Stochastic Models
in the Life Sciences
and Their Methods of Analysis

Other World Scientific Titles by the Author

Dynamical System Models in the Life Sciences and Their Underlying Scientific Issues
ISBN: 978-981-3143-33-3
ISBN: 978-981-3143-70-8 (pbk)

Stochastic Models
in the Life Sciences
and Their Methods of Analysis

Frederic Y M Wan

University of California, Irvine, USA

World Scientific

NEW JERSEY · LONDON · SINGAPORE · BEIJING · SHANGHAI · HONG KONG · TAIPEI · CHENNAI · TOKYO

Published by

World Scientific Publishing Co. Pte. Ltd.

5 Toh Tuck Link, Singapore 596224

USA office: 27 Warren Street, Suite 401-402, Hackensack, NJ 07601

UK office: 57 Shelton Street, Covent Garden, London WC2H 9HE

British Library Cataloguing-in-Publication Data
A catalogue record for this book is available from the British Library.

STOCHASTIC MODELS IN THE LIFE SCIENCES AND THEIR METHODS OF ANALYSIS

ISBN 978-981-3274-60-0

For any available supplementary material, please visit
https://www.worldscientific.com/worldscibooks/10.1142/11108#t=suppl

Desk Editor: Cheryl Heng

Typeset by Stallion Press
Email: enquiries@stallionpress.com

Printed in Singapore

To my grandparents Wen Ru Cai (溫如才) and Wen Cai Shi (溫蔡氏) who brought me up.

Contents

Preface

In the early spring of 2001, I was fortunate to meet with two colleagues from the School of Biological Sciences at the University of California, Irvine (UCI), Arthur D. Lander and J. Lawrence Marsh, through my Mathematics colleague Qing Nie. Together, we embarked on several projects in developmental biology. These projects led to a number of publications with much more work suggested by the research results. Recruiting graduate students and postdoctoral fellows to participate was an obvious way to expand the team research effort. However, candidates with mathematics degrees typically do not have an adequate background in biology and those in the biological sciences are usually not sufficiently prepared in the mathematics of differential equations and the related scientific computing. The dilemma motivated us to start a (gateway) graduate program at UCI to train a group of students each year who would have the necessary background in both mathematics and biology to participate in research in the biological sciences involving modeling, analysis and computation <http://mcsb.uci.edu/aboutmcsb>.

While mathematical modeling has long been an important aspect of the physical sciences and engineering, biology has historically been more of an empirical science. The molecular revolution in the 20th century has provided the field with an abundance of new factual information but with limited insight to the biochemical processes at work in biological organisms. In a 2009 report entitled *"A New Biology for 21st Century"* <http://www.nap.edu/catalog/php?record_id=12764>, the National Research Council (NRC) recommends a new approach to biology that "depends on greater integration within biology, and closer collaboration with physical, computational, and earth scientists, mathematicians and engineers be used to find solutions...". Given the centrality of mathematical models in the areas designated for collaboration, it is implicit in this recommendation that mathematical modeling should be exploited for significant advances in biology.

Pioneering efforts on mathematical modeling in the biological sciences have been reported in [46, 48, 49, 53, 54] and references therein. Modeling activities began to flourish in the 1990s and became intensified and more mechanistic (as opposed to phenomenological) in the new century with an interdisciplinary group of UCI faculty having a substantive role in quantitative studies of biological phenomena involving spatial dynamics. An award from the Howard Hughes Medical Institute (HHMI) Interdisciplinary Initiative enabled the group to initiate a gateway graduate program in Mathematical and Computational Biology (MCB) in 2007. The program has been sustained since then by an NIH award that designates the Center for Complex Biological Systems (CCBS) as a National Center of Excellence in Systems Biology at UCI <http://ccbs.uci.edu/>, several NIH training grants for supporting graduate students and for conducting short courses on spatial dynamics approach to biology for interested researchers, and the recent award of the NSF-Simons Foundation Center for Multi-scale Cell Fate Research. Approximately four scores of students have received their Ph.D. degrees since the inception of the MCB Program. In 2015, a stand alone interdisciplinary graduate (M. Sc. and Ph. D.) program in Mathematical and Computational Systems Biology (MCSB) was established to augment the MCB Gateway Program. It offers a more flexible curriculum to complement the departmental curriculum for Ph.D. degrees in anyone of the 10 academic departments affiliated with the MCB Gateway.

The present volume and its already published companion volume, *Dynamical system models in the life sciences* [72], are based on expanded versions of the course notes for the three quarter course sequence entitled Mathematical and Computational Biology offered by the Mathematics Department at UCI to meet a core "modeling, analysis and computation" requirement in the MCB and MCSB curriculum. Teaching mathematical modeling is much more challenging than teaching mathematics; there is not a recipe or algorithm for the development of a mathematical model for any specific phenomenon. This is even more so for the life sciences where there is not a counterpart of Newton's laws in mechanics. Courses and text books on general mathematical modeling typically consist of 1) case studies of specific phenomena, or 2) exposition of different mathematical methods with discussion of specific mathematical models as examples of their applications. For the MCB modeling sequence, we organize the teaching by the central issue of interest pertaining to the phenomenon being investigated, whether it is about the evolution of the phenomenon with time, its possible equilibrium configurations, the stability of these equilibria, the possibility of bifurcation to different behavior with minor changes of system characteristics, etc., see [72]. Our approach to presenting course material is motivated by the fact that analytical and computational tools needed for analyzing models are usually dictated by the issue they try to address as illustrated in [72]. With the modeling cycle and the scientific issues already discussed in some detail in [72], this second volume will focus on models where uncertainty plays a substantial role and the relevant methods for analyzing them.

Biological processes are evolutionary in nature and often evolve in a noisy environment or in the presence of uncertainty. Such evolving phenomena are necessarily modeled mathematically by stochastic differential/difference equations. Stochastic models have long been recognized as essential for a true understanding of many biological phenomena. Yet, there is a dearth of teaching material for students and researchers interested in models involving stochastic differential equations (SDE), notwithstanding the addition of some recent texts on stochastic modeling in the life sciences such as [2, 14, 79]. The reason may well be the demanding mathematical pre-requisites needed to "solve" SDE. The principal goal here is to provide a palatable path to a working knowledge of SDE based on the premise that familiarity with the basic (but challenging) elements of a stochastic calculus for random processes is unavoidable.

This volume begins with familiar topics such as Markov chains (MC) and other discrete sample space stochastic models. An effort is made however to engage readers in the mathematical proofs involved (e.g., the proof of existence and uniqueness of a steady state for a *regular* MC) to prepare them for the mathematical complexity and rigor of a calculus for such processes. We then undertake a substantive discussion of several birth and death type models that involve continuous time but discrete sample space. The last of these, a one-dimensional random walk model transitions the discussion to continuous (time and) sample space models and the notion of *stochastic process*. To ease the readers' effort to learn the indispensable elements of a (mean square) calculus for stochastic processes, we first examine models of biological phenomena that can be solved by transformation theory (originally developed for the study of electrical signal transmissions) and Liouville equation in dynamical system theory. These techniques allow us to work with a fixed instant of time and thereby further delay an explicit discussion of stochastic processes. At the same time, they show how SDE arise naturally in modeling of life science phenomena and, as a consequence, why there is a need for a calculus for their (stochastic) solution.

In presenting a calculus for stochastic processes, we limit the discussion to one based on mean square convergence. As we are mainly interested in analyzing models involving SDE, readers will only be expected to grasp the basics on limit, differentiation and integration with their mathematical justifications often omitted when they do not contribute to a better understanding of the calculus. The rest of the volume is devoted to methods for extracting information from SDE. Acquiring a working knowledge of methods for SDE is an arduous but unavoidable task to be proficient in mathematical modeling of life science phenomena. Reliance on numerical simulations of stochastic models would tantamount to dependence on results from a few repetitions of the same experiment. Neither validates a hypothesis; they only support (or refute) one.

While advances in SDE have been more recent compared to the conventional ODE and PDE, there is still a lot more valuable information than we can convey in this volume. As such we will have to be

- Selective in our choice of material and topics in SDE proper,
- Judicious in offering proofs of the mathematical results,

- Sparse in biological applications that require a great deal of background information, and
- Constrained by author's comfort level for different aspects of life sciences and SDE.

Nevertheless, the development of the MCB course curriculum has been steadfastly on course with the NRC Commission's recommended utilization of methodologies successful in other areas of science and engineering to advance the biological sciences.

The arduous effort to acquire a working knowledge of SDE is made more rewarding by the following three important features of this volume:

(1) Choosing for discussion modeled phenomena that allow for introduction and applications of advanced stochastic methods such as moments of first exit time, matched asymptotic expansions and the Fokker-Planck-Kolmogorov equations.
(2) Inclusion of many research results on stochastic models and their analysis not yet appeared in any mathematics or mathematical biology texts.
(3) Specific illustrations of several possible new approaches to known classes of problems that offer new research opportunities for life scientists.

Many important analytical and computational techniques for SDE feature reduction of the stochastic problem to a conventional (deterministic) problem in differential equations from which information about the modeled phenomenon can be extracted. As such, a good working knowledge of conventional ODE and PDE is indispensable to take advantage of these methods for SDE. Even stochastic simulation methods are not exempted from this requirement. For this volume to be reasonably self-contained, we have included review sections for topics such as eigenfunction expansions, Green's function and method of characteristics and others for ODE and PDE.

As I complete this second volume on the material for a MCB/MCSB program core requirements, I am reminded once more how fortunate I have been to be involved in the MCB/CCBS/MCSB projects. They have provided me an exciting opportunity for research and education to show that "there is life after administration." For the unanticipated opportunity, the congenial research collaboration, the unfailing support and warm friendship of Arthur, Larry and Qing for the last 18 years, I am most appreciative.

Frederic Y.M. Wan
Irvine, CA
June, 2018

Part 1
Discrete Stage Markov Chains

Chapter 1

Discrete Sample Space Probability

1.1 Terminology

It is human nature and necessity to make predictions. It is likely to rain tomorrow (so we should bring along rain gears when we go out). It is highly unlikely that I can fill an inside straight (so I should drop out from this poker hand before losing more money). These and other similar speculations reflect expectations of what is likely to happen based on past experience under similar (even better yet, identical) circumstances. The role of scientists and mathematicians is to quantify and make precise such vague predictions and others involving more complex observed phenomena and executable procedures. We begin to do this for phenomena and procedures with discrete outcomes, such as rain or no rain and success or failure in drawing one of the available cards (not already drawn previously) to fill an inside straight. The first step is to introduce some terminology and agree to their meaning throughout the ensuing developments.

1.1.1 *Events and sample space*

An **experiment** is a procedure that can be repeated or a phenomenon that recurs, possibly with different (well-defined) outcomes in either case even under the same "setting". Mechanical flipping of a coin (by some machine) under the same ambience to come up with a head (H) or a tail (T) is one such example. What constitutes "same setting" may vary from experiment to experiment and is to be specified in the context of the problem.

A **trial** of an experiment is one implementation or execution of the procedure or one observation of the phenomenon in question. All possible outcomes of a trial is called the **sample space** of the experiment denoted by S. The sample space of the coin flipping experiment consists of exactly two elements, $A_1 = H$ and $A_2 = T$ so that $S = \{A_1, A_2\} = \{H, T\}$. Throwing a dice (or a "die" if you prefer) with its faces numbered from 1 to 6 is another experiment with a sample space S consisting of six elements $\{1, 2, 3, 4, 5, 6\} \equiv \{A_1, A_2, \ldots, A_6\}$ in the set S.

An **elementary event** of an experiment is any one of the possible individual outcomes of the experiment. For coin flipping, H and T are the two elementary events of the experiment; for single dice rolling, $1, 2, 3, 4, 5$ and 6 are its six elementary events.

For some problems, we may be interested in combinations of elementary events. In the single die rolling experiment, we may be interested in betting on an outcome of an even number, i.e., the **event** $E_1 = \{2$ or 4 or $6\}$. Other combinations of elementary events of interest include $E_2 = \{$an odd number$\}$, $E_3 = \{1$ or $4\}$ $(= \{$a number in red$\}$ in a typically colored die), etc. An **event** E_i is any subset of the elementary events, i.e., any subset of the elements in the sample space S. When there are more than one distinctly different subsets of S, we may label them as E_i with different subscripts, as we did with the three subsets E_1, E_2 and E_3 defined above.

Among the events of an experiment, some are rather special and deserve to be singled out. One is an event that is certain to occur, called the **certain event**. The event $E = \{$either H or $T\}$ $(= \{H\} \cup \{T\}$ in set notations), is a certain event. The **complement** of an event E relative to its sample space S (or another event, denoted by E^c, consisting all elementary events in S that are not in E $(E^c = S - E)$. The complement of a certain event is an **impossible event** (or a **null event**) denoted by $\varnothing$. The event $\{$neither H nor $T\}$ cannot occur and is an impossible event.

The notion of a **union** of two events such as $E = \{H\} \cup \{T\}$ can be extended to a collection of events $\{E_1, E_2, E_3, \ldots, E_N\}$ denoted by: $E = E_1 \cup E_2 \cup E_3 \ldots = \cup_{i=1}^{N} E_i$. For example, if $E_1 = \{$an odd dice $\#\}$ and $E_2 = \{$a red dice $\#\}$, then $E_1 \cup E_2 = \{$any odd dice $\#$ or $4\}$.

The **intersection** E of two events such as $E_1 = \{2$ or 4 or $6\}$ and $E_3 = \{1$ or $4\}$, denoted by $E = E_1 \cap E_3$, requires both event to occur and hence $E = \{4\}$ for the specific example. The intersection of the N events $\{E_1, E_2, E_3, \ldots, E_N\}$ is the event $E = E_1 \cap E_2 \cap E_3 \cap \ldots \cap E_N = \cap_{i=1}^{N} E_i$.

1.1.2   *Finite, countable and continuous experiments*

In the early chapters of this volume, we will be concerned only with experiments that have a finite number of outcomes so that its sample space has only a finite number of elementary events. These experiments are said to have a *finite sample space*. Flipping a coin and drawing a card from the regular deck of 52 cards are experiments with a finite sample space. In later chapters, we consider more complex experiments or phenomena with a sample space that can be indexed by non-negative integers. An example would be an experiment with outcome of only even integers so that $S = \{A_0, A_1, A_2, A_3, \ldots, A_n, \ldots\}$ with elementary events $A_n = 2n$, $n = 0, 1, 2, 3, \ldots$. Such experiments and phenomena are said to be **countable** with a *countable sample space*.

Still distinctly different from finite and countable experiments are experiments with a sample space of elementary events that can be mapped one-to-one onto the real numbers (not just the integers). More will be said about these (**continuous**) experiments with a *continuous sample space* in later chapters.

1.2 Intuitive Probability for Finite Sample Space

1.2.1 *Relative frequencies and equiprobable sample space*

For an experiment with a finite number of elementary events that are equally likely to occur, their sample space is said to be **equiprobable**. If there are N elementary events in the sample space, then the fraction of a particular outcome occurring from a large number n of repeated trials is expected to be approximately $1/N$. That is, if you roll an *unbiased* die a large number of time, for example $n = 60,000$, the number of times a "5" turns up is expected to be close to $m_5 = 10,000$ with $m_5/n = 1/6$ in the limit as $n \to \infty$. Since the die is unbiased, the same would be true for any other face number so that $m_A/n = 1/6$ for any elementary event A.

More generally, **relative frequency** of an elementary event A from n trials of an experiment is m_A/n where m_A is the number of times A occurred. For an equiprobable sample space, we expect $m_A/n \to 1/N$ as $n \to \infty$. We call this limiting fraction of occurrence (for an infinite number of repeated trials) the **probability** of the (elementary) event A, denoted by $P(A)$.

Some elementary properties of $P(A)$ include:

(1) $0 \le P(A) \le 1$
(2) If $S = \{A_1, A_2, \ldots, A_N\}$ (where the A_i's are elementary events), then we have

$$P(A_1) + P(A_2) + \cdots + P(A_N) = \sum_{k=1}^{N} P(A_k) = 1.$$

(3) If $E = A_1 \cup A_2$, then $P(E) = P(A_1) + P(A_2)$.
(4) If all N elementary events $\{A_i\}$ in a sample space S are equally probable (from relative frequency data, by intuition, or by assumption), then $P(A_i) = 1/N$ for all $i = 1, 2, \ldots, N$.
(5) $P(S) = 1$ and $P(S^c) = P(\varnothing) = 0$.

1.2.2 *Basic properties of probabilities*

The elementary events of many experiments are known to be equally probable or may be idealized to be so to a good approximation. For many others however, we need to find a way to estimate $P(A_i)$ if a relative frequency calculation is not practical.

The following properties of probabilities are useful in making such estimates. The first of these is also sufficiently significant for theoretical developments to be designated as a theorem in most text:

Theorem 1. $P(E_1 \cup E_2) = P(E_1) + P(E_2) - P(E_1 \cap E_2)$

Proof. The proof is a combined application of properties (a), (b), (c) and the fact that the probability of an elementary event should not be counted more than once. □

Example 1. The probability of a "3" turn up in a roll of a fair dice is $P(3) = 1/6$. The probability of getting an even number is $P(2 \cup 4 \cup 6) = P(2) + P(4) + P(6) = 1/6 + 1/6 + 1/6 = 1/2$ since the three events (of rolling a 2, 4 or 6) are elementary event and are mutually exclusive. The probability of getting an even or a red number is $1/2 + 1/3 - 1/6 = 2/3$.

Corollary 1. *If E_1 and E_2 are mutually exclusive so that $E_1 \cap E_2 = \varnothing$, then $P(E_1 \cup E_2) = P(E_1) + P(E_2)$.*

Corollary 2. $P(E^c) = 1 - P(E)$.

Corollary 3. *If all N elementary events of a finite sample space S are equally probable, and if E is an event in S, then $P(E) = k_E/N$ where k_E is the number of distinct elementary events in E.*

Example 2. For a fair die, $P(2 \cup 4 \cup 6) = 3/6 = 1/2$ since $k_E = 3$.

1.2.3   *Expected value and variance*

Definition 1. When the elementary events of a finite sample space are expressed in terms of numerical values (as in rolling of a die), the events of such an experiment are specific realizations of a **random variable** X for the experiment. The **expected value** (aka **mean** and **statistical mean**) of the experiment, denoted by $E[X]$ or more simply μ, is defined to be

$$\mu = E[X] = A_1 P(A_1) + A_2 P(A_2) + \cdots + A_N P(A_N). \qquad (1.2.1)$$

Example 3. The six possible outcomes of rolling a die may be assigned the numerical values of $1, 2, \ldots, 6$, respectively. For a fair die with all faces equally probable so that $P(A_i) = 1/6$, the expected value of the experiment is $\mu = E[X] = 3.5$. (For an equiprobable sample space with elementary events that are numerical, the expected value is just the average or mean value of these events.)

Remark 1. An expected value of an experiment whose elementary events are not numerical values can also be calculated once we assign to each A_i a distinct numerical value a_i. For a fair coin, we may assign 0 to the event of a head turning up and 1 to a tail, each with probability $1/2$. In that case we have $\mu = E[X] = 1/2 \cdot a_1 + 1/2 \cdot a_2 = 1/2 \cdot (1 + 0) = 0.5$. Of course, we can also assign 1 to a head and 2 to a tail, in which case $\mu = 1.5$. When properly interpreted, that there can be a different expected value depending on the numerical assignments of the random variable is of no substantive significance.

We are also interested in higher order moments of the random variable

$$\mu_k = E[X^k] = A_1^k P(A_1) + A_2^k P(A_2) + \cdots + A_N^k P(A_N)$$

with $\mu_1 = \mu$.

Definition 2. The **variance** σ^2 of the experiment is defined by

$$\sigma^2 = E[(X - \mu)^2]$$
$$= (A_1 - \mu)^2 P(A_1) + (A_2 - \mu)^2 P(A_2) + \cdots + (A_N - \mu)^2 P(A_N).$$

with σ known as the **standard deviation** of the experiment.

Exercise 1. Show $\sigma^2 = E[X^2] - \mu^2$.

While the standard deviation σ is used as a measure of deviation from the statistical mean, another such measure is the **coefficient of variation**, denoted by c_v, CV or ρ, defined by

$$c_v = \frac{\sigma}{\mu}.$$

It should be noted that the quantity is not particularly meaningful unless the unit of measurement for the numerical-valued events are chosen so that they are non-negative (for otherwise μ may be zero or near zero.)

Exercise 2. Calculate $E[X^2]$, σ and c.v. for rolling an unbiased die.

1.3 The Binomial Distribution

A **Bernoulli trial** is an execution of an experiment that has only two outcomes, generically called *success* (*s*) and *failure* (*f*). When known, the probability for success is labelled p and the probability of failure is therefore $1 - p$. Flipping a coin is such an experiment and we may label the outcome of a head success and a tail failure (or vice versa). By executing the same Bernoulli trial repeatedly and independently with the two outcomes having the same two probabilities, p and $1 - p$ for each trial, we have a **Binomial Experiment**. (For independent trials, the outcome of any trial does not affect the outcome of other trials.) In many applications, we are interested in knowing the probability of a particular outcome of a sequence of n Bernoulli trials of the same Binomial experiment (with the same success probability p for every trial).

We demonstrate the richness of Binomial experiments with the simple example of rolling a fair die five times with the outcome of a "1" being a success and the other five outcomes being a failure. The probability of getting the sequence $\{f, f, s, f, s\}$

is the product probability

$$\text{Prob}\{f, f, s, f, s\} = \left(\frac{5}{6}\right)\left(\frac{5}{6}\right)\left(\frac{1}{6}\right)\left(\frac{5}{6}\right)\left(\frac{1}{6}\right) = \frac{125}{6^5}$$

given the probability of getting a "1" is $1/6$ and the probability of getting one of the other 5 numbers is $1 - 1/6 = 5/6$. If the order of occurrence of s and f does not matter, then the probability of getting two successes would increase to

$$\text{Prob}\{2 \text{ successes in 5 trials}\} = C(5, 2)\frac{125}{6^5} = \frac{5!}{2!3!}\frac{125}{6^5} = \frac{1250}{6^5}$$

since there are $C(5, 2) = 5!/(3!2!)$ ways in getting two successes in five trials.

More generally, when the probability of success for the binomial experiment is p, the probability of getting k successes in n repeated independent trials of the same binomial experiment is

$$\text{Prob}\{X = k\} = C(n, k)p^k(1 - p)^{n-k}$$

$$= \binom{n}{k}p^k(1 - p)^{n-k} = \frac{n!}{k!(n - k)!}p^k(1 - p)^{n-k}$$

where X is the random variable for the number of successes. The right hand side (in three different notations) is the **binomial distribution** with parameters n and p giving the discrete probability distribution of the number of successes in a sequence of n independent experiments.

Exercise 3. Six fair coins are flipped. Determine the probability for getting exactly three tails.

Exercise 4. Four cards are drawn from a typical deck of 52 cards with replacement after each draw. Calculate the probability of getting at least three aces.

1.4 Conditional Probability

Conditional probability is the probability of an event A occurring given that (by assumption, presumption, assertion or evidence) another event B has occurred. If the event of interest is A and the event B is known or assumed to have occurred, "the conditional probability of A given B", or "the probability of A under the condition B", is usually written as $P(A|B)$. For example, the probability that any given person has a cough on any given day may be only 5%. But if we know or assume that a person has a cold, then that person is much more likely to be coughing. The conditional probability of coughing given that you have a cold might be a much higher 75%.

The concept of *conditional probability* is one of the most fundamental and important concepts in probability theory. But conditional probabilities can be quite slippery and require careful interpretation. For example, there need not be a causal or temporal relationship between the events A and B.

$P(A|B)$ may or may not be equal to $P(A)$ (the unconditional probability of A). If $P(A|B) = P(A)$ (or its equivalent $P(B|A) = P(B)$), then events A and B are said to be **statistically independent**. In such cases, having knowledge about either event does not change our knowledge about the occurrence of the other event. Also, in general, $P(A|B)$ (the conditional probability of A given B) is not equal to $P(B|A)$. For example, if you have cancer you might have a 90% chance of testing positive for cancer. In this case the measured probability of event A (testing is positive) occurring is 90% if event B (having cancer) has occurred, $P(A|B) = 90\%$. On the other hand, even if you have tested positive for cancer, you may have only a 10% chance of actually having cancer because cancer is very rare. In this second case, what is being measured is the probability of the event B (having cancer) given that the event A (test is positive) has occurred, $P(B|A) = 10\%$. Falsely equating the two probabilities would cause various errors of reasoning such as the *base rate fallacy*. Also known as *base rate bias*, this common erroneous practice of ignoring generic, general (base rate) information when presented together with related specific information (information pertaining only to a certain case); human mind tends to ignore the former and focus on the latter.

Conditional probabilities can be correctly reversed using Bayes' theorem. Given two events A and B from a sample space with $P(B) > 0$, the conditional probability of A given B is defined as the quotient of the probability of the joint of events A and B and the probability of B:

$$P(A|B) = P(A \cap B)/P(B) \tag{1.4.1}$$

This may be visualized as restricting the sample space to B. The logic behind this equation is that if the outcomes are restricted to B, this set serves as the new sample space.

Note that (1.4.1) is a definition, not a theoretical result. We merely denote the quantity $P(A \cap B)/P(B)$ as $P(A|B)$ and call it the *conditional probability of A given B*. Recall that two events A and B are said to be statistically independent if $P(A|B) = P(A)$ and $P(B|A) = P(B)$. It follows from (1.4.1) that A and B are statistically independent if and only if

$$P(A \cap B) = P(A)P(B). \tag{1.4.2}$$

Chapter 2

Discrete Stage Regular Markov Chains

2.1 Introduction

In elementary probability theory, one way to establish the probability of different elementary events of a simple repeatable experiment is by the relative frequency ratio of these events after a large number of repeats. When the ratio for each elementary event settles down to a fixed number (and we call such an experiment *stationary* (*temporally homogeneous*), we adopt it as the probability for each elementary event accurate to the number of unchanged significant digits with further experiments. In this volume, we are interested instead in repetitions of an experiment with known probability for its elementary events (from a posteriori frequency ratio estimates or a priori attributions) under the same environment to see how these probabilities evolve with repetitions, starting with some initial probability distribution. It is implicitly assumed that the execution of each repetition of the experiment is completely random but the outcome possibly affected by memory of the past. The simplest of these are *Markov processes* whose outcome of the next trial is influenced only by the outcomes of recent past trial(s). We limit our discussion first to the class of Markov processes that have the following characteristics:

- The observations of the phenomenon (or the trials of the repeatable experiment) are made in discrete times, a generation, a year, a day, or a second at a time to be called "stages".
- There are only the same finite number of possible outcomes for each stage, and hence only N mutually exclusive elementary events.
- The experiment at each stage has very short memory with the outcome at a particular stage dependent only on the outcome of the previous stage and no others.
- The dependence of the outcome of the current stage on the outcome of the immediate past observation is linear.

The successive outcomes observed (at discrete instances of time) are known as the *states* of the experiment or observable phenomenon. The restricted class of such phenomena specified by the bulleted conditions are known as first order (finite) *discrete Markov chains (abbreviated as DMC or simply MC) with a finite number of states*. While many biological phenomena (including membrane channel opening and closing and gene sequencing) may be modeled as Markov chains for a deeper understanding of these phenomena, we begin with a simple mouse experiment to illustrates the nature and issues of such evolving processes.

2.2   A Simple Mouse Experiment

2.2.1   *The mathematical model of a mouse experiment*

A mouse is placed in a box with three compartments, I, II and III, that are completely indistinguishable to the mouse. Each compartment has two doors, one for exit to (or entrance from) each of the other two compartments. Suppose the mouse should choose a door at random to move from its current compartment to another and do so repeatedly. Intuitively, we expect the mouse to be in each compartment about one third of the times, i.e., equally often. However, we need a mathematical model that enables us to justify/validate this intuitive conclusion as well as provide other possible information about related experiments. More elaborate experiments of this type can be found in the work of Wecker studying prairie deer mice [76, 77]. The present version suffices for introducing the concept MC.

We take the three elementary events of this mouse experiment to be the mouse located in the three different compartments, respectively. We may let X be the *random variable* for the state of the mouse after one move and assign the numerical values for the three elementary events in the experiment's sample space to be $1, 2$ and 3 corresponding to the compartment I, II and III, respectively. As the experiment is repeated and the state of the mouse evolves from trial to trial, with the nth trial designated as stage n, two consecutive stages of the experiment correspond to the stage before and the stage after the mouse moving from one compartment to another. The corresponding state of the mouse at stage n is designated by the random variable X_n or $X(n)$ with the sample space of $\{1, 2, 3\}$ for all n.

Let $p_i(n)$ be the probability of $X_n = i$ (for the mouse in compartment corresponding to i):

$$p_i(n) = \text{Prob}\{X_n = i\}, \quad (i = 1, 2, 3 \text{ and } n = 1, 2, 3, \ldots).$$

We assume for the repeated trials of the mouse experiment that the state of the mouse (the location of the mouse in this case) at stage $n + 1$ depends only on its state at stage n (and not any of the earlier states). Such an evolving experiment is said to be a Markov process. In the mouse experiment, the stages are discrete moves that can be labelled by the positive integers (with $n = 0$ be the starting point before the first trial (move) of the experiment). We call such a Markov process a *discrete time Markov chain.* Since it has a finite sample space at each stage, it is called a discrete time finite state Markov chain (and abbreviated by MC).

There is more than one way for the mouse to get from its initial state to its current state i at stage n. To help analyze these ways and more generally the evolution of the state of an experiment, we let $p_{ij}(n)$ be the one step conditional probability of $X_{n+1} = i$ given $X_n = j$:

$$p_{ij}(n) = \text{Prob}\{X_{n+1} = i \mid X_n = j\}.$$

For the mouse experiment, we consider here the case where the one step transition probabilities $\{p_{ij}(n)\}$ do not change with stage so that $p_{ij}(n) = p_{ij}$ for all n. Such experiments or systems are said to be *time-invariant* (or *spatially homogeneous* when stages correspond to different locations in space). To the extent that they are probabilities, we have for a finite sample space of $N = 3$ elementary events

$$0 \le p_{ij} \le 1 \quad (i, j = 1, \ldots, N) \quad \text{and} \quad \sum_{i=1}^{N} p_{ij} = 1,$$

with the second condition characterizing the fact that the probabilities of the mouse ending up in one of all possible states should be a certainty. (Note that N may be ∞ for some Markov chains that have an infinite number of elementary events. An example would be to pick a nonnegative integer at random. Here, we consider only DTMC with only a finite sample space.)

We now consider a specific mouse experiment that is time-invariant with moving from its current compartment to either of the other two compartments equally likely. For example, the probability for the mouse to go from compartment II to compartment I on successive moves is $p_{12} = 0.5$. By the assumptions on the experiment, these probabilities do not change with stage. In that case we may summarize these probabilities in a transition matrix $M = [p_{ij}]$ for $i, j = 1, 2, 3$:

$$
\begin{array}{c}
X_{n+1} \backslash X_n \\
1 \\
2 \\
3
\end{array}
\quad
\begin{array}{ccc}
1 & 2 & 3 \\
\end{array}
\left[
\begin{array}{ccc}
0 & 0.5 & 0.5 \\
0.5 & 0 & 0.5 \\
0.5 & 0.5 & 0
\end{array}
\right] = [p_{ij}] = M.
\tag{2.2.1}
$$

The components $\{p_{1j}, p_{2j}, p_{3j}\}$ of column j of the transition matrix M are the probabilities of the mouse moving into the three different compartments from the compartment associated with that j^{th} column. They may be taken as components of a *probability vector* $\mathbf{p}_j$ with $p_{ij} \ge 0$ and $p_{1j} + p_{2j} + p_{3j} = 1$. The matrix M whose columns are all probability vectors is known as a *stochastic (or probability) matrix.* It should be noted that some authors prefer to work with the transpose of M, $P = M^T$, as the transition matrix (which is less natural for the matrix operations involved in the theory as we shall see).

Let $x_k(n)$, $k = 1, 2, 3$, be the probability of the mouse in compartment k at stage n for the MC, having started initially with $\mathbf{x}(0) = \mathbf{p}^o = (p_1^o, p_2^o, p_3^o)^T$. For example, if the mouse was initially at compartment II, then we would have $\mathbf{p}^o = (0, 1, 0)^T$. Evidently, the element $m_{ij}(n) = m_{ij}^{(n)}$ of the transition matrix $M^{(n)} = \left[m_{ij}^{(n)} \right]$ at stage n gives the probability for the mouse to get from compartment j to compartment i in one move at stage n. Since that probability has been stipulated to be p_{ij} independent of stage and is the same as that for the first move from the initial state we have $M^{(n)} = \left[m_{ij}^{(n)} \right] = [p_{ij}] = M$ for the time-invariant DTMC. Suppose we are interested in the probability of the mouse going from compartment j to compartment i in two moves instead. In principle, this can be accomplished in three ways: from compartment j to compartment ℓ and then to i, for $\ell = 1, 2$ or 3. In that case we have

$$p_{ij}^{(2)} = p_{i1}p_{1j} + p_{i2}p_{2j} + p_{i3}p_{3j} = \sum_{\ell=1}^{N=3} p_{i\ell}p_{\ell j}.$$

It follows that the 2-step transition matrix $M^{(2)}$ is given generally by

$$M^{(2)} = \left[\sum_{\ell=1}^{N} p_{i\ell}p_{\ell j} \right] = MM = M^2. \tag{2.2.2}$$

When applied to the mouse experiment, we have $p_{jj} = 0$, $j = 1, 2, 3$, so that there are in fact only two ways to get from state j to state i in two steps.

If p_{ij} should depend on n so that the system is **not** time-invariant, then we would have instead $\{p_{ij}(n)\}$ with

$$M^{(2)} = \left[m_{ij}^{(2)} \right] = \left[\sum_{k=1}^{N=3} m_{ik}(2)m_{kj}(1) \right]$$

and

$$M^{(3)} = \left[m_{ij}^{(3)} \right] = \left[\sum_{k=1}^{N=3} m_{ik}(3)m_{kj}^{(2)} \right] = \left[\sum_{\beta=1}^{N=3} \sum_{\alpha=1}^{N=3} m_{i\alpha}(3)m_{\alpha\beta}(2)m_{\beta j}(1) \right] = \prod_{i=1}^{N=3} M(i),$$

etc., with

$$M^{(n)} = \prod_{i=1}^{n} M(i). \tag{2.2.3}$$

Before we accept (2.2.2) as a transition matrix, we need to take care a small detail:

Exercise 5. With $M^{(k)} = \left[m_{ij}^{(k)} \right]$, show that $M^{(2)} = \left[m_{ij}^{(2)} \right]$ is a stochastic matrix.

2.3   Transition Matrix for a DMC

2.3.1   *Probability vectors and stochastic matrices*

In order to do more extensive analysis of MC, we now introduce more terminology.

Definition 3. A **probability vector** is a column vector $\mathbf{p} = (p_1, p_2, \ldots, p_N)^T$ with $0 \leq p_k \leq 1$ and $p_1 + p_2 + \cdots + p_N = 1$.

Definition 4. A **stochastic matrix** (or **probability matrix**) is a matrix M with each of its columns being a probability vector.

Lemma 1. *Multiplication of a probability vector of N components by an $N \times N$ stochastic matrix results in another probability vector.*

Proof. With

$$\mathbf{q} = M\mathbf{p} = \sum_{j=1}^{N} p_{ij} p_j$$

it is clear from $p_{ij} \geq 0$ and $p_j \geq 0$ that

$$q_i = \sum_{j=1}^{N} p_{ij} p_j \geq 0.$$

Moreover, we have

$$\sum_{i=1}^{N} q_i = \sum_{i=1}^{N} \sum_{j=1}^{N} p_{ij} p_j = \sum_{j=1}^{N} \sum_{i=1}^{N} p_{ij} p_j = \sum_{j=1}^{N} p_j = 1$$

so that $\mathbf{q}$ is a probability vector. $\qquad\square$

Theorem 2. *The product of two stochastic matrices is a stochastic matrix. In particular, any power of a stochastic matrix is a stochastic matrix.*

Proof. The proof is a straightforward application of Lemma 1. $\qquad\square$

Definition 5. A matrix M is a **power-positive matrix** if all elements of M^k are positive for all $k \geq k_p \geq 0$.

2.3.2 *Evolution of probability distribution*

For a Markov chain with N distinct states, let $x_k(n)$, $k = 1, 2, \ldots, N$ be the probability of the kth elementary event to occur at stage n. For the mouse experiment, $x_k(n)$, $k = 1, 2$ or 3, would be the probability for the mouse to be in compartment k at stage n. Suppose the first elementary event is known to have occurred initially and taken to be stage $n = 0$, corresponding to the mouse having been put in compartment I at the start of the mouse experiment. In that case, we have $\mathbf{x}(0) = (x_1(0), \ldots, x_N(0))^T = (1, 0, 0, \ldots, 0)^T$ and the probabilities for the different elementary events to occur at the next trial are components of

$$\mathbf{x}(1) = M\mathbf{x}(0) = (p_{11}, p_{21}, \ldots, p_{N1})^T.$$

To formulate an efficient method for determining $\mathbf{x}(n)$ for general and large n, we recall, at the risk of being repetitious, the situation for the mouse experiment once more. For that problem, the transition probabilities is given by (2.2.1) so that

$$\mathbf{x}(1) = M\mathbf{x}(0) = (0, 0.5, 0.5)^T.$$

For the same mouse to be in compartment I again after two moves, it can get there in two ways, from compartment I to II to I or from compartment I to III back to I. The probability of the mouse returning to compartment I after two moves is $0.5 \times 0.5 + 0.5 \times 0.5 = 0.5$. In contrast, there is only one way for the mouse to be in one of the other two compartments after two moves: It can only get to compartment II by first getting to compartment III (and no other way). Similarly, it can get to compartment III only by first getting to compartment II. In each case, the probability for either to happen is $0.5 \times 0.5 = 0.25$. We get the same result by multiplying the initial probability distribution (which is $(1, 0, 0)^T$ since the mouse is known to be in compartment I) by the transition matrix twice:

$$\mathbf{x}(2) = M\mathbf{x}(1) = M\{M\mathbf{x}(0)\} = \begin{bmatrix} 0 & 0.5 & 0.5 \\ 0.5 & 0 & 0.5 \\ 0.5 & 0.5 & 0 \end{bmatrix}^2 \begin{pmatrix} 1 \\ 0 \\ 0 \end{pmatrix}$$

$$= \begin{bmatrix} 0 & 0.5 & 0.5 \\ 0.5 & 0 & 0.5 \\ 0.5 & 0.5 & 0 \end{bmatrix} \begin{pmatrix} 0 \\ 0.5 \\ 0.5 \end{pmatrix} = \begin{pmatrix} 0.5 \\ 0.25 \\ 0.25 \end{pmatrix}. \qquad (2.3.1)$$

We may alternatively write the same result above in the form

$$\mathbf{x}(2) = M\mathbf{x}(1) = M\{M\mathbf{x}(0)\} = M^{(2)} \begin{pmatrix} 1 \\ 0 \\ 0 \end{pmatrix}$$

where

$$M^{(2)} = M^2 = \begin{bmatrix} 0 & 0.5 & 0.5 \\ 0.5 & 0 & 0.5 \\ 0.5 & 0.5 & 0 \end{bmatrix}^2 = \begin{bmatrix} 0.50 & 0.25 & 0.25 \\ 0.25 & 0.50 & 0.25 \\ 0.25 & 0.25 & 0.50 \end{bmatrix}. \qquad (2.3.2)$$

In a more general setting of the mouse experiment, the initial position of the mouse is not known with certainty but has a known probability distribution $\mathbf{x}(0) = \mathbf{p} = (p_1, p_2, p_3)^T$ instead (with the special case of the mouse known to be in compartment I corresponding to $p_1 = 1$ and $p_2 = p_3 = 0$). In that case, we have the following probability distribution of the mouse in different compartment after one move:

$$\mathbf{x}(1) = M\mathbf{x}(0) = \left(\sum_{j=1}^{3} p_{1j}p_j, \sum_{j=1}^{3} p_{2j}p_j, \sum_{j=1}^{3} p_{3j}p_j \right)^T$$

$$= \frac{1}{2} \left(p_2 + p_3, \; p_1 + p_3, \; p_1 + p_2 \right)^T.$$

Similar results can be obtained for the probability distribution for the three elementary events after $2, 3, \ldots$ or n moves.

Now for a general Markov chain with N elementary events, the transition matrix $M = [p_{ij}]$ would be an $N \times N$ matrix. The probability of the elementary events after $1, 2, \ldots, n$ trials is formally the same except for summing from 1 to N instead:

$$\mathbf{x}(1) = M\mathbf{x}(0) = \left(\sum_{j=1}^{N} p_{1j}p_j, \sum_{j=1}^{N} p_{2j}p_j, \ldots, \sum_{j=1}^{N} p_{\ell j}p_j \right)^T \tag{2.3.3}$$

$$\mathbf{x}(2) = M\mathbf{x}(1) = M^2\mathbf{x}(0) \equiv M^{(2)}\mathbf{x}(0), \quad \mathbf{x}(3) = M^{(3)}\mathbf{x}(0), \ldots, \tag{2.3.4}$$

$$\mathbf{x}(n) = M^{(n)}\mathbf{x}(0), \quad M^{(n)} = M^n. \tag{2.3.5}$$

Evidently, information essential for determining the probability distribution at the nth stage is completely contained in the nth order transition matrix $M^{(n)} = M^n$. This observation does not seem particularly helpful for a general stochastic matrix as M^n becomes intractable even after just three or four trials. However, when viewed as the evolution of the state vector $\mathbf{x}(n)$ of a discrete dynamical system,

$$\mathbf{x}(n+1) = M\mathbf{x}(n), \quad \mathbf{x}(0) = \mathbf{p}, \tag{2.3.6}$$

the problem of extracting useful information from the transition matrix becomes considerably simpler.

Suppose the $N \times N$ transition matrix M is an non-defective matrix (with a full set of N linear independent eigenvectors $\{\mathbf{v}^{(k)}\}$ and $\{\lambda_k\}$ the associated eigenvalues). Let $V = \left[\mathbf{v}^{(1)}, \mathbf{v}^{(2)}, \ldots, \mathbf{v}^{(N)} \right]$ be the *modal matrix* of the eigenvectors of M. In that case, we have

$$MV = V\Lambda, \quad \Lambda = \begin{bmatrix} \lambda_1 & 0 & \cdot & \cdot & 0 \\ 0 & \lambda_2 & 0 & \cdot & \cdot \\ \cdot & 0 & \cdot & \cdot & \cdot \\ \cdot & \cdot & \cdot & \cdot & 0 \\ 0 & \cdot & \cdot & 0 & \lambda_\ell \end{bmatrix}$$

where Λ is the diagonal matrix with the eigenvalues of M on the main diagonal. It follows that

$$M = V\Lambda V^{-1}, \quad M^{(n)} = M^n = V\Lambda^n V^{-1} \tag{2.3.7}$$

where

$$\Lambda^n = \begin{bmatrix} \lambda_1^n & 0 & \cdot & \cdot & 0 \\ 0 & \lambda_2^n & 0 & \cdot & \cdot \\ \cdot & 0 & \cdot & \cdot & \cdot \\ \cdot & \cdot & \cdot & \cdot & 0 \\ 0 & \cdot & \cdot & 0 & \lambda_\ell^n \end{bmatrix}.$$

In the form (2.3.7), the power transition matrix M^n is no longer so intractable. But because M is a stochastic matrix, we can say a whole lot more.

Lemma 2. $\lambda = 1$ *is always an eigenvalue of the transition matrix M of a Markov chain with an associated eigenvector determined up to a multiplicative factor* **c**.

Proof. Let $A = M - I$. Then all the rows of A sum to give a zero row. Hence zero is an eigenvalue of A or $\lambda = 1$ is an eigenvalue of M with eigenvector $c\mathbf{v}$ where c is any non-zero constant. □

Lemma 3. *All eigenvalues of a stochastic matrix must be ≤ 1 in magnitude.*

Proof. Suppose λ is an eigenvalue of M and $|\lambda| > 1$. Let $\{\lambda, \mathbf{y}\}$ be an eigen-pair of M^T (since λ is also an eigenvalue of M^T as shown in an exercise in Assignment I). Let $\max_{i=1}^N [|y_i|] = Y_j > 0$. Then it follows from $M^T \mathbf{y} = \lambda \mathbf{y}$ that

$$|\lambda| Y_j = \left| \sum_{\ell=1}^N p_{\ell j} y_\ell \right| \leq Y_j \sum_{k=1}^N p_{\ell j} = Y_j,$$

so that $|\lambda| \leq 1$, contradicting the initial assumption $|\lambda| > 1$. □

2.3.3 Chapman–Kolmogorov equations

As we saw in the mouse experiment, there is more than one way for the mouse to get from its initial state to its current state i at stage n. To help analyze more generally the evolution of a Markov chain, we consider again the transition matrix $M = [p_{ij}]$ of a time-invariant MC with

$$0 \leq p_{ij} \leq 1 \quad \text{and} \quad \sum_{i=1}^N p_{ij} = 1.$$

Now for an experiment that was in state j at stage k, let $p_{ij}^{(n-k)}$ be the probability that the MC is in state i at the n stage, $n > k \geq 0$:

$$p_{ij}^{(n-k)} = \text{Prob}\{X_n = i \mid X_k = j\}.$$

For $k = 0$ so that the MC started at state j initially, we have

$$p_{ij}^{(n)} = \text{Prob}\{X_n = i \mid X_0 = j\}.$$

It is important to note the difference between $p_{ij}^{(n)}$ (which is for getting from j to i in n stages) and $p_{ij}(n)$ (which is getting from j to i in one move starting at stage n) even if we are to consider mainly the case of time-invariant systems. By elementary probability theory, $p_{ij}^{(n)}$ may be taken to be the sum of the probabilities of getting to each of the N possible states at an intermediate stage k and from there to state i:

$$p_{ij}^{(n)} = \sum_{k=1}^{N} \text{Prob}\{X_n = i, X_k = \ell \mid X_0 = j\}.$$

With

$$\text{Prob}\{X_n = i, X_k = \ell \mid X_0 = j\} = \text{Prob}\{X_n = i \mid X_k = \ell, \ X_0 = j\}$$
$$\cdot \text{Prob}\{X_k = \ell \mid X_0 = j\},$$

we may rewrite the expression for $p_{ij}^{(n)}$ as

$$p_{ij}^{(n)} = \sum_{k=1}^{N} \text{Prob}\{X_n = i \mid X_k = \ell, \ X_0 = j\} \cdot \text{Prob}\{X_k = \ell \mid X_0 = j\}$$

$$= \sum_{k=1}^{N} \text{Prob}\{X_n = i \mid X_k = \ell\} \cdot \text{Prob}\{X_k = \ell \mid X_0 = j\}$$

where we have made use of the fact that the experiment is a Markov process so that it only depends on its most recent past.

From this, we get the following Chapman–Kolmogorov relation:

Theorem 3.

$$p_{ij}^{(n)} = \sum_{\ell=1}^{N} p_{i\ell}^{(n-k)} p_{\ell j}^{(k)}. \tag{2.3.8}$$

The Chapman–Kolmogorov relation (2.3.8) holds for any intermediate stage k, $0 < k < n$. Its continuous state counterpart will be important for subsequent developments of more general Markov processes.

For time-invariant systems, it trivially implies the following matrix version of the Chapman–Kolmogorov relation:

Corollary 4. *In terms of the transition matrix M of the relevant MC, the Chapman–Kolmogorov relation may be written as*

$$M^{(k)} = M^{(k-\ell)} M^{(\ell)}.$$

Proof. For time-invariant systems, the desired relation follows from the expression for the transition matrix

$$M^{(k)} = M^k = M^{k-\ell} M^\ell = M^{(k-\ell)} M^{(\ell)}. \tag{2.3.9}$$

For the more general case of time-varying systems, we have from (2.2.3)

$$M^{(k)} = \Pi_{i=1}^k M(i) = M(k)M(k-1)\cdots M(2)M(1)$$

$$= \Pi_{\alpha=1}^\ell M(\alpha) \cdot \Pi_{\beta=\ell}^k M(\beta) = M^{(\ell)} M^{(k-\ell)}. \tag{2.3.10}$$

$\square$

2.4 Regular Markov Chains

2.4.1 *Convergence to a steady state*

While some entries of the transition matrix M of the mouse experiment, notably the diagonal elements, are zeros, entries of the powers M^n of that transition matrix are all positive for $n \geq 2$ (see (2.3.1) for M^2 for example). This is not always the case for other transition matrix. Two simple examples are

$$I = \begin{bmatrix} 1 & 0 \\ 0 & 1 \end{bmatrix}, \quad S = \begin{bmatrix} 0 & 1 \\ 1 & 0 \end{bmatrix}. \tag{2.4.1}$$

Definition 6. A matrix M is a **power-positive** if all elements of M^k are positive for all $k \geq k_p \geq 1$.

Definition 7. A Markov Chain is **regular** if its $N \times N$ transition matrix M is a power-positive probability matrix.

The following theorem is the principal result for regular MC. Its proof will be given in an appendix of this chapter.

Theorem 4. *Starting with any initial probability vector* $\mathbf{x}(0) = \mathbf{p}$ *($\neq \mathbf{0}$ and with non-negative components $p_i \geq 0$), the vector sequence* $\{\mathbf{x}(1), \mathbf{x}(2), \mathbf{x}(3), \ldots\}$ *of successive stages of a regular Markov chain converges to a limiting vector* $c\mathbf{v} = c(v_1, v_2, \ldots, v_N)^T$ *which can be re-scaled to a probability* $\mathbf{x}_\infty$ *by setting* $c = 1/\sum_{k=1}^N v_k$.

Proof. (see Appendix of this section). $\square$

For a given initial probability vector $\mathbf{p}$, the limiting distribution vector $\mathbf{x}_\infty$ is clearly a probability vector by Lemma 1. The limiting vector $\mathbf{x}_\infty$ will be shown to be the same for all initial distribution and hence independent of $\mathbf{p}$ (see Theorem 5).

It follows that $\mathbf{x}_\infty$ is unique and asymptotically stable (analogous to the asymptotic stability of a critical point of a dynamical systems).

As a vector difference equation $\mathbf{x}_{n+1} - \mathbf{x}_n = (M - I)\mathbf{x}_n$, the limit distribution of the regular Markov chain corresponds to a fixed point of the evolving process:

$$\mathbf{x}_\infty = M\mathbf{x}_\infty. \tag{2.4.2}$$

In the language of matrix theory, $\mathbf{x}_\infty$ is necessarily an eigenvector associated with the eigenvalue $\lambda = 1$ of the transition matrix M. The eigen-pair $\{1, c\mathbf{v}\}$ is known to exist for MC in general; Theorem 5 below assures that $c\mathbf{v}$ may be taken as the unique probability $\mathbf{x}_\infty$).

2.4.2 The eigenvalues of a transition matrix

For a regular Markov chain with a power-positive transition matrix, we can also show that $\lambda = 1$ is the only eigenvalue of unit magnitude; hence all eigenvalues but $\lambda_1 = 1$ are inside the unit circle, i.e., $|\lambda_k| < 1$, $1 < k \leq \ell$. We take M to be positive to reduce the details of the proof.

Lemma 4. *For a regular MC, there is no complex eigenvalues with unit modulus, i.e., $|\lambda| \neq 1$.*

Proof. Suppose there should be a complex eigenvalue μ with $|\mu| = 1$ and $\mathbf{v} = \mathbf{u} + i\mathbf{w}$ its associated eigenvector with both $\mathbf{u}$ and $\mathbf{w}$ real. With $\mathbf{x}_\infty$ the limiting probability vector (= eigenvector for $\lambda = 1$) for the regular MC with transition matrix M and c a sufficiently large positive number so that both $\mathbf{u} + c\mathbf{x}_\infty$ and $\mathbf{w} + c\mathbf{x}_\infty$ are both positive vectors (to meet the requirement of Theorem 4), we have

$$M^n(\mathbf{u} + i\mathbf{w} + c(1 + i)\mathbf{x}_\infty) = \mu^n(\mathbf{u} + i\mathbf{w}) + c(1 + i)\mathbf{x}_\infty$$

given

$$M(\mathbf{u} + i\mathbf{w} + c(1 + i)\mathbf{x}_\infty) = \mu(\mathbf{u} + i\mathbf{w}) + c(1 + i)\mathbf{x}_\infty$$

As $n \to \infty$, we are left with

$$\lim_{n \to \infty} \{M^n(\mathbf{u} + i\mathbf{w})\} = \mu^n(\mathbf{u} + i\mathbf{w}).$$

This contradicts the eigen-pair assumption unless $\mu = 1$. $\qquad\square$

2.4.3 Uniqueness and stability of the steady state

Altogether, all eigenvalues of M, except for $\lambda_1 = 1$, are inside the unit circle with $|\lambda| < 1$. While we expect $\lambda_1 = 1$ to be associated with a limiting probability distribution $\mathbf{x}_\infty$ assured by Theorem 4, we still need to show that $\mathbf{x}_\infty$ is unique and independent of $\mathbf{x}(0)$.

Theorem 5. *Suppose that a regular Markov chain satisfies the initial value problem (IVP) defined by (2.3.6) with a limiting probability distribution $\mathbf{x}_\infty$. Then $\mathbf{x}_\infty$ is independent of the initial distribution $\mathbf{x}(0) = \mathbf{p}$.*

Proof. For simplicity, we prove the theorem for $M > O$ (and leave the more general case as an exercise). Let probability vectors $\mathbf{x}_\infty$ and $\mathbf{y}_\infty$ be two limiting distributions corresponding to two initial distributions $\mathbf{p}$ and $\mathbf{q}$ (which may be different or the same). Let $\mathbf{z}_\infty = \mathbf{x}_\infty - \alpha \mathbf{y}_\infty$ with α chosen so that $\mathbf{z}_\infty$ has at least one zero component with all the others positive. Since $\mathbf{x}_\infty$ and $\mathbf{y}_\infty$ are both fixed points of M, we have

$$ M\mathbf{z}_\infty = M\mathbf{x}_\infty - \alpha M\mathbf{y}_\infty = \mathbf{x}_\infty - \alpha \mathbf{y}_\infty = \mathbf{z}_\infty. $$

But by a previous exercise, we have, (with $M > O$) $M\mathbf{z}_\infty > 0$, contradicting the fact that $\mathbf{z}_\infty$ on the right hand side has a zero component unless $\mathbf{z}_\infty = \mathbf{x}_\infty - \alpha \mathbf{y}_\infty = \mathbf{0}$ or $\mathbf{x}_\infty = \alpha \mathbf{y}_\infty$. In that case, we must have $\alpha = 1$ since $\mathbf{x}_\infty$ and $\mathbf{y}_\infty$ are both probability vectors. It follows that there can only be the same limiting distribution $\mathbf{x}_\infty$ for any two initial distributions (different or not).   □

Corollary 5. *The limiting distribution $\mathbf{x}_\infty$ is unique and asymptotically stable.*

Proof. The corollary is a consequence of the fact that $\mathbf{x}_\infty$ is independent of the initial distribution.   □

We summarize below the properties of a regular MC defined by $\mathbf{x}(n + 1) = M\mathbf{x}(n)$:

(1) Its $N \times N$ transition matrix M is power positive, i.e., $M^k > O$ for all $k \geq k_p \geq 1$.
(2) If $\mathbf{x}(0) = \mathbf{p}$ is a probability vector, then $\mathbf{x}(n) = M^n \mathbf{p}$ is a **positive** probability vector when n is sufficiently large.
(3) $\mathbf{x}(n) \rightarrow$ a limiting (steady state) distribution $\mathbf{x}_\infty$ which is **independent** of the initial distribution $\mathbf{x}(0) = \mathbf{p}$. Hence $\mathbf{x}_\infty$ is unique and asymptotically stable.
(4) With $M\mathbf{x}_\infty = \mathbf{x}_\infty$, the limit distribution $\mathbf{x}_\infty$ is a fixed point of M and can be determined by the eigenvector $\mathbf{v}^{(1)}$ of M for the eigenvalue $\lambda_1 = 1$ with $\mathbf{x}_\infty = c\mathbf{v}^{(1)}$ where c is chosen so that $\mathbf{x}_\infty$ is a probability vector.
(5) The transient distribution $\mathbf{x}(n)$ can be found by solving the linear first order difference equation system.
(6) Except for $\lambda_1 = 1$, all other eigenvalues of the transition matrix M are with less than unit modulus, i.e., $|\lambda_k| < 1$, $1 < k \leq N$. (In particular, there are no complex eigenvalues with a unit modulus.)

2.4.4 Evolution and steady state of the mouse experiment

2.4.4.1 Steady state of the mouse experiment

The second stage transition matrix M of the mouse experiment was found in (2.3.1) to be

$$M_2 = M^2 = \begin{bmatrix} 0 & 0.5 & 0.5 \\ 0.5 & 0 & 0.5 \\ 0.5 & 0.5 & 0 \end{bmatrix}^2 = \begin{bmatrix} 0.5 & 0.25 & 0.25 \\ 0.25 & 0.5 & 0.25 \\ 0.25 & 0.25 & 0.5 \end{bmatrix}.$$

M is therefore a power-positive matrix. Starting with an initial probability vector, we are assured by Theorem 4 that the nth stage probability vector $\mathbf{x}(n) = M^n\mathbf{p}$ converges to some probability vector $\mathbf{x}_\infty$ with

$$M\mathbf{x}_\infty = \mathbf{x}_\infty.$$

A non-trivial solution exists for this homogeneous system since $\lambda_1 = 1$ is an eigenvalue. Written as

$$(M - I)\mathbf{x}_\infty = \begin{bmatrix} -1 & 0.5 & 0.5 \\ 0.5 & 0 & 0.5 \\ 0.5 & 0.5 & -1 \end{bmatrix} \begin{pmatrix} v_1 \\ v_2 \\ v_3 \end{pmatrix} = \mathbf{0},$$

this homogeneous system has the following unique solution $c(1, 1, 1)^T$ up to a multiplicative factor c. We can choose c to re-scale the solution to give a probability vector $\mathbf{x}_\infty$:

$$\mathbf{x}_\infty = c(1, 1, 1)^T = \frac{1}{3}(1, 1, 1)^T.$$

The result confirms Theorem 5 that the long term (steady state) behavior is independent of the initial condition.

2.4.4.2 Evolution of the probability distribution

If we are interested in the evolution of the probability distribution $\mathbf{x}(n)$ as n increases, we may make use of (2.3.5) and the decomposition of M^n in (2.3.7) to get

$$\mathbf{x}(n) = M^n\mathbf{x}(0) = M^n\mathbf{p} = V\Lambda^n V^{-1}\mathbf{p}$$

where V is the modal matrix (whose columns are the N distinct eigenvectors) of M and $\mathbf{p}$ is the prescribed probability distribution for $\mathbf{x}(0)$. The expression for $\mathbf{x}(n)$ leads to the conventional approach to solving the IVP for homogeneous systems of linear difference equations (2.3.6):

Theorem 6. *When M is not defective, the solution of the IVP (2.3.6) is given by*

$$\mathbf{x(n)} = c_1 \mathbf{v}^{(1)} \lambda_1^n + \cdots + c_N \mathbf{v}^{(N)} \lambda_N^n$$

where $\left\{ \mathbf{v}^{(k)}, \lambda_k \right\}$ are the eigen-pairs of M and $\mathbf{c} = (c_1, \ldots, c_N)^T$ is given by $\mathbf{c} = V^{-1}\mathbf{p}$ where $V = \left[\mathbf{v}^{(1)}, \ldots, \mathbf{v}^{(N)} \right]$ is the modal matrix of M.

For the mouse experiment, M is as given in (2.2.1) with three known eigen-pairs to form

$$\Lambda = \begin{bmatrix} 1 & 0 & 0 \\ 0 & 2^{-1} & 0 \\ 0 & 0 & 2^{-1} \end{bmatrix}, \quad V = \begin{bmatrix} 1 & 1 & 0 \\ 1 & 0 & 1 \\ 1 & -1 & -1 \end{bmatrix}, \quad V^{-1} = \frac{1}{3}\begin{bmatrix} 1 & 1 & 1 \\ 2 & -1 & -1 \\ -1 & 2 & -1 \end{bmatrix}.$$

It follows that

$$\mathbf{x}(n) = \frac{1}{3} V \begin{bmatrix} 1 & 0 & 0 \\ 0 & 2^{-n} & 0 \\ 0 & 0 & 2^{-n} \end{bmatrix} V^{-1}\mathbf{p}$$

which is a simpler expression for calculating $\{x_i(n)\}$ compared to computing the nth power of M. The gain in computational efficiency with the help of matrix diagonalization increases geometrically as n and N increase.

In the limit as $n \to \infty$, we have

$$\lim_{n \to \infty} [\mathbf{x(n)}] = \frac{1}{3} V \begin{bmatrix} 1 & 0 & 0 \\ 0 & 0 & 0 \\ 0 & 0 & 0 \end{bmatrix} V^{-1}\mathbf{p} = \frac{1}{3}\begin{bmatrix} 1 & 1 & 1 \\ 1 & 1 & 1 \\ 1 & 1 & 1 \end{bmatrix}\mathbf{p}$$

$$= \frac{1}{3}\begin{pmatrix} p_1 + p_2 + p_3 \\ p_1 + p_2 + p_3 \\ p_1 + p_2 + p_3 \end{pmatrix} = \frac{1}{3}\begin{pmatrix} 1 \\ 1 \\ 1 \end{pmatrix}$$

with the right hand side independent of the initial probability vector $\mathbf{p}$ as assured by Theorem 5.

Exercise 6. (Generations of Red Oak and White Pine) Consider a huge forest comprising of red oaks and white pines, with either one or the other type of trees found at any particular tree location inside this forest. Both types of trees have (roughly) the same life span. When a tree dies, the replacement tree grown in that location may be either of the same type or the other type. Available data show that when a red oak (R) die, it is equally likely replaced by another red oak or by a white pine (W) but the replacement for a white pine is more likely (by a ratio of 3 to 1) to be a red oak. Of interest is how does the forest evolve and what is the distribution of red oaks and white pines after many generations.

Treat the replacement process as a Markov chain, with the index n corresponding to one tree generation. The sample spaces of different generations of trees in any location consist of two outcomes *red oak* ($X_1(n)$) and *white pine* ($X_2(n)$).

We denote by $x_1(n)$ and $x_2(n)$ the two components of the probability vector $\mathbf{x}(n) = (x_1(n), x_2(n))^T$ (with the superscript T indicating the *transpose* of a matrix) of the tree at any particular tree location be red oak and white pine, respectively.

a) Formulate the transition matrix M for the transition of $\mathbf{x}(n)$ to $\mathbf{x}(n+1)$.
b) Calculate $M^{(2)} = M^2$ and $M^{(3)} = M^3$.
c) Determine the eigen-pairs of M and use it to solve the IVP with $\mathbf{x}(0) = (p_r, p_w)^T$.
d) Determine the limiting probability distribution vector $\mathbf{x}_\infty$ by solving (2.4.2).

Exercise 7. (Social Mobility) From data compiled by government census, it is known that a fraction of the offsprings of families in a particular income group becomes significantly more wealthy and another fraction becomes significantly less well off with the rest not doing any better or worse. To gain some insight to the properties of regular Markov chains, suppose we simply divide up families into high (X_1), middle (X_2) and low (X_3) income group. Suppose data over many generations show that 3/5 of families in high income group remain in that bracket after one generation, 3/10 drop to the middle income bracket and 1/10 drops more drastically to the low income bracket. Correspondingly, only 1/10 of the families in the middle income bracket moves up to the high income bracket, another 1/10 drops to the low income group while the remaining 4/5 remain in the same middle income group after one generation. The fractional changes in low income families after one generation to high and middle income bracket are 1/10 and 1/5, respectively (while the remaining families stay in the low income bracket).

Assume for the present discussion that these fractional changes remain the same from generation to generation.

a) Formulate the transition matrix M for the transition of $\mathbf{x}(n)$ to $\mathbf{x}(n+1)$.
b) Calculate $M^{(2)} = M^2$ and $M^{(3)} = M^3$.
c) Determine the eigen-pairs of M and use it to solve the IVP with $\mathbf{x}(0) = (p_1, p_2, p_3)^T$.
d) Determine the limiting probability distribution vector $\mathbf{x}_\infty$ by solving (2.4.2).

2.5 DNA Mutation

2.5.1 *The double helix*

The central dogma of biology is from DNA to RNA to proteins, the latter constitute fundamental units for all parts and functions of living organisms. DNA are therefore the basic building blocks for these organisms. DNA molecules encode genetic information and these molecules (with their genetic information) are copied and passed on from parents to an offspring. Though highly accurate, the copying process is not immune from errors leading to genetic mutation and consequently the evolution of living organisms. To study biological evolution, we should know how errors incur in the DNA copying process, copying errors that result in genetic mutation.

The 1962 Nobel Prize in Physiology or Medicine was awarded to James Watson, Francis Crick and Maurice Wilkins for their discovery of the double helix structure of DNA molecules. In 1953, Watson and Crick saw an x-ray pattern of a crystal of the DNA molecule made by Rosalind Franklin and Maurice Wilkins; it gave Watson and Crick enough information to make an accurate model of the DNA molecule. Their model showed a twisted double helix with little rungs connecting the two helical strands (see Fig. 2.1). At each end of a rung of the double helix ladder is one of four possible molecular subunits (called nucleobases or simply bases): *adenine* (A), *guanine* (G), *cystosine* (C), and *thymine* (T).

The shape of *adenine* is complementary to *thymine*; they are bound together consistently at opposite ends of a rung of the DNA ladder through a hydrogen bond to form a nucleotide. The nucleobase *guanine* is structurally similar to *adenine* and complementary to *cytosine*. The *guanine-cytosine* pair are also bound together consistently at the opposite ends of a DNA ladder rung. In other words, we always find either A paired with T or G paired with C (but neither A nor T with C or G). Thus, once we know the base at one side of a rung, we can deduce the base at the other end of the same rung. For example, if along one strand of the DNA ladder, we have a base sequence

$$ATTAGAGCGCGT,$$

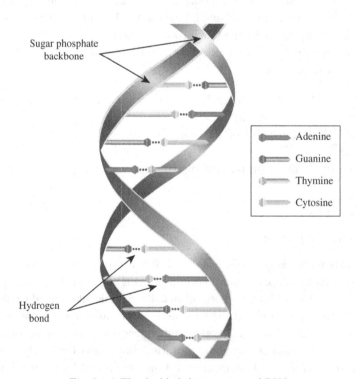

Fig. 2.1. The double helix structure of DNA

then the corresponding sequence along the other strand opposite to the same stretch must be

$$TAATCTCGCGCA.$$

(Because of their structural similarity, cytosine and thymine are called *pyrimidines* while adenine and guanine are called *purines*.) As such, a DNA molecule (or a segment of it) is specified by a sequence of letters that is some combination of the four letters A, T, C and G.

Once the model was established, its structure hinted that DNA was indeed the carrier of the genetic code and thus the key molecule of heredity, developmental biology and evolution. Heredity requires genetic information be passed on to an offspring. This is accomplished during cell division when the twisted double-helix DNA molecule ladder unzips into two separate strands. One new molecule is formed from each half-ladder and, due to the required pairings, this gives rise to two identical daughter copies from each parent molecule. Though elaborate safeguards are in place to ensure fidelity in the replication, errors still occur though infrequently.

2.5.2 *Mutation due to base substitutions*

The most common type of errors during the replication process is a replacement of one base by another at a certain site of the strand sequence. For instance, if the sequence in the parent DNA molecule *ATTAGAGC* should become *ATTACAGC* in the offspring DNA, then there is a *base substitution* $G \to C$ at the fifth site of the sequence. The base substitution replaces a pyrimidine by another pyrimidine (or one replaces a purine by another purine) is known as *transition*. A base substitution of a purine by a pyrimidine (or a pyrimidine by a purine) is called a *transversion*. Error types other than base substitution are also possible. These include deletion, insertion and inversion of a section of the sequence. Their occurrence are much less frequent than base substitutions and will be ignore in the discussion herein to focus only on mutations due to base substitutions.

For the restricted problem of mutations by base substitutions only (and for other types of mutations), one issue of interest is the amount of mutation that has occurred after a large number of generations of offspring from a DNA sequence, especially over an evolutionary time scale. If data for all generations involved are not available or not used, then the difference between the original sequence of generation n and the sequence of generation $n + m$ does not necessarily give an accurate estimate of the amount of mutation that has taken place since one or more back mutation might have taken place. If the sequence in the parent DNA molecule *ATTAGAGC* should become *ATTACAGC* four generations later, then one base substitution of $G \to C$ at the fifth site of the sequence is only one possible kind of mutation. It may also be a sequence of base substitutions such as $G \to C \to G \to G \to C$ or $G \to C \to T \to G \to C$ with the former having one back mutation while the latter involves no back mutation at all. A more sophisticated approach is needed to determine the amount of mutation involved between an ancestral DNA site and the same site of its descendent DNA molecule after n generations. In the remainder of this section, we describe how discrete Markov chain models have been used for this purpose.

2.5.3 *Markov chain model*

Given a particular nucleobase at a particular site of an initial (ancestral) DNA sequence, knowing the odds that it would mutate to another nucleobase after replication would allow us to better estimate its evolution. For the nucleobase A for example, it would be helpful to know the probability (or the fraction of times) p_{AG} that it would be replaced by the nucleobase G, the probability p_{AC} of being replaced by C and the probability p_{AT} of being replaced by T. Since deletion is not allowed, that the fraction of times that A remains unchanged would have to be $1 - p_{AG} - p_{AC} - p_{AT} = 1 - \Sigma_A$. Similar odds would be needed for the nucleobases G, C and T. We arrange all these quantities as a matrix M,

$$
\begin{array}{c}
\\
A_{n+1} \\
G_{n+1} \\
C_{n+1} \\
T_{n+1}
\end{array}
\begin{array}{cccc}
A_n & G_n & C_n & T_n \\
\left[\begin{array}{cccc}
1 - \Sigma_A & p_{AG} & p_{AC} & p_{AT} \\
p_{GA} & 1 - \Sigma_G & p_{GC} & p_{GT} \\
p_{CA} & p_{CG} & 1 - \Sigma_C & p_{CT} \\
p_{TA} & p_{TG} & p_{TC} & 1 - \Sigma_T
\end{array}\right] = M,
\end{array}
$$

for the following vector relation for the evolution of different bases at a particular site of the DNA sequence:

$$
\mathbf{x}(n+1) = M\mathbf{x}(n)
$$

where

$$
\mathbf{x}_n = (A, G, C, T)_n^T,
$$

with a component of $\mathbf{x}_n$ being the fraction of time (or the probability of) the particular nucleobase occupying the same site at the nth generation. The numerical values of the transition matrix entries $\{p_{ij}\}$ are key to the evolution of the organism; they are normally estimated from available replication data over many generations.

2.5.4 *Equal opportunity substitution*

To illustrate the use of the Markov chains to learn more about the genetic mutation, we consider the consequences of a highly speculative, one-parameter *equal opportunity* model for the transition matrix M. In this version of the Jukes-Cantor (1969) model, we stipulate that substitution of a nucleobase by any of the other three bases are equally likely with probability $\alpha/3$ where α is a number to be specified (or estimated from available data). It ranges from 10^{-8} mutations per site per year for mitochondrial DNA to 0.01 mutations per site per year for the influenza A virus DNA. The corresponding transition matrix M then becomes

$$M = \begin{bmatrix} 1-\alpha & \alpha/3 & \alpha/3 & \alpha/3 \\ \alpha/3 & 1-\alpha & \alpha/3 & \alpha/3 \\ \alpha/3 & \alpha/3 & 1-\alpha & \alpha/3 \\ \alpha/3 & \alpha/3 & \alpha/3 & 1-\alpha \end{bmatrix}. \tag{2.5.1}$$

For such a transition matrix, an ancestral sequence with A at a particular site (so that $\mathbf{x}_0 = (1,0,0,0,0)^T$ would have any one of the four nucleobases at the same site after one replication with probabilities given by

$$\mathbf{x}_1 = M\mathbf{x}_0 = \begin{pmatrix} 1-\alpha \\ \alpha/3 \\ \alpha/3 \\ \alpha/3 \end{pmatrix}.$$

In other words, a substitution by anyone of the three nucleobase after one replication is equally likely at $\alpha/3$ fraction of the time while the site remains occupied by A at $1-\alpha$ fraction of the time.

Exercise 8. Obtain the four distinct eigen-pairs for the transition matrix for the Jukes-Cantor Equal Opportunity Model. (Hint: Try $\lambda = 1 - 4\alpha/3$.)

2.6 Linear Difference Equations

From the various DTMC models discussed in the previous sections of this chapter, we see that many biological phenomena may be modeled by *linear* difference equations after suitable idealization and simplifications. It is therefore desirable to know some general mathematical techniques for solving such equations. In this section, we summarize some methods and approaches that have been frequently found useful in applications.

2.6.1 *A single first order linear equation*

A general single first order linear equation may be taken in the form

$$x(n+1) = \mu(n)x(n) + q(n) \tag{2.6.1}$$

where $\mu(n)$ and $q(n)$ are known scalars, generally functions of n. Even if $q(n) = \mathbf{0}$, Theorem 6 does not apply as long as $\mu(n)$ varies with n. On the other hand, (2.6.1) is effectively a recurrence relations giving successive $x(n)$ in terms of the same quantity at earlier stages; it is then straightforward to deduce the following result:

Proposition 1. *The unique solution of the IVP*

$$x(n + 1) = \mu(n)x(n), \quad x(0) = p$$

is

$$x(n) = p\, \Pi_{k-0}^{(n-1)} \mu(k).\qquad\qquad(2.6.2)$$

Proof. The solution follows upon writing

$$x(n) = \frac{x(n)}{x(n-1)} \frac{x(n-1)}{x(n-2)} \cdots \frac{x(2)}{x(1)} \frac{x(1)}{x(0)} x(0)$$

$$= \mu(n-1)\mu(n-2)\cdots\mu(2)\mu(1)\mu(0)x(0)$$

Uniqueness is left to an exercise. $\square$

Corollary 6. *For a constant coefficient* $\mu(n) = \mu(0) = \mu_0$, *(independent of n), the solution (2.6.2) reduces to the expected result:*

$$x(n) = p\mu_0^n,\qquad\qquad(2.6.3)$$

(with $\lambda = \mu_0$ and $c_1 = p$ in Theorem 6).

Proposition 2. *The unique solution of the IVP*

$$x(n + 1) = \mu(n)x(n) + q(n), \quad x(0) = p$$

is

$$x(n) = \Pi_{k=0}^{n-1}\mu(k) \left[p + \sum_{i=0}^{n-1} \frac{q(i)}{\Pi_{j=0}^{i}\mu(j)} \right]$$

for $n = 1, 2, \ldots$.

Proof. We prove the simpler case of a constant $\mu(n) = \mu_0$ (and leave the general case as an exercise). In this simpler case, we have

$$x(1) = \mu_0 x(0) + q(0),$$

$$x(2) = \mu_0 x(1) + q(1) = \mu_0 \left[\mu_0 x(0) + q(0)\right] + q(1)$$
$$= \mu_0^2 x(0) + \mu_0 q(0) + q(1),$$

By induction, we get

$$x(n) = \mu_0^n x(0) + \mu_0^{n-1} q(0) + \mu_0^{n-2} q(1) + \cdots + \mu_0 q(n-2) + q(n-1)$$

$$= \mu_0^n x(0) + \sum_{k=0}^{n-1} \mu_0^{n-1-k} q(k) = \mu_0^n \left[p + \sum_{k=0}^{n-1} \mu_0^{-(k+1)} q(k) \right]. \tag{2.6.4}$$

Again, a uniqueness proof is straightforward (by setting $z(n) = x(n) - y(n)$, the differences of two solutions should more than one exist.) □

While the results above can be seen as consistent counterparts of the corresponding results for a single first order linear ODE, it is fair to say that they are merely more compact expressions of the direct consequences of the respective recurrence relations after n steps (see also exercises in Assignment II for further exposition of this observation). There is really no creativity or ingenuity in the steps leading to them. On the other hand, we should be grateful for the fact that we can always get the solution of a problem involving difference equations because of recursive nature of the problem and the computing power available today.

2.6.2 *Linear systems with constant coefficients*

Similar to the ODE counterpart, single higher order difference equations and a system of more than one linear difference equations are more compactly written in terms of a state vector as we did for Markov chains in chapters 2 and 3:

$$\mathbf{x}(n+1) = M(n)\mathbf{x}(n) + \mathbf{q}(n), \quad \mathbf{x}(0) = \mathbf{p} \tag{2.6.5}$$

for $n = 0, 1, 2, \ldots$. Taken in the form (2.6.5), $\mathbf{x}(n)$, $\mathbf{q}(n)$ and $\mathbf{p}$ are N vectors and M is a known $N \times N$ matrix. Among the vectors, $\mathbf{q}(n)$ and $\mathbf{p}$ are prescribed and $\mathbf{x}(n)$ is to be determined starting with some initial state (distribution) $\mathbf{x}(0) = \mathbf{p}$. If $\mathbf{q}(n) = \mathbf{0}$, the linear system is said to be *homogeneous*. If $M(n)$ does not depend on n (so that it is M for all n), then the system is said to be of *constant coefficients or time-invariant*. The matrix M is said to be *non-defective* if it has a full set of eigenvectors. The general solution $\mathbf{x}^{(h)}(n)$ of linear homogeneous systems

($\mathbf{q}(n) = \mathbf{0}$) with a non-defective constant (transition) matrix M is given by Theorem 6. Briefly, we may write the solution as

$$\mathbf{x}^{(h)}(n) = c_1 \mathbf{v}^{(1)} \lambda_1^n + c_2 \mathbf{v}^{(2)} \lambda_2^n + \cdots + c_N \mathbf{v}^{(N)} \lambda_N^n$$

where $\{\lambda_k, \mathbf{v}^{(k)}\}$ are the eigen-pairs of M.

If $\mathbf{q}(n) = \mathbf{0}$ in the actual problem, the constants $\{c_1, c_2, \ldots, c_N\}$ are determined by the initial condition $\mathbf{x}(0) = \mathbf{p}$ with $\mathbf{x}(n) = \mathbf{x}^{(h)}(n)$ in that case.

If $\mathbf{q}(n) \neq \mathbf{0}$ in the actual problem, then the general solution of linear inhomogeneous systems with forcing and a non-defective constant matrix M may be obtained by the method of variation of parameters or, for some simple forcing term $\mathbf{q}(n)$, the method of undetermined coefficients. The latter is very similar to its ODE counterpart and will not be discussed here. The method of variation of parameters requires some slight modification and will be discussed in a later section. For a *defective* matrix with a multiple eigenvalue for which there are fewer than N eigenvectors, the sure fire method of solution would be to reduce M to Jordan normal form by a suitable similarity transformation analogous to what was done for ODE (see [72] for example).

When M varies with n, then Theorem 6 does not apply though the method of variation of parameters (see last subsection) continues to apply if we have a complete set of (complementary) solutions for the corresponding homogeneous equation. Techniques for finding complementary solutions for linear equations with variable coefficients can be developed similar to their counterparts in ordinary differential equations. However, the solutions obtained by such methods are no more attractive than a repeated execution of the recurrence relation implied by the difference equation.

Theorem 7. *The unique solution of the IVP (2.3.6) may be taken in the form*

$$\mathbf{x}(n) = \Pi_{k=0}^{n-1}[M_k] \left\{ p + \sum_{j=0}^{n-1} \Pi_{k=j}^{0}[M_k]^{-1} q_j \right\}$$

where we have set for brevity $M^{(k)} = M_k$, $\mathbf{q}(k) = \mathbf{q}_k$ *and*

$$\Pi_{j=m}^{k}[M_j] = M_k M_{k-1} \cdots M_{m+1} M_m. \tag{2.6.6}$$

Proof. For $n = 0$ and $n = 1$, we have

$$\mathbf{x}_1 = M_0 \mathbf{x}_0 + \mathbf{q}_0 = M_0 \mathbf{p} + \mathbf{q}_0$$

$$\mathbf{x}_2 = M_1 \mathbf{x}_1 + \mathbf{q}_1 = M_1[M_0 \mathbf{p} + \mathbf{q}_0] + \mathbf{q}_1 = M_1 M_0 \mathbf{p} + M_1 \mathbf{q}_0 + \mathbf{q}_1$$

$$\mathbf{x}_3 = \Pi_{k=0}^{2}[M_k]\mathbf{p} + \Pi_{k=0}^{2}[M_k] \sum_{m=0}^{2-k} \Pi_{m=0}^{2-k}[M_m]^{-1} \mathbf{q}_{2-k}$$

upon observing the notation (2.6.6). By induction, we get for general n

$$\mathbf{x}_n = M_{n-1}\mathbf{x}_{n-1} + \mathbf{q}_{n-1}$$

$$= \Pi_{k=0}^{n-1}[M_k] \left\{ \mathbf{p} + \sum_{m=0}^{n-1-k} \Pi_{m=0}^{n-1-k}[M_m]^{-1}\mathbf{q}_{n-1-k} \right\}. \qquad \Box$$

2.6.3 *Reduction of order*

For many problems, we do not have a sufficient number of complementary solutions. This is certainly the case of a defective matrix; but there are others. For the equation

$$x_{n+2} - (n+1)x_{n+1} + nx_n = 0,$$

we see by inspection that $x_n = 1$ is a solution. But it is not so easy to spot the second complementary solution which is needed to solve an IVP. Note that this is a linear equation of variable coefficients and can be rewritten as a first order vector difference equation by setting $y_n = x_{n+1}$ and writing the given linear difference equation as

$$\mathbf{x}_{n+1} = \begin{pmatrix} x \\ y \end{pmatrix}_{n+1} = \begin{bmatrix} 0 & 1 \\ -n & n+1 \end{bmatrix} \begin{pmatrix} x \\ y \end{pmatrix}_n = M(n)\mathbf{x}_n.$$

While Theorem 7 applies, a more informative solution can be obtained by the method of *reduction of order*, a difference equation analogue of the same method for ODE.

Consider the general inhomogeneous second order linear equation

$$y_{n+2} + \beta_n y_{n+1} + \alpha_n y_n = f_n \qquad (2.6.7)$$

with the initial conditions

$$y_0 = p_0, \quad y_1 = p_1. \qquad (2.6.8)$$

Suppose we know one complementary solution x_n of the homogeneous equation so that

$$x_{n+2} + \beta_n x_{n+1} + \alpha_n x_n = 0. \qquad (2.6.9)$$

To find a second complementary solution or, better yet, to find the complete solution for the IVP, we set, as in the ODE case,

$$y_n = u_n x_n$$

where u_n is an unknown function of n. In terms of u_n, the original difference equation (2.6.7) becomes

$$x_{n+2}u_{n+1} + u_{n+1}[x_{n+2} + \beta_n x_{n+1} + \alpha_n x_n] - \alpha_n x_n u_n = f_n,$$

where $w_n = u_{n+1} - u_n$ and where we have added and subtracted terms that sum up to zero. Now the combination inside brackets vanishes since x_n is a complementary solution so that the equation above simplifies to

$$x_{n+2}w_{n+1} = \alpha_n x_n w_n + f_n. \tag{2.6.10}$$

Since the known complementary solution x_n does not vanish, (2.6.10) is a linear first order difference equation for w_n and its solution is given by Theorem 1 with $w_0 = u_1 - u_0 = (p_1/x_1) - (p_0/x_0)$. Having w_n, we can solve another linear first order difference equation $u_{n+1} = u_n + w_n$ for u_n with $u_0 = p_0/x_0$. The product solution $y_n = u_n x_n$ evidently satisfies the initial condition on y_n with $y_0 = x_0 u_0 = p_0$ and $y_1 = x_1 u_1 = p_1$. Since (2.6.7) is the most general form of a linear second order difference equation, we have effectively formulated the method for solving the IVP for any second order linear difference equation with variable coefficients once we have one complementary solution.

For linear difference equation of order n, the method of reduction order reduces the solution process to solving a difference equation of order $n-1$. The simplification may or may not lead to a complete solution for the original equation.

2.6.4 *Variation of parameters*

When all n linearly independent complementary solutions of a linear system are known, the *method of variation of parameters* is usually more effective for a complete solution of the original system.

For the second order linear difference equation (2.6.7) with two linearly independent complementary solutions $x_n^{(k)}$, $k = 1$ and 2, we write the solution of the original IVP as

$$y_n = u_n x_n^{(1)} + v_n x_n^{(2)}. \tag{2.6.11}$$

Since we have introduced two new unknown functions u_n and v_n and there is only a single equation for determining y_n, we have the flexibility of specifying a second condition for the two new unknowns. We take it to be

$$\left(u_{n+1} - u_n\right) x_{n+1}^{(1)} + \left(v_{n+1} - v_n\right) x_{n+1}^{(2)} = 0 \tag{2.6.12}$$

so that

$$y_{n+1} = u_n x_{n+1}^{(1)} + v_n x_{n+1}^{(2)}.$$

In that case, the original second order linear inhomogeneous difference equation (2.6.7) simplifies to

$$\left(u_{n+1} - u_n\right) x_{n+2}^{(1)} + \left(v_{n+1} - v_n\right) x_{n+2}^{(2)} = f_n, \tag{2.6.13}$$

where we have made use of the fact that $x_n^{(1)}$ and $x_n^{(2)}$ are complementary solutions. The two equations (2.6.12) and (2.6.13) can be solved for $u_{n+1} - u_n$ and $v_{n+1} - v_n$

to give

$$u_{n+1} - u_n = \frac{x_{n+1}^{(2)}}{\Delta_n} \frac{f_n}{\alpha_n}, \quad v_{n+2} - v_{n+1} = -\frac{x_{n+1}^{(1)}}{\Delta_n} \frac{f_n}{\alpha_n} \tag{2.6.14}$$

where

$$\Delta_n = x_{n+1}^{(1)} x_n^{(2)} - x_{n+1}^{(2)} x_n^{(1)}. \tag{2.6.15}$$

The two relations (2.6.14) are two uncoupled first order linear inhomogeneous difference equations with a constant coefficient for which the solution (2.6.4) applies with $\mu_0 = 1$ and (2.6.8) provides the needed initial conditions. The resulting expressions for u_n and v_n are then inserted into (2.6.11) to complete the solution process.

Exercise 9. Prove the following relations for the Wronskian Δ_n of the second order linear ODE (2.6.7):

$$a) \ \ \Delta_{n+1} = -\alpha_n \Delta_n, \quad b) \ \ \Delta_n = c_0 \Pi_{n=0}^{n-1} \alpha_n$$

with $\alpha_0 = 1$ and c_0 a constant to be specified by an appropriate known initial condition (which is not zero if the two complementary solutions are linearly independent).

For higher order linear systems taken in the form of a first order vector difference equation

$$\mathbf{y}_{n+1} = A_n \mathbf{y}_n + \mathbf{f}_n, \tag{2.6.16}$$

the method of variation of parameters also applies by writing the solution as

$$\mathbf{y}_n = \Psi_n \mathbf{u}_n \tag{2.6.17}$$

where Ψ_n is the fundamental matrix of the complete set of linearly independent complementary solutions.

Exercise 10. Upon inserting (2.6.17) into (2.6.16), obtain the following first order linear system with constant coefficients for $\mathbf{u}_n$:

$$\mathbf{u}_{n+1} = \mathbf{u}_n + \Psi_{n+1}^{-1} \mathbf{f}_n$$

with $\mathbf{u}_0$ known from the relation (2.6.17) for $n = 0$

$$\mathbf{u}_0 = \Psi_0^{-1} \mathbf{y}_0$$

where $\mathbf{y}_0$ is the prescribed initial data.

2.7   Appendix A — Proof of a Basic Existence Theorem for Regular MC

Below is a proof of Theorem 4. With no loss in generality, we give a proof for the case $M > 0$.

Theorem 8. *For a regular Markov chain with an $N \times N$ transition matrix $M = [p_{ij}]$ and any initial probability distribution $\mathbf{x}(0) = \mathbf{p}$, there exists a probability vector $\mathbf{x}_\infty$ to which the sequence of probability vectors $\{\mathbf{x}(n) = M^n\mathbf{p}\}$ converges as $n \to \infty$.*

Proof. Form $\mathbf{q}^T M^n \mathbf{p} = \mathbf{p}^T (M^T)^n \mathbf{q}$ for any multiple of an arbitrary probability vector $\mathbf{q} \neq \mathbf{0}$. Set $\mathbf{w}(n) = (M^T)^n \mathbf{q}$. It suffices to show that $\mathbf{w}(n)$ converges as $n \to \infty$ for any vector $\mathbf{q} \neq \mathbf{0}$ with $q_i \geq 0$. Observe that $\mathbf{w}(n)$ satisfies the difference equation

$$\mathbf{w}(n+1) = M^T \mathbf{w}(n), \quad \mathbf{w}(0) = \mathbf{q}.$$

For each n, let $u(n)$ be the largest component of $\mathbf{w}(n)$ and $v(u)$ the smallest. Since

$$w_i(n+1) = \sum_{j=1}^{N} p_{ji} w_j(n)$$

and since $p_{ij} \geq 0$ and $\sum_{j=1}^{N} p_{ji} = 1$ for $i = 1, 2, \ldots, N$, it follows that

$$u(n+1) \leq u(n), \quad v(n+1) \geq v(n).$$

Therefore $\{u(n)\}$ is a monotone decreasing sequence bounded from below by zero; and $\{v(n)\}$ is a monotone increasing sequence bounded from above by 1, respectively. Hence, both converge to a limit, denoted by u_∞ and v_∞, respectively. Our goal is to show $u_\infty = v_\infty$.

With no loss in generality, let $u(n)$ be the first component of $\mathbf{w}(n)$, i.e., $u(n) = w_1(n)$. In that case, we have (for some $i \neq 1$)

$$v(n+1) = w_i(n+1) = \sum_{j=1}^{N} p_{ji} w_j(n) = \sum_{j=2}^{N} p_{ji} w_j(n) + (p_{11} - d)w_1(n) + du(n)$$

$$\geq \sum_{j=2}^{N} p_{ji} v(n) + (p_{11} - d)v(n) + du(n) = (1 - d)v(n) + du(n).$$

where $d = \min[p_{ij}] \leq 1/2$ for $n \geq 2$ (and $d > 0$ since M is a positive matrix). Similarly, we have also

$$u(n+1) \leq (1 - d)u(n) + dv(n).$$

Combining these two inequalities gives

$$u(n+1) - v(n+1) \leq (1 - 2d)\left[u(n) - v(n)\right] \leq (1 - 2d)^n \left[u(0) - v(0)\right].$$

Keeping in mind that $0 < d \leq 1/2$ so that $0 \leq 1 - 2d < 1$, the difference $[u(n) - v(n)] \to 0$ as $n \to \infty$ so that $u_\infty = v_\infty$ in the limit.

Thus, not only the two sequences $\{u(n)\}$ and $\{v(n)\}$ converge, they both converge to the same limit resulting in

$$\lim_{n \to \infty} \mathbf{w}(n) = (M^T)^n \mathbf{q} = w_\infty (1, 1, \ldots, 1)^T \tag{2.7.1}$$

where we have denoted by w_∞ the two equal limits u_∞ and v_∞. Since $\mathbf{q}$ is finite and arbitrary, the power matrix $(M^T)^n = (M^n)^T$ converges and $M^n \to$ a well-defined $\bar{M}$. (The kth column of the limiting matrix $\bar{M}$ corresponds to the limiting vector (2.7.1) for $\mathbf{q} = (\delta_{1k}, \delta_{2k}, \ldots, \delta_{Nk})^T$.)   □

Chapter 3

Discrete Stage Absorbing Markov Chains

3.1 Introduction

Regular Markov chains constitute an important class of discrete stochastic processes in applications. There are however other Markov chains that are not regular. Two simple counter-examples were given in (2.4.1). One class of MC prevalent in science and engineering that is not regular is known as *absorbing Markov chains*. In this chapter, we examine some of the properties and applications of this important class of Markov chains.

Definition 8. A state in a Markov chain is an **absorbing state** if it is impossible to leave it. A state is a **transient state** if it is not an absorbing state.

Definition 9. A Markov chain is said to be an **absorbing MC** if (i) it has at least one absorbing state, and (ii) from every state it is possible go to an absorbing state (not necessarily in one step).

Remark 2. Of the two counter-examples in (2.4.1), the first with $M = I$ is absorbing while the second is neither absorbing (why?) nor regular (why?).

Theorem 9. *For an absorbing Markov chain, it is a certainty that the process ends up in one of the absorbing states.*

Proof. (sketched) From a non-absorbing state S_j, let n_j be the minimum number of steps required to reach an absorbing state. Let p_j be the probability that starting from S_j the process does not reach an absorbing state in n_j steps. (Note that $p_j < 1$, why?) Let $n = \max[n_j]$ and $p = \max[p_j]$. The probability of not reaching an absorbing state in n steps is less than p, in $2n$ steps is less than p^2, etc. In general, the probability of not reaching an absorbing state in $k \cdot n$ steps is less than p^k. Since $p < 1$ (given that the Markov chain is absorbing), the probability of not reaching an absorbing state tends to zero as $k \to \infty$.   □

For an absorbing Markov chain, there are at least three interesting problems:

(1) What is the probability of the process ending up in a particular absorbing state?
(2) On the average, how "long" would it take for the process to reach an absorbing state starting from a non-absorbing state (also known as a *transient state*)?
(3) On the average, how many times does the process pass through each non-absorbing state before ending in an absorbing state?

For answers to these questions, we first examine a few well known examples of this class of Markov chains.

3.2   Gambler's Ruin

3.2.1   *The model*

The classic example of an absorbing Markov chain is the gambler's ruin (or success) in playing the following (fair) coin tossing game. Suppose Player 1 has $2 and Player 2 has $3 at the start of the game. Each time a fair coin is tossed and a head (*H*) turns up, Player 1 takes $1 from Player 2. If a tail (*T*) turns up, Player 1 gives Player 2 $1. The game ends when one of the players loses all his/her capital ($). The obvious question of interest is what happens to this game eventually? Intuitively, Player 1 who has less initial capital will probably lose. More specifically, if the game is played many times, it is expected that Player 1 would lose more often than not. How can we substantiate this expectation, that is, how can we prove it mathematically? How do we formulate a mathematical model that allows us to analyze this simple game so that we can answer questions such as the probability of Player 1 winning (or losing)?

In this section, we formulate a mathematical model of this gambler's ruin problem and introduce mathematical techniques to deduce the desired answers. In the process, we see how the mathematical tools used can also be extended to handle similar but more complicated absorbing Markov chain problems. We will also see how other life science problems can also be modeled and analyzed as absorbing Markov chains.

We begin by letting the *state* $S(n)$ (or S_n interchangeably) be the total of Player 1's capital after n tosses, designated as *stage n*. There are six possible states corresponding to the six elementary events {$0, $1, $2, $3, $4, $5} at each stage. The transition matrix M for the game remains unchanged from stage to stage and is given by

$$
\begin{array}{c}
\textit{Stage } n \\
Stage\ (n+1)\quad
\begin{array}{c|cccccc}
 & \$0 & \$1 & \$2 & \$3 & \$4 & \$5 \\
\hline
\$0 & 1 & \frac{1}{2} & 0 & 0 & 0 & 0 \\
\$1 & 0 & 0 & \frac{1}{2} & 0 & 0 & 0 \\
\$2 & 0 & \frac{1}{2} & 0 & \frac{1}{2} & 0 & 0 \\
\$3 & 0 & 0 & \frac{1}{2} & 0 & \frac{1}{2} & 0 \\
\$4 & 0 & 0 & 0 & \frac{1}{2} & 0 & 0 \\
\$5 & 0 & 0 & 0 & 0 & \frac{1}{2} & 1
\end{array}
\end{array} \equiv M = [p_{ij}]
$$

More specifically, let $\mathbf{x}(n) = (x_1(n), x_2(n), x_3(n), \ldots)^T$ be the distribution vector of the probability of Player 1's capital among the different elementary events. Then the corresponding distribution vector in the next stage $\mathbf{x}(n + 1)$ is given by the evolutionary relation

$$\mathbf{x}(n + 1) = M\mathbf{x}(n). \tag{3.2.1}$$

M is clearly a probability matrix since $p_{ij} \geq 0$ and $\sum_{i=1}^{N} p_{ij} = 1$ where $N = 6$ in our particular example of the Gambler's ruin.

The transition relation (3.2.1) between the probability vectors of consecutive stages is again a system of linear difference equations. The solution of the initial value problem (consisting of the difference equation (3.2.1) and the initial probability vector $\mathbf{x}(0) = \mathbf{p}$ as shown in (3.2.3)) for this system provides complete information about the evolution of the game with stages. However, some useful observations can be made even before solving the IVP.

(1) $\lambda = 1$ is again seen to be an eigenvalue of the transition matrix M above since the rows of the matrix $M - I$ sum up to a zero row.
(2) M is **not** a power positive matrix (see an assigned exercise).
(3) There is more than one equilibrium states since $\mathbf{x}^{(1)} = (1, 0, 0, 0, 0, 0)^T$ and $\mathbf{x}^{(2)} = (0, 0, 0, 0, 0, 1)^T$ are both fixed points of (3.2.1).

As such, the Markov chain with transition matrix M is *not* a regular Markov chain. To the extent that it has absorbing states that can be reached from all transient states (and hence an absorbing Markov chain), we expect the behavior of this MC will be different from those of regular Markov chains.

3.2.2   *Solution of IVP*

To motivate some further developments that will uncover these differences, we work out presently a simpler version of the same problem with Player 1 having \$1 and Player 2 having \$2 so that there are four elementary events {\$0, \$1, \$2, \$3} in the sample space at each stage. The transition matrix for this game is given by

$$
M = [m_{ij}] = \begin{bmatrix} 1 & \frac{1}{2} & 0 & 0 \\ 0 & 0 & \frac{1}{2} & 0 \\ 0 & \frac{1}{2} & 0 & 0 \\ 0 & 0 & \frac{1}{2} & 1 \end{bmatrix}.
\tag{3.2.2}
$$

For the solution of the linear system of difference equations

$$
\mathbf{x}(n+1) = M\mathbf{x}(n), \quad \mathbf{x}(0) = \mathbf{p}
\tag{3.2.3}
$$

we need the general superposition theorem for linear difference equations:

Theorem 10. *If* $\mathbf{y}^{(1)}(n)$ *and* $\mathbf{y}^{(2)}(n)$ *are two solutions of the (homogeneous linear vector difference) equation* $\mathbf{x}(n+1) = M\mathbf{x}(n)$, *then so is* $\mathbf{y}(n) = c_1\mathbf{y}^{(1)}(n) + c_2\mathbf{y}^{(2)}(n)$ *for any two constants* c_1 *and* c_2.

Proof. (left as an exercise)   □

To make use of this theorem, we seek the eigen-pairs of the matrix M with the eigenvalues being the roots of $|M - \lambda I| = (\lambda - 1)^2(\lambda^2 - \frac{1}{4}) = 0$. Unlike regular Markov chains, $\lambda = 1$ is a double root of the characteristic equation for the eigenvalues. Fortunately, M still has the full set of eigenvectors. It is straightforward to determine the four eigen-pairs to be $\{1, \mathbf{v}^{(1)}\}, \{1, \mathbf{v}^{(2)}\}, \{\frac{1}{2}, \mathbf{v}^{(3)}\}$ and $\{-\frac{1}{2}, \mathbf{v}^{(4)}\}$ where

$$
\mathbf{v}^{(1)} = \begin{pmatrix} 1 \\ 0 \\ 0 \\ 0 \end{pmatrix}, \quad \mathbf{v}^{(2)} = \begin{pmatrix} 0 \\ 0 \\ 0 \\ 1 \end{pmatrix}, \quad \mathbf{v}^{(3)} = \begin{pmatrix} 1 \\ -1 \\ -1 \\ 1 \end{pmatrix}, \quad \mathbf{v}^{(4)} = \begin{pmatrix} 1 \\ -3 \\ 3 \\ -1 \end{pmatrix},
$$

with the eigenvectors determined up to a multiplicative constant. We may then use Theorem 10 to write the (general) solution of the system $\mathbf{x}(n+1) = M\mathbf{x}(n)$ with the transition matrix (3.2.2) as

$$
\mathbf{x}(n) = c_1\mathbf{v}^{(1)}(1)^n + c_2\mathbf{v}^{(2)}(1)^n + c_3\mathbf{v}^{(3)}\left(\frac{1}{2}\right)^n + c_4\mathbf{v}^{(4)}\left(-\frac{1}{2}\right)^n,
\tag{3.2.4}
$$

where the constants $\{c_k\}$ are determined by the initial probability distribution $\mathbf{x}(0) = \mathbf{p} = (p_1, p_2, p_3, p_4)^T$:

$$
\mathbf{x}(0) = \begin{bmatrix} 1 & 0 & 1 & 1 \\ 0 & 0 & -1 & -3 \\ 0 & 0 & -1 & 3 \\ 0 & 1 & 1 & -1 \end{bmatrix} \begin{pmatrix} c_1 \\ c_2 \\ c_3 \\ c_4 \end{pmatrix} = \begin{pmatrix} p_1 \\ p_2 \\ p_3 \\ p_4 \end{pmatrix}.
$$

The solution of the linear system for $\{c_k\}$ above is

$$\mathbf{c} = \left(\frac{1}{3}(3p_1 + 2p_2 + p_3), \frac{1}{3}(p_2 + 2p_3 + 3p_4), -\frac{1}{3}(p_2 + p_3), \frac{1}{6}(p_3 - p_2) \right)^T. \quad (3.2.5)$$

In the limit as $n \to \infty$, we get

$$\lim_{n \to \infty} \mathbf{x}(n) = \frac{1}{3} \begin{pmatrix} 3p_1 + 2p_2 + p_3 \\ 0 \\ 0 \\ p_2 + 2p_3 + 3p_4 \end{pmatrix} = \mathbf{x}_\infty, \quad (3.2.6)$$

which may be written as

$$\mathbf{x}_\infty = \begin{bmatrix} 1 & \frac{2}{3} & \frac{1}{3} & 0 \\ 0 & 0 & 0 & 0 \\ 0 & 0 & 0 & 0 \\ 0 & \frac{1}{3} & \frac{2}{3} & 1 \end{bmatrix} \begin{pmatrix} p_1 \\ p_2 \\ p_3 \\ p_4 \end{pmatrix} = M_\infty \mathbf{p}. \quad (3.2.7)$$

The limiting solution (3.2.6) of the IVP provides the answer to the first question posed in the introductory paragraph of this section on Absorbing Markov chains. It gives the probability for the process ending up in a particular absorbing state. As originally stated, Player 1 started with \$1 initially, i.e., $\mathbf{p} = (0, 1, 0, 0)^T$. This means $\mathbf{x}_\infty = (2/3, 0, 0, 1/3)^T$. In other words, if the game is played over and over again, Player 1 would lose twice as often as he/she would win it.

Other related observations from the fixed point solution are:

- The situation is symmetric for the two players. If Player 1 has \$2 (and Player 2 has only \$1) initially, then we have $\mathbf{p} = (0, 0, 1, 0)^T$ and $\mathbf{x}_\infty = (1/3, 0, 0, 2/3)^T$ so that Player 2 now loses twice as often as she/he would win.
- In the two trivial cases where one player has all \$3 initially, then we have either $\mathbf{p} = (1, 0, 0, 0)^T$ or $\mathbf{p} = (0, 0, 0, 1)^T$ resulting in $\mathbf{x}_\infty = (1, 0, 0, 0)^T$ or $\mathbf{x}_\infty = (0, 0, 0, 1)^T$ giving the obvious outcome of the player losing or winning with probability 1.
- If Player 1's asset is not known but all four different amounts are equally probable so that $\mathbf{p} = \frac{1}{4}(1, 1, 1, 1)^T$, then we have $\mathbf{x}_\infty = (\frac{1}{2}, 0, 0, \frac{1}{2})^T$. Player 1 equally likely wins or loses.

Evidently, long term behavior for other initial distributions can also be easily read off the equilibrium (fixed point) solution (3.2.6). As such, the actual solution (3.2.4) together with (3.2.5) for the IVP (3.2.3) contains more information than we usually need. It seems reasonable to avoid expending effort to solve the IVP (by obtaining all the eigen-pairs and inverting the corresponding modal matrix) for information often not really needed. If all we need are the consequences of not starting in an absorbing state, the two rows of zeros in (3.2.6) and (3.2.7) are superfluous. In the next few sections, we develop a more efficient approach to get the long term behavior and related useful information about the game without solving the IVP or even the fixed point. This efficient approach turns out to be applicable broadly to all absorbing Markov chain problems.

3.2.3   Canonical transition and absorbing matrix

To obtain the useful information without solving the IVP, we begin by re-ordering the elementary events so that the non-absorbing states take more of the center stage. This is done by re-labelling the elementary events. Up to now, we have the following labeling for the four elementary events $\{E_k\}$:

$$E_1 \quad E_2 \quad E_3 \quad E_4$$
$$\$0 \quad \$1 \quad \$2 \quad \$3$$

To highlight the non-absorbing states, we re-order them by re-labeling as shown below:

$$E_1^* \quad E_2^* \quad E_3^* \quad E_4^*$$
$$\$1 \quad \$2 \quad \$0 \quad \$3$$

with $E_1^* = E_2$, $E_2^* = E_3$, $E_3^* = E_1$ and $E_4^* = E_4$. For this new ordering of elementary events, the transition matrix for the corresponding probability distribution vector $\mathbf{y}(n)$ is now changed to

$$\mathbf{y}(n+1) = M^*\mathbf{y}(n) \tag{3.2.8}$$

where

$$M^* = \begin{bmatrix} 0 & 0.5 & 0 & 0 \\ 0.5 & 0 & 0 & 0 \\ 0.5 & 0 & 1 & 0 \\ 0 & 0.5 & 0 & 1 \end{bmatrix} \equiv \begin{bmatrix} R & O \\ Q & I \end{bmatrix} \tag{3.2.9}$$

$$R = \begin{bmatrix} 0 & 0.5 \\ 0.5 & 0 \end{bmatrix}, \quad Q = \begin{bmatrix} 0.5 & 0 \\ 0 & 0.5 \end{bmatrix}, \quad \mathbf{y}(0) = \mathbf{p}^* = \begin{pmatrix} \mathbf{p}_t^* \\ \mathbf{p}_a^* \end{pmatrix} \tag{3.2.10}$$

with $\mathbf{p}^*$ being the re-ordered initial distribution vector and the two sub-vectors $\mathbf{p}_t^*$ and $\mathbf{p}_a^*$ corresponding to the initial distribution for the non-absorbing states and absorbing states, respectively.

As consequences of all the absorbing states being relegated to the last half of the grouping of elementary events, we have in the second group column the **identity** matrix I and the **zero** matrix O, reflecting the fact that the game ends as it cannot leave any of these states once in that state. Also with the first part of group members being in *transient* states, the elements of $R = [r_{ij}]$ and $Q = [q_{ij}]$ are necessarily less than unity, i.e.,

$$0 \le r_{ij} < 1, \quad 0 \le q_{ij} < 1.$$

A transition matrix with ordering of elementary events that results in the form (3.2.9) is said to be in *canonical form* or *canonical transition*.

The canonical transition relation (3.2.8) requires

$$\mathbf{y}(2) = M^*\mathbf{y}(1) = [M^*]^2\,\mathbf{y}(0) = [M^*]^2\,\mathbf{p}^* = \begin{bmatrix} R^2 & O \\ Q(R+I) & I \end{bmatrix}\mathbf{p}^*$$

and, by induction,

$$\mathbf{y}(n) = [M^*]^n\,\mathbf{p}^* = \begin{bmatrix} R^n & O \\ Q(R^{n-1} + R^{n-2} + \cdots + R + I) & I \end{bmatrix}\mathbf{p}^*.$$

As $n \to \infty$, $\mathbf{y}(n)$ approaches the following limit:

$$\lim_{n\to\infty}[\mathbf{y}(n)] = \lim_{n\to\infty}\begin{bmatrix} R^n & O \\ Q\sum_{k=0}^{n-1} R^k & I \end{bmatrix}\mathbf{p}^* = \begin{bmatrix} O & O \\ A & I \end{bmatrix}\begin{pmatrix} \mathbf{p}_t^* \\ \mathbf{p}_a^* \end{pmatrix} \tag{3.2.11}$$

With

$$(I - R)\sum_{k=0}^{n-1} R^k = I - R^n,$$

we get

$$\lim_{n\to\infty}\sum_{k=0}^{n-1} R^k = \lim_{n\to\infty}\left[(I - R)^{-1}(I - R^n)\right] = (I - R)^{-1}.$$

The *absorption matrix* A can therefore be re-written as

$$A = Q(I - R)^{-1}. \tag{3.2.12}$$

It follows that (3.2.11) is simplified to

$$\lim_{n\to\infty}[\mathbf{y}(n)] = \begin{pmatrix} \mathbf{0} \\ A\mathbf{p}_t^* + \mathbf{p}_a^* \end{pmatrix} = \begin{pmatrix} \mathbf{0} \\ A\mathbf{p}_t^* \end{pmatrix} + \begin{pmatrix} \mathbf{0} \\ \mathbf{p}_a^* \end{pmatrix}. \tag{3.2.13}$$

The result above is consistent with Theorem 9 and delineates explicitly why the lower half of the vector relation above is the truly informative part of the result.

For the transition matrix (3.2.2), the lower half of the limiting distribution is

$$\lim_{n\to\infty}\begin{bmatrix} y_3(n) \\ y_4(n) \end{bmatrix} = A\mathbf{p}_t^* + \mathbf{p}_a^* = A\begin{pmatrix} p_1^* \\ p_2^* \end{pmatrix} + \begin{pmatrix} p_3^* \\ p_4^* \end{pmatrix} = A\begin{pmatrix} p_2 \\ p_3 \end{pmatrix} + \begin{pmatrix} p_1 \\ p_4 \end{pmatrix}. \tag{3.2.14}$$

It is the principal information sought for the problem. The 2×2 absorption matrix A is easily calculated to be

$$A = Q(I - R)^{-1} = \begin{bmatrix} 0.5 & 0 \\ 0 & 0.5 \end{bmatrix} \begin{bmatrix} 1 & -0.5 \\ -0.5 & 1 \end{bmatrix}^{-1} = \begin{bmatrix} 2/3 & 1/3 \\ 1/3 & 2/3 \end{bmatrix}$$

and therewith

$$\lim_{n \to \infty} \begin{bmatrix} y_3(n) \\ y_4(n) \end{bmatrix} = \lim_{n \to \infty} \begin{pmatrix} x_1(n) \\ x_4(n) \end{pmatrix} = \frac{1}{3} \begin{pmatrix} 3p_1 + 2p_2 + p_3 \\ p_2 + 2p_3 + 3p_4 \end{pmatrix} \tag{3.2.15}$$

which was just a condensed version of the complete solution (3.2.6) with all the inessentials omitted. Moreover, we now obtain it by performing simple algebraic operations on a smaller (2×2) matrices R and Q and not having to solve any matrix eigenvalue problem for the larger (and generally much larger) original transition matrix.

Exercise 11. Consider the following coin-tossing process involving (the same) two coins, a dime and a quarter. The rules of the game require that the dime is tossed next if a head turns up and the quarter next for a tail. Due to the different engraved patterns on the two sides, the coins are **not** fair coins. Since the patterns on the coin faces are different, the probability of getting a head for the dime is p_d (obtained as the limit of repeated tossing of the same coin with each outcome independent of the outcomes of past tosses). Similarly, the probability of getting a head for the quarter is p_q. Correspondingly, the probability of getting a tail is $1 - p_d$ for the dime and $1 - p_q$ for the quarter, respectively. We are interested here in whether the dime or quarter would be tossed next after $(n - 1)$ tosses. Mathematically, we would like to know the probabilities of getting to toss each of the two coins at the nth toss.

The next three exercises have already been mentioned informally in Chapter 1 and are stated explicitly for possible assignments:

Exercise 12. Show that the eigenvalues of M^T is the same as that of a transition matrix M.

Exercise 13. If $\mathbf{x}(0) = \mathbf{p}$ is a probability vector, then so is $\mathbf{x}(n) = M^n \mathbf{p}$.

Exercise 14. If $M > O$, show $\mathbf{y} = M\mathbf{x} > 0$ for any probability vector $\mathbf{x}$. (A matrix $M > O$ means that all elements of M are positive. A vector $\mathbf{p} > \mathbf{0}$ means all components of the vector are positive.)

Exercise 15. Prove Theorem 10 (corresponding to a superposition principle for *complementary solutions* of linear difference equations).

3.3 Expected Transient Stops to an Absorbing State

There is more than improved computational efficiency and reduction of unessentials to the alternate form of the limiting distribution given in (3.2.15). The absorption matrix actually provides answers to the two remaining questions posed at the end of the paragraph after Theorem 9 (the first already answered by the limit distribution (3.2.6) through the solution of the IVP or (3.2.15) with the help of the absorption matrix (3.2.12)). We show below how the absorption matrix also provides the answer to the question: *Starting from one of its transient states, how many transient stops does the absorbing chain make on the average before reaching an absorbing state?* (Or how long does the evolving chain survives on the average before being trapped in an absorbing state?) The answer to the other question will also be obtained in the process.

Suppose the given absorbing Markov chain is in a transient state E_j initially. Let s_{ij} be the *number of stops on the average* (known as the expected number of stops) the absorbing chain makes at a particular transient state E_i before it reaches an absorbing state. If $i \neq j$, the chain can reach the state E_i on the first trial with probability p_{ij}. (Since the sample space is $\{0, 1\}$ for zero or one stop, p_{ij} is also the mean or expected number of stops at E_i in one trial.) If not, it may reach E_j in the second trial with probability $\sum_{\alpha=1}^{m_n} p_{i\alpha} p_{\alpha j}$ by passing through any one intermediate transient state E_α, for $\alpha = 1, 2, \ldots, m_n$ (with m_n = the number of transient states) on the first trial. If not, it may reach E_i in the third trial with probability $\sum_{\alpha=1}^{m_n} \sum_{\beta=1}^{m_n} p_{i\alpha} p_{\alpha\beta} p_{\beta j}$ and so on. If $i = j$, the chain is already in E_i with probability 1. Altogether, starting from in E_j, the expected number of stops the chain makes in E_i is therefore

$$s_{ij} = \delta_{ij} + p_{ij} + \sum_{\alpha=1}^{m_n} p_{i\alpha} p_{\alpha j} + \sum_{\alpha=1}^{m_n} \sum_{\beta=1}^{m_n} p_{i\alpha} p_{\alpha\beta} p_{\beta j} + \sum_{\alpha=1}^{m_n} \sum_{\beta=1}^{m_n} \sum_{\gamma=1}^{m_n} p_{i\alpha} p_{\alpha\beta} p_{\beta\gamma} p_{\gamma j} + \cdots$$

$$(3.3.1)$$

where δ_{ij} is the Kronecker delta ($= 1$ if $i = j$ and 0 otherwise). (Again, what we calculated was also the sum of different expected numbers of stops to get from transient state j to transient state i.) It gives the number of times, on the average, the absorbing chain dwells in the particular non-absorbing state E_i when the chain starts from E_j.

As we allow i and j to range over all the transient states, the relation (3.3.1) may be written in terms of the two $m_n \times m_n$ matrices $S = [s_{ij}]$ and $R = [p_{ij}]$ as

$$S = I + R + R^2 + R^3 + \cdots$$

To simplify the expression for S, we form $RS = R + R^2 + R^3 + \cdots = S - I$ to get $I = (I - R)S$ or

$$S = (I - R)^{-1}.$$

The expected number of stops (from all initial transient states) the chain makes in E_i is the sum of s_{ij} over j:

$$S_i = \sum_{\alpha=1}^{m_n} s_{i\alpha}.$$

Thus, the (expected) *transient stop matrix* S provides the answer to another of our original three questions.

For the remaining question, let

$$\bar{S}j = \sum_{k=1}^{m_n} s_{kj} = s_{1j} + s_{2j} + s_{3j} + \cdots + s_{m_n j}. \tag{3.3.2}$$

Evidently, starting at the transient state E_j, the sum $\bar{S}_j$ is the expected number of transient stops incurred by the absorbing chain prior to reaching an absorbing state. This expected number is the sum of the jth column of the matrix $(I - R)^{-1}$.

3.4 Birth and Death Processes

3.4.1 *General birth and death processes*

3.4.1.1 *The model*

To learn more about absorbing Markov chains, we consider here a DTMC model of the birth and death process applicable to typical populations in the life sciences. We limit discussion to populations in habitats with a finite carrying capacity of N individuals. Elapsed time between any two consecutive stages are taken to be so short that only one elementary event, one birth with probability b or one death with probability d, occurs. Generally, the birth and death rate vary from stage to stage. For many populations, they would depend on the size of the population as the rate of growth in the deterministic theory of population growth (such as logistic growth). We will focus on the former but not the latter type of population and denote the birth and death rate at stage i by b_i and d_i (instead of just b and d). It is reasonable to expect (and we will assume)

$$b_j > 0, \quad d_k > 0, \quad s_i \equiv b_i + d_i \leq 1 \tag{3.4.1}$$

for all $0 \leq k \leq N$ and $0 \leq j \leq N - 1$ with $b_N = 0$.

For a particular stage n, let the random variable $X(n)$ be the population size at that stage with X assuming different size from 0 to the carrying capacity N. Thus, there are $N + 1$ elementary events, $\{0, 1, 2, \ldots, N\}$. Let p_{ij} be the transition probability of $X(n + 1) = i$ given $X(n) = j$:

$$p_{ij} = P[X(n + 1) = i \mid X(n) = j] \tag{3.4.2}$$

Under the reasonable assertion $p_{00} = 1$ and $p_{i0} = 0$ for $i > 0$ (as offsprings can only be generated by an existing population), we should have the following transition matrix

$$M = \begin{bmatrix} 1 & d_1 & 0 & \cdot & \cdot & & 0 & 0 \\ 0 & 1-s_1 & d_2 & \cdot & \cdot & & 0 & 0 \\ 0 & b_1 & 1-s_2 & \cdot & \cdot & & & 0 \\ 0 & 0 & b_2 & \cdot & \cdot & & & \cdot \\ & & 0 & \cdot & \cdot & & & \cdot \\ \cdot & \cdot & \cdot & \cdot & \cdot & & & \cdot \\ \cdot & \cdot & \cdot & \cdot & 0 & & & \cdot \\ \cdot & \cdot & \cdot & \cdot & d_{N-2} & 0 & & \cdot \\ 0 & \cdot & \cdot & \cdot & 1-s_{N-2} & d_{N-1} & 0 & \\ 0 & 0 & \cdot & \cdot & b_{N-2} & 1-s_{N-1} & d_N \\ 0 & 0 & 0 & \cdot & 0 & b_{N-1} & 1-d_N \end{bmatrix}. \tag{3.4.3}$$

To see how we obtain the various transition matrix elements, we denote by m_{ij} the i, j - elements of the matrix $M = [m_{ij}]$ with $i, j = 1, \ldots, N+1$. We note that the entry $m_{23} = p_{12}$ requires a death to reduce the population size 2 at stage n by 1 to get a population size 1 at stage $n + 1$. The element $m_{43} = p_{32}$ requires a birth to increase the population size from 2 to 3. The probability of a death to occur at population size 2 is d_2 while the probability of a birth is b_2. The diagonal element $m_{33} = p_{22}$ corresponds to a probability of no change in the population size $i = 2$ from stage n to stage $n + 1$, or $m_{33} = p_{22} = 1 - (d_2 + b_2)$. With a finite carrying capacity, the population does not exceed some maximal size N and whence $p_{(N+1)N} = b_N = 0$

Given the restrictions (3.4.1), the columns of M are probability vectors. However, the DTMC characterized by M is not regular since the first column of M^k remains $(1, 0, \ldots, 0)^T$ for all $k > 0$. Furthermore, starting the process with the probability distribution $\mathbf{p}$, the transitional relation

$$\mathbf{x}(n+1) = M\mathbf{x}(n), \quad \mathbf{x}(0) = \mathbf{q} \tag{3.4.4}$$

(with $q_i = p_{i-1}, i = 1, 2, \ldots, N+1$) characterizes the evolution of the state (probability) distribution $\mathbf{x}(n) = (x_1(n), x_2(n), \ldots, x_{N+1}(n))^T$ with stage. For the special case of $\mathbf{p} = (1, 0, \ldots, 0)^T$, all future stage state will remain unchanged.

3.4.1.2 *Expected time to extinction*

We know with certainty that population modeled by this DTMC eventually goes extinct since the MC will eventually reach its only absorbing state of zero population. Consequently, it is not necessary to calculate $A\mathbf{p}_t^* + \mathbf{p}_a^*$ to determine the probabilities of reaching the chain's only absorbing state (see (3.2.14)). Of interest instead is the expected time for absorption starting from one of its transient states. This can be determined from Eq. (3.3.2) for $S_i, i = 1, \ldots, N$, which gives the expected number of stops before absorption with the MC starting from the transient state i. (It should be noted $S_0 = 0$ given that the chain is already in the absorbing state.) With one unit of elapsed time between consecutive stops, S_i is also the expected time prior to moving into the only absorbing state, starting from transient state i.

With

$$S_i = (1, 1, \ldots, 1)(I - R)^{-1} \equiv \mathbf{1}^T (I - R)^{-1} = \mathbf{1}^T S, \qquad (3.4.5)$$

it is in principle a matter of simply calculating the matrix $(I - R)^{-1}$ for our particular $N \times N$ matrix R with

$$I - R = \begin{bmatrix} s_1 & -d_2 & 0 & \cdot & \cdot & & \cdot & & 0 & 0 \\ -b_1 & s_2 & -d_3 & \cdot & \cdot & & \cdot & & & 0 \\ 0 & -b_2 & s_3 & \cdot & \cdot & & & & & \cdot \\ \cdot & 0 & -b_3 & \cdot & \cdot & & \cdot & & & \cdot \\ \cdot & & \cdot & & \cdot & & -d_{N-2} & 0 & & \cdot \\ \cdot & & \cdot & & \cdot & & s_{N-2} & -d_{N-1} & 0 & \\ 0 & \cdot & & \cdot & \cdot & & -b_{N-2} & s_{N-1} & -d_N \\ 0 & 0 & & \cdot & \cdot & & 0 & -b_{N-1} & d_N \end{bmatrix},$$

where $s_k = b_k + d_k$. While the calculation of $(I - R)^{-1}$ is possible, explicit analytical expressions of the elements of $S = [s_{ij}]$ seem complicated and tedious to obtain. It is therefore a considerable accomplishment that Nisbet and Gurney [39] succeeded in deriving the following rather elegant result for $\{S_k\}$, $k = 1, 2, \ldots, N$:

Theorem 11. *Starting from an initial population of size k, the expected time to extinction S_k is given by*

$$S_1 = \frac{1}{d_1} + \sum_{i=2}^{N} \frac{b_1 \cdots b_{i-1}}{d_1 \cdots d_i}, \qquad (3.4.6)$$

$$S_k = S_1 + \sum_{j=1}^{k-1} \left[\frac{b_1 \cdots b_j}{d_1 \cdots d_j} \sum_{i=j+1}^{N} \frac{b_1 \cdots b_{i-1}}{d_1 \cdots d_i} \right], \qquad (k = 2, \ldots, N). \qquad (3.4.7)$$

As mentioned earlier, it is decidedly nontrivial to deduce (3.4.6)-(3.4.7) from $\mathbf{1}^T (I - R)^{-1}$, the right hand side of (3.4.5). For the case where the birth and death probabilities are time-invariant so that $b_k = b$ $(1 \le k \le N - 1)$, $b_N = 0$ and $d_k = d$ for $(1 \le k \le N)$, the derivation of the expression for S_k can be simplified by an alternative method of solution.

We wish to determine S_k, the expected time for the chain to be in various transient states before exiting starting from state k. At state k, the chain can get to state $k + 1, k$ or $k - 1$ at the next stage with probability $b_k, (1 - d_k - b_k)$ and d_k, respectively (and no others). At this next stage, the expected time to extinction starting from the state at that stage would be $(1 + S_{k+1}), (1 + S_k)$ and $(1 + S_{k-1})$, respectively. It follows that the expected time to extinction starting from the k state is given by

$$S_k = b_k(1 + S_{k+1}) + d_k(1 + S_{k-1}) + (1 - b_k - d_k)(1 + S_k)$$

or

$$d_k S_{k-1} - (b_k + d_k)S_k + b_k S_{k+1} = -1, \quad (k = 1, 2, \ldots, N - 1) \tag{3.4.8}$$

and

$$S_0 = 0, \quad d_N (S_{N-1} - S_N) = -1 \tag{3.4.9}$$

since $b_N = 0$ in view of limiting capacity. The relations (3.4.8)-(3.4.9) constitute another approach for determining S_k, $k = 1, 2, \ldots, N$. A sketch of the proof of Theorem 11 using this approach can be found in [2].

3.4.1.3 Time-invariant systems

For time-invariant birth and dead rate constants, the linear difference equation is of constant coefficients:

$$dS_{k-1} - (b + d)S_k + bS_{k+1} = -1 \quad (k = 1, 2, \ldots, N - 1)$$

The two complementary solutions of this equation are $x_n^{(1)} = 1$ and $x_n^{(2)} = (d/b)^n$ so that the general solution can be obtained by the method of variation of parameters to be in the form

$$S_k = x_k^{(1)} u_k + x_k^{(2)} v_k = u_k + \left(\frac{d}{b}\right)^k v_k. \tag{3.4.10}$$

The two quantities u_k and v_k are determined by specializing (2.6.14)-(2.6.15) for our problem to obtain

$$u_k = u_0 + \frac{1}{d - b}, \quad v_k = v_0 + \frac{1}{d - b} \frac{d}{b} \sum_{m=0}^{k-2} \left(\frac{b}{d}\right)^m$$

with the two constants u_0 and v_0 determined by the two auxiliary conditions (3.4.9). However, the auxiliary conditions are seen to be one at each end of the solution domain giving rise to a two point boundary value problem (BVP) instead of the more familiar IVP. While we can still use the two boundary conditions to determine these two constants to complete the solution process, the following alternative method of solution also applies to time-varying systems.

With death and birth rate, d and b, independent of stages, the system (3.4.8)–(3.4.9) becomes

$$dS_{k-1} - (b + d)S_k + bS_{k+1} = -1 \quad (k = 1, 2, \ldots, N - 1) \tag{3.4.11}$$

and

$$S_0 = 0, \quad S_N = \frac{1}{d} + S_{N-1}. \tag{3.4.12}$$

Starting from $k = 1$, we get from (3.4.11)

$$S_2 = -\frac{1}{b} + \left(1 + \frac{d}{b}\right) S_1 \tag{3.4.13}$$

$$S_3 = -\frac{d}{b^2} + \left(1 + \frac{d}{b} + \frac{d^2}{b^2}\right) S_1 \tag{3.4.14}$$

$$\cdots$$

$$S_k = S_1 + \sum_{n=1}^{k-1} \left(\frac{d}{b}\right)^n \left[S_1 - \frac{1}{d} - \sum_{m=2}^{n} \frac{b^{m-1}}{d^m}\right]. \tag{3.4.15}$$

From the expression for general k, we obtain

$$S_N - S_{N-1} = \left(\frac{d}{b}\right)^{N-1} \left[S_1 - \frac{1}{d} \sum_{m=0}^{N-2} \left(\frac{b}{d}\right)^m\right]; \tag{3.4.16}$$

but from the second boundary condition in (3.4.12), we have

$$S_N - S_{N-1} = \frac{1}{d}. \tag{3.4.17}$$

Equating the right hand sides of (3.4.16) and (3.4.17), we obtain

$$S_1 = \frac{1}{d} \sum_{m=0}^{N-1} \left(\frac{b}{d}\right)^m = \frac{1 - (b/d)^N}{d - b} \tag{3.4.18}$$

which is nonnegative independent of sign of $d - b$. With (3.4.18), the corresponding expressions for S_k for $k = 2, \ldots, N$ can be obtained from (3.4.13), (3.4.14) and (3.4.15) and seen to be special cases of the general results of Theorem 11. In fact, the proof in [2] for that theorem, for the general case where birth and death rate constants vary with stage, proceeds similarly: solving for S_k, $k = 2, 3, \ldots, N$, in terms of S_1 and use the end condition in (3.4.12) as the condition for determining S_1 as given in (3.4.6). The remaining $S_k, k = 2, 3, \ldots, N$, follow from the intermediate relations already obtained for them in terms of S_1 to give (3.4.7).

3.5 A Simplified Infectious Disease Problem

3.5.1 *A simplified (SIS) model*

As another application of the general results for absorbing MC, we consider in this section a simple model for the spread of an infectious disease. For this model, some members of the (healthy but) susceptible population $S(n)$ at stage n becomes infected by stage $n + 1$ to add to the infected population $I(n + 1)$ at stage $n + 1$.

This is one way the infected population changes in size. There are two other ways:

- Members of the infected population recover with probability r (and become susceptible since they do not acquire immunity from the disease).
- Members of the infected population die with probability d_I (from the disease or another cause).

On the other hand, there are five ways the susceptible population can change in size:

- Members can be infected with probability $\bar{\mu}$ and lost to that population.
- Member can die with probability d_S.
- New members are born to the susceptible population with probability b_S.
- New members are born (disease free) to the infected population with probability b_I.
- Members of the infected population recovered from the disease (with probability r) but not immune from being infected again.

To simplify the solution process, we make the assumptions $d_S = b_S = b$ and $d_I = b_I = b$. They may not be particularly realistic but such simplifications are consistent with the general modeling practice in our first attempt to gain some insight to the evolution of the two populations. The total population is assumed fixed at N for all time so that we may take $\bar{\mu} = \mu/N$.

We take the duration of consecutive stages to be so short that only one addition or deletion of the infected population occurs. Let m_{ij} be the probability of the infected population to become size i at stage $n+1$ when it was of size j at stage n. Evidently, we have

$$m_{(i-1)i} = \text{Prob}\,\{I(n+1) = i-1 \mid I(n) = i\} = (d_I + r)i = (b+r)i$$

$$m_{(i+1)i} = \text{Prob}\,\{I(n+1) = i+1 \mid I(n) = i\} = \bar{\mu}(N-i) \equiv \frac{\mu}{N}i(N-i) \equiv \bar{m}_i$$

$$m_{ii} = \text{Prob}\,\{I(n+1) = i \mid I(n) = i\} = 1 - (b+r) - p_I(i)$$

$$m_{00} = 1, \quad m_{k0} = 0, \quad m_{ji} = 0$$

for all $i, k \geq 1$ and $j \neq i-1, i, i+1$. The last of these relations is the consequence of the "exactly one change between consecutive stages" assumption and the prescription $\bar{\mu} = \mu i/N$ reflects the fact that the probability of infection should depend on the size of the infected population.

Starting with an initial probability distribution $\mathbf{p}$ for the $N+1$ elementary events $\{0, 1, 2, \ldots, N\}$, we have the IVP (3.4.4) governing the evolution of this distribution with stages. The $(N+1) \times (N+1)$ transition matrix for the MC is given by

$$M = \begin{bmatrix}
1 & s & 0 & \cdot & \cdot & \cdot & 0 & 0 & 0 \\
0 & 1-\bar{s}_1 & 2s & 0 & \cdot & \cdot & \cdot & 0 & 0 \\
0 & \bar{m}_1 & 1-\bar{s}_2 & 3s & \cdot & \cdot & \cdot & 0 & 0 \\
0 & 0 & \bar{m}_2 & 1-\bar{s}_3 & \cdot & \cdot & \cdot & 0 & 0 \\
\cdot & \cdot & 0 & \bar{m}_3 & \cdot & \cdot & \cdot & \cdot & \cdot \\
\cdot & \cdot & \cdot & \cdot & \cdot & \cdot & \cdot & \cdot & \cdot \\
\cdot & \cdot & \cdot & \cdot & \cdot & \cdot & 0 & \cdot & \cdot \\
0 & \cdot & \cdot & \cdot & \cdot & \cdot & \cdot & (N-1)s & 0 \\
0 & 0 & \cdot & \cdot & \cdot & 0 & \bar{m}_{N-2} & 1-\bar{s}_{N-1} & Ns \\
0 & 0 & 0 & \cdot & \cdot & 0 & 0 & \bar{m}_{N-1} & 1-Ns
\end{bmatrix}$$

where

$$s = b + r, \quad \bar{s}_i = is + \bar{m}_i, \quad \bar{m}_i = \frac{\mu}{N} i(N-i).$$

Similar to the general birth and death processes, we consider parameter values in the range $0 < i(b+r) + \bar{m}_i \le 1$, $0 \le i \le N$ and $0 < \mu \le 1$, the last of these to ensure that $\mu I(n)/N$ is in fact a fraction.

3.5.2 Probability of absorption

We are interested in the eventual spread or disappearance of the infection disease. This can be done by determining the absorption matrix to be used together with the initial probability distribution. To obtain the absorption matrix, we partition the transition matrix into

$$M = \begin{bmatrix} 1 & Q \\ O & R \end{bmatrix} \tag{3.5.1}$$

where

$$Q = (s, 0, \ldots, 0),$$

$$R = \begin{bmatrix}
1-\bar{s}_1 & 2s & 0 & \cdot & \cdot & \cdot & 0 & 0 \\
\bar{m}_1 & 1-\bar{s}_2 & 3s & \cdot & \cdot & \cdot & 0 & 0 \\
0 & \bar{m}_2 & 1-\bar{s}_3 & \cdot & \cdot & \cdot & 0 & 0 \\
\cdot & 0 & \bar{m}_3 & \cdot & \cdot & \cdot & \cdot & \cdot \\
\cdot & \cdot & \cdot & \cdot & \cdot & \cdot & \cdot & \cdot \\
\cdot & \cdot & \cdot & \cdot & \cdot & \cdot & 0 & \cdot \\
\cdot & \cdot & \cdot & \cdot & \cdot & \cdot & (N-1)s & 0 \\
0 & \cdot & \cdot & \cdot & 0 & \bar{m}_{N-2} & 1-\bar{s}_{N-1} & Ns \\
0 & 0 & \cdot & \cdot & 0 & 0 & \bar{m}_{N-1} & 1-Ns
\end{bmatrix}$$

and O is the $N \times 1$ zero matrix. Similar to the derivation of the limiting behavior of M^* for the Gambler's Ruin problem, we obtain from the partition of the present transition matrix (3.5.1)

$$\lim_{n \to \infty} M^n = \begin{bmatrix} 1 & Q(I-R)^{-1} \\ O & 0 \end{bmatrix} = \begin{bmatrix} 1 & A \\ O & O \end{bmatrix}$$

so that

$$\lim_{n \to \infty} \mathbf{x}(n) = \mathbf{x}_\infty = \begin{bmatrix} 1 & A \\ O & O \end{bmatrix} \mathbf{p} = \begin{bmatrix} 1 & A \\ O & O \end{bmatrix} \begin{pmatrix} q_a \\ \mathbf{q}_t \end{pmatrix} = q_a + A\mathbf{q}_t$$

where q_a is the first component of $\mathbf{p}$, $\mathbf{q}_t$ is the $N \times 1$ vector of the remaining components of $\mathbf{p}$ and

$$A = Q(I-R)^{-1}$$

is the $1 \times N$ absorption matrix. Evidently, the probability of the spreading infection dying out depend on the values of the parameters r, b, μ, N and the initial probability distribution.

Chapter 4

Discrete Stage Nonlinear Markov Processes

4.1 Mendelian Genetics and Difference Equation

While much more can be said about first (or higher) order Markov chains, we wish to turn our attention now to the more difficult problem pertaining to modeling and analysis of nonlinear discrete stochastic processes. This chapter starts with one well-known problem of this type, the simplest type of genetics problem in the form of difference equations, to illustrate the complexity of these processes. Some techniques for solving nonlinear difference equations will be presented at the end of this chapter.

Genetics is a branch of biological science that investigates the mechanism for passing physiological traits from one generation to the next. Genetic analysis predates Gregor Mendel, but Mendel introduced a number of innovations to the science of genetics and enabled him to formulate laws that provide the theoretical basis of our understanding of the genetics of inheritance. Briefly, Mendel concluded that the hereditary determinants are of a particulate nature and called them **genes**.

To start, we limit our discussion to the inheritance of physical traits with two possible manifestations (appearances if you can see them visually) or **phenotypes:** Either you are an albino or you are not; either the surface of a developed biological organism is wrinkled or it is smooth. (Other physical traits such as the color of your eyes have more than two phenotypes.) Each parent has a pair of genes (**gene pair**) in each cell for each trait of interest. Each of these two genes can be of one of two **allele** types, denoted by D (for dominant) and R (for recessive), though A and a are sometimes used instead by some writers. So a gene pair can be one of the three **genotypes**: $\mathbf{D} = (D, D)$, $\mathbf{H} = (D, R) = (R, D)$ and $\mathbf{R} = (R, R)$.

Below are some terminology in Mendelian genetics:

- **Allele** is a specific form of a gene in the gene-pair (with two possible alleles for our two phenotype trait (e.g., "wrinkled" and "smooth" are the two alleles for the surface appearance of an organ).
- **Allelic pair** is the combination of two alleles which comprise a gene pair.
- **Homozygote** is an individual which has only one allele in the gene (and allelic) pair; e.g., DD is homozygous dominant and RR is homozygous recessive.
- **Heterozygote** is an individual whose gene-pair (of a particular physical trait) has two different alleles (hybrid); e.g., the DR gene pair.
- **Genotype** is a specific allelic combination for a certain gene or set of genes.
- An allele of a gene is **dominant** if an organism of genotype **D** is indistinguishable from one of the genotype **H**.
- An allele of a gene is **recessive** if an organism of genotype **R** appear to be different from one of the genotype **H**.

For example, for the gene controlling sickle cell anemia, an individual with (R, R) pair would show severe anemia while neither a **D** (dominant) genotype nor a **H** (hybrid) genotype shows such a trait. The genotypes themselves are called *dominant*, *hybrid* and *recessive*, respectively.

In the simplest setting, a gene is drawn for the gene pair of a trait from each of two parents, male and female, to form the genotype of the offspring for this trait. Allele frequencies in a population is to be the same across generations; the static allele frequencies effectively assumes: no mutation (the alleles don't change), no migration or emigration (no exchange of alleles between populations), infinitely large population size, and no selective pressure for or against any genotypes. Genotype frequencies are also to be static while mating is random.

Together, the two alleles comprise the gene pair. With each offspring gene pair containing one member of each parent's gene pair in accordance with Mendel's two laws of genetics:

Mendel's First Law — *The Law of Segregation*: For the pair of alleles an offspring has of some gene, one is a copy of a randomly chosen one in the father, and the other is a copy of a randomly chosen one in the mother.

Mendel's Second Law — *The Law of Independent Assortment*: Each allele of a parent's allelic pair has an equal chance to be the one copied for the offspring, and that the copying of alleles to different offspring or from different parents are independent.

(Today, we know that some genes are in fact "linked" and are inherited together, but for the most part Mendel's laws have proved surprisingly robust.)

Let $(d, 2h, r)$ be the probability of the offspring be of the dominant, hybrid and recessive genotype, respectively. The distribution of the three probabilities evidently depends of the genotypes of the two parents. Below is a table of distributions for the different combinations of parent genotypes with the explanations given in bullets to follow:

$$
\begin{array}{cccc}
m \backslash f & (D,D) & (D,R) & (R,R) \\
(D,D) & (1,0,0) & (\frac{1}{2},\frac{1}{2},0) & (0,1,0) \\
(D,R) & (\frac{1}{2},\frac{1}{2},0) & (\frac{1}{4},\frac{1}{2},\frac{1}{4}) & (0,\frac{1}{2},\frac{1}{2}) \\
(R,R) & (0,1,0) & (0,\frac{1}{2},\frac{1}{2}) & (0,0,1)
\end{array}
\qquad (4.1.1)
$$

- If the genotypes of both parents are dominant, denoted by $\mathbf{D}_m$ and $\mathbf{D}_f$ (with the subscript m and f indicating male and female, respectively), it is certain that the genotype of the offspring will be dominant and hence $(d, 2h, r) = (1, 0, 0)$.
- Similarly the genotype of the offspring will be recessive with $(d, 2h, r) = (0, 0, 1)$, if the genotypes of the parents are $\mathbf{R}_m$ and $\mathbf{R}_f$.
- If the genotypes of the two parents are i) $\mathbf{D}_m$ and $\mathbf{H}_f$, respectively, or ii) $\mathbf{D}_f$ and $\mathbf{H}_m$, respectively, then the offspring genotype would be $\mathbf{D}$ (with probability $1 \cdot \frac{1}{2} = \frac{1}{2}$) or $\mathbf{H}$ (with probability $1 \cdot \frac{1}{2} = \frac{1}{2}$), corresponding to the entry $(\frac{1}{2}, \frac{1}{2}, 0)$ in the first super- and sub-diagonal position of the matrix (4.1.1).
- The offspring genotype from parents of genotypes iii) $\mathbf{R}_m$ and $\mathbf{H}_f$, respectively, or ii) $\mathbf{R}_f$ and $\mathbf{H}_m$, respectively, are similarly $\mathbf{R}$ and $\mathbf{H}$, both with probability $1 \cdot \frac{1}{2} = \frac{1}{2}$, giving the entry $(0, \frac{1}{2}, \frac{1}{2})$ in the last super- and sub-diagonal position of the table (4.1.1).

This leaves the complicated case of both parents being of the hybrid genotype, $\mathbf{H}_m$ and $\mathbf{H}_f$, each containing one allele for the dominant phenotype and one for the recessive phenotype. To get a dominant genotype offspring, we need a dominant allele from each parent. Each of these occurs with a probability of $\frac{1}{2}$ resulting in a probability of $d = \frac{1}{4}$ the event of a dominant offspring. Similarly, the probability of a recessive genotype offspring is $r = \frac{1}{4}$. For a hybrid offspring, we can get it in two ways: a dominant allele from the "male" parent and a recessive allele from the "female" parent and the mirror image of this construction. Each of these two elementary events is with a probability of $\frac{1}{4}$ totaling to a probability of $2h = \frac{1}{2}$ for a hybrid genotype offspring. Altogether, the probability distribution for the genotype offspring of a pair of hybrid (genotype) parents is $(d, 2h, r)^T = (\frac{1}{4}, \frac{1}{2}, \frac{1}{4})^T)$ as shown in the center box of the table above.

4.2 Hardy–Weinberg Stability Theorem

In evolutionary biology, we are interested more than just the next generation off-springs but multi-generational evolution of the genotypes of offsprings. Starting with an initial distribution of $\boldsymbol{\delta}_0 = (d_0, 2h_0, r_0)^T$, the genotype of subsequent generations is **not** determined by a Markov Chain, $\boldsymbol{\delta}_{n+1} \neq M\boldsymbol{\delta}_n$, where we now use a subscripted variable, x_n, instead of the previous form of $x(n)$. Given the nth generation probability distribution vector $\boldsymbol{\delta}_n = (d_n, 2h_n, r_n)^T$, a dominant offspring in the $(n+1)^{th}$ generation can come from combinations of dominant or hybrid parents. The probability of a dominant allele from the male parent of the nth generation is $1 \cdot d_n + \frac{1}{2} \cdot (2h_n)$; the same is true for the female parent. Together, they give the probability of $d_{n+1} = (d_n + h_n)^2$ for a *dominant* genotype offspring. Similarly, the probability of a *recessive* genotype offspring is $r_{n+1} = (r_n + h_n)^2$. On the other hand, we can get a hybrid offspring in two ways. One is to get a dominant allele from the male parent with probability $(d_n + h_n)$ and a recessive allele from the female parent with a probability $(r_n + h_n)$ so that the probability of a hybrid genotype offspring from this combination is $(d_n + h_n)(r_n + h_n)$. Now, we can also get a hybrid offspring through a dominant allele from the female parent and a recessive allele from the male parent with the same probability so that $2h_{n+1} = 2(d_n + h_n)(r_n + h_n)$. These observations are summarized as the following systems of three *difference equations*:

$$d_{n+1} = (d_n + h_n)^2,$$
$$2h_{n+1} = 2(d_n + h_n)(r_n + h_n), \qquad (4.2.1)$$
$$r_{n+1} = (r_n + h_n)^2.$$

Unlike Markov chains, the difference equations that govern the probability distribution $\boldsymbol{\delta}_n = (d_n, 2h_n, r_n)^T$ are not linear and solutions of the form $c\lambda^n$ is generally not applicable. In fact, there are much less general methods for analyzing the solution of nonlinear difference equations.

For the present nonlinear system (4.2.1), we note that the solution $(d_1, 2h_1, r_1)$ is obtained from (4.2.1) for $n = 0$ to be

$$d_1 = (d_0 + h_0)^2,$$
$$2h_1 = 2(d_0 + h_0)(r_0 + h_0), \qquad (4.2.2)$$
$$r_1 = (r_0 + h_0)^2.$$

Next, we get from (4.2.1) with $n = 1$

$$d_2 = (d_1 + h_1)^2 = \left[(d_0 + h_0)^2 + (d_0 + h_0)(r_0 + h_0) \right]^2$$
$$= (d_0 + h_0)^2 \left[(d_0 + h_0) + (r_0 + h_0) \right]^2 = (d_0 + h_0)^2. \qquad (4.2.3)$$

Similarly, we have

$$2h_2 = 2(d_1 + h_1)(r_1 + h_1)$$
$$= 2 \left[(d_0 + h_0)^2 + (d_0 + h_0)(r_0 + h_0) \right] \left[(r_0 + h_0)^2 + (d_0 + h_0)(r_0 + h_0) \right]^2$$
$$= 2(d_0 + h_0)(r_0 + h_0),$$

and

$$r_2 = (r_1 + h_1)^2 = \left[(r_0 + h_0)^2 + (d_0 + h_0)(r_0 + h_0) \right]^2$$
$$= (r_0 + h_0)^2 \left[(d_0 + h_0) + (r_0 + h_0) \right]^2 = (r_0 + h_0)^2.$$

Upon repeating the calculations for $\boldsymbol{\delta}_3, \boldsymbol{\delta}_4, \ldots$, we have the following celebrated Hardy–Weinberg stability theorem in (two allele -) Mendelian genetics:

Theorem 12. *Given the initial probability distribution* $\boldsymbol{\delta}_0 = (d_0, 2h_0, r_0)^T$, *the subsequent probability distribution* $\boldsymbol{\delta}_n$ *is invariant after one generation with*

$$d_n = (d_0 + h_0)^2, \quad 2h_n = 2(d_0 + h_0)(r_0 + h_0), \quad r_n = (r_0 + h_0)^2 \quad (n \geq 1).$$

Proof. (by induction)   □

In the language of difference equations, the probability distribution of genotypes for our model evolves into a steady state. In the case of a *regular* Markov chain starting with an initial distribution that is not the steady state, its expected steady state is approached through a converging process and reached only in the limit as $n \to \infty$. For the present simple Mendelian model of genetic evolution, the equilibrium configuration is reached in two generations and does not change thereafter. The development and attainment of an equilibrium genotype distribution in this model is remarkably rapid and its implication is of great significance. It is very much consistent with the physical traits in a population being very stable. However, if it were completely stable, there would be no changes in physical traits, and there would be no evolution.

Fortunately, the Hardy–Weinberg equilibrium distribution is, in the language of dynamical systems, stable but not asymptotically stable. Suppose at some stage N (>1), there is a perturbation from the equilibrium distribution so that we have $\boldsymbol{\delta}_N^* = (d_N^*, 2h_N^*, r_N^*)^T$ instead of $\boldsymbol{\delta}_N = \big((d_0 + h_0)^2, 2(d_0 + h_0)(r_0 + h_0), (r_0 + h_0)^2 \big)^T$. By the Hardy–Weinberg stability theorem, the genotype probability distributions for all future generations would be

$$\boldsymbol{\delta}_{N+k}^* = \left((d_N^* + h_N^*)^2, 2(d_N^* + h_N^*)(r_N^* + h_N^*), (d_N^* + h_N^*)^2 \right)^T \quad (k \geq 1).$$

In other words, the genotype distribution quickly reaches another steady state configuration. If $\boldsymbol{\delta}_N^*$ is close to $\boldsymbol{\delta}_N$, then the new equilibrium configuration $\boldsymbol{\delta}_{n+k}^*$ would be close to $\boldsymbol{\delta}_{N+k} = \boldsymbol{\delta}_N$. In that sense, the Hardy–Weinberg equilibrium configuration is stable but not asymptotically stable; the perturbed genotype distribution does not evolve and return to the equilibrium configuration before perturbation. Thus, Hardy–Weinberg is compatible with the view that evolution is a process for physical traits to change from some existing state apparently with a high degree of stability.

To see what may be responsible for the observed evolution, it is important to make explicit the assumptions in the simple Mendelian model of genetics that led to the Hardy–Weinberg law. These include

- The controlling genes have only two trait alleles.
- Genotypes are formed by different combinations of the two alleles, one from each of the two parents.
- The population is bisexual with the same distribution of genotypes in both.
- Generations are discrete.
- A pair of male and female parents is selected at random in each generation to produce an offspring.
- The offspring genotype is determined by an allele from a randomly selected gene from each parent.

Clearly, Hardy–Weinberg law may not apply when anyone of these assumptions is violated. Additional biological processes that are implicitly excluded from the model that led to the Hardy–Weinberg law include

- mutation
- nonrandom mating (inbreeding, selective breeding, assortative mating, etc.)
- natural selection
- gene flow
- genetic drift

While some of these exclusions are consequences of the assumptions listed earlier, they are mentioned explicitly because of their importance as biological processes that may lead to evolutionary changes. We examine a few of these in next few sections.

4.3   Selective Breeding I

Instead of random mating, suppose only the dominant genotype of one parent is allowed to breed. For example, a plant flower consists of both the male (with pollens corresponding to sperms) and female (center of the stamen) parts. A honey bee usually does the transfer of pollens to complete the cycle. As such, flowers are said to self-fertilize. A flower grower may retain only pollens of genotype for more brilliant flower colors for the bee to transfer. With male parent gene to have both dominant alleles, the sample space for the offspring genotype consists of only two elementary events $\{(D_m, D_f), (D_m, R_f)\}$ with a recessive genotype offspring being an impossibility (and hence $r_n = 0$ for all $n > 0$). The probability d_{n+1} of a *dominant* offspring genotype is then $1 \cdot (d_n + h_n)$ and the probability $2h_{n+1}$ of a *hybrid* offspring is $1 \cdot (r_n + h_n)$ with $r_{n+1} = 0 \cdot (r_n + h_n) = 0$ for $n > 0$.

For $n = 0$, we have

$$d_1 = (d_0 + h_0), \quad 2h_1 = (r_0 + h_0), \quad r_1 = 0 \tag{4.3.1}$$

with no restriction on the known (prescribed) initial distribution $\delta_0 = (d_0, 2h_0, r_0)^T$ other than it being a probability vector. For larger n, we have

$$d_2 = (d_1 + h_1) = \left[(d_0 + h_0) + \frac{1}{2}(r_0 + h_0)\right] = 1 - \frac{1}{2}(r_0 + h_0),$$

$$2h_2 = (r_1 + h_1) = \frac{1}{2}(r_0 + h_0), \quad r_2 = 0,$$

and

$$d_3 = (d_2 + h_2) = 1 - \frac{1}{4}(r_0 + h_0),$$

$$2h_3 = (r_2 + h_2) = \frac{1}{4}(r_0 + h_0), \quad r_3 = 0.$$

By induction, we get for $n > 0$ the following result:

Proposition 3. *For selective breeding with the gene of one breeding parent to have two dominant alleles, the components of the genotype distribution at the $(n+1)^{th}$ stage is given by*

$$d_{n+1} = (d_n + h_n) = 1 - \frac{1}{2^n}(r_0 + h_0), \tag{4.3.2}$$

$$2h_{n+1} = (r_n + h_n) = \frac{1}{2^n}(r_0 + h_0), \quad r_{n+1} = 0. \tag{4.3.3}$$

Proof. (by induction) □

In the limit as $n \to \infty$, we get

$$\delta_n \to (1, 0, 0)^T$$

as we would expect (when there are no other changes in the genetic environment).

It is worth mentioning that the relations (4.3.2) and (4.3.3) constitute a set of linear difference equations and hence amenable to an explicit solution of the form $c\lambda^n$ with the constant λ to be determined by the method Section 3 of Chapter 2. Equation (4.3.2) and the first equation of (4.3.3) with $r_n = 0$ (by the second equation in (4.3.3)) may be written in matrix form,

$$\delta_{n+1} = \begin{pmatrix} d_{n+1} \\ 2h_{n+1} \end{pmatrix} = \begin{bmatrix} 1 & \frac{1}{2} \\ 0 & \frac{1}{2} \end{bmatrix} \begin{pmatrix} d_n \\ 2h_n \end{pmatrix} \quad (n \geq 1) \tag{4.3.4}$$

with the initial conditions (4.3.1)

$$\delta_1 = \begin{pmatrix} d_1 \\ 2h_1 \end{pmatrix} = \begin{pmatrix} d_0 + h_0 \\ r_0 + h_0 \end{pmatrix}. \tag{4.3.5}$$

A solution proportional to λ^n, i.e., $\delta_n = \mathbf{x}\lambda^n$, is possible. The linear system of two difference equations requires the constant λ to be a root of the quadratic equation

$$2\lambda^2 - 3\lambda + 1 = 0,$$

namely $\lambda_1 = 1$ and $\lambda_2 = \frac{1}{2}$. Superposition of the two linearly independent solutions corresponding to the two roots and the auxiliary conditions at $n = 0$ give

$$d_{n+1} = d_0 + 2h_0 \left(1 - \frac{1}{2^{n+1}}\right) + r_0 \left(1 - \frac{1}{2^n}\right) = 1 - \frac{1}{2^n}(r_0 + h_0)$$

$$2h_{n+1} = \frac{1}{2^n}(r_0 + h_0)$$

which is the same as previously obtained.

4.4 Gene Frequencies

Let

$$p_n = d_n + h_n, \quad q_n = h_n + r_n. \tag{4.4.1}$$

Evidently, p_n and q_n are, respectively, the frequency of the dominant gene and recessive gene. For the two allele Mendelian model in the first section of this chapter, the evolution of genotype distribution may be rewritten as

$$d_{n+1} = p_n^2, \quad 2h_{n+1} = 2p_n q_n, \quad r_{n+1} = q_n^2.$$

It follows that

$$p_{n+1} = p_n^2 + p_n q_n = p_n, \quad q_{n+1} = q_n^2 + p_n q_n = q_n,$$

for $n = 0, 1, 2, \ldots$ with $p_{n+1} + q_{n+1} = p_n + q_n = \cdots = p_0 + q_0 = 1$. Whatever the initial gene frequency distribution, it remains the same thereafter.

Proposition 4. *In the two allele Mendelian model of population genetics, gene frequency distribution is conserved.*

4.5 Selective Breeding II

To illustrate the necessary care needed in the delineation of gene frequencies and genotypes, we consider here a different kind of selective breeding that is the opposite to the one discussed previously. In the new problem, those individuals of recessive genotype do not participate in the reproductive process. For example, they may expire prior to reproductive age or may simply be prohibited from participating in reproduction.

Suppose we start with a genotype distribution of $\delta_0 = (d_0, 2h_0, r_0)$. By the Mendelian model without the participation of the recessive genotype in reproduction, we now have the following genotype distribution for the next generation (instead of (4.2.1) for $n = 1$):

$$d_1 = (d_0 + h_0)^2, \quad 2h_1 = 2(d_0 + h_0)h_0, \quad r_1 = h_0^2. \tag{4.5.1}$$

Before we proceed to calculate the genotype distribution of the next generation, it is important to note that

$$d_1 + 2h_1 + r_1 = (d_0 + h_0)^2 + 2(d_0 + h_0)h_0 + h_0^2,$$
$$= (d_0 + 2h_0)^2 = (1 - r_0)^2.$$

Evidently, given that not the entire gene pool was allowed to participate in reproduction, the parts of the pool allowed to participate do not add up to the whole. To focus on the parts of the pool allowed to participate in reproduction as the whole pool for the reproduction of the next generation, we set

$$\delta_0^* = (d_0^*, 2h_0^*, r_0^*) = \left(\frac{d_0}{1 - r_0}, \frac{2h_0}{1 - r_0}, 0 \right),$$

with the frequency distribution taken to be that for the genes allowed to participate in reproduction

$$p_0 = d_0^* + h_0^* = \frac{d_0 + h_0}{1 - r_0}, \quad q_0 = r_0^* + h_0^* = \frac{h_0}{1 - r_0} \tag{4.5.2}$$

and

$$p_0 + q_0 = \frac{d_0 + 2h_0}{1 - r_0} = 1.$$

With recessive genotype individuals not participating in the reproductive process, the next generation's genotype distribution is appropriately given by

$$d_1 = (d_0^* + h_0^*)^2 = p_0^2, \quad r_1 = (r_0^* + h_0^*)^2 = q_0^2,$$
$$2h_1 = 2(d_0^* + h_0^*)(r_0^* + h_0^*) = 2p_0 q_0,$$

with

$$d_1 + 2h_1 + r_1 = \left(\frac{d_0 + h_0}{1 - r_0} \right)^2 + 2 \left(\frac{d_0 + h_0}{1 - r_0} \right) \frac{h_0}{1 - r_0} + \left(\frac{h_0}{1 - r_0} \right)^2$$
$$= \left(\frac{d_0 + 2h_0}{1 - r_0} \right)^2 = 1. \tag{4.5.3}$$

The gene pool for the second stage is not what is available from $\{d_1, 2h_1, r_1\}$ as $\mathbf{R}_1$ does not participate in the reproduction. So we would have to re-calibrate as we did before stage 1 by setting

$$\delta_1^* = (d_1^*, 2h_1^*, r_1^*) = \left(\frac{d_1}{1 - r_1}, \frac{2h_1}{1 - r_1}, 0 \right).$$

In general, we re-calibrate at stage n

$$\boldsymbol{\delta}_n^* = (d_n^*, 2h_n^*, r_n^*) = \left(\frac{d_n b}{1 - r_n}, \frac{2h_n}{1 - r_n}, 0 \right)$$

before determining $\boldsymbol{\delta}_{n+1} = (d_{n+1}, 2h_{n+1}, r_{n+1})$ from

$$d_{n+1} = (d_n^* + h_n^*)^2, \quad 2h_{n+1} = 2 \left(h_n^* \right) \left(d_n^* + h_n^* \right), \quad r_{n+1} = \left(h_n^* \right)^2 \tag{4.5.4}$$

and then re-calibrate for stage $n + 1$, etc.

Lemma 5.

$$d_{n+1} + 2h_{n+1} + r_{n+1} = 1 \quad (n \geq 0)$$

Proof. The initial distribution is a probability vector so that $d_0 + 2h_0 + r_0 = 1$ and consequently $d_1 + 2h_1 + r_1 = 1$ by (4.5.3). Suppose $d_k + 2h_k + r_k = 1$ holds for $k = n$; then we have

$$d_{n+1} + 2h_{n+1} + r_{n+1} = (d_n^* + h_n^*)^2 + 2 \left(h_n^* \right) \left(d_n^* + h_n^* \right) + \left(h_n^* \right)^2$$

$$= (d_n^* + 2h_n^*)^2 = \left(\frac{d_n + 2h_n}{1 - r_n} \right)^2$$

$$= \left(\frac{1 - r_n}{1 - r_n} \right)^2 = 1.$$

$\square$

For the nth generation, we now take the gene frequency to be that of the genes allowed to participate:

$$p_n = d_n^* + h_n^* = \frac{d_n + h_n}{1 - r_n}, \quad q_n = h_n^* = \frac{h_n}{1 - r_n}.$$

with

$$d_{n+1} = (d_n^* + h_n^*)^2 = p_n^2,$$

$$2h_{n+1} = 2 \left(h_n^* \right) \left(d_n^* + h_n^* \right) = 2p_n q_n, \tag{4.5.5}$$

$$r_{n+1} = \left(h_n^* \right)^2 = q_n^2.$$

We are interested in the evolution of (p_n, q_n) or $\boldsymbol{\delta}_n$ as n increases and the possibility of a limiting behavior as $n \to \infty$. When the system (4.5.4) is used to express the components of $\boldsymbol{\delta}_{n+1}$ in terms of the components of $\boldsymbol{\delta}_n$, we have a system of three highly nonlinear difference equations for $\{d_n, 2h_n, r_n\}$. However, working with the gene frequencies $\{p_n, q_n\}$ instead of $\{d_n, 2h_n, r_n\}$ results in considerable simplification as seen from the following proposition

Proposition 5. *The evolution of gene frequencies for $n > 0$ (in the selective breeding model of this section) is given by*

$$q_{n+1} = \frac{q_n}{1 + q_n}, \quad p_{n+1} = \frac{1}{1 + q_n}. \tag{4.5.6}$$

Proof.

$$p_{n+1} = d_{n+1}^* + h_{n+1}^* = \frac{d_{n+1} + h_{n+1}}{1 - r_{n+1}} = \frac{p_n^2 + p_n q_n}{1 - q_n^2} = \frac{p_n}{1 - q_n^2} = \frac{1}{1 + q_n},$$

$$q_{n+1} = h_{n+1}^* = \frac{h_{n+1}}{1 - r_{n+1}} = \frac{p_n q_n}{1 - q_n^2} = \frac{p_n q_n}{1 - q_n^2} = \frac{q_n}{1 + q_n}. \qquad \square$$

The determination of the two gene frequencies p_n and q_n is uncoupled with first order difference equation for q_n in (4.5.6) and the initial condition for q_0 in (4.5.2) defining an IVP for a single first order difference equation. Though still nonlinear, a single first order difference equation is considerably simpler than (4.5.4) in terms of $\{d_n, 2h_n, r_n\}$. As in the case of single first order ODE, there are some techniques available for an explicit solution for a single first order nonlinear difference equation.

The solution technique for the present problem is to transform the nonlinear difference equation for q_n into a single linear difference equation by setting $q_n^{-1} = x_n$ to get

$$q_{n+1} = \frac{q_n}{1 + q_n} = \frac{1}{1 + q_n^{-1}}$$

so that

$$x_{n+1} = 1 + x_n,$$

where $x_n = 1/q_n$ and $x_0 = 1/q_0 = (1 - r_0)/h_0$. The solution for IVP for x_n is immediate. With $x_1 = 1 + x_0$, $x_2 = 1 + x_1 = 2 + x_0$, it can be proved by induction

$$x_n = n + x_0 = \frac{n h_0 + (1 - r_0)}{h_0}.$$

From the calculations above, we have the following result for the evolution of the gene frequency distribution:

Proposition 6. *The frequency of the recessive gene pool for the Selective Breeding II model decreases to zero as $n \to \infty$ with*

$$q_n = \frac{1}{n + x_0} = \frac{h_0}{n h_0 + (1 - r_0)} \sim \frac{1}{n}.$$

Correspondingly, the frequency of the dominant gene pool increases slowly toward unity:

$$p_n = \frac{n + x_0 - 1}{n + x_0} = \frac{(n-1)h_0 + (1 - r_0)}{n h_0 + (1 - r_0)} \sim 1 + O\left(\frac{1}{n}\right).$$

with $\{p_n, q_n\} \to \{1, 0\}$ as $n \to \infty$.

4.6 Mutation

Under the appropriate idealized conditions, we were led to the Hardy–Weinberg law of Section 2 which predicts genetic stability for the population after a generation. The theoretical prediction turns out to be quite consistent with the observed persistency in the heredity of traits. However, changes do occurs naturally, albeit very infrequently and/or very slowly. This happens even under selective breeding of Section 3 where only the dominant genotype of one parent is allowed to breed; a recessive genotype occurs on rare occasions, when the modeling result predicts that it should not. We simply call this observed process a gene **mutation** (with the biological details of how it takes place discussed briefly in the section on DNA mutation in Chapter 2).

For a simple model of the phenomenon, we consider the situation that whenever a dominant gene is transmitted, there is a small probability α ($0 < \alpha \ll 1$) that the gene will mutate to a recessive gene. We suppose that mutation occurs after selection of the dominant gene from a parent. Otherwise, we retain all the hypotheses of the Mendelian (*panmixia*) model. In that case, the Mendelian model (4.2.1) governing the evolution of genotypes is modified by a reduction of the dominant gene frequency and an increase in the recessive gene frequency. The modification is most simply done by working with dominant and recessive gene frequencies, p_n and q_n, introduced in the last section.

The dominant offspring genotype probability d_{n+1} is the product of the (available) dominant gene frequency from the two parents both now reduced to $(1-\alpha)p_n$ by a loss αp_n due to mutation

$$d_{n+1} = [(1-\alpha)p_n]^2. \tag{4.6.1}$$

The recessive gene frequency available for the offsprings is enhanced by mutation from q_n to $\alpha p_n + q_n$ thereby giving

$$2h_{n+1} = 2(1-\alpha)p_n(\alpha p_n + q_n), \tag{4.6.2}$$

$$r_{n+1} = (\alpha p_n + q_n)^2. \tag{4.6.3}$$

Note that

$$d_{n+1} + 2h_{n+1} + r_{n+1} = (p_n + q_n)^2 = 1$$

so that gene frequency is conserved.

Instead of solving for the genotype distribution, we form

$$p_{n+1} = d_{n+1} + h_{n+1} = (1-\alpha)p_n\left[(1-\alpha)p_n + (\alpha p_n + q_n)\right] = (1-\alpha)p_n.$$

It follows immediately that

$$p_n = p_0(1-\alpha)^n$$

where p_0 is the initial dominant gene frequency. Correspondingly, the recessive gene frequency is

$$\alpha p_n + q_n = \alpha p_n + (1 - p_n) = 1 - (1 - \alpha)p_n$$
$$= 1 - p_0(1 - \alpha)^{n+1} \equiv 1 - \pi_n.$$

In terms of $\pi_n = p_0(1 - \alpha)^{n+1}$, we have

$$d_{n+1} = \pi_n^2, \quad 2h_{n+1} = 2\pi_n(1 - \pi_n), \quad r_{n+1} = (1 - \pi_n)^2.$$

When there is no mutation so that $\alpha = 0$ and therewith $\pi_n = p_0$, the above results reduce to the Hardy–Weinberg scenario with

$$d_{n+1} = p_0^2, \quad 2h_{n+1} = 2p_0q_0, \quad r_{n+1} = q_0^2.$$

For $0 < \alpha < 1$, we have $\pi_n = p_0(1 - \alpha)^{n+1} \to 0$ so that

$$d_{n+1} \to 0, \quad 2h_{n+1} \to 0, \quad r_{n+1} \to 1.$$

if the particular type of mutation is the only evolutionary process at work (which fortunately is not).

4.7   A Nonlinear Infectious Disease Model

Another candidate for a nonlinear Markov process model is the spread of a (nonfatal) infectious disease where the susceptible population $S(n)$ at stage n may be infected but does not become immune when they recovered. Typically, a fraction of the susceptible $S(n)$ becomes infected to increase the infected population at the next stage, while the recovery of a fraction of the infected reduces it. At the same time, each population would also be changed by their rates of birth and death. For the infected population, these activities result in the evolutionary change

$$I(n + 1) = \bar{\mu}S(n)I(n) + (1 - d_I - r)I(n) \tag{4.7.1}$$

where d_I and r are the fraction (or the probability) of the infected population that dies and recovers, respectively, and where the first term corresponds to the infection of the susceptible by first order interaction between the two populations. For the susceptible, we have

$$S(n + 1) = -\bar{\mu}S(n)I(n) + (1 - d_S + b_S)S(n) + (b_I + r)I(n) \tag{4.7.2}$$

where we have assumed that the disease is not hereditary (so that no one is born infected).

To gain some insight to the evolution of the two populations, we make the (not particularly realistic) assumptions $d_S = b_S = b$ and $d_I = b_I = b$ in order to simplify the solution process. Since birth rates now equal death rates, the total population must be fixed at N for all times. We may then take $\bar{\mu} = \mu/N$ and write the two relations (4.7.1) and (4.7.2) as

$$I(n+1) = \frac{\mu}{N} S(n) I(n) + (1 - b - r) I(n) \qquad (4.7.3)$$

$$S(n+1) = \left(1 - \frac{\mu}{N} I(n)\right) S(n) + (b + r) I(n). \qquad (4.7.4)$$

As in the general birth and death processes, we consider parameter values in the range $0 < b + r \leq 1$ and, to ensure that $\mu I(n)/N$ is in fact a fraction, $0 < \mu \leq 1$.

Since the total population is conserved, we have $S(n) + I(n) = N$. This is re-affirmed by adding the two equations (4.7.3) and (4.7.4) to get

$$S(n+1) + I(n+1) = S(n) + I(n) = S(0) + I(0) = N.$$

With $S(n) = N - I(n)$, the relation (4.7.3) becomes an equation for $I(n)$ alone:

$$I(n+1) = \left\{1 + \mu - b - r - \frac{\mu}{N} I(n))\right\} I(n).$$

Exercise 16. For an infectious disease that has a very short completion time (either death or recovery), available clinical data show that $3/4$ of the healthy individuals in month n remain healthy the following month while the remaining $1/4$ become infected. Half of those who are currently infected will be cured the following month; half of the remaining half (or $1/4$) will die while the rest will remain infected. Of course, those who die will remain dead thereafter. The evolution of the population fractions may be treated as a Markov chain of three states: H (healthy), I (infected) and D (dead). a) Obtain the transition matrix M for the evolution of the three states. b) Is this a regular DMC? c) Does the DMC evolves into a steady state?

Exercise 17. Suppose $7/16$ of those currently infected recover by the next stage and the rest dies while those recovered from the infectious disease acquire immunity from further infection thereafter. Formulate a model for the evolution of the four (fraction) states (*Healthy* $H(n)$, *Infected* $I(n)$, *Recovered* $R(n)$ and *Dead* $D(n)$) by constructing the transition matrix M. Of interest is the eventual fate of the population after long time. a) Identify all the absorbing states, and b) determine the steady state solution $\mathbf{x}(\infty) = \mathbf{x}_\infty$.

4.8 Single Nonlinear Difference Equations

As seen from previous sections, many biological phenomena may be modeled by *nonlinear* difference equations after suitable idealization and simplifications. It is therefore desirable to know some general mathematical techniques for solving such equations. Regrettably, there are not many such general techniques as there are for linear difference equations. In fact, there are arguably fewer such techniques than those available for nonlinear ordinary differential equations. In this section, we briefly summarize the few general approach that have been found useful and effective. Most of these are ways to reduce different kinds of nonlinear difference equation(s) to linear equations for which we have already developed general methods of solution in terms of known functions in Chapter 2.

4.8.1 *Taking logarithms*

Consider the nonlinear first order difference equation

$$x_{n+1} = x_n^2, \quad x_0 = p > 0 \tag{4.8.1}$$

again using the subscript notation to conserve space, i.e., $x_n = x(n)$. Let $y_n = \ln(x_n)$ and take the natural logarithm of both sides to get

$$y_{n+1} = 2y_n, \quad y_0 = \ln(p).$$

The new equation for y_n is a linear difference equation and can be solved by the usual assumed solution of the form $y_n = c\lambda^n$ to get $\lambda = 2$ and therewith

$$y_n = 2^n \ln(p) = \ln(x_n)$$

or

$$x_n = p^{2^n} \quad (n = 0, 1, 2, 3, \ldots).$$

For the difference equation (4.8.1), calculating x_n recursively from (4.8.1) leads to the same expression for x_n. The method described in this subsection does demonstrate how some nonlinear difference equations (including some with variable coefficients) become linear upon taking the logarithm of both side of the equations. The resulting equation may be more tractable as shown by the example above.

4.8.2 *Algebraic and trigonometric identities*

Known functional identities may be used sometimes to simplify nonlinear difference equations and reduce them to a more tractable form. Below are some examples.

Example 4.

$$x_{n+1} = 2x_n^2 - 1, \quad x_0 = p \quad (0 < p < 1). \tag{4.8.2}$$

For this nonlinear first order difference equation, by setting $x_n = \cos(\theta_n)$, we may re-write the difference equation (4.8.2) as

$$\cos(\theta_{n+1}) = 2\cos^2(\theta_n) - 1, \quad \cos(\theta_0) = \cos(p). \tag{4.8.3}$$

Though still nonlinear, the new difference equation (4.8.3) may be simplified further by the trigonometric identity $\cos(2x) = 2\cos^2(x) - 1$ to give

$$\cos(\theta_{n+1}) = \cos(2\theta_n), \quad \cos(\theta_0) = \cos(p).$$

The equation for θ_n above is satisfied if

$$\theta_{n+1} = 2\theta_n + 2k\pi, \quad k = 0, 1, 2, 3, \ldots$$

which is a first order linear equation. With $\theta_0 = \cos^{-1}(p) + 2m\pi$, we get

$$\theta_1 = 2\theta_0 + 2k\pi = 2\{\cos^{-1}(p) + 2m\pi\} + 2k\pi = 2\cos^{-1}(p) + 2\pi(2m + k)$$

so that

$$x_1 = \cos(\theta_1) = \cos(2\cos^{-1}(p) + 2\pi(2m + k)) = \cos(2\cos^{-1}(p)).$$

By induction, we get

$$\theta_n = 2^n \cos^{-1} p + 2j\pi$$

for any integer j. It follows that

$$x_n = \cos(2^n \cos^{-1}(p)).$$

Example 5.

$$x_{n+1} = 2x_n^2 - 1, \quad x_0 = q \quad (|q| \geq 1). \tag{4.8.4}$$

Given $|q| \geq 1$, the transformation $x_n = \cos(\theta_n)$ is no longer appropriate. Instead, we set $x_n = \cosh(\theta_n)$ which satisfies a similar identity $\cosh(2x) = 2\cosh^2(x) - 1$ and reduces the given equation to the more tractable form of

$$\cosh(\theta_{n+1}) = \cosh(2\theta_n),$$

for which the solution is $\theta_{n+1} = 2\theta_n$. With $\theta_0 = \cosh^{-1} q$ and $|q| \geq 1$, the solution for θ_n is

$$\theta_n = 2^n \theta_0 = 2^n \cosh^{-1} q.$$

These example suggests that many nonlinear difference equations become solvable upon making use of known mathematical identities such as the two above and others.

Exercise 18. Solve $x_{n+1}^2 = 4x_n^2 (1 - x_n^2)$, $x_0 = q$ ($|q| \leq 1$).

4.8.3 *Raising the order*

Given $x(n)$, $n = 0, 1, 2, \ldots$, the *difference operator* D operating on $x(n)$ is defined by

$$D[x(n)] = x(n + 1) - x(n) \tag{4.8.5}$$

for n in the range where $x(n) \equiv x_n$ is defined. Evidently, the difference operation is the discrete counterpart of differentiation for continuously differentiable functions. Similar to the ODE case, an intractable nonlinear difference equation may become solvable after further differencing to get a higher order (but hopefully simpler) equation. Below are some successes by this approach.

Example 6.

$$x(n)[x(n+1) - x(n)] = (n+1)[x(n+1) - x(n)]^2 + 1, \quad x(0) = p. \qquad (4.8.6)$$

The appearance of the difference $x(n+1) - x(n)$ suggests that something may be gained by setting

$$y(n) = x(n+1) - x(n)$$

and rewriting the given difference equation (4.8.6) as

$$x(n)y(n) = (n+1)[y(n)]^2 + 1. \qquad (4.8.7)$$

Superficially, the substitution has only made the problem worse; we now have one equation for two unknowns. To rid of the unknown $x(n)$ in (4.8.7), we apply the difference operator to both side of the new equation to get

$$x(n)[y(n+1) - y(n)] = [(n+2)y(n+1) + (n+1)y(n)][y(n+1) - y(n)] \qquad (4.8.8)$$

(see Assignment II for the needed tools to derive the result above). The relation (4.8.8) may be satisfied in two ways:

$$i) \quad y(n+1) = y(n)$$

or

$$ii) \quad x(n) = (n+2)y(n+1) + (n+1)y(n).$$

Each of these is a single linear difference equation and can be solved by the method of the previous section (exercise).

Example 7.

$$x(n+1) = \alpha - \frac{\beta}{x(n)}$$

Another way to increase the order of a given difference equation (in hope of simplification) is to set $x(n) = z(n+1)/z(n)$. Upon making this substitution, the equation above becomes

$$z(n+2) - \alpha z(n+1) + \beta z(n) = 0$$

which is *linear and solvable with its solution* in the form $z(n) = c_1 \lambda_1^n + c_2 \lambda_2^n$ where λ_1 and λ_2 are the two roots of

$$\lambda^2 - \alpha\lambda + \beta = 0.$$

The solution for $x(n)$ is then given by

$$x(n) = \frac{z(n+1)}{z(n)} = \lambda_1 \frac{1 + c\rho^{n+1}}{1 + c\rho^n}, \quad \rho = \frac{\lambda_2}{\lambda_1}. \tag{4.8.9}$$

Given the initial data $x(0) = q$, we have

$$c = \frac{\lambda_1 - q}{q - \lambda_1 \rho}.$$

For cases with two real unequal characteristic roots, we may let $|\rho| < 1$ with no loss in generality. In that case, we have $x(n)$ tends to the steady state solution λ_1 as $n \to \infty$. On the other hand, with $x(n) \to x$ as $n \to \infty$, the original nonlinear difference equation for the limiting solution x becomes

$$x = \alpha - \frac{\beta}{x} \quad \text{or} \quad x = \frac{1}{2}\left\{\alpha \pm \sqrt{\alpha^2 - 4\beta}\right\} = \begin{pmatrix} \lambda_1 \\ \lambda_2 \end{pmatrix}$$

where we have assumed $|\lambda_2/\lambda_1| < 1$ to be concrete. Evidently, the difference equation has two steady state solution by this steady state calculation. That the solution (4.8.9) tends to λ_1 suggests that this steady state is asymptotically stable while the other steady state solution is unstable. The need to confirm this supposition means that stability analysis is also an important aspect of the solution process, especially when we wish to bypass the exact solution of the IVP. To the extent that the general method of linear stability analysis so useful for this purpose is rather similar to that for ordinary differential equations, we will not discuss this topic here.

4.8.4   Other ad hoc substitutions

Example 8.

$$x(n)x(n+1) + 1 = \mu(n)\left[x(n+1) - x(n)\right], \quad x(0) = p. \tag{4.8.10}$$

While there is no standard recipe for solving such an equation, an unusual (and rather ingenious) substitution $x(n) = \tan(y_n)$ transforms (4.8.10) into

$$\cos(y_{n+1} - y_n) = \mu(n)\sin(y_{n+1} - y_n)$$

or

$$y_{n+1} - y_n = \cot^{-1}(\mu(n)).$$

The equation for y_n is linear and its solution is

$$y_n = \tan^{-1}(x_0) + \sum_{k=0}^{n-1} \cot^{-1}(\mu(k))$$

and therewith

$$x_n = \tan\left(\tan^{-1}(x_0) + \sum_{k=0}^{n-1} \cot^{-1}(\mu(k))\right).$$

4.8.5 *Evolving a few stages*

For some nonlinear models, none of the methods mentioned above would simplify the problem. In some cases, using the difference equation to calculate the unknown for a few stages sometimes gives some insight to the actual solution. The following nonlinear difference equation is one of the simplest example of this kind of situations:

Example 9.

$$x_{n+1} = x_n^{1/3}, \quad x_0 = p$$

For $p = 1$, it is clear from

$$x_1 = x_0^{1/3} = 1, \quad x_2 = x_1^{1/3} = 1$$

that $x_n = 1$ for all $n > 0$. For $p > 1$, we have

$$x_1 = x_0^{1/3} = p^{1/3} \quad (1 < p^{1/3} < p), \quad x_2 = x_1^{1/3} = p^{1/9} \quad (1 < p^{1/9} < p^{1/3} < p).$$

We infer from these $x_n \to 1$ from above as $n \to \infty$. Similarly, for $p < 1$, we have

$$x_1 = x_0^{1/3} = p^{1/3} \quad (p < p^{1/3} < 1), \quad x_2 = x_1^{1/3} = p^{1/9} \quad (p < p^{1/3} < p^{1/9} < 1),$$

so that $x_n \to 1$ from below as $n \to \infty$. Altogether, we have for $p > 0$, the solution of the IVP tends to 1 as $n \to \infty$. The solution for the complementary case of $p < 0$ can also be seen to tend to -1 as $n \to \infty$. For the remaining case of $p = 0$, the exact solution is $x_n = 0$ for all n.

Thus the nonlinear difference equation has three steady states $-1, 0$ and 1. A linear stability analysis (similar to that for ODE) shows that 0 is an unstable steady state while the other two are (locally) asymptotically stable. With the convergence to one of the asymptotically stable steady state depends on the initial data, the given nonlinear difference equation exhibits the characteristics of bi-stability.

Similar to MC models, not all nonlinear Markov processes would tend to a steady state. Below is an example of such a Markov process:

Example 10.

$$x(n+1) = \frac{1}{1 - x(n)}, \quad x(0) = p \quad (\neq 1).$$

With no obvious technique for such an equation, we again calculate a few $x(n)$ as we did for the the previous example (and for the Hardy–Weinberg model). Here we have

$$x(1) = \frac{1}{1-p}, \quad x(2) = \frac{1}{1 - x(1)} = \frac{p-1}{p} = 1 - \frac{1}{p}, \quad x(3) = p.$$

Evidently, the process evolves cyclically: $p \to (1-p)^{-1} \to 1 - p^{-1} \to p \to \cdots$. In contrast to Hardy–Weinberg, the present nonlinear Markov process does not reach a steady state. It is an example how a Markov process does not converge as $n \to \infty$.

Part 2
Continuous Time Markov Chains

Chapter 5

Continuous Time Birth and Death Type Processes

Up to now, the biological phenomena evolving with uncertainty encountered in this volume all share several distinctive features. Most conspicuous of these is the Markovian property of the processes whose evolution with time t depends only on its recent past. It is a feature that persists among the evolving processes treated in this volume since most biological organisms have relatively short term memory and models built on this assumption are more tractable mathematically and simpler for numerical simulations. Two other features common to the random processes in the previous chapters are:

1) The sample space of the phenomena modeled is countable (and hence may be indexed by the integers) with elementary events $\{E_1, E_2, \ldots\}$).
2) The processes evolve with outcomes recorded only in stages indexed by n with a finite time interval between consecutive stages, $\Delta_n = t_{n+1} - t_n$ (not necessarily with same actual elapsed time as illustrated by the mouse experiment).

Other phenomena with inherent uncertainty exist that do not share these two features. In Part II of this volume, we examine some important biological phenomena that are Markovian with a discrete sample space but evolves continuously in time. This new type of Markov processes is known as *continuous time Markov chains* (or CTMC). At each instant of (continuous) time, the state of a CTMC may be taken as a random variable. We will be interested at this time (and in the next chapter) in the probability distributions of such random variables with discrete sample space at different moments of time. The later chapters of this volume will be devoted to discussions of Markov processes with continuous sample space evolving continuously with time.

5.1 The Poisson Process

5.1.1 *Stationary processes*

To develop mathematical methods for analyzing continuous time Markov chains, let us focus on a specific problem of harvesting fish over an interval of time $(0, t)$ in a large body of water containing N fish (with N sufficiently large to be taken as ∞ for some purposes). Suppose $X(t)$ is the number of fish caught by time t with no fish caught by the starting time $t = 0$ so that $X(0) = 0$. Let $P_k(t)$ be the probability of k fish caught by time t:

$$P_k(t) = \text{Prob}\{X(t) = k\} \tag{5.1.1}$$

for some non-negative integer k (since we cannot catch less than zero fish). For δt sufficiently small so that no more than one fish would be caught in the interval $(t, t + \delta t)$ for any time t. For N very large but still sparsely located in a very large body of water (e.g., the large population of Atlantic cod or halibut inhabiting Georges Bank off the coast of Nova Scotia), it seems reasonable to assume that the probability of catching one fish in the interval $(t, t + \delta t)$ is proportional to the elapse time, say $\lambda \delta t$. The significance of this assumption is that the probability does not depend on $t_1 = t$ or $t_2 = t + \delta t$, only on the elapse time $t_2 - t_1 = \delta t$. A randomly evolving process with this property is known as a *stationary process*.

5.1.2 *Independent increments*

We are interested in the probability of catching k fish by time t under the assumptions of fish harvesting in a large body of water being a stationary random process with *independent increments*. For a fixed t, the number of fish caught by time t, $X(t)$, is a random variable with a discrete sample space $\{0, 1, \ldots, N\}$. As t ranges over an interval (a, b), the collection of random variables $\{X(t)\}$ constitutes a *stochastic process*. For any sequence $0 \leq \tau_1 < t_1 \leq \tau_2 < t_2 \leq \cdots \leq \tau_k < t_k \leq \cdots$, the increments $\{\Delta X_k = X(t_k) - X(\tau_k)\}$ are random variables. The stochastic process $X(t)$ is with *independent increments* if $\{\Delta X_k\}$ are mutually independent random variables.

For the specific probability of $\lambda \delta t$ for catching one fish in a sufficiently small interval δt (and probability zero for catching more than one fish during that interval), the probability of catching no fish $P_0(t + \delta t)$ by the time $t + \delta t$ is

$$P_0(t + \delta t) = \text{Prob}\{X(t + \delta t) = 0\}$$

$$= \text{Prob}\{\Delta X(t) = 0, \ X(t) - X(0) = 0\}$$

$$= \text{Prob}\{\Delta X(t) = 0, \ X(t) = 0\}$$

where $\Delta X(t) = X(t + \delta t) - X(t)$. Since the two intervals $(0, t)$ and $(t, t + \delta t)$ do not overlap, we expect (and assume) the two random variables to be statistically independent so that $X(t)$ is with independent increments. We may then rewrite $P_0(t + \delta t)$ as

$$P_0(t + \delta t) = \text{Prob}\{\Delta X(t) = 0\} \cdot \text{Prob}\{X(t) = 0\}$$

$$= (1 - \lambda \delta t) P_0(t) \qquad (5.1.2)$$

with $\text{Prob}\{\Delta X(t) = 0\} = \text{Prob}\{X(t + \delta t) - X(t) = 0\} = 1 - \lambda \delta t$ (possibly plus negligibly small terms that tend to zero faster than δt).

Upon re-arranging (5.1.2) to read

$$\frac{1}{\delta t} \{P_0(t + \delta t) - P_0(t)\} = -\lambda P_0(t),$$

we get in the limit as $\delta t \to 0$

$$\frac{dP_0}{dt} = -\lambda P_0.$$

Together with the initial condition $P_0(0) = 1$, we have

$$P_0(t) = e^{-\lambda t}. \qquad (5.1.3)$$

For $k = 1$, $P_1(t + \delta t)$ is the probability of catching one fish by $t + \delta t$ with

$$P_1(t + \delta t) = \text{Prob}\{X(t + \delta t) = 1\}$$

$$= \text{Prob}\{\Delta X(t) = 0, \ X(t) = 1\} + \text{Prob}\{\Delta X(t) = 1, \ X(t) = 0\}$$

$$= (1 - \lambda \delta t) P_1(t) + \lambda \delta t P_0(t)$$

either exactly or up to terms of higher order in δt. After re-arranging the last relation to get

$$\frac{1}{\delta t} \{P_1(t + \delta t) - P_1(t)\} = -\lambda \{P_1(t) - P_0(t)\} + O(\delta t^\sigma),$$

for some $\sigma > 0$, we take the limit as $\delta t \to 0$ to get

$$\frac{dP_1}{dt} = -\lambda P_1 + \lambda P_0(t) = -\lambda P_1 + \lambda e^{-\lambda t}.$$

With the initial condition $P_1(0) = 0$, we get by the method of integrating factor

$$P_1(t) = \lambda t e^{-\lambda t}.$$

For a larger k, we have

$$P_k(t + \delta t) = \text{Prob}\{X(t + \delta t) = k\}$$

$$= \sum_{i=0}^{k} \text{Prob}\{\Delta X(t) = i, \ X(t) = k - i\}$$

$$= (1 - \lambda \delta t) P_k(t) + \lambda \delta t P_{k-1}(t) + \sum_{i=2}^{k} \text{Prob}\{\Delta X(t) = i\} \cdot P_{k-i}(t).$$

But $\text{Prob}\{\Delta X(t) = i\}$ for $i \geq 2$ is negligibly small (assumed either to be zero or tending to zero faster than δt). The expression for $P_k(t + \delta t)$ is simplified to

$$P_k(t + \delta t) = (1 - \lambda \delta t) P_k(t) + \lambda \delta t P_{k-1}(t) + O(\delta t^m) \quad (m > 1)$$

or in the limit as $\delta t \to 0$

$$\frac{dP_k}{dt} = -\lambda P_k + \lambda P_{k-1}(t)$$

for $k = 2, 3, 4, \ldots$, with $P_k(0) = 0$.

In particular, the ODE for $P_2(t)$ is

$$\frac{dP_2}{dt} = -\lambda P_2 + \lambda^2 t e^{-\lambda t}$$

so that the solution of the corresponding IVP is

$$P_2(t) = \frac{1}{2!} (\lambda t)^2 e^{-\lambda t}.$$

For $k > 2$, we get sequentially for $N = \infty$

$$P_k(t) = \frac{1}{k!} (\lambda t)^k e^{-\lambda t}.$$

The collection of probabilities $\{P_k(t)\}$, $k = 0, 1, 2, \ldots$, is known as the *Poisson (probability) distribution* which gives the probability of catching k fish by time t, $k = 0, 1, 2, \ldots$, for our fish harvesting problem starting with no fish caught at $t = 0$.

5.1.3 *Other probabilistic information*

5.1.3.1 *The expected value*

Having $\{P_k(t)\}$, we can calculate for $N = \infty$ the expected value of the catch from

$$E[X(t)] = \sum_{k=0}^{\infty} k P_k(t) = \sum_{k=1}^{\infty} \frac{1}{(k-1)!} (\lambda t)^k e^{-\lambda t}$$

$$= [\lambda t] e^{-\lambda t} \sum_{k=1}^{\infty} \left[\frac{1}{(k-1)!} (\lambda t)^{k-1} \right] = \lambda t.$$

5.1.3.2 Second moment and variance

Correspondingly, we have the second moment

$$E[X^2(t)] = \sum_{k=0}^{\infty} k^2 P_k(t) = e^{-\lambda t} \sum_{k=1}^{\infty} \frac{k}{(k-1)!} (\lambda t)^k$$

$$= (\lambda t)^2 e^{-\lambda t} \sum_{k=2}^{\infty} \frac{1}{(k-2)!} (\lambda t)^{k-2} + (\lambda t) e^{-\lambda t} \sum_{k=1}^{\infty} \frac{1}{(k-1)!} (\lambda t)^{k-1}$$

$$= (\lambda t)^2 + \lambda t$$

and the variance of a Poisson process

$$\sigma^2(t) = E[X^2(t)] - E^2[X(t)] = \lambda t.$$

5.1.3.3 Probability generating function

The (*probability*) *generating function* (pgf) of a sequence of probabilities $\{P_k(t)\}$ is defined to be

$$G(t, x) = \sum_{k=0}^{\infty} P_k(t) x^k. \tag{5.1.4}$$

For the Poisson distribution, we have

$$G(t, x) = = \sum_{k=0}^{\infty} \frac{1}{k!} (x\lambda t)^k e^{-\lambda t} = e^{\lambda t(x-1)}.$$

From this expression, we obtain

$$G(t, 1) = \sum_{k=0}^{\infty} P_k(t) = 1$$

so that the total probability is 1 as we would expect. We can also obtain the expected number of fish caught:

$$E[X(t)] = \sum_{k=0}^{\infty} k P_k(t) = \left[\frac{\partial G}{\partial x}\right]_{x=1} = \lambda t$$

as found earlier.

5.1.4 *Waiting time*

Let t_k be the time when the kth fish is caught so that $X(t_k) = k$ and denote the *holding time* $\tau_k = t_{k+1} - t_k$ when no new fish is caught during the interval (t_k, t_{k+1}). From the Poisson distribution for our stationary process with independent increments, we have from (5.1.3)

$$Q_k(t) = \text{Prob}\{\text{no fish caught in the interval } (t_k, t_k + t)\} = e^{-\lambda t}$$

so that $Q_k(t)$ is the probability that the number of fish caught remains k for the duration t (in the time interval $(t_k, t_k + t)$).

For δt sufficiently small, we also have

$$\text{Prob}\{\text{one fish caught in the interval } (t_k + t, t_k + t + \delta t)\} = \lambda \delta t$$

except for small terms that tend to zero faster than δt. Combining these two observations, we get

$$\text{Prob}\{\text{holding time lies in the interval } (t_k + t, t_k + t + \delta t)\}\}$$
$$= \text{Prob}\{t \le \tau_k < t + \delta t\} = e^{-\lambda t} \lambda \delta t + o(\delta t)$$

with the holding time τ_k independent of t_k. Its limit for infinitessimal time increment is

$$\text{Prob}\{t \le \tau_k < t + dt\} = e^{-\lambda t} \lambda dt.$$

The expected waiting time between k^{th} and $(k+1)^{th}$ catch is then

$$E[\tau_k] = \int_0^\infty t e^{-\lambda t} \lambda dt = \frac{1}{\lambda}.$$

The *expected waiting time to m catches* is the sum

$$T_m = \sum_{k=0}^m E[T_k] = \frac{m}{\lambda}.$$

In general, the mean waiting time may depend on the state of the process at various stages. In that case, the probability $\text{Prob}\{t \le \tau_k < t + dt\}$ would be state dependent and the determination of the expected waiting time $E[\tau_k]$ would be more complicated.

5.2 Pure Birth and Pure Death Processes

5.2.1 *Pure birth processes*

Suppose at the initial time $t = 0$, there is a population of N individuals each capable of giving birth to one new member at a time to add to the population. For this section we are interested in the case where no one dies during the time period of interest so that the population size can only increase. Let $P_k(t)$ be the probability for the population $X(t)$ to be of size k at time t:

$$P_k(t) = \text{Prob}\{X(t) = k\}.$$

Evidently, we have $P_k(t) = 0$, $k \le N - 1$ for all t, $P_N(0) = 1$ and $P_j(0) = 0$ for $j > N$. As in the previous section, we have

$$P_N(t + \delta t) = \text{Prob}\{X(t) = N, \Delta X(t) = 0\}$$

where $\Delta X(t) = X(t + \delta t) - X(t)$. Note that the option of $\{X(t) = N - 1$ and $\Delta X(t) = 1\}$ is not available since we started with an initial population of size N (and there is no death) so that $X(t)$ cannot be smaller than N.

As before, we take δt sufficiently small so that no more than one birth can occur in the time interval $(t, t + \delta t)$ (or more properly, the probability of more than one birth is negligibly small tending to zero faster than δt) with the (stationary) probability of an individual giving birth being $\lambda \delta t$. For a population of size k, there would be k individuals who can give birth to one new member; hence the probability of having one birth in an interval $(t, t + \delta t)$ would be $k\lambda \delta t$. This stipulation results in a simple pure birth model. A more general birth model would replace $k\lambda \delta t$ by $\lambda_k \delta t$ with λ_k depending on k.

5.2.1.1 *The Kolmogorov differential equations*

The intervals $(0, t)$ and $(t, t + \delta t)$ do not overlap. The independent increment assumption will again be made to rewrite $P_N(t + \delta t)$ as

$$\begin{aligned} P_N(t + \delta t) &= P_N(t) \cdot \text{Prob}\{\Delta X(t) = 0\} \\ &= (1 - N\lambda \delta t)P_N(t) \end{aligned}$$

with

$$\text{Prob}\{\Delta X(t) = 0\} = 1 - N\lambda \delta t$$

since there are N individuals capable of giving one birth with probability $\lambda \delta t$ each. As $\delta t \to 0$, we get

$$\frac{dP_N}{dt} = -N\lambda P_N, \quad P_N(0) = 1$$

so that

$$P_N(t) = e^{-\lambda N t}. \tag{5.2.1}$$

For $k > N$, we have

$$P_k(t + \delta t) = \text{Prob}\{X(t) = k, \Delta X(t) = 0\}$$
$$+ \text{Prob}\{X(t) = k - 1, \Delta X(t) = 1\} + o(\delta t)$$

with $o(\delta t) \to 0$ faster than δt (given the "at most one birth" stipulation for the elapsed time δt). That $(0, t)$ and $(t, t + \delta t)$ being independent increments leads to

$$P_k(t + \delta t) = (1 - k\lambda\delta t) P_k(t) + (k - 1) \lambda\delta t P_{k-1}(t) + o(\delta t).$$

As $\delta t \to 0$, we get

$$\frac{dP_k}{dt} = -k\lambda P_k + (k - 1) \lambda P_{k-1}(t), \quad P_k(0) = 0 \qquad (5.2.2)$$

for $k - N + 1, N + 2, \dots$. These IVP for the system of Kolmogorov ODE can be solved sequentially starting with $k = N + 1$ having found $P_N(t)$ in (5.2.1) above.

Exercise 19. From the IVP (5.2.2) with $P_N(t)$ given by (5.2.1), deduce

$$P_{N+m}(t) = \binom{N + m - 1}{m} e^{-\lambda N t} \left(1 - e^{-\lambda t}\right)^m$$
$$= \frac{(N + m - 1)!}{(N - 1)!m!} e^{-\lambda N t} \left(1 - e^{-\lambda t}\right)^m$$

for $k = N, N + 1, N + 2, \dots$.

5.2.1.2 *Probability generating function*

The probability distributions for the pure birth process can be obtained by solving the IVP (5.2.2) sequentially as we did for the Poisson process (see Exercise above). For pedagogical purposes, we obtain presently the same results by way of the probability generating function $G(t, x)$ for the problem as previously defined in (5.1.4).

Multiply both sides of (5.2.2) by x^k and sum over k to get

$$\sum_{k=0}^{\infty} \frac{dP_k}{dt} x^k = \sum_{k=0}^{\infty} -k\lambda P_k x^k + \sum_{k=0}^{\infty} (k - 1) \lambda P_{k-1}(t) x^k.$$

Keeping in mind $P_k(t) = 0$ for $k < N$, we may write this relation as

$$\frac{\partial G}{\partial t} = -\lambda x \frac{\partial G}{\partial x} + \lambda x^2 \frac{\partial G}{\partial x}$$

or

$$\frac{\partial G}{\partial t} + \lambda x (1 - x) \frac{\partial G}{\partial x} = 0.$$

The solution of this first order PDE can be obtained by the method of characteristics (see an appendix of this chapter). The method calls for solving the following system of two ODE:

$$\frac{dx}{dt} = \lambda x \, (1 - x), \quad \frac{dG}{dt} = 0$$

or

$$x(t) = \frac{c_0 e^{\lambda t}}{1 + c_0 e^{\lambda t}}, \quad G(t) = G_0. \tag{5.2.3}$$

The initial conditions are $x(0) = x_0$ and $G(0) = x_0^N$ (with x_0 yet unspecified) given $P_N(0) = 1$ and $P_k(0) = 0$ for all $k \neq N$. They require

$$c_0 = \frac{x_0}{1 - x_0}, \quad G_0 = x_0^N.$$

We can now solve (5.2.3) for x_0 in terms of x and t to get

$$x_0 = \frac{xe^{-\lambda t}}{1 - x(1 - e^{-\lambda t})}$$

and then make use of the result to eliminate x_0 from $G(t, x) = x_0^N$ to get

$$G(t, x) = \left[\frac{xe^{-\lambda t}}{1 - x(1 - e^{-\lambda t})} \right]^N = e^{-\lambda N t} \left[x^{-1} - (1 - e^{-\lambda t}) \right]^{-N}$$

$$= e^{-\lambda N t} \sum_{k-0}^{\infty} \binom{N + k - 1}{k} (1 - e^{-\lambda t})^k x^{k+N}$$

$$= e^{-\lambda N t} \sum_{k-0}^{\infty} \frac{(N + k - 1)!}{k!(N - 1)!} (1 - e^{-\lambda t})^k x^{k+N}.$$

From this expression for $G(t, x)$, we get

$$P_{N+k}(t) = \frac{(N + k - 1)!}{k!(N - 1)!} e^{-\lambda N t} (1 - e^{-\lambda t})^k$$

as found previously in Exercise (19).

The expected value of the population size for an initial population size N is

$$E[X(t)] = \left[\frac{\partial G}{\partial x} \right]_{x=1} = \left[(-N) e^{-\lambda N t} \left[x^{-1} - (1 - e^{-\lambda t}) \right]^{-N-1} (-x^{-2}) \right]_{x=1}$$

$$= N e^{-\lambda N t} \left[e^{-\lambda t} \right]^{-N-1} = N e^{\lambda t}.$$

Exercise 20. Deduce the following expression for the variance $\sigma^2(t)$:

$$\sigma^2(t) = N e^{2\lambda t} (1 - e^{-\lambda t}).$$

5.2.1.3   *SI model for epidemics*

Epidemics of infectious diseases may be analyzed at different levels of complexity. We consider here only the simplest model to indicate how the modeling and analysis of Pure Birth processes also apply to this case. For this simplest model (denoted by SI in the literature), a population of size N consists of two groups: the susceptible $S(t)$ and the infectious $I(t)$ with $S(t) + I(t) = N$. Typically a fraction rS of the susceptible becomes infected through contact with the infectious. It is assumed that r depends on the size of the infectious. The SI model would have $r = \lambda I$ so that $rS = \lambda I(N - I)$.

For a stochastic SI epidemic model based on the assumption that the spread of the epidemic is a stationary process with independent increments, we again let $P_k(t)$ be the probability that $I(t) = k$. For δt sufficiently short so that no more than one susceptible is infected during the period of $(t.t + \delta t)$, we have

$$P_k(t + \delta t) = P_k(t) \cdot \text{Prob}\{\Delta X(t) = 0\} + P_{k-1}(t) \cdot \text{Prob}\{\Delta X(t) = 1\}$$

$$= \left(1 - \frac{\lambda}{N}k(N - k)\delta t\right) P_k(t) + \frac{\lambda}{N}(k - 1)(N - k + 1)\delta t P_{k-1}(t)$$

except for terms tending to zero faster than δt. In the limit as $\delta t \to 0$, we obtain

$$\frac{dP_k}{dt} = -\frac{\lambda}{N}(N - k)kP_k + \frac{\lambda}{N}(k - 1)(N - k + 1)P_{k-1}, \quad P_k(0) = 0.$$

The corresponding generating function is determined by

$$\frac{\partial G}{\partial t} = \sum_{k=0} -\frac{\lambda}{N}(N - k)kP_k x^k + \sum_{k=0} \frac{\lambda}{N}(k - 1)(N - k + 1)P_{k-1}x^k.$$

To simplify our discussion, we consider the case $N >> k$ (or the probability of being infected remaining unchanged with the size of the susceptible population) so that we may replace $(N - k)/N$ by 1 to get

$$\frac{\partial G}{\partial t} = \sum_{k=0} -k\lambda P_k x^k + \sum_{k=0}(k - 1)\lambda P_{k-1}x^k$$

$$= \left(-\lambda x + \lambda x^2\right)\frac{\partial G}{\partial x} = -\lambda x(1 - x)\frac{\partial G}{\partial x}.$$

The solution for the pure birth problem applies to this approximate version of the SI problem.

5.2.2   *Pure death*

The features of the problem of only death and no birth is somewhat different from that of pure birth; the modeling and analysis however are similar. Suppose at the initial time $t = 0$, there is a population of N non-reproducing individuals so that the population size only decreases with death. Let $P_k(t)$ again be the probability for the population $X(t)$ to be of size k at time t:

$$P_k(t) = \text{Prob}\{X(t) = k\}.$$

Evidently, we have $P_k(t) = 0$, $k > N$ for all t (since the population does not increase), $P_N(0) = 1$ and $P_j(0) = 0$ for $j < N$. Similar to the previous section, we have

$$P_k(t + \delta t) = \text{Prob}\{X(t) = k, \Delta X(t) = 0\} + \text{Prob}\{X(t) = k + 1, \Delta X(t) = -1\}$$

where $\Delta X(t) = X(t + \delta t) - X(t)$. As before, we take δt sufficiently small so that no more than one death can occur in the time interval $(t, t + \delta t)$ (or the probability of more than one death is negligibly small tending to zero faster than δt) with the (stationary) probability of an individual death being $\mu \delta t$. For a population of size k, there would be k individuals who may die; hence the probability of having one death in an interval $(t, t + \delta t)$ would be $k\mu \delta t$. This stipulation results in a simple pure death model. A more general death model would replace $k\mu \delta t$ by $\mu_k \delta t$ with μ_k being a function of k.

The intervals $(0, t)$ and $(t, t + \delta t)$ do not overlap. The independent increments assumption will again be made to rewrite $P_k(t + \delta t)$ as

$$P_k(t + \delta t) = P_k(t) \cdot \text{Prob}\{\Delta X(t) = 0\} + P_{k+1}(t) \cdot \text{Prob}\{\Delta X(t) = -1\}$$
$$= (1 - k\mu \delta t)P_k(t) + (k + 1)\mu \delta t P_{k+1}(t) + o(\delta t).$$

As $\delta t \to 0$, we get the Kolmogorov differential system

$$\frac{dP_k}{dt} = -k\mu P_k + (k + 1)\mu P_{k+1}, \quad P_k(0) = 0, \tag{5.2.4}$$

$k = 0, 1, \ldots, N - 1$, since we start with a population of size N initially, and

$$\frac{dP_N}{dt} = -N\mu P_N, \quad P_N(0) = 1$$

with the population size not more than N at any time so that $P_{N+1}(t) = 0$. The last equation gives immediately the following probability for a population of size N at time t:

$$P_N(t) = e^{-\mu N t}. \tag{5.2.5}$$

Exercise 21. From the IVP (5.2.4) with $P_N(t)$ given by (5.2.5), deduce $P_m(t)$, $m = N - 1, N - 2, \ldots, 1, 0$.

5.2.2.1 Probability generating function

The probability distribution for the pure death model can be obtained by solving the IVP (5.2.4) sequentially (see Exercise above). We obtain presently the same result by way of the pgf $G(t, x)$ for the problem as previously defined in (5.1.4). Multiply both sides of (5.2.4) by x^k and sum over k to get

$$\sum_{k=0} \frac{dP_k}{dt} x^k = \sum_{k=0} -k\mu P_k x^k + \sum_{k=0} (k+1)\mu P_{k+1} x^k$$

or

$$\frac{\partial G}{\partial t} - \mu(1-x)\frac{\partial G}{\partial x} = 0.$$

The solution of this first order PDE can be obtained by the method of characteristics that solves the following system of two characteristic ODE:

$$\frac{dx}{dt} = -\mu(1-x), \quad \frac{dG}{dt} = 0$$

to get

$$1 - x(t) = (1 - x_0)e^{\mu t}, \quad G(t) = x_0^N$$

given $P_N(0) = 1$ and $P_k(0) = 0$ for all $k \neq N$. We solve the first relation for x_0 to get

$$x_0 = 1 - \{1 - x\}e^{-\mu t},$$

and use the result to eliminate x_0 from the second to obtain

$$G(t, x) = \{1 - (1-x)e^{-\mu t}\}^N = e^{-\mu Nt}\{x + (e^{\mu t} - 1)\}^N.$$

Expanding the right hand side as a Taylor series in x leads to

$$G(t, x) = e^{-\mu Nt} \sum_{k=0}^{N} \binom{N}{k} (e^{\mu t} - 1)^{N-k} x^k \qquad (5.2.6)$$

$$= (1 - e^{-\mu t})^N \sum_{k=0}^{N} \binom{N}{k} (e^{\mu t} - 1)^{-k} x^k.$$

From (5.2.6), we get

$$P_k(t) = \frac{N!}{k!(N-k)!}(1 - e^{-\mu t})^{N-k}e^{-k\mu t}.$$

5.2.2.2 Expected population size and variance

From the generating function, we get also the following expected value for the pure death model:

$$E[X(t)] = \sum_{k=0}^{N} kP_k(t) = \left[\frac{\partial G}{\partial x}\right]_{x=1} = Ne^{-\mu t}.$$

Exercise 22. Deduce from the generating function (5.2.6) the following expression for the variance $\sigma^2(t)$ for the pure death model:

$$\sigma^2(t) = Ne^{-\mu t}(1 - e^{-\mu t}).$$

5.3 Simple Birth and Death Models

Exercise 23. Simple Birth and Death Model

Combine the ideas and methods of the last two subsections to formulate a model for a population initially of size N that allows for both birth and death with similar underlying assumptions such as stationarity and independent increments.

(a) Deduce the following Kolmogorov ODE for the probability $P_k(t)$ for the population to be of size k at time t:

$$\frac{dP_k}{dt} = \mu(k+1)P_{k+1} - (\lambda + \mu)\, k P_k + \lambda\, (k-1)\, P_{k-1}(t).$$

(b) Derive the corresponding pgf.

Exercise 24. Queueing Processes

A checkout queue with customers arriving and departing may be modeled as a continuous time Markov process similar to the birth and death process above. With the assumptions of stationarity and independent increments, derive the system of Kolmogorov ODE for this problem.

Exercise 25. SIS Model of Infectious Disease

The SI model for the spread of an infectious disease among a population of size N may be extended to allow for recovery without developing immunity to the disease. As such, the recovered individuals can be infected again immediately.

(a) Under similar assumptions for the corresponding discrete time model including susceptible and infected individuals born and deceased with the same probability b, show that $\{P_n(t)\}$ satisfy the following Kolmogorov ODE:

$$\frac{dP_k}{dt} = (b+\mu)(k+1)P_{k+1} - \left\{(b+\mu) + \frac{\lambda}{N}(N-k)\right\} k P_k$$

$$+ \frac{\lambda}{N}(k-1)(N-k+1)P_{k-1},$$

with $P_k(t) = 0$ for $k < 0$ and $P_k(0) = \delta_{kN}$.
(b) For large N, we have a good approximation with the following simpler ODE:

$$\frac{dP_k}{dt} = (b+\mu)(k+1)P_{k+1} - \{(b+\mu) + \lambda\} k P_k + \lambda\,(k-1)\,P_{k-1},$$

with the same initial conditions and constraints for $k < 0$.

In either case, $(b+\mu)\,\delta t$ corresponds to the probability of infected being removed from that group in the interval $(t, t+\delta t)$ by recovery (μ) and death (equal to birth b) in that class.

5.4   Other Birth and Death Processes

5.4.1   *First passage time*

Starting from a certain (initial) state $X(0) = E_0$, the first passage time of a random process $X(t)$ to reach another state E_f of the process is called the first passage time for that state and is denoted by $T_{f,0}$ with $X(T_{f,0}) = E_f$. Since the time it takes to reach the prescribed end state is uncertain, $T_{f,0}$ is a random variable.

For the fish harvesting problem, the different elementary events or states are the number of fish, k, caught, $k = 0, 1, 2, \ldots$. For the SI model for the spread of an infectious disease, the states are the number of individuals infected. In both models, the total population is fixed to be of size N so that the total fish caught and the total infected individuals cannot exceed N. For the fish harvesting model, the state would remain at N once it is reached (exercise). As such, $X(t) = N$ is an absorbing state of an absorbing Markov chain for that model.

We now broaden the definition of $Q_k(t)$ introduced at the end of the section on Poisson process to mean

$$Q_k(t) = \text{Prob}\{X(t) \text{ remains in state } E_k \text{ in the interval in } (t_k, t_k + t)\}$$

where t_k is the time for the process reaching state E_k. Note that E_k may not be k as was in the case of fish harvesting. At the start of the process, we have

$$Q_k(0) = 1$$

since no time has elapsed to allow for any change.

For a sufficiently small δt, we also have

$$\text{Prob}\{X(t) \text{ no longer at } E_k \text{ in the interval } (t_k + t, t_k + t + \delta t)\} = \lambda(E_k)\delta t$$

except for small terms that tend to zero faster than δt. Combining these two observations, we get

$$Q_k(t + \delta t) = Q_k(t)\{1 - \lambda(E_k)\delta t\} + o(\delta t)$$

with the parameter λ dependent on E_k. Its limit for infinitessimal time increment is

$$\frac{dQ_k}{dt} = -\lambda(E_k)Q_k, \quad Q_k(0) = 1.$$

The solution of this IVP is

$$Q_k(t) = e^{-\lambda(E_k)t}.$$

5.4.1.1   *Expected time to target*

It follows that

$$E[\tau_k] = \int_0^\infty t Q_k(t)\lambda(E_k)\,dt$$

$$= \int_0^\infty e^{-\lambda(E_k)t}\lambda(E_k)t\,dt = \frac{1}{\lambda(E_k)}$$

and

$$E[T_N] = \sum_{k=1}^{N-1} \frac{1}{\lambda(E_k)}. \tag{5.4.1}$$

5.4.1.2 Expected time to extinction for SI epidemic model

For the SI epidemic model, the expected time for all N individuals to be infected can be calculated from (5.4.1) with $\lambda(E_k) = k(N-k)\mu/N$ so that

$$E[T_N] = \frac{N}{\mu} \sum_{k=1}^{N-1} \frac{1}{k(N-k)} = \frac{1}{\mu} \sum_{k=1}^{N-1} \left(\frac{1}{k} + \frac{1}{N-k} \right) = \frac{2}{\mu} \sum_{k=1}^{N-1} \frac{1}{k}.$$

5.5 Surname Survival

5.5.1 A simple branching process

In a patriarchal society, a family name in the population is carried on through male descendants of the family. When a man dies, he leaves behind K number of male children with his surname. As K varies from individual to individual, it is a random variable with a prescribed or estimated probability

$$q_k = \text{Prob}\{K = k\}. \tag{5.5.1}$$

Suppose for a sufficiently small δt, the probability of death for a male member of the population during the time interval $(t, t + \delta t)$ is $\mu \delta t$ up to negligibly small terms that tend to zero faster than δt. Then for a population of n individuals at time t, the probability of one death in the time interval $(t, t + \delta t)$ is again $n\mu \delta t$ except for negligibly small terms. In formulating a model for this simple branching process, we have incorporated underlying assumptions (such as stationarity and independent increments) similar to those for the previous models in this chapter.

5.5.2 The mathematical model

Let $P_n(t)$ be the probability of a surname population of size n at time t. The assumptions adopted enable us to write

$$P_0(t + \delta t) = P_0(t) + q_0\mu\delta t P_1(t) + o(\delta t),$$
$$P_1(t + \delta t) = \{1 - \mu\delta t\} P_1(t) + q_1\mu\delta t P_1(t) + 2q_0\mu\delta t P_2(t) + o(\delta t),$$
$$P_2(t + \delta t) = \{1 - 2\mu\delta t\} P_2(t) + q_2\mu\delta t P_1(t) + 2q_1\mu\delta t P_2(t) + 3q_0\mu\delta t P_3(t) + o(\delta t).$$

In each case, the first term on the right corresponds to the probability of a current population of the same size n with no death during the elapsed time δt. Each remaining term is the probability of one death from a population of size k that leaves $n - k + 1$ descendants. All other possible scenarios not explicitly listed are either impossibilities (e.g., a population of zero member at time t followed by a population of 1 after the elapsed time δt) or of negligibly small probabilities that tend to zero faster than δt (e.g., a current population with two or more deaths) summarized by the $o(\delta t)$ term.

In general, the probability of a population of n members at $t + \delta t$ would be a sum of the probabilities of all possible scenarios:

$$P_n(t + \delta t) = \{1 - n\mu\delta t\} P_n(t) + (n + 1)q_0\mu\delta t P_{n+1}(t) + n q_1 \mu\delta t P_n(t)$$
$$+ \cdots + q_n\mu\delta t P_1(t) + o(\delta t)$$
$$= \{1 - n\mu\delta t\} P_n(t) + \sum_{k-0}^{n}(n + 1 - k)q_k\mu\delta t P_{n+1-k}(t).$$

By taking the limit of $\delta t \to 0$, we get for the general case of a population initially of size N

$$\frac{dP_n}{dt} = -n\mu P_n(t) + \sum_{k-0}^{n}(n + 1 - k)q_k\mu P_{n+1-k}(t), \quad P_n(0) = \delta_{nN} \quad (5.5.2)$$

where $\delta_{nN} = 0$ for $n \neq N$ and $\delta_{NN} = 1$.

5.5.3   *Probability generating function*

While the IVP (5.5.2) is for a linear ODE system with constant coefficients, it is an infinite system of coupled equations to be solved simultaneously. In practice, there is a limit to the size of the population so that $P_n(t)$ is negligibly small for large n. It is therefore not unreasonable to take $P_n(t) = 0$ for $n > N$ leaving us with a coupled system of N ODE for N unknowns: $P_1(t), P_2(t), \ldots, P_N(t)$, that can be solved to get an approximate solution.

Instead of pursuing such a solution, we develop a method for an exact solution for the probability generating function:

$$F(t, x) = \sum_{n=0}^{\infty} P_n(t)x^n$$

(Since we must have a non-zero initial population size for the problem to be meaningful, we stipulate $P_n(t) = 0$ for $n < 0$.)

Upon multiplying both sides of (5.5.2) by x^n and summing over n, we get

$$\frac{\partial F}{\partial t} = -\mu x \frac{\partial F}{\partial x} + \mu \sum_{n=0}^{\infty}\sum_{k-0}^{n}(n + 1 - k)P_{n+1-k}(t)x^{n-k}q_k x^k$$
$$= -\mu x \frac{\partial F}{\partial x} + \mu \sum_{k=0}^{\infty}\sum_{n=k}^{\infty}(n + 1 - k)P_{n+1-k}(t)x^{n-k}q_k x^k \quad (5.5.3)$$

with the previous stipulation of $P_m(t) = 0$ for $m < 0$ consistent with the interchange of order of summation. In the re-arranged form, the relation (5.5.3) may be written as

$$\frac{\partial F}{\partial t} = -\mu x \frac{\partial F}{\partial x} + \mu \frac{\partial F}{\partial x} \sum_{k=0}^{\infty} q_k x^k$$

or

$$\frac{\partial F}{\partial t} + \mu \{x - q(x)\} \frac{\partial F}{\partial x} = 0 \qquad (5.5.4)$$

where

$$q(x) = \sum_{k=0}^{\infty} q_k x^k$$

is the pgf of the probability distribution $\{q_k\}$.

5.5.4 *One child policy*

The relation (5.5.4) for the pgf $F(t,x)$ is a first order linear partial differential equation (PDE) and can again be solved by the method of characteristics. To illustrate, we consider the specific case of a linear descendent pgf:

$$q(x) = q_0 + q_1 x, \quad q_0 = 1 - q_1$$

which may be a reasonable characterization of a one child policy. In that case, we have as the characteristic equations of (5.5.4)

$$\frac{dx}{dt} = \mu q_0 \{x - 1\}, \quad \frac{dF}{dt} = 0$$

with

$$x(0) = x_0, \quad F(0, x_0) = x_0^N$$

assuming the population being N at $t = 0$ so that $P_N(0) = 1$ and $P_n(0) = 0$ for $n \neq N$. The solution of the characteristic equations is

$$\frac{x - 1}{x_0 - 1} = e^{q_0 \mu t}, \quad F(t, x) = x_0^N.$$

Upon solving for x_0 in terms of x and t and then using the resulting expression for x_0 in $F(t, x)$, we obtain

$$F(t, x) = x_0^N = \left\{1 + e^{-q_0 \mu t}(x - 1)\right\}^N.$$

From the solution for $F(t, x)$, we note

$$F(t, 1) = 1, \quad \left[\frac{\partial F}{\partial x}\right]_{x=1} = N e^{-q_0 \mu t} = \sum_{n=1}^{\infty} n P_n(t)$$

with the second expression being the expected population size at time t. With $q_0 \mu > 0$, the mean population size tends to zero as $t \to \infty$, suggesting the population (of the same surname) is expected to extinct eventually. In fact, from

$$\lim_{t \to \infty} [F(t, x)] = 1,$$

the population becomes extinct eventually with probability 1 (given that the coefficients of x^n of $F(t, x)$ for all $n > 0$ vanish in the limit).

5.5.4.1 Two child policy

In contrast to a population subject to a one child policy and destined for eventual extinction, consider one with

$$q(x) = q_0 + q_2 x^2, \quad q_0 = 1 - q_2$$

corresponding to a special case of a two-child policy (for a population whose members would have either two children or none). In that case, we have as the characteristic equations of (5.5.4)

$$\frac{dx}{dt} = \mu \{x - q(x)\} = -\mu q_2 (x - 1)(x - x_1), \quad \frac{dF}{dt} = 0$$

where $x_1 = q_0/q_2$ augmented by the initial conditions

$$x(0) = x_0, \quad F(0, x_0) = x_0^N$$

assuming the population being N at $t = 0$ so that $P_N(0) = 1$ and $P_n(0) = 0$ for $n \neq N$. The solution of the characteristic equations is

$$\frac{x - 1}{x_0 - 1} \frac{x_0 - x_1}{x - x_1} = E_d(t), \quad F(t, x) = x_0^N.$$

where

$$x_1 = \frac{q_0}{q_2}, \quad E_d(t) = e^{-(q_2 - q_0)\mu t}.$$

Upon solving for x_0 in terms of x and t and then using the result in the expression for $F(t, x)$, we obtain

$$F(t, x) = x_0^N = \left\{ \frac{x_1 (x - 1) - (x - x_1) E_d(t)}{(x - 1) - (x - x_1) E_d(t)} \right\}^N$$

with

$$F(t, 1) = 1, \quad \left[\frac{\partial F}{\partial x} \right]_{x=1} = N e^{(q_2 - q_0)\mu t} = \sum_{n=1}^{\infty} n P_n(t) = E[X(t)].$$

Unlike the previous (one-child policy) example, whether the expected population tends to zero with time now depends on the relative magnitude of q_0 and q_2. If $q_0 > q_2$ so that it is more probable for a member of the population to die without a descendent, then the mean population decays exponentially in time and tends to zero as $t \to \infty$. Furthermore, we have

$$F(\infty.x) = 1 = P_0(\infty)$$

so that the population eventually becomes extinct with probability 1.

On the other hand, if $q_2 > q_0$ so that it is less probable for a member of the population to die without a descendent, then the mean population grows exponentially in time. It is still possible for the population to extinct eventually but only with probability

$$F(\infty.x) = x_1^N = \left(\frac{q_0}{q_2}\right)^N < 1 \tag{5.5.5}$$

that decreases as the initial population size increases.

In the special case of $q_2 = q_0$ so that $x_1 = 1$, the solution of the characteristic equations simplifies to

$$x_0 = \frac{x(2 - \mu t) + \mu t}{2 + \mu t - x\mu t}, \quad F(t, x) = x_0^N.$$

The pgf for the problem is therefore

$$F(t, x) = x_0^N = \left(\frac{x(2 - \mu t) + \mu t}{2 + \mu t - x\mu t}\right)^N$$

with

$$F(t.1) = 1, \quad \left[\frac{\partial F}{\partial x}\right]_{x=1} = N = E[X(t)].$$

The mean population size for this special case is N uniformly for all t, unchanged from the initial population size. Similar to the case $q_2 > q_0$, there is still a finite probability that the population would extinct eventually. This is given by

$$P_0(t) = F(t, 0) = \left(\frac{\mu t}{2 + \mu t}\right)^N$$

which tends to 1 as $t \to \infty$.

5.5.5 A general $q(x)$

For a general $q(x)$, we have as the characteristic equations of (5.5.4)

$$\frac{dx}{dt} = \mu \{x - q(x)\}, \quad \frac{dF}{dt} = 0 \tag{5.5.6}$$

augmented by the initial conditions

$$x(0) = x_0, \quad F(0, x_0) = x_0^N$$

assuming the population being N at $t = 0$ so that $P_N(0) = 1$ and $P_n(0) = 0$ for $n \neq N$. The solution of the characteristic equations is

$$\mu t = \int_{x_0}^{x} \frac{dz}{z - q(z)}, \quad F(t, x) = x_0^N.$$

For further progress without specifying $q(x)$, we note for x sufficiently close to 1, we have (except for higher order terms in $z - 1$)

$$z - q(z) = z - 1 - \mu_q(z - 1), \quad \mu_q = q'(1),$$

so that the exact solution may be approximated by

$$\mu t = \int_{x_0 - 1}^{x - 1} \frac{dy}{y - \mu_q y} = \frac{1}{1 - \mu_q} \ln\left(\frac{x - 1}{x_0 - 1}\right),$$

or

$$\frac{x_0 - 1}{x - 1} = e^{(\mu_q - 1)\mu t}, \quad F(t, x) = x_0^N = \left[1 + (x - 1)e^{(\mu_q - 1)\mu t}\right]^N$$

with

$$F(t, 1) = 1, \quad E[X(t)] = \left[\frac{\partial F}{\partial x}\right]_{x=1} = Ne^{(\mu_q - 1)\mu t}.$$

It follows that the mean population $E[X(t)]$ decreases or grows with time depending on the magnitude of the mean number of descendants per individual μ_q with

$$\lim_{t \to \infty} E[X(t)] = \begin{cases} 0 & (\mu_q < 1) \\ \infty & (\mu_q > 1) \end{cases}.$$

For the case $\mu_q < 1$, we have also

$$\lim_{t \to \infty} F(t, x) = 1 = P_0(\infty) \quad (\mu_q < 1).$$

Hence, eventual extinction is certain (with probability 1) in this case.

The situation for $\mu_q > 1$ is a little more complicated. The following exercises helps achieve a corresponding result on eventual extinction similar to (5.5.5) for the $q(x) = q_0 + q_2 x^2$ case (when $q_2 > q_0$):

Exercise 26. If the mean number of descendants per individual $\mu_q = q'(1) > 1$ (and $q(0) = q_0 > 0$), show that $q(x) = x$ has one and only one root ρ in $(0, 1)$ and there is no such root if $\mu_q \leq 1$.

Exercise 27. For $\mu_q > 1$, prove: a) $x \leq x_0 < \rho$ if $0 \leq x < \rho$; and b) $\rho < x_0 \leq x$ if $\rho < x < 1$.

With these two results, we can prove the following lemma:

Lemma 6. *If $\mu_q > 1$, then $x_0 \to \rho$ as $t \to \infty$.*

Proof. When $\mu_q > 1$, we know from the second exercise above that the only two roots of $q(x) = x$ are 1 and $\rho < 1$. We may then write

$$\frac{1}{q(x) - x} = \frac{\rho}{q_0 (1 - \rho)} \left\{ \left(\frac{1}{x - 1} - \frac{1}{x - \rho} \right) + g(x) \right\}$$

where $g(x)$ has no singularities in $[0, 1]$ and $g(0) = 0$. The first characteristic equation (5.5.6) for the PDE (5.5.4) may then be integrated to get

$$-\frac{q_0}{\rho} (1 - \rho) \mu t = \ln \left(\frac{x - 1}{x_0 - 1} \frac{x_0 - \rho}{x - \rho} \right) + G(x, x_0)$$

with $G(x_0, x_0) = 0$. As $t \to \infty$ (with $G(x, x_0)$ having no singularities in the interval $[0, 1]$ and therefore dominated by the logarithmic term), we have

$$\mu q_0 (1 - \rho) t \sim -\rho \ln \left(\frac{x - 1}{x_0 - 1} \frac{x_0 - \rho}{x - \rho} \right)$$

since $G(x, x_0)$ has no singularity in $[0, 1]$. Upon solving for x_0 in terms of x and t, we obtain

$$\frac{x - 1}{x_0 - 1} \frac{x_0 - \rho}{x - \rho} \sim e^{-\mu q_0 (1 - \rho) t / \rho}$$

which can be solved to obtain

$$x_0 \sim \frac{\rho (x - 1) - (x - \rho) e^{-\mu q_0 (1 - \rho) t / \rho}}{(x - 1) - (x - \rho) e^{-\mu q_0 (1 - \rho) t / \rho}}.$$

Hence, we have $x_0 \to \rho$ as $t \to \infty$. □

It follows from the lemma above

$$\lim_{t \to \infty} F(t, x) = \rho^N = P_0(\infty) \quad (\mu_q > 1)$$

so that even when $\mu_q > 1$ (with the mean population size grows with time), the actual population may still extinct with a finite probability.

5.6 Chapman–Kolmogorov Equation

Recall that in the discussion of the Poisson process, we defined $P_k(t)$ to be the probability for the random variable $X(t)$ to be in state k with the stipulation that it was at state 0 initially (no fish had been caught up to $t = 0$). So instead of (5.1.1), we should have defined it to be

$$P_k(t) = \text{Prob}\{X(t) = k \mid X(0) = 0\}.$$

We now define $p_k(t)$ to be the probability of the random variable $X(t)$ to be in state k at time t (starting with any possible state initially, not just 0 as it is for $P_k(t)$):

$$p_k(t) = \text{Prob}\{X(t) = k\}$$

and let

$$p_{kj}(t, s) = \text{Prob}\{X(t) = k \mid X(s) = j\}.$$

We will be mainly interested in temporally stationary processes

$$p_{kj}(t, s) = p_{kj}(t - s) = \text{Prob}\{X(t - s) = k \mid X(0) = j\}$$

since such processes depend only on the elapsed time $t - s$ between the two instants involved and not the reference time s or the final time t (in which case the reference time may be replaced by the starting time). Evidently, we have

$$\sum_{k=0}^{N} p_{kj}(t) = 1 \quad (t \geq 0)$$

since the probabilities of getting from state j to any of the sample states must add up to certainty. Hence, $P(t) = [p_{ij}(t)]$ is a probability (stochastic) matrix.

The sum above over all possible sample states (with N finite or infinite) is not the same as $p_m(t)$. The latter is the sum of all probabilities of getting to state m from the different sample space states. That sum could in fact be very small (but not zero for otherwise the state would not be in the sample space) if there should be a strong bias toward other sample states as destinations. The event of getting from any sample state to any state of the same sample space is a certainty, hence

$$\sum_{m=0}^{N} p_m(t) = 1.$$

Analogous to the DTMC case, the transition probabilities $p_{ij}(t)$ satisfies the Chapman–Kolmogorov relation:

$$p_{ij}(t) = \sum_{k=0}^{N} p_{ik}(t - s)p_{kj}(s) \quad (0 < s < t < \infty)$$

since

$$p_{ij}(t) = \text{Prob}\{X(t) = i \mid X(0) = j\}$$

$$= \sum_{k=0}^{N} \text{Prob}\{X(t) = i, \ X(s) = k \mid X(0) = j\}$$

$$= \sum_{k=0}^{N} \text{Prob}\{X(t) = i \mid X(s) = k, \ X(0) = j\}\text{Prob}\{X(s) = k \mid X(0) = j\}$$

$$= \sum_{k=0}^{N} \text{Prob}\{X(t) = i \mid X(s) = k\}\text{Prob}\{X(s) = k \mid X(0) = j\}$$

$$= \sum_{k=0}^{N} \text{Prob}\{X(t-s) = i \mid X(0) = k\}\text{Prob}\{X(s) = k \mid X(0) = j\}$$

$$= \sum_{k=0}^{N} p_{ik}(t-s)p_{kj}(s).$$

As indicated earlier, the analogue of this Chapman–Kolmogorov equation for stochastic processes with continuous sample space (to be discussed in later chapters) will be important in developing the mathematical tools for analyzing these processes.

5.7   Appendix — The Method of Characteristics

Similar to ODE, a first order PDE is an equation involving an unknown of several independent variables and its various *first* partial derivatives and possibly the independent variables themselves. In the case of two independent variables x and y, a general first order PDE for an unknown function $z(x, y)$ is a relation

$$F(x, y, z, z_{,x}, z_{,y}) = 0, \tag{5.7.1}$$

where $z_{,t} = \partial z/\partial t$. For example, the equation

$$(z_{,x})^2 + x\sin(z_{,y}) - z = 0 \tag{5.7.2}$$

is a nonlinear first order PDE. In view of the squaring of $z_{,x}$ and the appearance of $\sin(z_{,y})$, the usual superposition principle does not hold. Given the form of the nonlinearity of interest herein, we limit our discussion here to *quasilinear first order* PDE in which the first partials of the unknown appear only linearly. For the two-variable case, such a *PDE* must be of the form

$$P(x, y, z)z_{,x} + Q(x, y, z)z_{,y} = R(x, y, z) \tag{5.7.3}$$

where $P(x, y, z)$, $Q(x, y, z)$ and $R(x, y, z)$ do not depend on the partial derivatives of the unknown but may depend on the unknown itself. We describe in this appendix the solution for a single quasi-linear PDE by the *method of characteristics*.

As evident from the developments in the previous chapters, the solution of a PDE is completely specified only with additional appropriate auxiliary conditions. For a first order PDE such as (5.7.3), we expect from our experience with ODE that one auxiliary condition along a boundary curve would generally suffice. The actual situation pertaining to auxiliary conditions, however, is more complex as we shall see in later chapters.

5.7.1   *The simplest first order PDE*

We begin with the following simple first order quasilinear PDE:

$$P_0 z_{,x} + Q_0 z_{,y} = 0. \tag{5.7.4}$$

When P_0 and Q_0 are constants, the equation is *linear* and *with constant coefficients*. We are interested in the case $P_0Q_0 \neq 0$ for otherwise (5.7.4) would reduce to an ODE (or triviality if $P_0 = Q_0 = 0$). Now a linear ODE with constant coefficients would admit exponential solutions. For (5.7.4), such solutions would be exponential in both x and y in the form

$$z(x, y; \alpha, \beta) = Ce^{\alpha x + \beta y} \tag{5.7.5}$$

where C, α and β are parameters to be determined by the PDE. Unlike the ODE case, these constants are not restricted to a few special numbers; they can be any real or complex number as long as they satisfy the relation

$$P_0\alpha + Q_0\beta = 0. \tag{5.7.6}$$

A solution (5.7.5) for any constant C which may vary with the free parameter α (or β) is said to be a *particular solution* of the PDE. Superposition of all particular solutions by "summing" (or integrating) over all possible α (or β) subject to (5.7.6) leads to a more general solution of the form

$$z(x, y) = \int C(\alpha, \beta)e^{\alpha x + \beta y} d\alpha \equiv f(mx + y), \quad m = -\frac{Q_0}{P_0} \tag{5.7.7}$$

for any (piecewise) continuously differentiable function $f(\cdot)$ (given the superposition principle holds for the linear PDE).

We may arrive at the same conclusion (5.7.7) by substituting $f(\xi) = f(mx + y)$ for $z(x, y)$ directly into the PDE to get

$$(P_0m + Q_0)\frac{df}{d\xi} = 0.$$

Hence, the PDE (5.7.4) is satisfied by any (non-constant, piecewise) continuously differentiable function $f(\xi) = f(mx + y)$ if

$$m = -Q_0/P_0.$$

Similar to ODE problems, we need auxiliary conditions to determine the still unknown function $f(\cdot)$. Suppose we are interested in the solution of (5.7.4) for $y > 0$ that fits the prescribed initial data

$$z(x, 0) = \sin(x), \quad (-\infty < x < \infty). \tag{5.7.8}$$

In that case, we must have

$$z(x, 0) = f(mx) = \sin(x), \quad (-\infty < x < \infty)$$

or, equivalently,

$$f(\xi) = \sin(\xi/m) \quad -\infty < \xi < \infty.$$

With $\xi = mx + y = y - Q_0 x / P_0$, we get

$$z(x,y) = f(\xi) = \sin(\xi/m) = \sin(x - P_0 y / Q_0), \qquad (5.7.9)$$

for $-\infty < x - P_0 y / Q_0 < \infty$. It may be verified by direct substitution that $z(x,y)$ as given in (5.7.9) satisfies the PDE (5.7.4) and the initial condition (5.7.8) and is well-defined for $y > 0$ (as well as $y < 0$ if it should be of interest).

More generally, the solution for the same PDE with the initial condition

$$z(x,0) = f_0(x), \qquad (-\infty < x < \infty),$$

is

$$z(x,y) = f_0(x + y/m) = f_0(x - P_0 y / Q_0)$$

in some region of the x, y-plane. For (5.7.4), the $y - Q_0 x / P_0 = $ constant lines are rather special as seen from the following properties:

- The solution $z(x,y)$ is constant along these lines. In particular, for any fixed η, we have $z(x,y) = f_0(\eta))$ for any point (x,y) along the straight line $y - Q_0 x / P_0 = \eta$.
- A finite jump discontinuity in the initial data $f_0(x)$ at $x = x_0$ is propagated along the **characteristic base curve** $x - P_0 y / Q_0 = x_0$ in the solution. For example, for the initial data

$$z(x,0) = f_0(x) = \begin{cases} \sin(x) & (x < 0) \\ \cos(x) & (x > 0) \end{cases},$$

the solution for $y > 0$ is

$$z(x,y) = f_0(mx + y) = \begin{cases} \sin(x - P_0 y / Q_0) & (x < P_0 y / Q_0) \\ \cos(x - P_0 y / Q_0) & (x > P_0 y / Q_0) \end{cases},$$

where the discontinuity in the initial data propagating along the straight line $y - Q_0 x / P_0 = 0$.
- Suppose the initial data is prescribed along such a straight line, say

$$z(x,y) = \sin(x), \quad \text{along } y - Q_0 x / P_0 = c.$$

This prescribed auxiliary condition requires

$$z(x,y) = f(mx + y) = f(c) = \sin(x)$$

which is not possible since $f(c)$ is a constant and cannot be made to fit the prescribed data which varies with locations along the line $y - Q_0 x / P_0 = c$ where the data is prescribed. The only exception is when the prescribed data is a constant $f_0(x) = c_0$ so that $f(c) = c_0$. The condition fixes $f(\cdot)$ along $y - Q_0 x / P_0 = c$ and nowhere else. As such, the problem for (5.7.4) generally has no solution if a non-constant z is prescribed along any of the special lines $y - Q_0 x / P_0 = c$. When the prescribed data for the unknown is constant along such a line, the solution is not completely specified by the auxiliary condition away from the particular line.

Because of these and other rather unusual properties of the solution of the PDE, the lines $y - Q_0 x/P_0 = c$ are called **characteristic base curves** Γ_0 of the PDE (5.7.4). The space curve Γ traced out by $(x, y, Z(x, y))$ for all (x, y) along a characteristic base curve Γ_0 is called a **characteristic** of the PDE. Evidently, the solution surface $z = Z(x, y)$ is built up from a collection of characteristics each emanating from a different point along some curve C, which should not be a characteristic curve itself by the observation in the third bullet above.

The situation is not different if we seek a solution in the first quadrant of the x, y-plane with the initial data (5.7.8) at $y = 0$ now prescribed only for $x > 0$ augmented by a similar condition along $x = 0$, say

$$z(0, y) = y \cos(y).$$

In that case, the prescribed condition along $y = 0$ and $x > 0$ determines

$$z(x, y) = \sin(x - P_0 y/Q_0) \text{ only for } x - P_0 y/Q_0 > 0.$$

At the same time, the prescribed condition along $x = 0$ and $y > 0$ requires

$$z(0, y) = f(y) = y \cos(y), \quad (y > 0).$$

This relation determines

$$z(x, y) = f(\xi) = \xi \cos(\xi) = (y - Q_0 x/P_0) \cos(y - Q_0 x/P_0)$$

for $y - Q_0 x/P_0 > 0$. Altogether, we have as the final solution of the initial value problem with initial data prescribed along the positive x-axis and positive y-axis

$$z(x, y) = f_0(mx + y) = \begin{cases} \sin(x - P_0 y/Q_0) & (x < P_0 y/Q_0) \\ (y - Q_0 x/P_0) \cos(y - Q_0 x/P_0) & (x > P_0 y/Q_0). \end{cases}$$

Note that the solution is continuous along and across the characteristic base line, Γ_0: $Q_0 x = P_0 y$, where $z(x, y) = 0$.

Suppose the data along positive y-axis is changed to $z(0, y) = \cos(y)$. Then the corresponding solution of the problem is

$$z(x, y) = f_0(mx + y) = \begin{cases} \sin(x - P_0 y/Q_0) & (x < P_0 y/Q_0) \\ \cos(y - Q_0 x/P_0) & (x > P_0 y/Q_0). \end{cases}$$

The discontinuity in the initial data at $x = y = 0$ is now propagated along the characteristic base line, Γ_0: $Q_0 x = P_0 y$.

Exercise 28. Suppose P_0, Q_0 and R_0 are constants in the linear PDE

$$P_0 z_{,x} + Q_0 z_{,y} = R_0 z. \tag{5.7.10}$$

(1) Determine the relation between α and β for a particular solution of the PDE of the form (5.7.5).
(2) Obtain the solution of the PDE that fits the initial data $z(x, 0) = \sin(x)$.

5.7.2 A quasi-linear equation

If P and Q in (5.7.3) are no longer constants but depend on x, y and z, it is easily verified by direct substitution that the first order PDE can no longer be satisfied by a function $f(mx + y)$ for some constant m even if $R \equiv 0$. We need a different approach for the more general problem. For this purpose, it is useful to recall that the solution for this equation can be viewed as a surface in the (x, y, z) space and may be written as

$$z = Z(x, y) \quad \text{or} \quad u(x, y, z) = c, \tag{5.7.11}$$

for some function $Z(\cdot, \cdot)$ and some constant c. With $z = Z(x, y)$ and $u_{,t} = \partial u / \partial t$ for a function u of x, y and possibly z, we have the following expressions for the two partial derivatives of $u(x, y, Z(x, y))$ with respect to x and y:

$$[u_{,x}]_{y \text{ fixed}} = [u_{,x}]_{(y,z) \text{ fixed}} + [u_{,z}]_{(x,y) \text{ fixed}} [z_{,x}]_{y \text{ fixed}} = 0,$$

$$[u_{,y}]_{x \text{ fixed}} = [u_{,y}]_{(x,z) \text{ fixed}} + [u_{,z}]_{(x,y) \text{ fixed}} [z_{,y}]_{x \text{ fixed}} = 0,$$

(keeping in mind $u(x, y, z) = \text{constant}$). From these two relations, we obtain

$$z_{,x} = -\frac{[u_{,x}]_{(y,z) \text{ fixed}}}{[u_{,z}]_{(x,y) \text{ fixed}}}, \qquad z_{,y} = -\frac{[u_{,y}]_{(x,z) \text{ fixed}}}{[u_{,z}]_{(x,y) \text{ fixed}}}.$$

The PDE (5.7.3) can then be written as

$$P [u_{,x}]_{(y,z) \text{ fixed}} + Q [u_{,y}]_{(x,z) \text{ fixed}} + R [u_{,z}]_{(x,y) \text{ fixed}} = 0. \tag{5.7.12}$$

The relation (5.7.12), viewed as the scalar product of two vectors

$$\nabla u \cdot (P\mathbf{i} + Q\mathbf{j} + R\mathbf{k}) = 0,$$

shows that the vector ∇u is normal to the vector $\mathbf{T} = P\mathbf{i} + Q\mathbf{j} + R\mathbf{k}$. Since ∇u is normal to the surface defined by $u(x, y, z) = c$, it follows that the vector $\mathbf{T}$ is tangent to the solution surface. Starting from a point $\mathbf{r}_0 = x_0\mathbf{i} + y_0\mathbf{j} + z_0\mathbf{k}$ in the (x, y, z) space, the tangent vector traces out a space curve C in that space. The position vector $\mathbf{r}(\xi)$ of this space curve is determined by the IVP

$$\frac{d\mathbf{r}}{d\xi} = \mu\mathbf{T} = \mu(P\mathbf{i} + Q\mathbf{j} + R\mathbf{k}), \qquad \mathbf{r}(\xi = 0) = \mathbf{r}_0 \tag{5.7.13}$$

where μ depends on the choice of the independent variable ξ. In particular, when ξ is arc length measured from the initial point $\mathbf{r}_0$, then μ must be chosen so that $\mu\mathbf{T} = \mu(P\mathbf{i} + Q\mathbf{j} + R\mathbf{k})$ is of unit length. With $\mathbf{T}$ not depending on ξ, the factor μ can be set to 1 if ξ is not arc length (and will be assumed as such in the subsequent development).

With $\mu = 1$, the vector equation (5.7.13) corresponds to three scalar first order ODE system

$$\frac{dx}{d\xi} = P, \qquad \frac{dy}{d\xi} = Q, \qquad \frac{dz}{d\xi} = R. \tag{5.7.14}$$

known as the characteristic equations for the first order PDE. The solution is a family of curve $\mathbf{r}(\xi, \mathbf{c}) = (x(\xi, \mathbf{c}), y(\xi, \mathbf{c}), z(\xi, \mathbf{c}))$ where $\mathbf{c} = (c_1, c_2, c_3)$ are three constants of integration to fit the appropriate initial data for the characteristic system. Note that the vector $\mathbf{c}$ is related, but not identical, to the parameter c in (5.7.11). An initial point

$$x(0) = x_0, \qquad y(0) = y_0, \qquad z(0) = z_0, \tag{5.7.15}$$

fixes the three constants $\mathbf{c} = \mathbf{c}(x_0, y_0, z_0)$ and the solution of (5.7.14) is a space curve on the surface that passes through the initial point. Other prescribed initial points generate similar characteristic curves. Typically, the solution of the PDE (5.7.3) is required to assume a prescribed value at each point along a given *initial base curve* $\Gamma_0 = \{x_0(\eta), y_0(\eta), \ a \leq \eta \leq b\}$ in the x, y-plane. The prescribed value of the unknown is generally different at different points on Γ_0 and hence to be denoted by $z_0(\eta)$. Thus, the solution of the PDE (5.7.3) subject to the initial condition

$$x(\xi = 0) = x_0(\eta), \qquad y(\xi = 0) = y_0(\eta), \qquad z(\xi = 0) = z_0(\eta) \tag{5.7.16}$$

for $a \leq \eta \leq b$, may be written as $\mathbf{r}(\xi, \mathbf{c}) = \phi(\xi, \eta)\,\mathbf{i}_x + \psi(\xi, \eta)\,\mathbf{i}_y + \zeta(\xi, \eta)\,\mathbf{i}_z = \mathbf{r}(\xi, \eta)$ or

$$x = \phi(\xi, \eta), \qquad y = \psi(\xi, \eta), \qquad z = \zeta(\xi, \eta), \tag{5.7.17}$$

where η varies over the domain (a, b) for the initial curve. Evidently, rhe relations (5.7.17) is a parametric representation of a surface in space consisting of solution curves generated by the ODE (5.7.14) emanating from different points on the *initial curve* $\Gamma = \{x_0(\eta), y_0(\eta), z_0(\eta), \ a \leq \eta \leq b\}$. The solution curves of (5.7.17) are called the *characteristic curves* (or simply *characteristics)* of the PDE (5.7.3).

   If possible, we may solve the first two relations of (5.7.17) for ξ and η in terms of x and y and use the results to eliminate ξ and η from the last relation, $z = \zeta(\xi, \eta)$, to get

$$\xi = S(x, y), \quad \eta = T(x, y), \tag{5.7.18}$$
$$z = \zeta(S(x, y), T(x, y)) \equiv Z(x, y), \tag{5.7.19}$$

with $Z(x, y)$ being the solution surface of the PDE (5.7.3) that emanates from the initial curve Γ.

5.7.3   *A single linear first order PDE*

The method of solution for a first order quasi-linear PDE (5.7.3) developed in the previous subsection is special form of the *method of characteristics*. We illustrate the application of this method to several linear first order PDE taken to be in the form

$$P(x, y)z_{,x} + Q(x, y)z_{,y} = R_0(x, y) + R_1(x, y)z \tag{5.7.20}$$

where the coefficients P, Q, R_0 and R_1 do not involve the unknown z or its partial derivatives.

Example 11.

$$P_0 z_{,x} + Q_0 z_{,y} = 0, \quad z(x,0) = \sin(x) \quad (-\infty < x < \infty).$$

For this previously solved problem, the initial curve may be taken in the form $\Gamma = \{(\eta, 0, \sin(\eta)), -\infty < \eta < \infty\}$ and the relevant system of ODE for the characteristics is $x' = P_0$, $y' = Q_0$, and $z' = 0$. The solution of the IVP for the characteristic ODE is

$$x = P_0 \xi + \eta, \qquad y = Q_0 \xi + 0, \qquad z = \sin(\eta).$$

Upon solving the first two relations for ξ and η and use the results to eliminate η from the last, we get

$$\xi = \frac{y}{Q_0}, \qquad \eta = x - \frac{P_0}{Q_0} y, \qquad z = \sin\left(x - \frac{P_0}{Q_0} y\right)$$

which is identical to the one given in (5.7.9) obtained by an elementary method. The latter applies only to linear PDE with constant coefficients.

Remark 3. If the initial curve for the same PDE is prescribed to be $P_0 z_{,x} + Q_0 z_{,y} = 0$, $z(x,0) = \sin(x)$ or

$$\Gamma : \{x_0(\eta) = P_0 \eta, \quad y_0(\eta) = Q_0 \eta, \quad z_0(\eta) = \sin(\eta)\}, \quad (-\infty < \eta < \infty) \quad (5.7.21)$$

in parametric form, the solution of the characteristic ODE is

$$x = P_0 (\xi + \eta), \qquad y = Q_0 (\xi + \eta), \qquad z = \sin(\eta).$$

In this case, it is not possible to solve for ξ and η in terms of x and y to get $z = Z(x,y)$. Observe that the initial curve Γ is itself a characteristic curve. This confirms a previous observation that initial data should not be a characteristic curve, i.e., the initial value of the unknown $Z(x,y)$ should not be prescribed on a characteristic base curve Γ_0, which is $Q_0 x - P_0 y = c_0$ in this case.

Example 12.

$$y z_{,x} + x z_{,y} = 0, \quad z(x,0) = f_0(x) \quad (-\infty < x < \infty).$$

The initial curve is Γ may be taken in the form $\{(\eta, 0, f_0(\eta)), -\infty < \eta < \infty\}$ and the relevant system of ODE for the characteristics is $x' = y$, $y' = x$, and $z' = 0$. The solution of the IVP for the characteristic ODE is

$$x^2 - y^2 = x_0^2 - y_0^2 = \eta^2, \quad z = z_0(\eta) = f_0(\eta)$$

from which the following solution follows:

$$z = f_0\left(\pm\sqrt{x^2 - y^2}\right).$$

Evidently, the solution is constant along a characteristic base curve Γ_0: $x^2 - y^2 = \eta^2$ for any particular η and does not have a real-valued solution for $x^2 < y^2$.

Example 13.

$$xz_{,x} + yz_{,y} = z \quad \text{with}$$

$$\Gamma : \{x_0 = \eta, \; y_0 = \eta^2, \; z_0 = f_0(\eta), \; -\infty < \eta < \infty\}.$$

The relevant system of ODE for the characteristics is $x' = x$, $y' = y$, and $z' = z$. The solution of the IVP for the characteristic ODE is

$$x = \eta e^\xi, \qquad y = \eta^2 e^\xi, \qquad z = f_0(\eta)e^\xi.$$

From the first two equations, we obtain

$$\eta = \frac{y}{x}, \qquad e^\xi = \frac{x^2}{y}.$$

Upon elimination of η and ξ, the expression for z becomes

$$z = f_0\left(\frac{y}{x}\right)\frac{x^2}{y}.$$

5.7.4 *A single quasi-linear first order equation*

The method of characteristics developed for a single linear PDE in the previous subsection also applies to the general single first order quasi-linear PDE

$$P(x, y, z)z_{,x} + Q(x, y, z)z_{,y} = R(x, y, z)$$

where P, Q and R now may depend on the unknown z in a general way (but not on its first derivatives). Consider the example

$$xzz_{,x} + yz_{,y} = z^2$$

The corresponding characteristic system is

$$x' = xz, \quad y' = y, \quad z' = z^2.$$

Unlike linear first order equations, the first two characteristic ODE $dx/d\xi = P$ and $dy/d\xi = Q$ for a quasilinear first order PDE can no longer be solved separately for x and y independent of the third equation. For the present problem, this coupling of the three characteristic ODE poses no impediment to obtaining the following exact general solution:

$$y = c_2 e^{\xi}, \quad z = \frac{1}{c_3 - \xi}, \quad x(c_3 - \xi) = c_1.$$

where c_1, c_2 and c_3 are three constants of integration.

Suppose the initial curve Γ defined by $z = 1/x$ along the line $y = x$ in the positive first quadrant. This initial curve may be written in parametric form as

$$x_0(\eta) = \eta, \quad y_0(\eta) = \eta, \quad z_0(\eta) = \frac{1}{\eta} \quad (\eta > 0).$$

Then, the corresponding initial conditions determine the three constants of integration to be

$$c_2 = \eta, \quad c_3 = \eta, \quad c_1 = \eta^2,$$

and therewith the following parametric representation of the solution surface:

$$x = \frac{\eta^2}{\eta - \xi}, \quad y = \eta e^{\xi}, \quad z = \frac{1}{\eta - \xi}. \tag{5.7.22}$$

While it does not appear possible to solve the first two relations for ξ and η in terms of x and y, it is possible to solve the first and third relations above for ξ and η in terms of x and z to get

$$\eta^2 = \frac{x}{z}, \quad \xi = \sqrt{\frac{x}{z}} - \frac{1}{z}.$$

These two expressions may be used to eliminate ξ and η from the expression for y to get

$$y = \sqrt{\frac{x}{z}} e^{\sqrt{\frac{x}{z}} - \frac{1}{z}}.$$

The above example shows that the implementation of the method of characteristics for a general quasilinear first order PDE is straightforward since a parametric representation (such as (5.7.22) for the example above) is perfectly adequate for further mathematical analysis, numerical evaluation and graphical presentation given the computing capacity available today. It behooves us to have a working knowledge of this method as it will be found useful for many other stochastic models for the life sciences in later chapters of this volume.

Chapter 6

Spread of Chlamydia and Stochastic Optimization

6.1 Stochastic Models for the Development of C. Trachomatis

As a cause for ocular infection in humans, the bacterium *Chlamydia trachomatis* is the world's leading cause of preventable blindness. This condition, known as *trachoma*, currently affects 84–150 million people in the world, causing blindness to 8 million people (see [33, 71] and references cited therein). The pathogenic bacterium is known to have an unusual intracellular developmental cycle involving conversion among different forms of the bacterium within a cytoplasmic inclusion (see Fig. 6.1 for a cartoon of the evolving cytoplasmic inclusion at an instant in time prior to lysing). The cycle starts with the (endocytosed units of the) bacteria infecting a human cell. The *reticulate body* (RB) form of the bacteria repeatedly divides by binary fission, asynchronously differentiates and eventually converts into *elementary body* (EB) form units that do not divide but survive host cell lysing to infect other cells. Until recently, these developmental events have not been quantified because conventional electron-microscopy only visualizes sections of the large chlamydial inclusion. Using a novel three-dimensional electron microscopy approach, the lab of Tan and Suetterlin has been able to obtain quantitative data on proliferation and conversion that have been replicated quite consistently by a stochastic population growth model examined in [33]. Instead of leaving the life cycle essentially as a consequence of an assigned probability, the possibility of some Darwinian force at work for the spread of the infectious disease is examined in [71]. To this end, we task the RB-to-EB conversion rate as an instrument chosen to maximize the (mean) terminal EB population. The deterministic optimal control type models [71] that delineate how fast the infectious Chlamydia bacteria may spread and how this optimal spread of the disease may be accomplished is for a prescribed terminal time. For each of these models, the theoretically optimal RB-to-EB conversion strategy

36 h.p.i.

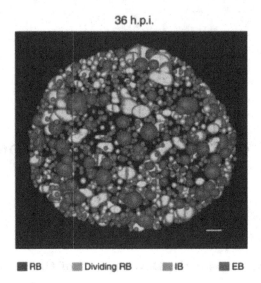

RB   Dividing RB   IB   EB

Fig. 6.1. A cartoon of Chlamydial inclusion in an infected cell.

for maximum spread have been found for different initial infecting RB population sizes and different ranges of system parameter values.

In this chapter, we formulate new models with more realistic features toward the same goal of determining the maximal spread of the infectious disease. Among them are the uncertainty of division and proliferation rate and the cause of host cell lysing that determines the terminal time of the proliferation-conversion cycle. For these models, we abbreviate the developmental cycle to involve only two forms of the bacteria, the reticulate body form RB and the elementary body form EB. RB units proliferate but do not survive outside the cytoplasmic inclusion; they can convert to smaller EB units that infect other cells once outside the cytoplasmic inclusion. After (an initial group of endocytosed) EB units entering a new host cell, they immediately turn into RB to form the "*initial infecting RB population*" of size N (the number of initial RB units) and begin to proliferate, repeating the life cycle as shown in Fig. 6.2. Of interest is the optimal conversion of some or all RB units into EB at each instant of time t to maximize the EB population (under some form of uncertainty) at a terminal time T specified by a conditions that determines host cell lysing.

We begin with a simple birth-and-death type probabilistic model that focuses on the complementary choice of division of RB (birth) and conversion of RB to EB (death) over a finite time T before the host cell lyses. In this first model, we consider only time-invariant probabilities for division and RB-to-EB conversion. The optimal conversion probability is sought to maximize the expected EB population at a terminal time (related to host cell lysing) determined by a threshold of weighted sum of the (expected) RB and EB populations. The optimal ratio from this simple birth and death model is confirmed by the corresponding result for the

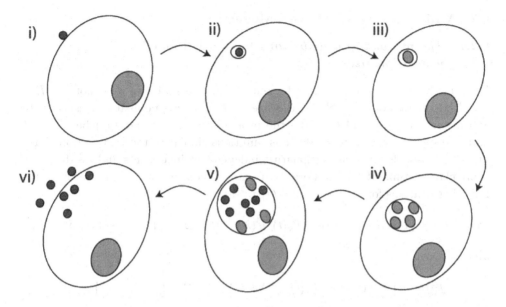

Fig. 6.2.   Life cycle of Chlamydia Trachomatis.

corresponding deterministic optimization problem (with a time-invariant conversion rate constant).

It is generally expected that a larger (expected) terminal EB population (than that induced by a time-invariant conversion-to-duplication probability ratio) would result if we allow the conversion probability to vary with time. If the capacity to convert is sufficiently high (but still finite), we may allow the RB population to grow by not converting for an initial period of time before converting at a high rate (higher than the natural RB growth rate) to result in a larger terminal EB population. A straightforward modification of the birth-and-death process model above has been investigated by the method of optimal control previously applied to simpler C. trachomatis models in [71, 72]. Interested readers are referred to [8] for the rather surprising results for the new time-varying conversion probability model. Here, we examine instead another more realistic probabilistic model where the only knowledge of the uncertain terminal time is a prescribed *probability density function* (pdf) suggested by empirical data on host cell lysing. The task is to maximize the expected terminal EB population, resulting in a novel stochastic optimization problem. With the bacterial growth dynamics linear in both state and control variables, the optimal conversion strategy is expected to be bang-bang when the conversion rate is not constrained to be less than the natural RB growth rate. The only question then is the location of the switch point between the two corner controls. That problem turns out to be rather mathematically challenging. A complete analysis of the problem is presented herein to show why mathematical analysis is indispensable for a complete and conclusive solution of many stochastic optimization problems.

6.2   A Birth and Death Process Model

6.2.1   *The probability distributions for the RB and EB population sizes*

For a birth and death type model for our Chlamydia problem, we denote by $R(t)$ and $E(t)$ the random variable for the size of RB and EB population, respectively, at time t. Let $P_k(t)$ and $Q_j(t)$ be the probability of $R(t)$ and $E(t)$ be of size k and j, respectively, at time t. With assumptions similar to those for various birth and death models (such as stationarity, independent increments and a sufficiently short elapsed time δt for no more than one division or conversion), we have the following relations for $P_k(t + \delta t)$ and $Q_j(t + \delta t)$:

$$P_k(t + \delta t) = \{1 - k(\lambda_D + \lambda_C)\delta t\} P_k(t) + (k + 1)\lambda_C \delta t P_{k+1}(t) + (k - 1)\lambda_D \delta t P_{k-1}(t)$$

and

$$Q_j(t + \delta t) = \{1 - \lambda_C \delta t P_1(t) - 2\lambda_C \delta t P_2(t) - 3\lambda_C \delta t P_3(t) - \cdots\} Q_j(t)$$
$$+ \lambda_C \delta t \{P_1(t) + 2P_2(t) + 3P_3(t) + \cdots\} Q_{j-1}(t)$$

except for negligibly small terms relative to terms proportional to δt. In these two relations, $\lambda_C \delta t$ and $\lambda_D \delta t$ are the probability of one conversion and one duplication, respectively, during the elapsed time δt. As a first model (that is to be refined later), we let both λ_C and λ_D be independent of time and population size.

In the limit as δt tends to zero, we get

$$\frac{dP_k}{dt} = -k(\lambda_D + \lambda_C)P_k + (k + 1)\lambda_C P_{k+1} + (k - 1)\lambda_D P_{k-1}, \qquad (6.2.1)$$

$$\frac{dQ_j}{dt} = \lambda_C (Q_{j-1} - Q_j) \sum_{i=0}^{\infty} iP_i, \qquad (6.2.2)$$

augmented by the initial conditions $P_k(0) = \delta_{kN}$ and $Q_j(0) = 0$, $k, j = 0, 1, 2, \ldots$, given that the process starts with N RB and no EB.

6.2.2   *Solution by generating functions*

The ODE system for $P_k(t)$ can be reformulated as a problem for the corresponding generating function $\sum P_k(t)x^k \equiv G(x, \tau)$ with $G(x, \tau)$ determined by

$$G_{,\tau} + \left\{(1 + \rho)x - \rho - x^2\right\} G_{,x} = 0, \qquad G(x, 0) = x^N, \qquad (6.2.3)$$

where

$$\rho = \frac{\lambda_C}{\lambda_D}, \quad \tau = \lambda_D t, \quad G_{,z} = \frac{\partial G}{\partial z} \quad (z = \tau, x).$$

The IVP for this first order PDE (6.2.3) is solved by the *method of characteristics* to give

$$G(x, \tau) = x_0^N, \quad x_0 = \frac{(x - \rho) e^{-(1-\rho)\tau} + \rho(1 - x)}{(x - \rho) e^{-(1-\rho)\tau} + (1 - x)}. \tag{6.2.4}$$

Note that

$$\left[\sum_{k=0}^{\infty} P_k(t) x^k \right]_{x=1} = G(1, \tau) = \left[\frac{(1 - \rho) e^{-(1-\rho)\tau}}{(1 - \rho) e^{-(1-\rho)\tau}} \right]^N = 1$$

and

$$[G(x, \tau)]_{\tau=0} = \left(\frac{x_0 (1 - \rho)}{1 - \rho} \right)^N = x_0^N$$

as they should be. More importantly, we have for a fixed t the following expression for the *expected value* of the random variable $R(t) \equiv R_t$ for a fixed t:

$$E[R_t] = \sum_{k=0}^{\infty} k P_k(t) = [G_{,x}(x, \tau)]_{x=1} = N e^{(1-\rho)\tau} \equiv \bar{R}_t(\tau; \rho) \tag{6.2.5}$$

with

$$\lim_{t \to \infty} [\bar{R}_t(\tau; \rho)] = \begin{cases} 0 & (\lambda_D < \lambda_C) \\ N & (\lambda_D = \lambda_C) \\ \infty & (\lambda_D > \lambda_C). \end{cases}$$

The expressions $E[R_t]$ and $\bar{R}_t(\tau; \rho)$ will be used interchangeably for the expected RB population size as a function of time t with $\tau = \lambda_D t$ and $\rho = \lambda_C / \lambda_D$. Similar expressions will also be used for the expected EB population size below.

Turning now to the ODE system for $\{Q_j(t)\}$, we also formulate the IVP for the corresponding generating function $F(x, \tau) = \sum Q_j(t) x^j$ to get

$$F_{,\tau} - N(x - 1) e^{(1-\rho)\tau} F = 0, \quad F(x, 0) = 1.$$

The equation for $F(x, t)$ above is a first order linear ODE that can be solved with the help of an integrating factor to get

$$F(x, \tau) = e^{I_x(\tau)}, \quad I_x(\tau) = \frac{\rho N(x - 1)}{1 - \rho} \left[e^{(1-\rho)\tau} - 1 \right].$$

We can obtain $\{Q_j(t)\}$ by expanding $F(x, \tau)$ as a Taylor series in x.

Of more immediate interest is the expected value for the random variable $E_t = E(t)$ for the EB population for a fixed t:

$$E[E_t] = [F_{,x}(x, \tau)]_{x=1} = \frac{\rho N}{1 - \rho} \left[e^{(1-\rho)\tau} - 1 \right] \equiv \bar{E}_t(\tau; \rho). \tag{6.2.6}$$

6.2.3   Maximum expected EB population at terminal time

6.2.3.1   Host cell lysis time

The infected host cell has only a finite carrying capacity for the two RB and EB populations. Suppose it lyses when the total C. trachomatis bacteria reaches that carrying capacity, denoted by P_c, for which the host cell and its bacterial inclusion membranes cannot sustain. Let T denote the *terminal time* when the host cell lyses. We have at that instant T

$$\left\{\bar{E}_t(\tau;\rho) + m\,\bar{R}_t(\tau;\rho)\right\}_{t=T} = P_c, \tag{6.2.7}$$

In (6.2.7), the factor m (>1) reflects the fact that the size of a typical RB unit is substantially larger than that of an EB unit. That the host cell lyses at P_c and any larger bacteria population determines the terminal time T as a function of the conversion and replication probability λ_c and λ_D and the initial *infecting RB population* N.

If λ_D and N should be known or prescribed, we may also consider (6.2.7) as a condition for determining the terminal time T as a function of the conversion-to-duplication probability ratio $\rho = \lambda_C/\lambda_D$. Even for this one free parameter case, it is still rather tedious to try to solve for T in terms of ρ (with N and λ_D fixed). In the subsequent development for finding a maximum point $\rho_{\max}$ that maximizes the expected EB population at terminal time, our approach is to determine $\rho_{\max}$ and the corresponding normalized terminal time $\tau_{\max} = \tau_T(\rho_{\max})$ simultaneously.

6.2.3.2   Maximum point for expected terminal EB population

With N and λ_D prescribed and $\tau_T = \lambda_D T$, the problem of maximizing $\bar{E}_t(\tau_T;\rho)$ subject to the constraint (6.2.7) will be solved by the method of Lagrange multipliers. By this approach, we seek a maximum point (τ_T^*, ρ^*) of the augmented objective function

$$J(\rho, \tau_T; \lambda) = \bar{E}_t(\tau_T;\rho) + \lambda\left\{\bar{E}_t(\tau_T;\rho) + m\,\bar{R}_t(\tau_T;\rho) - P_c\right\} \tag{6.2.8}$$

(subject to the constraint (6.2.7)) where λ is a Lagrange multiplier.

Any maximum point must be a stationary point of (6.2.8) with

$$\left[\frac{\partial J}{\partial \tau_T}\right]_{(\rho^*,\tau_T^*)} = 0 \quad \text{and} \quad \left[\frac{\partial J}{\partial \rho}\right]_{(\rho^*,\tau_T^*)} = 0.$$

The first condition requires

$$(1+\lambda)\rho^* + m\lambda(1-\rho^*) = 0 \tag{6.2.9}$$

since $e^{(1-\rho)\tau_T^*}$ does not vanish. Upon observing (6.2.9), the second condition requires

$$(1+\lambda)\left\{e^{(1-\rho)\tau_T^*} - 1\right\} = 0. \tag{6.2.10}$$

since $\tau_T^* = 0$ is not acceptable (it being a minimum point with $\bar{E}_t = 0$), the two conditions (6.2.9) and (6.2.10) yield a single stationary point

$$\lambda = -1, \quad \rho^* = 1 \tag{6.2.11}$$

with τ_T^* determined by (6.2.7) to be

$$\tau_T^* = \frac{P_c}{N} - m. \tag{6.2.12}$$

6.2.3.3 *The maximum expected EB population*

The corresponding maximum expected EB at terminal time is then

$$\bar{E}_t(\tau_T^*; \rho^*) = \lim_{\rho \to 1} \left\{ \rho N \frac{e^{(1-\rho)\tau_T^*} - 1}{1 - \rho} \right\}$$

$$= N \lim_{\rho \to 1} \left\{ \tau_T^* e^{(1-\rho)\tau_T^*} \right\} = N\tau_T^* = P_c - mN \tag{6.2.13}$$

We highlight this outcome in the following proposition:

Proposition 7. *For a prescribed combination of λ_D and N, the optimal (time-invariant) RB to EB conversion probability is $\lambda_c = \lambda_D$ resulting in a maximum expected terminal EB population of $P_c - mN$ attained at the terminal time*

$$T_{\max} = \frac{P_c - mN}{\lambda_D N}. \tag{6.2.14}$$

Before leaving the completed solution for the present birth and death process model, two important observations should be noted. The first of these pertains to the optimal expected terminal EB population (6.2.13) and the time required to reach it given in (6.2.14). We note that, for later comparison with the corresponding result using other approaches, the expression for $\bar{E}_t(\tau_T^*; \rho^*)$ is independent of the conversion capacity, only on the total C. trachomatis population for lysing, P_c, the initial infecting RB population N and the weight factor m in the lysing condition (6.2.7). Though the optimal expected terminal EB population is large, the time it takes to reach it with the constant conversion rate $\lambda_C = \lambda_D$ is very long. In normalized time αT, it is effectively $O(P_c/N) \gg 1$, two orders of magnitude larger than the replication time of RB units and seemingly not consistent with observations reported in [33]. While more empirical work may be conducted to address this issue, a more refined mathematical model that capture the more than two states of the C. trachomatis during its life cycle may also be another way to provide a more meaningful comparison.

6.2.4   On the host cell lysing condition

A second observation to be noted pertains to the threshold criterion (6.2.7) for host cell lysing. It is a special case of the more general lysing criterion

$$L\left(\bar{E}_T, \bar{R}_T\right) = P_c, \tag{6.2.15}$$

where

$$\bar{E}_T = \bar{E}_t(\tau_T; \rho), \quad \bar{R}_T = \bar{R}_t(\tau_T; \rho)$$

and $L\left(\cdot, \cdot\right)$ is some differentiable function of two variables. As long as

$$1 = \frac{\partial L}{\partial \bar{E}_T} < \frac{\partial L}{\partial \bar{R}_T},$$

any change due to the more general condition (6.2.15) would be of quantitative nature. Here, we examine instead other possible modifications of the threshold condition that are qualitatively different from (6.2.7).

The condition (6.2.7) with $m > 1$ corresponds to the basic fact that RB particles are bigger and the membrane inclusion has only a finite size-capacity for the total Chlamydia particles. Empirical evidence suggests that there are additional factors contributing to the lysis of the host cell. Among these is the fact that EB particles secrete chemicals that enhance host cell lysing. To incorporate this additional effect, we may again limit the threshold criterion to a linear condition of the form (6.2.7) for a fixed total Chlamydia population threshold P_c but modify it to read

$$n\,\bar{E}_T + m\,\bar{R}_T = P_c, \tag{6.2.16}$$

with the factor $n > 1$ characterizing the effects of the secreted chemicals. Depending on the potency of the lysis inducing EB chemicals, the ratio $m/n > 0$ may be >1 or <1. In that case, the results obtained for (6.2.7) again applies but with m and P_c replaced by m/n and P_c/n. In particular, we have as the maximum expected EB,

$$\bar{E}_T = \frac{1}{n}\left(P_c - mN\right), \tag{6.2.17}$$

to be attained at the terminal time

$$T_{\max} = \frac{P_c - mN}{n\lambda_D N}. \tag{6.2.18}$$

With $n > 1$ (given that the EB secreted chemicals promote host cell lysis), we see from (6.2.14) and (6.2.17) that the host cell would lyse sooner with a smaller expected terminal EB population.

Research findings in the Tan–Suetterlin Lab also suggest the possibility of RB particles exhibiting inhibitory effects on host cell lysing to prolong their own existence. In the context of the linear threshold condition (6.2.16), such inhibitory effects may be incorporated by modifying (6.2.16) to read

$$n\langle E(\tau_T; \rho)\rangle + (m - i)\langle R(\tau_T; \rho)\rangle = P_c, \tag{6.2.19}$$

for some $i > 0$. The results (6.2.17) and (6.2.18) again apply with m replaced by $m - i$. It follows that for a given pair of (m, n) values, the host cell would lyse later and the expected terminal EB population would be larger if $i > 0$.

When warranted, the incorporation of the two new effects into the lysing threshold condition (6.2.7) can again be extended to the nonlinear form (6.2.15) with suitable conditions on the partial derivatives of L with respect to E_T and R_T, respectively. However, we should do this only when we have validated the order of magnitude of the long lysing time for typical life cycles of C. trachomatis.

6.2.5 *Time-dependent conversion probability*

The simple birth and death type probabilistic model for the developmental cycle of C. trachomatis formulated in (6.2.1) and (6.2.2) is limited by its inherent linearity and time-invariant growth and conversion probabilities. One direction for a richer model is to ask whether or not we may attain a faster spread of the infectious bacteria by allowing the conversion probability to vary with time as shown by the empirical data of [33]. A straightforward change for the birth-and-death process model above to allow for λ_C to depend on t has been investigated by the method of optimal control previously applied to simpler C. trachomatis models in [71, 72]. Interested readers are referred to [8] for the rather surprising results for such a time-varying conversion probability model. Instead, we will examine in the next section another more realistic probabilistic model with *time-varying conversion probability* but without a prescribed host cell lysing criterion. To reflect the findings of [33], the terminal time T in this model of the development cycle is treated as a random variable with a prescribed *probability density function* (pdf) suggested by the empirical data reported in [33].

6.3 Uncertain Host Death

6.3.1 *The growth of RB and EB population*

We wish to investigate here a birth and death process model of C. trachomatis with a time-varying conversion probability $\lambda_C(\tau)$ for attaining a faster spread of the infectious bacteria (than $\lambda_C(\tau) = \lambda_D$ for the time-invariant λ_C case). Our experience with the case of a known T in [8, 71, 72] suggests that the appropriate method of solution for this type of problem is to make use of the optimal control methodology. To do so, we henceforth change to the more conventional optimal control notations, writing $u(t), \alpha, E_T, E(t)$ and $R(t)$ for $\lambda_c(t), \lambda_D, \bar{E}_T, \bar{E}_t(\tau; \rho)$ and

$\bar{R}_t(\tau; \rho)$, respectively. In terms of the new notations, the following proposition gives the growth dynamics of the expected RB and EB population:

Proposition 8. *The rates of growth of the expected RB population $R(t)$ and the expected EB population $E(t)$ are given by the IVP,*

$$R' = (\alpha - u)R, \quad R(0) = N \tag{6.3.1}$$

and

$$E' = uR, \quad E(0) = 0, \tag{6.3.2}$$

respectively.

Proof. Upon multiplying (6.2.1) by k and summing from 0 to ∞, we obtain

$$R' = \sum_{k=0}^{\infty} \left\{ \alpha \left[-k^2 P_k + k(k-1)P_{k-1} \right] + u \left[-k^2 P_k + k(k+1)P_{k+1} \right] \right\}$$

$$= \sum_{j=1}^{\infty} \left\{ \alpha \left[-j^2 P_j + j(j+1)P_j \right] + u \left[-j^2 P_j + j(j-1)P_j \right] \right\}$$

$$= \alpha \sum_{j=1}^{\infty} j P_j - u \sum_{j=1}^{\infty} j P_j = (\alpha - u)R$$

proving the validity of the ODE in (6.3.1). The initial condition in (6.3.1) follows from

$$\sum_{j=1}^{\infty} j P_j(0) = N$$

given $P_j(0) = \delta_{jN}$. The second IVP can be derived similarly from (6.2.2) and $Q_j(0) = \delta_{j0}$. □

6.3.2 *Prescribed probability density function for lysis time*

Empirical data collected in [33] show that the lysis time varies. A different but realistic approach to investigate the spread of C. trachomatis would be to work with the limited available data on the variation of the host cell lysis time. A model on this approach is to allow for the considerable uncertainty on the terminal time T to obtain over a fixed interval evident from the data available [33]. In other words, we do not know with certainty when the host cell lyses or what actually causes it to lyse. Instead, we only have some statistics on the host cell lysis time T allowing us to estimate or postulate a probability distribution or a probability density function $f_T(t)$ for the terminal time T of the life cycle. Since any host cell being considered is still alive, we should have as the pdf for the random variable T:

$$f_T(t) = U(t)f(t) = \begin{cases} 0 & (t < 0) \\ f(t) & (t > 0) \end{cases}$$

where $U(\cdot)$ is the Heaviside unit step function and

$$\int_{-\infty}^{\infty} f_T(t)dt = \int_0^{\infty} f(t)dt = 1.$$

In that case, the corresponding (expected) EB population at terminal time is also a random variable $E(T)$, a transformed random variable of the random variable T. It is a proven fact that the *expected value* of $E(T)$, denoted by E_T, is given in terms of the pdf $f_T(t)$ by

$$E_T = \int_0^{\infty} E(t)f(t)dt$$

From the linear growth dynamics (6.3.2) of EB, we have

$$E(t) = \int_0^t u(z)R(z)dz.$$

For maximum spread of the C. trachomatis bacteria, we choose the (per unit RB particle) conversion rate $u(t)$ to maximize the expected terminal *EB* population

$$E_T = \int_0^{\infty} f(t) \left[\int_0^t u(z)R(z)dz \right] dt. \tag{6.3.3}$$

Given the reality of a finite capacity for the one-way RB-to-EB conversion, the optimal choice $u_{op}(t)$ among all $u(t)$ ($\equiv \lambda_C(t)$) in the admissible set Ω of piecewise smooth (PWS) functions is required to satisfy the inequality constraints

$$0 \le u(t) \le u_{\max}. \tag{6.3.4}$$

Upon interchanging the order of integration, we may re-write the expression above as

$$E_T = \int_0^{\infty} F_c(t)u(t)R(t)dt \tag{6.3.5}$$

where

$$F_c(t) = \begin{cases} \int_t^{\infty} f(z)dz & (t \ge 0) \\ 1 & (t \le 0) \end{cases} \tag{6.3.6}$$

is the probability of the host cell NOT lysed at time t. Typical probability density functions include the *uniform density distribution (over the interval (T_1, T_2))*

$$f_T(t) = \frac{1}{\Delta}[H(t - T_1) - H(t - T_2)], \quad \Delta = T_2 - T_1 > 0, \tag{6.3.7}$$

and the *generalized inverse distribution density function*

$$f_T(t) = f(t)U(t - T_1) = \frac{T_1}{t^2}U(t - T_1) = \begin{cases} 0 & (t < T_1) \\ T_1/t^2 \equiv f(t) & (t > T_1). \end{cases} \qquad (6.3.8)$$

The two corresponding $F_c(t)$ are

$$F_c(t) = \begin{cases} 0 & (t \geq T_2) \\ \dfrac{1}{\Delta}[T_2 - t] & (T_1 \leq t \leq T_2) \\ 1 & (t \leq T_1) \end{cases} \qquad (6.3.9)$$

and

$$F_c(t) = \begin{cases} T_1/t & (t \geq T_1) \\ 1 & (t \leq T_1), \end{cases} \qquad (6.3.10)$$

respectively.

6.3.3 *Stochastic optimization*

As indicated earlier, we wish to investigate the possibility of a time-varying conversion probability $\lambda_C(\tau) = u(t)$ for attaining a faster spread of the infectious bacteria than the time-invariant λ_C case. In view of (6.3.5), we state the optimization problem as

$$\max_{u \in \Omega} \left\{ E_T = \int_0^\infty F_c(t)u(t)R(t)dt \right\} \qquad (6.3.11)$$

for the control $u(t)$ chosen from the admissible set Ω of piecewise smooth (PWS) functions that satisfy the constraints (6.3.4) and the (mean) RB population $R(t)$ determined by the IVP (6.3.1).

In the formulation above, the sample space of the random variable T is implicitly taken to be $(0, \infty)$ since the time when the host cell should lyse is uncertain and T may be infinite. In reality, all cells lyse in finite time within some upper bound (reflecting the fact that a host cell is of finite size with only a limited carrying capacity for the different forms of C. trachomatis bacteria). To be realistic and to simplify the solution process, we take the terminal time T upper bound to be finite for any life cycle in the subsequent development. In that case, all probability distribution $F_c(t)$ for a cell not yet lysed at time t is with compact support vanishing for $t < 0$ and $t > T_u$ (upper bound). The uniform density function (6.3.7) leading to the distribution (6.3.9) is a typical example of probability distribution with such compact support.

To apply the method of optimal control, we introduce the Hamiltonian

$$H_s(t) = u(t)R(t)F_c(t) + \lambda(t)R(t)\{\alpha - u(t)\}$$
$$= u(t)R(t)\{F_c(t) - \lambda(t)\} + \alpha\lambda(t)R(t) \tag{6.3.12}$$

The adjoint function (aka Lagrange multiplier) $\lambda(t)$ for the problem is to satisfy the adjoint ODE (see [5, 72])

$$\lambda' = u[\lambda - F_c(t)] - \alpha\lambda \tag{6.3.13}$$

for t in the bounded time interval $(0, T]$. In that case, the adjoint (Euler) boundary condition (BC) at the terminal time T is [72]

$$\lambda(T) = 0. \tag{6.3.14}$$

The Maximum Principle requires that we choose u to maximize H_s [5, 40]:

$$\max_{u \in \Omega}[H_s[u]] = \max_{u \in \Omega} R^*(t)[u(t)\{F_c(t) - \lambda^*(t)\} + \alpha\lambda^*(t)]$$
$$= H_s[u_{op}(t)] \quad (0 \le u \le u_{max}) \tag{6.3.15}$$

where $(\ )^* = (\ )_{u=u_{op}(t)}$. Since the Hamiltonian is linear in the control function $u(t)$, the optimal control is expected to be a combination of *singular controls* and *extreme values* of the admissible control over different time intervals.

6.3.4 Singular solution not applicable

Natural candidates for the optimal conversion rate corresponding are the stationary points of the Hamiltonian determined by:

$$\frac{\partial H_s}{\partial u} = R(t)\{F_c(t) - \lambda(t)\} = 0. \tag{6.3.16}$$

This stationarity condition requires

$$\lambda_S(t) = F_c(t) \tag{6.3.17}$$

given $R(t) > 0$. Note that the control $u(t)$ does not appear in the stationary condition (6.3.16) and hence is not determined by (6.3.16). For that reason the immediate consequence of the stationary condition (6.3.17) is known as a *singular solution* with the singular control $u_S(t)$ to be deduced from other requirements of the Maximum Principle when appropriate. For our singular solution, the adjoint DE requires

$$F'_c(t) = -\alpha F_c(t)$$

which is generally *not* satisfied by the probability distribution $F_c(t)$ of the host cell not lysing (and certainly not by the two illustrative examples in (6.3.9) and

(6.3.10)). Hence, the singular solution does not hold in any interval of the solution domain $[0, T]$ and we have the following negative result for our problem:

Proposition 9. *The singular solution is not applicable in any part of the solution domain $[0, T]$ for our uncertain terminal time T problem.*

6.3.5   Upper corner control adjacent to terminal time

It follows from Proposition 9 that the optimal control can only be a combination of the two *corner controls*: 0 and $u_{\max}$. With $R(t)$ being positive and converted only proportionately, $u(T) < u_{\max}$ is evidently not optimal when conversion is called for, since we can always choose a larger u (still within the admissible range) to convert some of the remaining RB for a larger terminal EB population. By continuity, the optimal control $u_{op}(t)$ continues to be $u_{\max}$ for some interval adjacent to T with the corresponding state and adjoint function denoted by $R_g(t)$ and $\lambda_g(t)$, respectively. We state this observation more formally as the next proposition:

Proposition 10. *The upper corner control maximizes the Hamiltonian (6.3.12) for some interval $(t_s, T]$ with t_s being the largest root of the switch condition*

$$S_g(t) \equiv F_c(t) - \lambda_g(t) = 0 \qquad (6.3.18)$$

nearest to T.

Proof. With the Euler BC $\lambda_g(T) = 0$, the Hamiltonian reduces to

$$H_s(T) = u(T)R_g(T)F_c(T)$$

i) For the general case of $F_c(T) > 0$, $H_s(T)$ is maximized by the upper corner control so that $u_{op}(T) = u_{\max}$ and

$$\lambda_g'(T) = -u_{\max}F_c(T) < 0.$$

Hence, $\lambda_g(t)$ is a decreasing function of time and is positive in some interval adjacent to T. By the continuity of the state and control functions, we have $u_{op}(t) = u_{\max}$ for some finite interval $(t_s, T]$ since the upper corner control continues to maximize the Hamiltonian

$$H_s(t) = u(t)R(t)\left[F_c(t) - \lambda_g(t)\right] + \alpha R(t)\lambda_g(t), \quad (\,t_s < t \leq T)$$

as long as $F_c(t) > \lambda_g(t)$. It ceases to maximize when $F_c(t) < \lambda_g(t)$ with the switch point t_s being the largest root of the switching condition (6.3.18).

ii) For the exceptional case $F_c(T) = 0$, $u_{op}(t)$ continues to be $u_{\max}$ for some interval adjacent to T since $\lambda_g'(t) = 0$ so that $\lambda_g(t) = 0$ in $(t_0, T]$ (with $t_0 = T_2$ for the uniform distribution case (6.3.9)). From there, the analysis of i) above applies.   □

6.3.6 Maximum conversion for the range $u_{max} \leq \alpha$

For the range $u_{max} \leq \alpha$ or, equivalently,

$$u_\alpha = u_{max} - \alpha \leq 0, \tag{6.3.19}$$

we have the following stronger result:

Proposition 11. *For $u_\alpha \leq 0$, $u_{op}(t)$ must be u_{max} for all t in $[0, T]$ with the optimal expected terminal EB population given by*

$$[E_T]_{op} = u_{max} N \int_0^\infty F_c(t) e^{-u_\alpha t} dt. \tag{6.3.20}$$

Proof. While a more formal proof can be given, it suffices to note that with the RB population not decreasing, converting at any smaller rate than u_{max} would result in less terminal EB and more terminal RB that can be converted with the higher conversion rate $u_{max}R$ any time in the interval $[0, T]$. □

To illustrate, we apply (6.3.20) to the uniform pdf (6.3.7) with the corresponding $F_c(t)$ given in (6.3.9). With $u(t) = u_{max} < \alpha$, the RB and EB population determined by the IVP (6.3.1) and (6.3.2), respectively, grow with time according to

$$R(t) = Ne^{-u_\alpha t}, \quad E(t) = \frac{u_{max}}{u_\alpha} N \left(1 - e^{-u_\alpha t}\right)$$

The resulting expected terminal EB population is given by (6.3.20) but also can be calculated directly from

$$[E_T]_{u(t)=u_{max}} = \int_0^\infty E(t) f_T(t) dt = \frac{u_{max}}{u_\alpha} N \int_0^\infty \left(1 - e^{-u_\alpha t}\right) f_T(t) dt$$

$$= \frac{u_{max} N}{u_\alpha \Delta} \int_{T_1}^{T_2} \left(1 - e^{-u_\alpha t}\right) dt = \frac{u_{max} N}{u_\alpha^2 \Delta} \left\{e^{-u_\alpha T_2} - e^{-u_\alpha T_1} + \Delta\right\}$$

where we have made use of the expression (6.3.7) for $f_T(t)$. In the limit as $u_{max} \to \alpha$ from below, we get by L'Hospital's rule

$$[E_T]_{u(t)=\alpha} = \frac{1}{2} \alpha N (T_1 + T_2). \tag{6.3.21}$$

The corresponding expected time to host cell lysis is given by

$$\int_0^\infty t f_T(t) dt = \frac{1}{\Delta} \int_{T_1}^{T_2} t dt = \frac{1}{2} (T_2 + T_1)$$

which is not large and does not seem unreasonable.

6.4 Uniform Density on a Finite Interval

Given Propositions 11 and 10, it remains to determine for the $u_\alpha = u_{max} - \alpha > 0$ range: *i*) the switch point t_s; *ii*) the optimal control in the complementary region

$[0, t_s)$, and *iii*) the optimal expected terminal EB population E_T. We illustrate the method of solution for these problems by working out the details for the uniform density function (6.3.7) in this section.

6.4.1 The adjoint function

For the uniform probability density function (6.3.7) and the corresponding distribution (6.3.9) for a cell not lysed at time t, the adjoint function for the upper corner control u_{max}, denoted by $\lambda_g(t)$, is determined by the terminal value problem [5, 70, 72]:

$$\lambda_g' = -\frac{\partial H_s}{\partial R} = -(\alpha - u_{max})\lambda_g - u_{max}F_c(t), \quad \lambda_g(T) = 0. \tag{6.4.1}$$

The solution for any finite T and any t in the interval $[0, T]$ is

$$\lambda_g(t) = \begin{cases} \Lambda_1(t; T) & (0 \le t < T \le T_1) \\ \Lambda_2(t; T, T_2) & (T_1 \le t < T \le T_2) \\ \Lambda_{12}(t; T, T_1, T_2) & (t < T_1 < T \le T_2) \\ 0 & (t \ge T_2) \end{cases} \tag{6.4.2}$$

where

$$\Lambda_1(t; T) = \frac{u_{max}}{u_\alpha}[1 - e^{-u_\alpha(T-t)}] \tag{6.4.3}$$

$$\Lambda_2(t; T, T_2) = \frac{u_{max}}{u_\alpha^2 \Delta} \left\{ \begin{array}{c} u_\alpha[(T_2 - t) - (T_2 - T)e^{-u_\alpha(T-t)}] \\ -[1 - e^{-u_\alpha(T-t)}] \end{array} \right\} \tag{6.4.4}$$

$$\Lambda_{12}(t; T, T_1, T_2) = \{\text{Exercise}\} \tag{6.4.5}$$

With $\lambda_g(t) = 0$ for $t \ge T_2$, the reticulate bodies have no "(shadow) value" beyond T_2. Since $F_c(t) = 0$ for $t \ge T_2$, the host cell has already lysed with probability 1 so that we would only be interested in the range of time $t < T_2$.

6.4.2 Lower corner control for $t < t_s$

In view of Proposition 11, we focus on the range $u_\alpha > 0$. In that case, we know from Proposition 10 that the upper corner control u_{max} is optimal for (t_s, T_2) where t_s is the largest zero of (6.3.18). If u_{max} should also be optimal for $t \lesssim t_s$ for the uniform pdf (6.3.7), then the ODE for the adjoint function at t_s (the zero of (6.3.18)) simplifies to

$$\lambda_g'(t_s) = -\alpha\lambda_g(t_s) = -\alpha F_c(t_s) < 0$$

so that $\lambda_g(t)$ is decreasing exponentially in for $t \lesssim t_s$. Since our particular $F_c(t)$ is either constant or monotone decreasing linearly in t, the quantity $F_c(t) - \lambda_g(t)$ is negative for $t < t_s$ at least in that range of t for $t \lesssim t_s$. It follows that the upper corner control is not optimal there. With the singular solution not applicable, this leaves us with the lower corner control as the only option. Indeed, we can prove the following stronger result:

Proposition 12. *For $u_\alpha > 0$, the lower corner control maximizes the Hamiltonian in $[0, t_s)$ so that the optimal control for the problem is the bang-bang control*

$$u_{op}(t) = \begin{cases} 0 & [0 < t < t_s) \\ u_{\max} & (t_s < t \le T], \end{cases} \tag{6.4.6}$$

Proof. For the lower corner control, the corresponding adjoint function, denoted by $\lambda_\ell(t)$, is determined by

$$\lambda_\ell' = -\alpha\lambda_\ell, \quad \lambda_\ell(t_s) = F_c(t_s).$$

The exact solution is

$$\lambda_\ell(t) = F_c(t_s)e^{-\alpha(t-t_s)}.$$

With $F_c(t)$ a linearly decreasing function of t, we have $F_c(t) < \lambda_\ell(t)$ for $t < t_s$ so that the lower corner control maximizes the Hamiltonian. $\square$

6.4.3 The switch point t_s

The process of determining the switch point t_s generally depends on the location of t_s since the expressions for $F_c(t_s)$ and $\lambda_g(t_s)$ vary with the range of t in which t_s is located (see (6.4.2)). As t_s is the zero of (6.3.18) nearest to T, we expect the determination of t_s by the condition

$$\lambda_g(t_s) = F_c(t_s) \tag{6.4.7}$$

to depend on the location of the (random variable) T in the interval $(0, T_2]$. In generally, the switch condition (6.4.7) takes different form in three different ranges of t_s ($< T$) as shown in the three subsections below:

6.4.3.1 $t_s < T \le T_1$

For this range, we have from (6.4.2) and (6.4.3)

$$\frac{u_{\max}}{u_\alpha}[1 - e^{-u_\alpha(T-t)}] = F_c(t_s) = 1$$

so that

$$t_s = T - \frac{1}{u_\alpha}\ln\left(\frac{u_{\max}}{\alpha}\right) \quad (T \le T_1). \tag{6.4.8}$$

As $u_{\max}$ increases (with α fixed), the switch point t_s tends to T as you would expect. At the other extreme, the switch point t_s in this range tends to $T - 1/\alpha$ as $u_{\max} \downarrow \alpha$ from above.

Since T is a random variable, it may assume any positive value. However, it has been endowed with a particular uniform distribution (6.3.7) for this example, the probability of $T < T_1$ is zero. As such, the result (6.4.8) for the location of the switch point is not particular relevant. With the corresponding expected value of T, denoted by $\overline{T}$, readily calculated to be

$$\overline{T} = \frac{1}{2}(T_2 + T_1) > T_1$$

we should consider t_s associated with $T_1 < T \leq T_2$, given the probability of $T > T_2$ is also zero.

6.4.3.2 $T_1 < t_s < T \leq T_2$

For $T_1 < T \leq T_2$, the switch point t_s may be anywhere between 0 and T depending on magnitude of $u_{\max}$ relative to the natural growth rate of RB α (as well as on other parameters). To the extent that t_s is the root of the condition (6.4.7) nearest to T, we consider first the case $t_s > T_1$. Upon observing (6.4.2) and (6.4.4), we have from $\Lambda_2(t_s, T, T_2) = F_c(t_s)$

$$\frac{\alpha u_\alpha}{u_{\max}} x = \left[1 - \frac{\alpha u_\alpha}{u_{\max}} \Delta_T\right] + [u_\alpha \Delta_T - 1]e^{-u_\alpha x}$$

where $x = T - t_s$ and $\Delta_T = T_2 - T$. As an equation for determining the switch point, the relation above may be re-arranged as

$$\beta_0 + \beta e^{-u_\alpha x} = \frac{\alpha u_\alpha}{u_{\max}} x \tag{6.4.9}$$

with

$$\beta = [u_\alpha(T_2 - T) - 1], \quad \beta_0 = 1 - \frac{\alpha u_\alpha}{u_{\max}} \Delta_T.$$

Graphing the two sides of (6.4.9) as functions of x with all other parameters fixed shows that the switch condition determines a unique positive root x^* that decreases with increasing $u_{\max}$. It follows that

$$t_s = T - x^* < T \tag{6.4.10}$$

increases with u_α. On the other hand, t_s may decreases below T_1 for sufficiently small u_α. With x^* monotone increasing as u_α decreases toward zero, t_s decreases to less than T_1 for all $u_\alpha < u_\alpha^*$ for some small u_α^*. (A rough estimate gives $u_\alpha^* \alpha \leq 1$.) For $u_\alpha < u_\alpha^*$ but still positive, t_s no longer falls in the range $t_s \geq T_1$ and the condition (6.4.9) for determining the switch point no longer applies. For u_α sufficiently small (relative to a prescribed set of other parameters of the problem such as Δ and $u_{\max}$), it is necessary to work with the third range of t_s in the next subsection.

6.4.3.3 $\ t_s < T_1 < T \leq T_2$

Even with T in the interval (T_1, T_2), we still have the possibility that t_s lies below $T_1(< T)$. In that case, the switch condition (6.4.7) for t_s requires

$$\Lambda_{12}(t_s, T, T_1, T_2) = 1. \tag{6.4.11}$$

Given (6.4.5), Eq. (6.4.11) may be written as

$$\frac{\alpha}{u_{\max}} = \gamma(T)e^{-u_\alpha(T-t_s)} \tag{6.4.12}$$

Unlike the case $t_s(T) > T_1$ (for $T > T_1$), we have an explicit solution of the switch point in terms of the terminal time T and other system parameters.

6.4.3.4 *The expected switch point for the optimal control*

Recall that the terminal time T is not known but a random variable with a given pdf. For the case of uniform pdf over the interval (T_1, T_2), the expected terminal time $\overline{T}$ was found to be $(T_1 + T_2)/2 > T_1$. Since T is a random variable, the optimal switch point t_s, being a function of T, is also a random variable. Its expected value $\overline{t}_s$ can be calculated from the relation

$$\overline{t}_s = \int_{-\infty}^{\infty} t_s(T)f_T(T)dT = \frac{1}{\Delta}\int_{T_1}^{T_2} t_s(T)(T_2 - T)dT$$

once we have the switch point $t_s(T)$ as determined by (6.4.10).

6.4.4 The optimal expected terminal EB population

6.4.4.1 *A local maximum*

For $u_\alpha > 0$, we know from Propositions 10 and 12 that the bang-bang control (6.4.6) maximizes the Hamiltonian with $t_s = t_s^*$ determined by (6.4.10). (If $u_\alpha \leq 0$, then the optimal strategy for maximizing the expected terminal EB population is to convert at the maximum rate possible from the start as the argument for Proposition 11 applies.) To the extent that maximizing Hamiltonian is not synonymous with maximizing E_T, we still need to prove that $u_{op}(t)$ maximizes E_T for $u_\alpha > 0$.

Proposition 13. *For $u_\alpha > 0$, the bang-bang control (6.4.6) maximizes the expected terminal EB population.*

Proof. From Propositions 10 and 12, we know already that $u_{op}(t)$ as given by (6.4.6) maximizes the Hamiltonian H_s with t_s in the interval (T_1, T_2). For our

relatively simple control problems, we may appeal to the uniqueness of the switch
point and that the optimal control is superior to not-converting any RB to EB at all.

□

6.4.4.2 The expected terminal RB population

For the optimal conversion rate $u_{op}(t)$ given in (6.4.6), we have from the growth
dynamics (6.3.1)

$$R_{op}(t) = \begin{cases} Ne^{\alpha t} & (0 \le t \le t_s) \\ Ne^{-u_\alpha t + u_{max} t_s} & (t_s \le t < T). \end{cases}$$

The associated expected EB population is given by (6.3.3)

$$E_T = u_{max} N \int_{-\infty}^{\infty} f_T(T) e^{u_{max} t_s(T)} \left(\int_{t_o(T)}^{T} e^{-u_\alpha t} dt \right) dT$$

$$= \frac{u_{max} N}{u_\alpha \Delta} \int_{T_1}^{T_2} e^{\alpha t_s} \{1 - e^{-u_\alpha (T - t_s)}\} dT \tag{6.4.13}$$

As we do not have an explicit solution for $t_s(T)$ in the range (T_1, T_2), the integral
(6.4.15) for the expected terminal EB population resulting from the bang-bang
control will have to be evaluated numerically for any given set of system parameters.

6.4.5 The optimal expected terminal time

Recall that the terminal time T is not known but a random variable with a given
pdf. For the case of uniform pdf over the interval (T_1, T_2), the expected terminal
time $\bar{T}$ is given by

$$\bar{T} = \int_{-\infty}^{\infty} t f_T(t) dt = \frac{1}{\Delta} \int_{T_1}^{T_2} t \, dt = \frac{1}{2}(T_1 + T_2).$$

6.5 Theory and Experimental Results

Given the rather unusual life cycle of C. trachomatis, we have embarked on some
modeling and analysis to investigate possible Darwinian force at work on the choice
for the rate of proliferation of the bacteria. In our first attempt in this direction,
we formulate and analyze several optimal control models to determine the optimal
RB-to-EB conversion rate that maximizes for these models the terminal EB population at a *known* terminal time when the host cell lyses [71, 72]. As applications
of the previously developed birth and death processes, we formulate and analyze in
this chapter two different and more realistic models that prescribe some constraints
on the host cell lysing.

The first is a birth and death type probabilistic model with time-invariant (per unit particle) probabilities for RB division and RB-to-EB conversion (in the context of standard birth and death processes) as an application of the mathematical machinery developed in the last chapter. An added feature to the usual birth and death models is a host cell capacity constraint for the total C. trachomatis population to be used for the determination of the terminal time. The condition is a weighted sum of the total RB and EB population that reflects the substantial size difference between the RB and EB units; it fixes the termination time for the RB proliferation and RB-to-EB conversion development cycle. The optimal RB-to-EB conversion probability α_C is found, rather surprisingly, to be equal to the RB growth rate α_D. More importantly, the theoretical findings from this simple model do not seem to be consistent with the experimental results of [33]. Among the inconsistency is the time-varying conversion rate shown in the empirical data collected.

Along with time-varying RB-to-EB conversion should be a larger optimal expected terminal EB population. As the causes contributing to the host cell lysing are complex and not completely known and the terminal time for the developmental cycle has been observed to vary (see [33]), it seems reasonable to investigate models with an uncertain terminal time T (when the host cell lyses). In that case, T would be a random variable with a prescribed probability density function (pdf) taken to be representative of the available collection of data on the host cell lysis time. A novel stochastic optimization problem is then formulated for such models when an optimal time-varying conversion probability $\lambda_C(t) \equiv u(t)$ is sought to maximize the terminal expected EB population. The stochastic optimal control problem for a pdf uniformly distributed over a finite time interval (T_1, T_2) (corresponding to a linear probability distribution in that period) is worked out in detail to show how the optimal control methodology may be used for stochastic optimization problems and how mathematical analysis (not just explicit analytical solution, numerical simulations and statistical inferences) can play an important role in the understanding phenomena in the biological sciences. When the system conversion capacity has an upper limit $u_{\max}$ $(\geq u(t))$ and $u_{\max} > \alpha = \lambda_D$, the optimal conversion turns out to be bang-bang for these new models featuring a random terminal time with a prescribed pdf. The corresponding optimal expected terminal EB population may be calculated numerically. However. while a bang-bang control is time-varying, it is not quite the multiple change pattern shown in [33]. A more in-depth discussion of consistency between the theoretical findings and empirical observations (as well as additional modeling work) can be found in [8].

Chapter 7

Random Walk, Diffusion and Heat Conduction

In this chapter, we begin with another application of the branching process to the problem of one-dimensional random walk along the x-axis. We then deduce from the relevant Kolmogorov ODE system the simplest diffusion equation in the limiting case of the drunkard's small step size δx tending to zero. This limiting partial differential equation is a special case of the general linear PDE for diffusive phenomena including the phenomenon of heat conduction (see [72] for example). To the extent that heat and temperature distribution and other diffusive phenomena are important in the life sciences, we take the appearance of the heat equation as an opportunity to review some basic theory and solution techniques for linear PDE. These theory and techniques will also be needed for the solution of stochastic partial differential equations in later chapters.

7.1 One-Dimensional Random Walk

7.1.1 *Unbiased random walk and diffusion approximation*

A drunkard leaves a tavern located at the origin $x = 0$ for home along the street taken to be the x-axis. He is so drunk that he would go in either direction on any step. Let δt be sufficiently short that at most one step is taken during that interval of elapsed time. The probability of him taking one step of size δx in the time interval $(t, t + \delta t)$ is $\beta \delta t$ except for negligibly small terms that tend to zero faster than δt. If a step is taken, then it is equally probable that it be to the left or to the right so that

$$\text{Prob}\{X \to X + \delta x\} = \text{Prob}\{X \to X - \delta x\} = \frac{1}{2}. \qquad (7.1.1)$$

Let $P_k(t)$ be the probability of the drunkard located at $k\delta x$ from the origin,

$$P_k(t) = \text{Prob}\{X(t) = k\delta x\},$$

where k may be positive, negative or zero. Then by the assumptions stated above, we have

$$P_k(t + \delta t) = (1 - \beta \delta t)\, P_k(t) + \frac{1}{2}\beta \delta t \,\{P_{k+1}(t) + P_{k-1}(t)\} + o(\delta t). \qquad (7.1.2)$$

In the limit as $\delta t \to 0$, we get the following Kolmogorov ODE system

$$\frac{dP_k}{dt} = -\beta P_k + \frac{\beta}{2} \{P_{k+1}(t) + P_{k-1}(t)\}, \quad (k = 0, \pm 1, \pm 2, \pm 3, \ldots). \qquad (7.1.3)$$

The system is supplemented by the initial conditions

$$P_0(0) = 1, \quad P_k(0) = 0 \quad (k \neq 0) \qquad (7.1.4)$$

given the drunkard started at the origin. It is reasonable to assume that

$$\lim_{k \to \pm \infty} P_k(t) = 0 \quad (0 < t < \infty), \qquad (7.1.5)$$

since the drunkard can only take a finite number of steps in finite time. (For this particular version of the problem, the drunkard is so drunk that he would not recognize his own home even if he should get there.)

7.1.2 *Diffusion approximation for small step size*

The solution of the Kolmogorov system is far from straightforward. In order to make progress in that direction, we consider the extreme case of drunkard's steps being so small to allow for a Taylor polynomial approximation of the unknown probability. For this purpose, we let $x = k\delta x$ and $u(x, t) = P_k(t) = P(k\delta x, t)$ and re-write (7.1.3) as

$$\frac{\partial u}{\partial t} = -\beta u + \frac{\beta}{2} \{u(x + \delta x, t) + u(x - \delta x, t)\}.$$

Upon expanding $u(x \pm \delta x, t)$ as a Taylor series about x to get

$$u(x \pm \delta x, t) = u(x, t) \pm u_{,x}(x, t)\delta x + u_{,xx}(x, t)(\delta x)^2 / 2 + \cdots$$

where $u_{,x}$ denotes partial derivative with respect to x. The Kolmogorov ODE for $u(x, t)$ becomes

$$\frac{\partial u}{\partial t} = \frac{\beta}{2}(\delta x)^2 \frac{\partial^2 u}{\partial x^2} + \cdots \qquad (7.1.6)$$

where the largest of the omitted terms (indicated by "$\cdots$") are proportional to $(\delta x)^4$. In the limit as $\delta x \to 0$, the right hand side tends to an unacceptable limit of zero for any finite β. It is unacceptable because $\beta\delta t$ is the probability of the drunkard taking a step during an interval of duration δt and is unreasonable to remain unchanged as the step size varies. For the limiting case of $\delta x \to 0$, the relation (7.1.6) requires that $\beta\,(\delta x)^2$ should be finite in order for the limiting relation to be meaningful. We set

$$\beta\,(\delta x)^2 = a.$$

In that case, we have for very small δx

$$\frac{\partial u}{\partial t} = \frac{a}{2}\frac{\partial^2 u}{\partial x^2} \qquad (7.1.7)$$

except for negligibly small term of the order of $(\delta x)^2$ or smaller with the relation (7.1.7) being exact in the limit as $\delta x \to 0$.

7.1.3 Fundamental solution

The relevant solution for the partial differential equation (7.1.7), known as one dimensional *diffusion* (or *heat*) *equation*, can be obtained by a number of different methods. Here we consider a similarity solution of the form

$$u(x,t) = f(y) = f\left(\frac{x}{\sqrt{t}}\right) \qquad (7.1.8)$$

motivated by the observation that the diffusion equation (7.1.7) is invariant after a change of scale by replacing x by λx and t by $\lambda^2 t$. With $y = x/\sqrt{t}$, the function $u(x,t) = f(y)$ is scale invariant in the same way as the diffusion equation, hence a possible form of solution for the PDE. Upon substitution of (7.1.8) into (7.1.7), we obtain

$$-y\frac{df}{dy} = a\frac{d^2 f}{dy^2}.$$

The first order ODE for df/dy is separable and can be solved to give

$$\frac{df}{dy} = c_1 e^{-y^2/2a} \quad \text{and} \quad u(x,t) = f(y) = c_0 + c_1 \int_{-\infty}^{y} e^{-z^2/2a}dz.$$

The previously noted vanishing condition (7.1.5) at $-\infty$ requires $c_0 = 0$ while the $x = k\delta x$ counterpart of the condition

$$\lim_{N \to \infty} \sum_{k=-N}^{N} P_k(t) = 1 \qquad (7.1.9)$$

requires

$$\lim_{N \to \infty} \int_{-N\delta x}^{N\delta x} u(x,t)dx = c_1 \int_{-\infty}^{\infty} e^{-y^2/2a}dy = c_1\sqrt{2\pi a} = 1.$$

It follows that the solution of (7.1.7) subject to (7.1.9) is

$$u(x,t) = \frac{1}{\sqrt{2\pi a}}\cdot \int_{-\infty}^{x/\sqrt{t}} e^{-y^2/2a}dy. \qquad (7.1.10)$$

In continuous probability theory, it is more convenient to work with the corresponding probability density function (pdf)

$$p(x,t) = \frac{\partial u(x,t)}{\partial x} = \frac{1}{\sqrt{2\pi at}} e^{-x^2/2at}. \tag{7.1.11}$$

It can be verified that $p(x,t)$ as given in (7.1.11) is also a solution of (7.1.7) and is known as the **fundamental solution** of that one-dimensional diffusion equation. Note that while $p(x,t)$ tends to 0 as $t \to 0$ as long as $x \neq 0$, the function is not well-defined at $x = 0$ and $t = 0$. However, for any small positive ε, $0 < \varepsilon \ll 1$, we have

$$\lim_{t \to 0} \int_{-\varepsilon}^{\varepsilon} \frac{1}{\sqrt{2\pi at}} e^{-x^2/2at} dx = \lim_{t \to 0} \int_{-\varepsilon/\sqrt{t}}^{\varepsilon/\sqrt{t}} \frac{1}{\sqrt{2\pi a}} e^{-y^2/2a} dy$$

$$= \int_{-\infty}^{\infty} \frac{1}{\sqrt{2\pi a}} e^{-y^2/2a} dy = 1.$$

Hence, $p(x,t)$ behaves like the Dirac delta function $\delta(x)$ in the limit as $t \to 0$.

For a drunkard starting his random walk at a location $x = \xi$ (instead of the origin), the corresponding fundamental solution is seen to be

$$G(x,t;\xi,0) = \frac{1}{\sqrt{2\pi at}} e^{-(x-\xi)^2/2at}.$$

If the initial position of the drunkard can be anywhere along the real line with a prescribed probability density function $p_0(x)$, it can be shown (see [69]) that the pdf $p(x,t)$ for the drunkard to be at position x at time $t > 0$ is given by

$$p(x,t) = \int_{-\infty}^{\infty} G(x,t;\xi,0)p_0(\xi)d\xi. \tag{7.1.12}$$

For that reason, $G(x,t;\xi,0)$ is also known as the *Green's function* for the diffusion equation on the real line; it is the solution of the IVP

$$\frac{\partial p}{\partial t} = \frac{a}{2}\frac{\partial^2 p}{\partial x^2} \quad (-\infty < x < \infty, \ t > 0) \tag{7.1.13}$$

$$p(x,0) = \delta(x-\xi) \quad (-\infty < x < \infty). \tag{7.1.14}$$

subject also to the limiting conditions

$$\lim_{|x| \to \infty} [p(x,t)] = 0 \quad (t > 0).$$

The method of separation of variables for linear PDE (instead of the method of similarity solution) was used in [69] to derive (7.1.12) in the context of one-dimensional heat conduction. For this volume to be self-contained, the eigenfunction expansion approach will be reviewed in the subsequent sections of this chapter as the solution process will be needed for other stochastic PDE problems to be discussed in later chapters. Reader familiar with the separation of variables approach may leapfrog to start on the second half of this volume for discussions on models with a continuous sample space.

7.1.4 Biased random walk

If instead of moving in either direction with equal probability as in (7.1.1), the drunkard has a slight preference to walk in one of the two directions (perhaps because some foot injury) so that

$$\text{Prob}\{X \to X + \delta x\} = p, \quad \text{Prob}\{X \to X - \delta x\} = m. \tag{7.1.15}$$

with $m + p = 1$. Though unnecessary to do so, we take for convenience

$$p = \frac{1}{2} + q, \quad m = \frac{1}{2} - q.$$

Again, we let $P_k(t)$ be the probability of the drunkard located at $k\delta x$ from the origin where k may be positive, negative or zero. Instead of (7.1.2), we have now

$$P_k(t + \delta t) = (1 - \beta \delta t) P_k(t) + \frac{1}{2}\beta \delta t \{mP_{k+1}(t) + pP_{k-1}(t)\} + o(\delta t). \tag{7.1.16}$$

In the limit as $\delta t \to 0$, we get the following new ODE system for $P_k(t)$:

$$\frac{dP_k}{dt} = -\beta P_k + \frac{\beta}{2}\{mP_{k+1}(t) + pP_{k-1}(t)\}, \quad (k = 0, \pm 1, \pm 2, \pm 3, \ldots). \tag{7.1.17}$$

The system is supplemented by the same initial conditions (7.1.4) and the conditions (7.1.5) at $\pm\infty$.

For very small step size δx, the same Taylor polynomial approximation of the unknown probability $P_{k\pm 1}(t)$ leads to the following approximation for (7.1.17)

$$\frac{\partial u}{\partial t} = -(p - m)\beta \delta x \frac{\partial u}{\partial x} + \frac{\beta}{2}(\delta x)^2 \frac{\partial^2 u}{\partial x^2} + \cdots \tag{7.1.18}$$

where the largest of the omitted terms (indicated by "$\cdots$") are proportional to $(\delta x)^3$. For a reasonable limit for (7.1.18) as $\delta x \to 0$, we continue to set $\beta(\delta x)^2 = a$ as before. To avoid having the first term on the right dominating for very small δx (in fact becoming unbounded as $\delta x \to 0$), we also need $p - m$ proportional to δx with

$$(p - m)\beta \delta x = b \tag{7.1.19}$$

independent of δx. In that case, we have for very small step size δx

$$\frac{\partial u}{\partial t} = -b\frac{\partial u}{\partial x} + \frac{1}{2}a\frac{\partial^2 u}{\partial x^2} \tag{7.1.20}$$

except for negligibly small term of the order of $(\delta x)^2$ or smaller, with the relation (7.1.20) being exact in the limit as $\delta x \to 0$.

The second order PDE (7.1.20) is a special case of the more general *Kolmogorov forward equation* for the transition probability density function of diffusive Markov processes. A great deal more will be said about the more general case, also known as the *Fokker–Planck equation*, in the second half of this volume. In other applications, the more limited (7.1.20) is also known as convection-diffusion equation, advection-diffusion equation, drift diffusion equation or transport equation with the term multiplied by b characterizing the effect of drift.

Exercise 29. Transform (7.1.20) into the conventional diffusion equation (7.1.7) by setting

$$u(x,t) = e^{\left(2abx - b^2 t\right)/2a} w(x,t)$$

and obtain an exact solution for (7.1.20).

7.2 Diffusion on a Bounded Domain

7.2.1 *Random walk on a finite interval*

For many problems, the domain for the one dimensional diffusion equation (7.1.7) may be only a part of the real line for a number of reasons. For the drunkard walking randomly in one dimension, his home may be located at a point ℓ ($= k_\ell \delta x$) > 0 and he, not too drunk, would stop his meander when he gets there. In that case, the stretch of the road $x > \ell$ would have no relevance to the problem if his starting location is $\xi < \ell$ since the part $x > \ell$ would never be reached. As with the case of ODE, we need an auxiliary condition at the finite end point ℓ of the road for the diffusion version of the problem. While we do not know the probability of the drunkard landing on $x = \ell$ at time $t > 0$, we can say with certainty that, once reached, he would not leave that location thereafter. As the drunkard would not take another step forward or backward, we have for the diffusion approximation model of the random walk problem

$$\left[\frac{\partial u(x,t)}{\partial x} \right]_{x=a} = 0. \tag{7.2.1}$$

When the derivative of the unknown of a PDE is prescribed at an end of the solution interval, the prescribed condition is known mathematically as a "Neumann" boundary condition for the PDE. The PDE (7.1.7) is for the unknown $u(x,t)$ and the end condition (7.2.1) is a condition on the derivative of $u(x,t)$; hence the prescribed end condition is a Neumann condition on $u(x,t)$. In neurobiology, this type of condition at the synaptic end of an axon is known as a "sealed end" condition for the PDE governing the unknown transmembrane depolarization voltage along the axon. We will have occasion to discuss more about modeling of nerve cells including their axons in later chapters of this volume.

We noted previously that the density function $p(x,t) = u_{,x}(x,t)$ also satisfies the same diffusion equation so that

$$\frac{\partial p}{\partial t} = \frac{a}{2}\frac{\partial^2 p}{\partial x^2}. \tag{7.2.2}$$

In terms of $p(x,t)$, the boundary condition (7.2.1) at $x = \ell$ now takes the form

$$p(\ell, t) = 0.$$

In particular, the prescribed end condition is on the unknown $p(x,t)$ itself. When the unknown of a PDE is prescribed at an end of the solution interval, the prescribed condition is known mathematically as a "Dirichlet" boundary condition for the PDE. In neurobiology, this type of condition at the synaptic end of an axon is known as a "killed end" condition for the PDE governing the unknown transmembrane depolarization voltage along the axon.

In general, the three PDE (7.1.7), (7.1.20) and (7.2.2) are all second order in the spatial variable x for the interval $-\ell < x < L$ and first order in the time variable t for $t > 0$. It will be seen from the development in the next few sections that we need to augment such PDE with two prescribed boundary conditions, one at each end of the finite spatial interval and an initial condition at some reference time t_0. Together, the PDE and the auxiliary conditions define an *initial-boundary value problem* (IBVP) of the unknown governed by the PDE.

7.2.2 *Heat conduction in an insulated wire*

We noted earlier that one dimensional diffusion equations such as (7.1.7) and (7.1.20) also appear in other applications including problems on heat conduction. A concise summary of this fact at a level adequate for our purpose can be found in [69, 72]. To the extent that heat conduction is simpler to visualize and the results easier to interpret, we indicate briefly below the relevant features of this natural phenomenon and the mathematical methods for heat conduction along a fine wire of cross-sectional area A and insulated around its cylindrical surface (see Fig. 7.1). For a very thin wire, it is expected that the temperature T of the wire is uniformly distributed over the cross-section of the wire and we will so assume. In that case, the temperature distribution T is only a function of x and t.

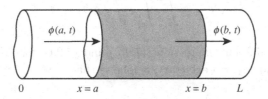

Fig. 7.1.   Heat flux in a thin wire.

The capacity of heat in a medium at time t may be quantified in units of heat energy per unit mass at a specific location of the medium and taken to be proportional to the temperature of the medium $T(\mathbf{x}, t)$. The "constant" of proportionality c_0 is known as the **specific heat** of the medium giving the amount of heat energy needed to raise the temperature of a unit mass of the medium one degree Kelvin at that spatial location of the wire and time. The amount of heat in the segment of wire spanning the interval $x_1 \leq x \leq x_2$ at time t is then given by

$$H(t) = \int_M c_0 T \, dm = \int_V c_0 T \rho \, dV = \int_{x_1}^{x_2} c_0 T \rho A \, dx$$

with

$$\Delta H = H(t_2) - H(t_1) = \left[\int_{x_1}^{x_2} c_0 T \rho A \, dx \right]_{t_1}^{t_2} = \int_{t_1}^{t_2} \int_{x_1}^{x_2} \frac{\partial}{\partial t} [c_0 T \rho A] \, dx \, dt, \quad (7.2.3)$$

for the heat energy increment ΔH accrued in a segment (x_1, x_2) of the wire over the time interval (t_1, t_2). In these relations, ρ is the mass density of the medium which (like T, A and c_0) may vary with location and time.

The same heat energy increment of the same wire segment may also be calculated in terms of the **heat flux density** Q, the amount of heat per unit area per unit time flowing across a cross-sectional area A. The amount of heat energy that passes through the cross-section A (positive in the direction of increasing x) over the time interval (t_1, t_2) is then given by

$$h(x) = \int_{t_1}^{t_2} A(x, t) Q(x, t) dt \quad (7.2.4)$$

The energy increment over the same time interval accrued in the wire segment between the two end cross-sections at x_1 and x_2 is then given by

$$\Delta H = h(x_1) - h(x_2) = \int_{t_1}^{t_2} [A(x_1, t)Q(x_1, t) - A(x_2, t)Q(x_2, t)] \, dt. \quad (7.2.5)$$

Assuming differentiability, ΔH may be written as

$$\Delta H = - \int_{t_1}^{t_2} \int_{x_1}^{x_2} \frac{\partial}{\partial x} (AQ) \, dx \, dt. \quad (7.2.6)$$

Equating the two different expressions (7.2.3) and (7.2.6) for ΔH and with both intervals (t_1, t_2) and (x_1, x_2) arbitrary, it follows that

$$\frac{\partial}{\partial t} (c_0 T \rho A) + \frac{\partial}{\partial x} (AQ) = 0. \quad (7.2.7)$$

The relation (7.2.7) is one equation for two unknowns, a second relation for determining $T(x,t)$ and $Q(x,t)$ is usually taken to be the Fourier's law. It postulates the heat flux Q to be proportional to the temperature gradient. For our one-dimensional problem, this takes the form

$$Q(x,t) = -K_0 \frac{\partial T}{\partial x} \equiv -K_0 T_{,x} \tag{7.2.8}$$

where K_0 is the coefficient of **thermal conductivity**. Upon using (7.2.8) to eliminate Q from (7.2.7), we obtain the following linear PDE for the temperature distribution in the thin wire:

$$\frac{\partial}{\partial t} [c_0 T \rho A] = \frac{\partial}{\partial x} \left[A K_0 \frac{\partial T}{\partial x} \right]. \tag{7.2.9}$$

For a wire with uniform properties so that all material and geometrical parameters are constants in space and time, the linear PDE (7.2.9) becomes

$$\frac{\partial T}{\partial t} = \frac{\nu}{2} \frac{\partial^2 T}{\partial x^2}, \quad \frac{\nu}{2} = \frac{K_0}{c_0 \rho}, \tag{7.2.10}$$

which is the same as (7.1.7) with a (to be reserved as a label for an end point of the wire in the subsequent development) replaced by ν. The diffusion constant $\nu/2$ is known as coefficient of **thermal diffusivity** in heat conduction.

7.2.3 *Inhomogeneous problem and auxiliary conditions*

Though it may give rise to the same diffusion equation in a special case, the phenomenon of heat conduction is considerably more complex and richer in content (than the one-dimensional random walk problem). To illustrate, let us examine the case of an electric current passing through the thin wire insulated along its cylindrical surface. The resistance inherent in the metallic wire material to the continuous current flow results in continuous new heat energy generation along the wire not included in the previous formulation. This addition of new energy may be characterized by a energy source intensity term that may vary with location and time. With this energy generation per unit mass per unit time denoted by $f(x,t)$, (7.2.5) is replaced by

$$\Delta H = h(x_1) - h(x_2) + \int_{t_1}^{t_2} \int_{x_1}^{x_2} f(x,t) \rho A \, dx \, dt,$$

leading to the inhomogeneous PDE

$$\frac{\partial}{\partial t} [c_0 T \rho A] = \frac{\partial}{\partial x} \left[A K_0 \frac{\partial T}{\partial x} \right] + \rho A f(x,t). \tag{7.2.11}$$

For a wire with uniform properties, this equation simplifies to

$$\frac{\partial T}{\partial t} = \frac{\nu}{2} \frac{\partial^2 T}{\partial x^2} + \frac{1}{c_0} f(x, t). \qquad (7.2.12)$$

In either case, the PDE holds for the entire length of the wire, naturally taken to locate along the interval $(0, \ell)$ of the x-axis. In this way, the domain of the IBVP is $0 < x < L$ and $t > 0$.

For this finite wire, a boundary condition is required at each end point. These conditions are much more transparent in the context of heat conduction. For example, we may prescribed

$$T(L, t) = T_\ell \qquad (7.2.13)$$

at the end point $x = L$, corresponding to keeping the temperature fixed at T_ℓ there. The case of a killed end corresponds to keeping the end temperature at $0°$. In general, we have a *Dirichlet (boundary) condition* when an unknown of problem is specified at an end point.

In the context of heat conduction, the case of a sealed end, at $x = 0$ for example, corresponds to the end being insulated so that there cannot be heat flow in or out of that cross section. The more general condition

$$T_{,x}(0, t) = q_0 \qquad (7.2.14)$$

is easily seen to be a prescribed heat flux $Q(0, t) = Q_0$ at the location $x = 0$ which Fourier's law translates into $T_{,x}(0, t) = -Q_0/K_0 \equiv q_0$. In general, we have a *Neumann (boundary) condition* when a (normal) derivative of an unknown of problem is specified at an end point.

A mixed boundary condition of the form

$$\alpha T(L, t) + \beta T_{,x}(L, t) = \gamma(t)$$

is also possible and appropriate at an end point $x = L$. It corresponds to some kind of heat leakage proportional to the temperature. In developing methods of solution for the IBVP, we will stay with the simpler killed and sealed end conditions. The treatment of a mixed end condition mainly requires more bookkeeping but otherwise involves natural extensions of the steps taken in the two simpler cases.

At some initial (reference) time before the onset of heat generation by the internal heat source $f(x, t)$, the thin wire is usually at the ambient temperature $T_0(x)$ of its surrounding environment so that

$$T(x, 0) = T_0(x) \quad (0 \le x \le L). \qquad (7.2.15)$$

Altogether, (7.2.11), (7.2.14), (7.2.13) and (7.2.15) constitute an IBVP for the evolving temperature distribution $T(x, t)$ in the thin wire.

For the purpose of developing a method of solution in the next section for the IBVP above, we work with the following cleaner version of this IBVP:

$$u_{,t} = u_{,xx} + f(x,t), \quad (0 < x < \ell,\ t > 0), \tag{7.2.16}$$

the boundary conditions

$$u_{,x}(0,t) = 0, \quad u(\ell,t) = 0, \quad (t > 0), \tag{7.2.17}$$

and the inhomogeneous initial condition

$$u(x,0) = U(x), \quad (0 \le x \le \ell). \tag{7.2.18}$$

This normalized version of the problem is the result of re-scaling the spatial and temporal variables (when all wire properties are uniform along the length of the wire)

$$t' = t/c_0, \quad x' = x/\sqrt{K_0/\rho},$$

and setting

$$\ell = L/\sqrt{K_0/\rho}, \quad u(x',t') = T(x,t), \quad f'(x',t') = f(x,t), \quad U(x') = T_0(x),$$

with the primes omitted from the new variables after re-scaling.

From the actual heat conduction problem, we see inhomogeneous end conditions such as

$$u_{,x}(0,t) = g_0(t), \quad u(\ell,t) = g_\ell(t), \quad (t > 0)$$

arise naturally for a more general IBVP with the homogeneous boundary conditions (7.2.17) being a special case. However, the more general case can in fact be reduced to the special case by setting

$$u(x,t) = w(x,t) + g_0(t)(x - \ell) + g_\ell(t).$$

The new unknown $w(x,t)$ now satisfies the PDE

$$w_{,t} = \frac{\nu}{2} w_{,xx} + \left\{ f(x,t) - \left[(x - \ell)\frac{dg_0}{dt} + \frac{dg_\ell}{st} \right] \right\},$$

with the initial condition

$$w(x,0) = U(x) - \{ g_0(0)(x - \ell) + g_\ell(0) \},$$

and the homogeneous end conditions

$$w_{,x}(0,t) = 0, \quad w(\ell,t) = 0 \quad (t > 0).$$

7.2.4 Superposition principles

The solution of the IBVP (7.2.16)–(7.2.18) is facilitated by a number of superposition principles analogous to those for linear ODE (see [4, 72] for example). For the purpose of formulating the few such principles to be used herein, we consider first the following special cases of this IBVP:

- The *(Time-Invariant) Steady State Problem* is the time independent solution $u^{(s)}(x)$ for the BVP defined by the ODE (7.2.16), after simplification by $u^{(s)}{}_{,t} = 0$, and the boundary conditions (7.2.17) without any reference to the initial condition (7.2.18). For this BVP, the forcing term f must be a function of spatial variable only to allow for the (steady state) solution to be independent of t. (There may also be time-varying steady state (in the form of periodic solutions) for the IBVP but these are not relevant for the present discussion.)
- The *Forced Problem* is the solution $u^{(f)}(x,t)$ of the *IBVP* (7.2.16)–(7.2.18) with $U(x) = 0$ for all $(0 \leq x \leq \ell)$.
- The *Initial Value Problem* is defined by the *IBVP* (7.2.16)–(7.2.18) with $f(x,t) = 0$ for all $(0 < x < \ell, t > 0)$ with its solution denoted by $u^{(i)}(x,t)$.

With $u^{(s)}(x)$, $u^{(f)}(x,t)$ and $u^{(i)}(x,t)$ as the solution for the *Steady State Problem*, the *Forced Problem* and the *Initial Value Problem*, respectively, we have the following superposition principles to help simplify the solution process for the *IBVP*.

Proposition 15. *(Superposition I)* $u(x,t) = u^{(f)}(x,t) + u^{(i)}(x,t)$ *is the solution of the original IBVP defined by* (7.2.16)–(7.2.18).

Proof. The *PDE* (7.2.16) is satisfied by $u^{(f)}(x,t) + u^{(i)}(x,t)$ since

$$u^{(f)}{}_{,t} = u^{(f)}{}_{,xx} + f(x,t), \quad u^{(i)}{}_{,t} = u^{(i)}{}_{,xx}.$$

The homogeneous boundary conditions (7.2.17) are satisfied by both $u^{(f)}(x,t)$ and $u^{(i)}(x,t)$; hence so does their sum. Finally, the initial condition (7.2.18) is also satisfied by $u(x,t)$ since

$$u(x,0) = u^{(f)}(x,0) + u^{(i)}(x,0) = 0 + U(x) = U(x). \qquad \square$$

Proposition 16. *(Superposition II)* *If* $f(x,t)$ *is independent of t and* $u^{(s)}(x)$ *is the solution of the corresponding Steady State Problem, then the solution of original IBVP defined by* (7.2.16)–(7.2.18) *can be written as* $u(x,t) = u^{(s)}(x) + u^{(I)}(x,t)$ *where* $u^{(I)}(x,t)$ *is the solution of the Initial Value Problem with inhomogeneous initial condition* $u^{(I)}(x,0) = U(x) - u^{(s)}(x) \equiv U_0(x)$.

Proof. For a source term $f(x)$, the steady state solution is easily seen to be:

$$u^{(s)}(x) = \int_x^\ell \int_0^\xi f(\eta) d\eta d\xi.$$

Upon substituting the exact solution $u(x,t) = u^{(s)}(x) + u^{(I)}(x,t)$ into the various relations that define the IBVP for the general problem, equations (7.2.16) and (7.2.17) give a PDE and two homogeneous boundary conditions for $u^{(I)}(x,t)$

$$u^{(I)},_t = u^{(I)},_{xx} \quad (0 < x < \ell,\ t > 0),$$

$$u^{(I)},_x (0,t) = 0, \quad u^{(I)}(\ell,t) = 0 \quad (t > 0).$$

The initial condition however becomes $u(x,0) = u_s(x) + u^{(I)}(x,0) = U(x)$ so that

$$u^{(I)}(x,0) = U(x) - u^{(s)}(x)$$

$$= U(x) - \int_x^\ell \int_0^\xi f(\eta)d\eta d\xi \equiv U_0(x). \tag{7.2.19}$$

$\square$

There are also problems with boundary conditions (7.2.17) replaced by the more general inhomogeneous mixed boundary conditions:

$$\alpha_0 u,_x (0,t) + \beta_0 u(0,t) = \gamma_0(t), \quad \alpha_\ell u,_x (\ell,t) + \beta_{\ell L} u(\ell,t) = \gamma_L(t), \tag{7.2.20}$$

(with proper restrictions on the coefficients $\alpha_k, \beta_k,\ k = 0, \ell$ to ensure the problem is well posed). In that case, we have the following new type of problem:

- The *Boundary Data Problem* is the solution of (7.2.16), (7.2.18) and the mixed inhomogeneous boundary conditions (7.2.20) (instead of homogeneous boundary conditions (7.2.17)) with $f(x,t) = U(x) = 0$.

With the solution of this problem denoted by $u^{(b)}(x,t)$, we have the following superposition principle:

Proposition 17. (*Superposition III*) $u(x,t) = u^{(f)}(x,t) + u^{(i)}(x,t) + u^{(b)}(x,t)$ *is the solution of the IBVP defined by* (7.2.16), (7.2.20) *and* (7.2.18).

Proof. (left as an exercise) $\square$

It should be evident from the development above that similar superposition principles can also be established for more complex linear PDE systems including the single linear PDE (7.2.11) for the thin wire with nonuniform geometrical and material properties. The latter case will be discussed in a later section of this chapter.

7.3 Fourier Series

7.3.1 *Product solutions and eigenvalue problem*

Given the various superposition principles established in the last section, we develop here a powerful method of solution for general linear PDE. The main idea of this method (and of the general approach to linear PDE on a finite solution domain) is to solve separately a number of special (and simpler) problems for ODE and combine their solutions to form the solution of the actual PDE problem. For the purpose of developing and illustrating this solution technique, we consider first the Initial Value Problem for the simple diffusion equation in one spatial dimension associated with (7.2.16)–(7.2.18), re-stated below for convenient reference:

$$u^{(i)}_{,t} = u^{(i)}_{,xx} \quad (0 < x < \ell, \ t > 0) \tag{7.3.1}$$

$$u^{(i)}_{,x}(0, t) = u^{(i)}(\ell, t) = 0, \quad u^{(i)}(x, 0) = U(x) \tag{7.3.2}$$

While the dependence of $u^{(i)}(x, t)$ on x and t can be in many different forms, we focus our effort by seeking a solution of the product form $u^{(i)}(x, t) = \Phi(x)\Psi(t)$ for some functions $\Phi(x)$ and $\Psi(t)$ to be determined by the PDE. For example, if $U(x) = \cos(\lambda x)$, then the product solution $u^{(i)}(x, t) = e^{-\lambda^2 t}\cos(\lambda x)$ would satisfy the PDE (7.3.1) as well as the first and third auxiliary condition in (7.3.2). The same solution would also satisfy the remaining (second) auxiliary condition if $\lambda = \pi/2\ell$ and thereby solves the Initial Value Problem.

For a more general $U(x)$, we consider the general product solution $u^{(i)}(x, t) = \Phi(x)\Psi(t)$ for which the governing PDE (7.3.1) becomes $\Phi(x)\Psi_{,t}(t) = \Phi_{,xx}(x)\Psi(t)$, or

$$\frac{\Phi_{,xx}(x)}{\Phi(x)} = \frac{\Psi_{,t}(t)}{\Psi(t)}. \tag{7.3.3}$$

With the left hand side only a function of x and the right hand side only a function of t, equality of the two sides is possible only if both sides are equal to the same constant, denoted by $-\lambda^2$ (with the negative sign imposing no special restriction since λ^2 is unknown and may be negative). In that case, we have from (7.3.3)

$$\Phi'' + \lambda^2\Phi = 0, \quad \dot{\Psi} + \lambda^2\Psi = 0, \quad (0 < x < \ell, \ t > 0) \tag{7.3.4}$$

where a prime, $(\ )'$, indicates differentiation with respect to x, a dot, $(\ )\dot{}$, indicates differentiation with respect to t. The general solutions for these two ODE are

$$\Phi(x) = A\cos(\lambda x) + B\sin(\lambda x), \quad \Psi(t) = Ce^{-\lambda^2 t} \tag{7.3.5}$$

where A, B and C are unknown constants of integration.

The two boundary conditions on $u^{(i)}(x, t)$ in (7.3.2) require

$$\Phi'(0) = 0, \quad \Phi(\ell) = 0. \tag{7.3.6}$$

The first condition in (7.3.6) is met by $B = 0$ so that $\Phi(x) = A\cos(\lambda x)$. The second condition in (7.3.6) requires

$$\Phi(\ell) = A\cos(\lambda \ell) = 0.$$

Since we want a *nontrivial* solution and therefore should have $A \neq 0$, the condition above is satisfied only if

$$\lambda = \lambda_n \equiv (2n - 1)\frac{\pi}{2\ell} \quad (n = 1, 2, 3, \ldots). \tag{7.3.7}$$

(In particular, $\lambda = 0$ is not admissible as it does not lead to a nontrivial solution.) Except for the special values $\{\lambda_n\}$ of the parameter λ given by (7.3.7), the homogeneous BVP for $\Phi(x)$ has only the trivial solution $\Phi(x) \equiv 0$. The set of special values $\{\lambda_n^2\}$ constitutes the *eigenvalues* (or the *characteristic values*) of the homogeneous *BVP* problem. The problem of finding all possible values $\{\lambda_n^2\}$ for which the homogeneous *BVP* has a nontrivial solution is called an *eigenvalue problem*; the different nontrivial solutions associated with the different eigenvalues are called *eigenfunctions* (or *characteristic functions*). For a particular λ_k^2 from the set specified by (7.3.7), we have the product solution

$$u_k(x,t) = \Phi_k(x)\Psi_k(t) = D_k \cos(\lambda_k x)e^{-\lambda_k^2 t}, \quad (k = 1,2,\ldots) \qquad (7.3.8)$$

where D_k is a yet unknown constant of integration ($= AC$ for λ_k).

Historically, the parameter λ (instead of λ^2) was first used in (7.3.4) as the eigenvalue parameter. Today both λ and λ^2 are being used in different texts and research articles. Since λ^2 is used in these notes, the eigenvalues are the collection $\{\lambda_n^2\}$ (and not $\{\lambda_n\}$).

7.3.2 *Superposition and orthogonality*

For the product solution $u_k(x,t)$ to be a solution of the *IBVP* (7.3.1)–(7.3.2), it must still satisfy the initial condition $u^{(i)}(x,0) = U(x)$, or, in the language of PDE theory, $u^{(i)}(x,0)$ should *fit* the *initial data* $U(x)$. With

$$u_k(x,0) = \Phi_k(x)\Psi_k(0) = D_k \cos(\lambda_k x),$$

$u_k(x,t)$ could satisfy the prescribed initial condition only if the initial data has exactly a $\cos(\lambda_k x)$ distribution in x, i.e., $U(x) = \bar{U}_k \cos(\lambda_k x)$ for some constant $\bar{U}_k$. For that special initial data, the Initial Value Problem is solved by setting $D_k = \bar{U}_k$.

For other forms of the initial data, $u_k(x,t)$ is not a solution for the problem of interest since, for any fixed k, it clearly could not fit the prescribed initial data. To obtain a solution for the more general distribution $U(x)$, we consider the combination

$$u_M(x,t) \equiv \sum_{m=1}^{M} D_m \cos(\lambda_m x)e^{-\lambda_m^2 t} \qquad (7.3.9)$$

of the collection of particular solutions $\{u_m(x,t)\}$ which satisfies the PDE (7.3.1) and the two boundary conditions in (7.3.2). The initial distribution of the new solution,

$$u_M(x,0) \equiv \sum_{m=1}^{M} D_m \cos(\lambda_m x),$$

is obviously more general and would fit more general initial data $U(x)$. But it is equally clear that for a fixed M, the combination $u_M(x,t)$ cannot fit all possible initial data, not even continuous data or differentiable data.

Take for example the initial data $U(x) = \cos(\lambda_{M+1}x)$. If $u_M(x,0) = \cos(\lambda_{M+1}x)$ for some selected set of constants $\{D_m\}$, then we would have

$$\sum_{m=1}^{M} D_m \cos(\lambda_m x) = \cos(\lambda_{M+1}x),$$

and therewith

$$\sum_{m=1}^{M} D_m \int_0^\ell \cos(\lambda_m x)\cos(\lambda_k x)dx = \int_0^\ell \cos(\lambda_{M+1}x)\cos(\lambda_k x)dx. \qquad (7.3.10)$$

However, we know from elementary calculus

$$\int_0^\ell \cos(\lambda_m x)\cos(\lambda_k x)dx = \begin{cases} 0 & if\ m \neq k \\ \ell/2 & if\ m = k. \end{cases} \qquad (7.3.11)$$

For any value $m \leq M$, the *orthogonality relation* (7.3.11) reduces (7.3.10) to

$$D_m = \frac{2}{\ell}\cdot(0) = 0, \quad (m = 1, 2, \ldots, M) \qquad (7.3.12)$$

(since $k, m \leq M < M+1$) so that (7.3.9) becomes

$$u_M(x,t) = 0 \quad (0 \leq x \leq \ell,\ t > 0) \qquad (7.3.13)$$

which clearly does not fit the initial data $U(x) = \cos(\lambda_{M+1}x)$. Hence, the expression (7.3.9) for $u_M(x,t)$ cannot be the solution for the Initial Value Problem with the initial data $U(x) = \cos(\lambda_{M+1}x)$.

The observation on the initial data $U(x) = \cos(\lambda_{M+1}x)$ suggests that we should consider an even more general solution, namely a linear combinations of all the particular product solutions as given by (7.3.8):

$$u^{(i)}(x,t) = \sum_{m=0}^{\infty} u_m(x,t) = \sum_{m=0}^{\infty} D_m \cos(\lambda_m x)e^{-\lambda_m^2 t}. \qquad (7.3.14)$$

Such a solution would at least cover all initial data of the form $U(x) = \cos(\lambda_j x)$. However, unlike the case of a finite combination of $\{u_m(x,t)\}$ such as (7.3.9), the infinite series solution (7.3.14) is meaningful only if it can be shown to converge for the relevant ranges of x and t of interest. Furthermore, it must converge to the prescribed initial data for all x in $(0, \ell)$. An infinite series $\sum_{m=0}^{\infty} a_m$ converges to "a" if

$$\left|\sum_{m=0}^{M} a_m - a\right| < \epsilon$$

for any chosen ϵ if M is sufficiently large.

Before discussion the convergence of the infinite series solution, we first make a few observations that are evident from (7.3.14):

- With $e^{-\lambda_m^2 t} \to 0$ as $m \to \infty$ for a fixed $t > 0$ and as $t \to \infty$ for a fixed $m > 0$, it suffices then to establish the convergence of the *Fourier series* solution at $t = 0$,

$$u^{(i)}(x,0) = \sum_{m=0}^{\infty} D_m \cos(\lambda_m x) = U(x), \qquad (7.3.15)$$

to the initial data.
- If equality in (7.3.15) holds and term-by-term integration is permitted, then applications of the orthogonality condition (7.3.11) leads to the uncoupled relations

$$D_m = \frac{2}{\ell} \int_0^\ell \cos(\lambda_m x) U(x) dx \quad (m = 1, 2, \ldots) \qquad (7.3.16)$$

for an explicit determination of the individual coefficients $\{D_m\}$. (In the absence of orthogonality among the eigenfunctions, we would have to solve an infinite system of simultaneous equations for $\{D_m\}$.) Note that the orthogonality relation (7.3.11) with $k \neq m$ is critical to the uncoupled determination of the Fourier coefficients D_m. The following example illustrates its application:

Example 14. If $U(x) = x$ $(0 \leq x \leq \ell)$, the Fourier coefficients $\{D_m\}$ can be obtained by evaluating the integral in (7.3.16) to get:

$$D_m = \frac{2}{\ell} \int_0^\ell \cos(\lambda_m x) x \, dx = \frac{2}{\lambda_m} \sin(\lambda_m \ell) - \frac{2}{\lambda_m^2 \ell}$$

$$= \frac{4\ell}{\pi} \left\{ \frac{(-)^{m-1}}{2m-1} - \frac{2}{\pi} \frac{1}{(2m-1)^2} \right\} \quad (m = 1, 2, \ldots). \qquad (7.3.17)$$

Hence, the Fourier series (7.3.15) is an alternating harmonic series and thus converges rather slowly (caused by the finite discontinuities of the periodically extended function $U^*(t)$ of $U(x)$).

7.3.3   Convergence

Regarding the convergence of the Fourier series solution (7.3.14) (with $\{D_m\}$ determined by (7.3.16)), the following facts and their proofs can be found in most texts on the theory of Fourier series such as [78]:

- The continuity of $U(x)$ is neither necessary nor sufficient for the convergence of the infinite series (7.3.15).
- If $U(x)$ is a piecewise smooth (PWS) function with a finite jump discontinuity at only a finite number of points in the interval $(0, \ell)$, the corresponding infinite

series (7.3.15), with the *Fourier coefficients* $\{D_m\}$ calculated from (7.3.16), converges to $U(x)$ at any location x in $(0, \ell)$ where the initial data is continuous and to the average of the two one-sided limits at points of finite jump discontinuity:

$$u^{(i)}(x, 0) = \frac{1}{2}\{U(x+) + U(x-)\}$$

$$= \frac{1}{2}\lim_{\epsilon \to 0}\{U(x + \epsilon) + U(x - \epsilon)\}. \qquad (7.3.18)$$

- Since the solution components $u_m(t) = D_m \cos(\lambda_m x)e^{-\lambda_m^2 t} \to 0$ as $t \to \infty$, it follows that the solution of the Initial Value Problem tends to zero for any initial data $U(x)$. As such, the solution of the Initial Value Problem for the diffusion equation is a transient phenomenon. The system (whether it is biological, chemical or mechanical) returns to quiescence (a resting state) eventually.

While we are mainly concerned with the infinite series solution in the $0 < x < \ell$, it turns out to be relevant to the discussion of convergence, especially at the end points $x = 0$ and $x = \ell$, to extend $U(x)$ to the entire real line in a way consistent with the behavior of the Fourier series solution $u^{(i)}(x, 0)$ beyond the basic interval $(0, \ell)$. For an appropriate extension of the initial data, we observe from the properties of $\cos(\lambda_m x)$ that $u^{(i)}(x, 0)$ as given in (7.3.15) is an even function of x with

$$u^{(i)}(x, 0) = \begin{cases} u^{(i)}(-x, 0) & (-\ell \leq x \leq 0) \\ -u^{(i)}(2\ell - x, 0) & (\ell \leq x \leq 2\ell) \\ -u^{(i)}(-2\ell - x, 0) & (-2\ell \leq x \leq -\ell) \end{cases}$$

and periodic with period 4ℓ for $-\infty < x < \infty$. Accordingly, we define for our problem the extended function $U^*(x)$ of the PWS $U(x)$ to be

$$U^*(x) = \begin{cases} U(x) & (0 < x < \ell) \\ U(-x) & (-\ell < x < 0) \\ -U(2\ell - x) & (\ell < x < 2\ell) \\ -U(-2\ell - x) & (-2\ell < x < -\ell) \end{cases} \qquad (7.3.19)$$

with the PWS $U^*(x)$ periodic of period 4ℓ for $-\infty < x < \infty$. The following fact can also be proved:

- The infinite series solution (7.3.14) converges to (the PWS) $U^*(x)$ wherever it is continuous and to the average of the two one-sided limits at points of simple jump discontinuities of $U^*(x)$ as in (7.3.18). The series is periodic with period 4ℓ.

7.3.4   *Time-dependent distributed heat sources*

With the solution $u^{(i)}(x, t)$ for the Initial Value Problem, we have by Proposition 16 also the solution of the IBVP if the internal distributed heat source of the wire is independent of time so that $f(x,t) = f(x)$. If $f(x,t)$ varies with time, we can still make use of the Fourier series technique to obtain the solution for the IBVP (7.2.16)–(7.2.18). Given the Superposition Principle I, we need only to solve the relevant *Forced Problem*. This is done by setting

$$\{u^{(f)}(x,t), f(x,t)\} = \sum_{n=0}^{\infty} \{u_n^{(f)}(t), f_n(t)\} \cos(\lambda_n x) \qquad (7.3.20)$$

where λ_n is given in (7.3.7) and

$$f_n(t) = \frac{2}{\ell} \int_0^{\ell} f(x,t) \cos(\lambda_n x) dx \quad (n = 1,2,3,\ldots). \qquad (7.3.21)$$

Upon substituting the expansion for $u^{(f)}$ in (7.3.20) into the PDE (7.2.16) for $u(x,t)$ and making use of the orthogonality condition (7.3.11), we obtain

$$\frac{du_n^{(f)}}{dt} + \lambda_n^2 u_n^{(f)} = f_n(t) \quad (n = 1,2,3,\ldots). \qquad (7.3.22)$$

The initial condition $u^{(f)}(x,0) = 0$ requires

$$u_n^{(f)}(0) = 0 \quad (n = 1,2,3,\ldots). \qquad (7.3.23)$$

The solution of the *IVP* (7.3.22)–(7.3.23) is

$$u_n^{(f)}(t) = \int_0^t e^{-\lambda_n^2(t-\tau)} f_n(\tau) d\tau \quad (n = 1,2,3,\ldots), \qquad (7.3.24)$$

which completes the solution process with $u(x,t) = u^{(f)}(x,t) + u^{(i)}(x,t)$.

Example 15. $f(x,t) = x \cos(t)$

For this forcing function we have from the results of the last example,

$$f_n(t) = D_n \cos(t)$$

where the Fourier coefficients $\{D_n\}$ are given in (7.3.17). Correspondingly, the unknown coefficients $\{u_n^{(f)}(t)\}$ are determined by (7.3.24) to be

$$u_n^{(f)}(t) = \frac{\lambda_n^2}{1 + \lambda_n^4} \left\{ \left[\cos(t) + \frac{1}{\lambda_n^2} \sin(t) \right] - e^{-\lambda_n^2 t} \left[\cos(t) - \frac{1}{\lambda_n^2} \sin(t) \right] \right\}$$

with

$$u_n^{(f)}(t) \rightarrow \frac{\lambda_n^2}{1 + \lambda_n^4} \left[\cos(t) + \frac{1}{\lambda_n^2} \sin(t) \right] \quad \text{as } t \rightarrow \infty.$$

In the limit as $t \to \infty$, we have

$$\lim_{t \to \infty} [u(x,t)] = \lim_{t \to \infty} \left[u^{(f)}(x,t) + u^{(i)}(x,t) \right]$$

$$= \sum_{n=0}^{\infty} \frac{\lambda_n^2}{1 + \lambda_n^4} \left[\cos(t) + \frac{1}{\lambda_n^2} \sin(t) \right] \cos(\lambda_n x)$$

which is a periodic function in time with period 2π. This example illustrates an earlier observation that there exist time-varying steady state solution for IBVP in PDE.

7.3.5   *Uniqueness*

In the preceding subsections, we showed that the solution of IBVP (7.2.16)–(7.2.18) can be obtained by a combination of the method of separation of variables and superposition principles. The general approach actually applies to problems for other linear PDE and more general auxiliary conditions (such as inhomogeneous mixed boundary conditions of the form

$$\alpha u(x_0, t) + \beta u_{,x}(x_0, t) = \gamma \tag{7.3.25}$$

where α, β and γ are known constants). As in the theory of ODE, we need to know if there are other solutions not found by the method of Fourier series. We illustrate a general approach to uniqueness for the heat equation by proving the following uniqueness theorem for the problem (7.2.16)–(7.2.18).

Theorem 13. *The solution of the IBVP (7.2.16)–(7.2.18) is unique.*

Proof. Suppose there are two solution $u_1(x,t)$ and $u_2(x,t)$. Let $u = u_1 - u_2$ be their difference. Then $u(x,t)$ is the solution of the homogeneous IBVP

$$u_{,t} = u_{,xx} \qquad (0 < x < \ell, \ t > 0), \tag{7.3.26}$$

$$u_{,x}(0,t) = u(\ell, t) = 0 \quad (t > 0), \tag{7.3.27}$$

$$u(x,0) = 0 \qquad (0 \le x \le \ell). \tag{7.3.28}$$

From

$$\int_0^\ell u u_{,t} \, dx = \int_0^\ell u u_{,xx} \, dx,$$

integrate by parts to get

$$\frac{\partial}{\partial t} \int_0^\ell \frac{u^2}{2} dx = [u u_{,x}]_{x=0}^\ell - \int_0^\ell (u_{,x})^2 dx$$

$$= - \int_0^\ell (u_{,x})^2 dx \le 0$$

where we have made use of the homogeneous boundary conditions (7.3.27) to eliminated the term $[uu_{,x}]^{\ell}_{x=0}$ on the right hand side. Now integrate both side with respect to t from 0 to τ to get

$$\int_0^{\ell} \frac{1}{2} [u(x,\tau)]^2 \, dx = -\int_0^{\tau} \int_0^{L} (u_{,x})^2 dx dt \leq 0 \tag{7.3.29}$$

where we have made use of the homogeneous initial condition (7.3.28) to eliminate the "constant" of integration. The right hand is non-positive while the left hand side is non-negative. Equality of these two terms holds only when both vanish. Since $\tau > 0$ is arbitrary, the vanishing of the right hand side requires $(u_{,x})^2 = 0$ or $u(x,t) = u(t)$ is a function of time only. In that case, (7.3.29) becomes

$$\int_0^{L} \frac{1}{2} [u(t)]^2 \, dx = 0,$$

which implies $u(x,t) = u(t) = 0$ or $u_1 = u_2$ and hence uniqueness. □

The technique for proving uniqueness above can also be used to establish uniqueness theorems for problems involving the heat equation but with more general boundary conditions (including the mixed boundary condition (7.2.20)) and for other PDE similar in type to the diffusion (heat) equation (known as *parabolic* PDE).

7.4 Sturm–Liouville Problems

7.4.1 *Regular Sturm–Liouville problems*

The method of separation of variables applies to many other linear PDE models resulting also in one or more eigenvalue problems similar to the one encountered earlier but often more complicated. For the simple one-dimensional diffusion problem with uniform properties previously analyzed, we were able to obtain explicit exact solution for the eigen-pairs. This is not always possible for other problems. For instance, when the coefficient of thermal conductivity K_0 is not spatially uniform, the governing PDE for the temperature distribution $T(x,t)$ in the straight thin wire now takes the form

$$c_0 \rho T_{,t} = [K_0(x) T_{,x}]_{,x} + \rho f(x,t). \tag{7.4.1}$$

For a product solution $T(x,t) = \phi(x) \psi(t)$ for the corresponding Initial Value Problem, the ODE for $\phi(x)$ is a special case of the general *self-adjoint* equation

$$L[\phi(x)] + \lambda^2 r(x) \phi(x) = 0, \quad (a < x < b) \tag{7.4.2}$$

where the linear differential operator $L[\phi]$ is defined by

$$L[\phi] = \left\{ \frac{d}{dx} \left(p(x) \frac{d\phi}{dx} \right) - q(x) \phi \right\}, \tag{7.4.3}$$

for some known functions $p(x), q(x)$ and $r(x)$. For the heat equation (7.4.1) (with $f(x,t) = 0$), we have $p(x) = K_0(x)$ and $q(x) = 0$. Note that we have changed the interval to (a, b) instead of $(0, \ell)$ to conform with the conventional specification of the *Sturm–Liouville problem*.

As usual, the ODE is augmented at each of the two ends of the solution interval by a boundary condition appropriate for the phenomenon being modeled. Each of these may be a Dirichlet, Neumann, or *mixed (boundary) condition*. In case where these conditions are *homogeneous*, they correspond to special cases of the following general (homogeneous) mixed boundary condition

$$\alpha_c \phi(c) + \beta_c \phi'(c) = 0 \qquad (7.4.4)$$

where c is an end point (a or b).

Definition 10. The eigenvalue problem for the second order self-adjoint ODE (7.4.2) with λ^2 as the eigenvalue parameter, together with one condition of the form (7.4.4) at each end of the solution interval (a, b), is known as a *Sturm–Liouville (S–L) problem*.

Definition 11. A Sturm–Liouville problem is *regular* if $p(x) > 0$, $r(x) > 0$ and $q(x) \geq 0$ are continuous functions and $p(x)$ is also differentiable in (a, b).

7.4.2 Some facts on regular S–L problems

While an exact solution of the ODE (7.4.2) is generally not possible, we can nevertheless establish many important properties for the eigen-pairs of Sturm–Liouville problems. Three of these properties known to be very useful in applications are presently established for regular S–L problems with a homogeneous Dirichlet condition at both ends:

$$\phi(a) = 0, \quad \phi(b) = 0. \qquad (7.4.5)$$

It should be evident that many of the results obtained continue to hold if one of the Dirichlet end condition is replaced by a homogeneous Neumann condition. Somewhat less obvious is the validity of the results for homogeneous mixed end conditions (7.4.4) as well.

Theorem 14. *If λ^2 is an eigenvalue of a regular S–L problem (with Dirichlet condition (7.2.4)), it must be simple, i.e., there is only one eigenfunction (up to a multiplicative factor) for the eigenvalue.*

Proof. Suppose there are two eigenfunction $\phi^{(1)}(x)$ and $\phi^{(2)}(x)$ for the eigenvalue λ^2 with

$$\mathcal{L}\left[\phi^{(i)}(x)\right] \equiv L\left[\phi^{(i)}(x)\right] + \lambda^2 r(x) \phi^{(i)}(x) = 0, \quad (i = 1, 2) \qquad (7.4.6)$$

where $L[\varphi]$ is the linear operator previously defined in (7.4.3).

Consider the bilinear form

$$B\left[\phi^{(1)}, \phi^{(2)}\right] \equiv \phi^{(2)} \mathcal{L}\left[\phi^{(1)}(x)\right] - \phi^{(1)} \mathcal{L}\left[\phi^{(2)}(x)\right] = 0. \qquad (7.4.7)$$

After some cancellation, we have

$$B\left[\phi^{(1)}, \phi^{(2)}\right] = \frac{d}{dx}\left[pW(x)\right] = 0, \qquad (7.4.8)$$

where

$$W(x) = \phi^{(2)}\left[\phi^{(1)}\right]' - \phi^{(1)}\left[\phi^{(2)}\right]'. \qquad (7.4.9)$$

It follows from (7.4.8) that

$$p(x)W(x) = C_0.$$

Given $\phi^{(i)}(a) = 0$, $i = 1, 2$, we have $W(a) = 0$ so that $C_0 = 0$ and therewith

$$W(x) = \phi^{(2)}\left[\phi^{(1)}\right]' - \phi^{(1)}\left[\phi^{(2)}\right]' = 0 \qquad (7.4.10)$$

keeping in mind that $p(x) > 0$.

Now, (7.4.10) may be re-arranged as

$$\frac{\left[\phi^{(1)}\right]'}{\phi^{(1)}} = \frac{\left[\phi^{(2)}\right]'}{\phi^{(2)}}$$

or

$$\phi^{(2)}(x) = c_0 \phi^{(1)}(x). \qquad \square$$

Corollary 7. *Theorem 14 holds if the homogeneous Dirichlet condition (7.4.5) at one end point is replaced by a mixed condition (7.4.4).*

Proof. (exercise) $\square$

Theorem 15. *If λ^2 is an eigenvalue of a regular Sturm–Liouville problem (with Dirichlet condition (7.2.4)), it must be real.*

Proof. Suppose that there is a complex eigenvalue $\lambda^2 = \Lambda$ with $\varphi(x)$ the associated eigenfunction. Then their complex conjugates Λ^* and $\varphi^*(x)$ also constitute an eigen-pair. By the bilinear relation (7.4.7), we have

$$(\Lambda - \Lambda^*) \int_a^b r(x)\, \varphi(x)\, \varphi^*(x)\, dx = \int_a^b \left\{\varphi^* L\left[\varphi\right] - \varphi L\left[\varphi^*\right]\right\} dx$$

$$= \int_a^b \left\{\varphi^* (p\varphi')' - \varphi \left(p\left[\varphi^*\right]'\right)'\right\} dx$$

$$= \left[\varphi^* (p\varphi') - \varphi \left(p\left[\varphi^*\right]'\right)\right]_a^b = 0$$

where we have made use of the homogeneous Dirichlet condition (7.4.5) in the last step. Since $\varphi(x)\varphi^*(x) = |\varphi(x)|^2 > 0$ and $r(x) > 0$, we must have $\Lambda^* = \Lambda$ so that λ^2 must be real. □

Corollary 8. *Theorem* 15 *holds if the homogeneous Dirichlet condition* (7.4.5) *at one end point is replaced by a mixed condition* (7.4.4).

Proof. (left as an exercise) □

Theorem 16. *The eigenfunctions for two different eigenvalues of a regular S–L problem are orthogonal relative to the weight function* $r(x)$, *i.e.*,

$$\int_a^b r(x)\varphi_m(x)\varphi_n(x)\,dx = 0 \quad (\lambda_m \neq \lambda_n). \tag{7.4.11}$$

Proof. (The proof is left as an exercise since it is similar to that for the one dimensional heat conduction problem.) □

7.5   The Rayleigh Quotient

7.5.1   *Variational formulation*

In addition to the three properties above, we can also prove that not only the eigenvalues of a regular S–L problem are real; they must also be non-negative. For that we need to introduce a quantity called the Rayleigh quotient for the problem.

Definition 12. The *Rayleigh quotient* of a regular Sturm–Liouville problem is the ratio

$$R[u] = \frac{\int_a^b p(x)\,[u']^2\,dx + \int_a^b q(x)\,[u]^2\,dx}{\int_a^b r(x)\,[u]^2\,dx} \equiv \frac{N[u]}{D[u]}, \tag{7.5.1}$$

where p, q and r are the coefficients of the ODE (7.4.2).

A function $u(x)$ that satisfies the prescribed boundary conditions (7.4.5) generally does not satisfy the differential equation (7.4.2) for any value λ^2 unless λ^2 is an eigenvalue and $u(x)$ is the corresponding eigenfunction.

Theorem 17. *If* $\{\lambda^2, u(x)\}$ *is an eigen-pair for a regular Sturm–Liouville problem, then* $\lambda^2 = R[u] \geq 0$.

Proof. Multiply (7.4.2) by $u(x)$ and integrate over (a, b). After integration by parts and application of the boundary conditions (7.4.5), we obtain

$$\lambda^2 \int_a^b r(x)u^2 dx - \int_a^b p(x)\,[u']^2\,dx - \int_a^b q(x)u^2 dx = 0.$$

Since $r(x) > 0$ and $u(x)$ is an eigenfunction (which is not identically zero by definition), the relation above may be re-written as $\lambda^2 = R[u]$. With $p(x) > 0$, $r(x) > 0$ and $q(x) \geq 0$, we have $R[u] \geq 0$ for any eigenfunction $u(x)$, the corresponding eigenvalue must be nonnegative. $\qquad\Box$

Remark 4. $\{\lambda_0 = 0,\ u_0(x) = 1\}$ is an eigen-pair for the eigenvalue problem $u'' + \lambda^2 u = 0$, $u'(0) = u'(1) = 0$. This is consistent with the relation $\lambda^2 = R[u]$ since $q(x) \equiv 0$ for the given ODE.

Corollary 9. *If the more restrictive homogeneous Dirichlet condition (7.4.5) is prescribed at one end point, then $\lambda^2 = R[u] > 0$ so that all eigenvalues of such regular S–L problems are positive.*

Proof. (left as an exercise) $\qquad\Box$

The following proposition is an application of the calculus of variations (see [70, 72]).

Proposition 18. *If $\hat{u}(x)$ is an extremal of $R[u]$ and satisfies a Dirichlet condition (7.2.4) at both end points a and b, then $\hat{u}(x)$ is a solution of the Sturm–Liouville problem (with Dirichlet condition (7.2.4)) for $\lambda^2 = N[u]/D[u]$.*

Proof. With $\delta R = (D\delta N - N\delta D)/D^2 = (\delta N - \lambda^2 \delta D)/D$ where Theorem 17 is used to set $N/D = \lambda^2$, we obtain

$$\delta R = \frac{2}{D} \left\{ \int_a^b p(x)\hat{u}'\delta u' dx + \int_a^b \{q(x) - \lambda^2 r(x)\}\hat{u}\delta u dx \right\}$$

$$= \frac{2}{D} [p(x)\hat{u}'\delta u]_{x=a}^b - \frac{2}{D} \int_a^b \{(p\hat{u}')' + (\lambda^2 r - q)\hat{u}\}\delta u dx.$$

With $\hat{u}(a)$ and $\hat{u}(b)$ as prescribed by (7.4.5) in order for $\hat{u}(x)$ to be an admissible comparison function, we have $\delta u(a) = \delta u(b) = 0$. The requirement $\delta R = 0$ for a stationary value of R leads to the Euler DE

$$(p\hat{u}')' + (\lambda^2 r - q)\hat{u} = 0$$

which is the Sturm–Liouville ODE (7.4.2). $\qquad\Box$

For many Sturm–Liouville problem for which there is not an explicit exact solution, the Rayleigh quotient formulation enables us to obtain accurate approximate solution for the eigen-pairs. To illustrate, consider the simple eigenvalue problem

$$u'' + \lambda^2 u = 0, \quad u(0) = u(1) = 0. \tag{7.5.2}$$

The C^2 comparison function $u_1(x) = x(1-x)$ satisfies both end conditions for u. The corresponding $R[u_1]$ is

$$R[u_1] = \frac{\int_0^1 (1-2x)^2 dx}{\int_0^1 x^2(1-x)^2 dx} = 10$$

(and there is no freedom in $u_1(x)$ for possible extremization). The result is very close to the exact solution $\lambda_1^2 = \pi^2 (= 9.8696\ldots)$. If we try to improve on the approximation by taking $u_2 = u_1 + cx^2(1-x)$, then we have

$$R[u_2] = \frac{\int_0^1 [1 - 2x + cx(2 - 3x)]^2 dx}{\int_0^1 [x(1-x) + cx^2(1-x)]^2 dx}$$

$$= \frac{(5 + 5c + 2c^2)/15}{(7 + 7c + 2c^2)/210} = \frac{14(5 + 5c + 2c^2)}{7 + 7c + 2c^2} = R(c).$$

For a stationary value of R, we require $dR(c)/dc = 4c(2+c)/(7+7c+2c^2)^2 = 0$. This shows that $R(c)$ attains its stationary value at $c = 0$ and at $c = -2$, with $R(0) = 10$ as the minimum again, now relative to the larger class of admissible comparison functions. ($R(-2) = 42$ is a maximum and, by the theory to be developed, is not relevant in estimating the smallest eigenvalue.) If we use a more symmetric function $\tilde{u}_2 = u_1 + cx^2(1-x)^2$, the minimum $R[\tilde{u}_2]$ would be $9.8697496\ldots$ which is accurate to 0.001%!

7.5.2   *Existence of a discrete spectrum of eigenvalues*

In all the theorems established in the previous subsections, we had to assume that one or more eigen-pairs exist. While they do exist for the eigenvalue problem associated with the heat conduction problem, we have yet to prove that they exist for a general regular S–L problem. The following theorem on the existence of an infinite sequence of eigenvalues can in fact be proved:

Theorem 18. *For a regular Sturm–Liouville problem with the Dirichlet end conditions (7.4.5), there exists an infinite sequence of positive simple eigenvalues $\{\lambda_m^2\}$ that can be ordered as $0 < \lambda_1^2 < \lambda_2^2 < \lambda_3^2 < \cdots$ with the sequence increasing without bound.*

While we will not prove the theorem here, it should be noted that we have already proved in the previous section the simplicity and positiveness of the eigenvalues (since Theorem 18 is for the restricted regular S–L problem with at least one

Dirichlet condition at an end). Hence, the eigenvalues, if they exist, should form an increasing sequence. It remains to prove that there exists a countably infinite number of eigenvalues and that λ_n^2 tends to ∞ as $n \to \infty$. The most difficult part of the proof is proving the existence of at least one eigen-pair. This is accomplished by proving that there exists a minimizer for $R[u]$. (For a proof of this fact, see Chapter 12 of [78].)

We can (but will not) prove also that an extremal $u_1(x)$ of $R[u]$ actually minimizes the Rayleigh quotient and $R[u_1]$ corresponds to the minimum eigenvalue λ_1^2. We state this as the following theorem:

Theorem 19. *The smallest eigenvalue λ_1^2 is the minimum value of $R[u]$ over all admissible C^2 comparison functions $u(x)$.*

Having found an approximate solution for the smallest eigenvalue with the help of the Rayleigh Quotient, the key to finding the other eigenvalues approximately is the orthogonality condition of Theorem 16, requiring two different eigenfunctions for the general Sturm–Liouville problem to satisfy (7.4.11). With this requirement, we can state (without proof) the following theorem which can be used to determine the second eigenvalue (usually approximately):

Proposition 19. *The second eigenvalue is determined by minimizing the functional $R[u]$ subject to the admissible comparison functions that are orthogonal to the first eigenfunction.*

Proof. (omitted) □

To illustrate this solution process, we consider again the simple Sturm–Liouville problem (7.5.2). Let $u_1(x)$ be the exact eigenfunction $\sin(\pi x)$ and take $u_2(x) = x(1 - x)(c - x)$. To get the second eigenvalue λ_2^2 by $R[u_2]$, we need to have u_2 orthogonal to u_1:

$$\int_0^1 u_2(x)u_1(x)dx = \int_0^1 x(1 - x)(c - x)\sin(\pi x)dx = 0$$

or

$$c\int_0^1 x(1 - x)\sin(\pi x)dx = \int_0^1 x^2(1 - x)\sin(\pi x)dx$$

which determines an appropriate constant $c = c^*$ with $u_2^*(x) = x(1 - x)(c^* - x)$. The corresponding approximate value for λ_2^2 is then given by $R[u_2^*]$.

As we have seen previously, an eigenfunction is determined up to a multiplicative factor. It is sometime convenient to choose the available constant to normalize the associated eigenfunction by

$$C[u] = \int_a^b r(x)u^2 dx = 1. \tag{7.5.3}$$

In that case, $R[u]$ becomes

$$J_1[u] = \int_a^b \left\{ p(x)\left[u'\right]^2 + q(x)u^2 \right\} dx \tag{7.5.4}$$

with the $u(x)$ normalized as in (7.5.3). If (7.5.3) is suppressed, the *Euler DE* of $J_1[u]$ as given in (7.5.4) generally has only one extremal: $\hat{u}(x) = 0$. This can be seen from cases when $p(x) = p_0 > 0$ and $q(x) = q_0 > 0$. To account for the constraint (7.5.3), we may use the method of Lagrange multipliers (see [72]) by working with the modified performance index

$$I[u] = J_1[u] + \lambda^2\{1 - C[u]\}$$
$$= \int_a^b \left\{ p(x)\left[u'\right]^2 + \left[q(x) - \lambda^2 r(x)\right] u^2 \right\} dx + \lambda^2$$

with $\delta I[u] = 0$ leading to (7.4.2) as the Euler DE. In this approach, the eigenvalue corresponds to the Lagrange multiplier which together with the extremal $\hat{u}(x)$ renders the performance index $J_1[u] - \lambda^2 C[u]$ stationary.

Part 3
Continuous State Random Variables

Chapter 8

Continuous Sample Space Probability

8.1 Random Variables

8.1.1 *Random variables and probability density functions*

Elementary probability theory limits discussion to experiments with finite and countable sample spaces. The actual elementary events involved may be numerical in nature (such as rolling a die) or non-numerical (such as flipping a coin). As pointed out previously, even non-numerical elementary events may be assigned numerical labels for the purpose of mathematical analysis. An appropriate assignment may depend on the nature of the quantitative analysis. When elementary events are numerical or numerically labelled, they are specific realizations of a **random variable** X for the sample space.

For an experiment that picks out a real number at random, the random variable X for this experiment may assume any number on the real line. Another experiment that measures the arc length between the point $(1, 0)$ in the (x, y) plane and a randomly selected point on the unit circle centered at the origin would have a sample space of elementary events consisting of all real numbers in the interval $[0, \pi]$. For these and other experiments, we need to extend the range of sample space for a random variable X to the real line, or an entire segment thereof. Random variables with an interval of the real line as their sample space are known as continuous (state) random variables.

As we often have to investigate the interaction among several such continuous random variables, the relevant analyses often become more compact and efficient by treating the interacting random variables as components of a vector random variable $\mathbf{X} = (X_1, X_2, \ldots, X_N)$. With the sample space of each component being the real line or a part thereof, the sample space of the vector random variable is then a part of or the entire *N-dimensional Euclidean space*.

One experiment with the real line as its sample space is the one dimensional random walk problem considered in the last chapter but with the limit of the step size tending to zero. For this limiting case, it would not be appropriate to assign a finite probability to each of the elementary events, i.e., to each point on the real line; otherwise, the sum of the probabilities for all elementary events would be unbounded in general. A theory for continuous sample space problems focuses instead on the probability of occurrence of a range of elementary events. In particular, we assign a probability $P(X \leq A) \equiv P_X(A)$ for the random variable X to assume all values less than or equal to A. (To be precise, $P_X(A)$ really means $P(-\infty \leq X \leq A)$; but we usually abbreviate it as $P(X \leq A)$ with the "$\geq -\infty$" part tacitly understood.) With all real numbers being elementary events in our sample space, we postulate

(1) $P(X \leq -\infty) = 0$
(2) $P(X \leq \infty) = 1$
(3) $P(y \leq X \leq z) = P(X \leq z) - P(X \leq y)$ (y and z being real numbers).

For most applications, we are usually interested in the **probability distribution function** $P_X(x) = P(X \leq x)$ since we want to know how the phenomenon being investigated varies with x. However for the purpose of analysis of the mathematical models, it is often more helpful to introduce and work with a **probability density function** $p_X(x)$, conventionally abbreviated as **pdf**, defined by

$$P_X(z) = \int_{-\infty}^{z} p_X(x)dx, \tag{8.1.1}$$

with

$$\frac{dP_X(z)}{dz} = p_X(z). \tag{8.1.2}$$

Sometimes, the subscript X in $p_X(x)$ is omitted when there is no other random variable under consideration. Evidently, properties (1) and (3) above are automatically satisfied by the definition of of the distribution function in terms of the integral of **pdf**. Property 2 imposes a constraint on $p_X(x)$:

$$P_X(\infty) = \int_{-\infty}^{\infty} p_X(x)dx = 1. \tag{8.1.3}$$

The probability theory for continuous random variables contains the theory for discrete and finite sample spaces as special cases. For example, in the experiment of rolling a die with the conventional sample space of six discrete elementary events of $\{1, 2, 3, 4, 5, 6\}$, the probability density function of an unbiased die would be

$$p(x) = \sum_{n=1}^{6} \frac{1}{6}\delta(x - n)$$

where $\delta(x-x^*)$ is the *Dirac delta function* characterized by the following properties:

$$(i) \quad \delta(x-x^*) = 0 \quad \text{for all } x \neq x^*,$$

$$(ii) \quad \int_{-\infty}^{\infty} \delta(x-x^*)dx = \int_{x^*-\epsilon}^{x^*+\epsilon} \delta(x-x^*)dx = 1,$$

$$(iii) \quad \int_{-\infty}^{\infty} f(x)\delta(x-x^*)dx = \int_{x^*-\epsilon}^{x^*+\epsilon} f(x)\delta(x-x^*)dx = f(x^*),$$

for any $f(x)$ continuous at x^*. In that case, we have for the fair die

$$P(X \leq 4) = \int_{-\infty}^{4+\epsilon} p(x)dx = \int_{-\infty}^{4+\epsilon} \frac{1}{6} \sum_{n=1}^{6} \delta(x-n)dx = \frac{2}{3}$$

and

$$P(X = 4) = P(4 - \epsilon \leq X \leq 4 + \epsilon) = \int_{4-\epsilon}^{4+\epsilon} \frac{1}{6} \sum_{n=1}^{6} \delta(x-n)dx = \frac{1}{6}.$$

8.1.2 Moments and characteristic functions

Given an (integrable) probability density function $p(x)$ for the random variable X, we introduce moments of X by the following definitions:

Definition 13. The **expectation** of a random variable X with a probability density function $p(x)$ is given by

$$E[X] = \int_{-\infty}^{\infty} xp(x)dx \equiv \mu. \tag{8.1.4}$$

For random variables with a discrete and finite sample space, the expression (8.1.4) reduces to the previous definition for that special case in (1.2.1) of Chapter 1. $E[X]$ is also known as the (statistical) *mean* or (statistical) *average* of the random variable X.

Definition 14. The nth moment of a random variable X with a probability density function $p(x)$ is

$$E[X^n] = \int_{-\infty}^{\infty} x^n p(x)dx \equiv \mu_n.$$

Evidently, μ_1 is the expected value μ of X.

Definition 15. The **variance** of a random variable X with a probability density function $p(x)$ is

$$Var[X] = \int_{-\infty}^{\infty} (x-\mu)^2 p(x)dx.$$

Proposition 20. $\qquad Var[X] = E[X^2] - E[X]^2 = \mu_2 - \mu^2.$

Proof. (exercise) $\qquad\qquad\qquad\qquad\qquad\qquad\qquad\qquad\qquad$ □

Just as μ is often used to denote $E[X]$, σ^2 is often used to denote $Var[X]$.

Corollary 10. $\qquad\qquad \mu_2 \geq \mu^2$

Proof. (exercise) $\qquad\qquad\qquad\qquad\qquad\qquad\qquad\qquad\qquad$ □

The nth moment of a random variable X is actually a special case of a more general moment of X. Let $g(\cdot)$ be a continuous and continuously differentiable function on the real line.

Definition 16. The expectation of $g(X)$, $E[g(X)]$, is defined to be

$$E[g(X)] = \int_{-\infty}^{\infty} g(x)p(x)dx.$$

The particular function $g(x) = e^{iux}$ where $i = \sqrt{-1}$ is the imaginary unit and u is a parameter that may assume any real value, is particularly significant in the theory of continuous probability. The expectation of e^{iux} is known as the *characteristic function* of the random variable X:

Definition 17. The characteristic function of a random variable X with a probability density function $p(x)$ is

$$\hat{p}(u) = E[e^{iuX}] = \int_{-\infty}^{\infty} e^{iux}p(x)dx. \qquad (8.1.5)$$

Evidently, $\hat{p}(u)$ is just the Fourier transform of $p(x)$. When $p(x)$ is absolutely integrable, we have from the theory of Fourier transforms the following inversion formula for the transform pair:

$$p(x) = \frac{1}{2\pi} \int_{-\infty}^{\infty} e^{-iux}\hat{p}(u)du. \qquad (8.1.6)$$

Proposition 21. *With* $\hat{p}^{(k)}(0) = \left[d^k\left\{\hat{p}(u)\right\}/du^k\right]_{u=0}$, *the characteristic function of the random variable* X *is related to the moments* $\{\mu_k\}$ *of the same random variable by*

$$\hat{p}(u) = \sum_{k=0}^{\infty} \frac{1}{k!}\hat{p}^{(k)}(0)u^k = \sum_{k=0}^{\infty} \frac{i^k\mu_k}{k!}u^k. \qquad (8.1.7)$$

Proof. Use the Taylor series for e^{iux} about the origin to write

$$\hat{p}(u) = \sum_{n=0}^{\infty} \frac{(iu)^n}{n!} \int_{-\infty}^{\infty} x^n p(x) dx = \sum_{n=0}^{\infty} \frac{(iu)^n}{n!} \mu_n. \qquad \square$$

8.1.3 *Some probability density functions*

Unlike experiments with a finite sample space, the probability distribution and pdf of continuous random variables are not readily known or estimated. Below are two most frequently encountered probability density functions and some of their elementary properties. A few others will be introduced through exercises.

8.1.3.1 *Normal (Gaussian) distribution*

The normal (or Gaussian) probability density function for the random variable X is defined by

$$p(x) = \frac{1}{\sqrt{2\pi}\sigma} e^{-(x-\mu)^2/2\sigma^2} \qquad (8.1.8)$$

where the two real-valued parameters μ and σ^2 will be shown to be the mean and variance of X (denoted by $X \sim N(\mu, \sigma^2)$) among the exercises below.

Proposition 22. *For $A > \mu$, we have*

$$P(X \leq A) = \int_{-\infty}^{A} p(x) dx = \frac{1}{2} + \int_{0}^{A} p(x) dx = \frac{1}{2} + \mathrm{erf}\left(\frac{A - \mu}{\sigma}\right) \qquad (8.1.9)$$

and

$$P(X \leq \infty) = 1.$$

Proof. (exercise) $\qquad \square$

Proposition 23. $\qquad E[X] = \mu, \quad and \quad Var[X] = \sigma^2.$

Proof. (exercise) $\qquad \square$

It follows from these

$$(i) \quad \int_{-\infty}^{\infty} x N(0, \sigma^2) dx = 0, \quad (ii) \quad \int_{-\infty}^{\infty} x^2 N(0, 1) dx = 1.$$

The verification of these two properties are also left as an exercise.

Proposition 24. *The characteristic function for the normal pdf* (8.1.8) *is*

$$\hat{p}(u) = E[e^{iuX}] = e^{i\mu u - \sigma^2 u^2}.$$

Proof. A relatively simple proof for $\mu = 0$ would be to expand e^{iux} as a Taylor series about $x = 0$, integrate term by term and then sum up the resulting series. □

A random variable X is said to be normally distributed with mean μ and variance σ^2 if it has a normal distribution (8.1.9) (and a pdf (8.1.8)).

8.1.3.2 *Poisson's distribution*

The Poisson's probability density function is defined by

$$p(x) = \sum_{k=0}^{\infty} \frac{\lambda^k e^{-\lambda}}{k!} \delta(x - k)$$

where λ is a real-valued parameter. Below are some its properties whose proofs are left as exercises:

Proposition 25. *For $N < A < N + 1$, we have*

$$P(X \leq A) = \int_{-\infty}^{A} p(x)dx = e^{-\lambda} \sum_{k=0}^{N} \frac{\lambda^k}{k!} \quad \text{with } P(X \leq \infty) = 1.$$

Proof. (exercise) □

Proposition 26. $E[X] = \lambda, \quad and \quad E[X^2] = \lambda^2 + \lambda.$

Proof. (exercise) □

The following related result follows from the proposition above:

Corollary 11. $Var[X] = \lambda.$

Proof. (exercise). □

Remark 5. The development above effectively rewrites the Poisson distribution derived early for countable sample spaces in terms of a pdf normally appropriate for continuous random variables.

8.2 Multivariate Random Variables

8.2.1 *Two random variables*

We start with two random variables X and Y and their non-negative *joint density function* $p(x, y)$ defined in the extended plane $\{|x| \leq \infty, |y| \leq \infty\}$. An example is the following joint Gaussian (normal) density function

$$p(x,y) = \frac{1}{2\pi\sigma_x\sigma_y\sqrt{1-r^2}}e^{-\frac{1}{2(1-r^2)}\left[\left(\frac{\xi}{\sigma_x}\right)^2 - 2r\frac{\xi}{\sigma_x}\frac{\eta}{\sigma_y} + \left(\frac{\eta}{\sigma_y}\right)^2\right]} \qquad (8.2.1)$$

with

$$\xi = x - \mu_x, \quad \eta = y - \mu_y, \quad \mu_z = E[Z], \quad \sigma_z^2 = Var[Z]$$

where μ_z and σ_z^2 are the *mean* and *variance* of the random variable Z and the parameter r, $-1 \le r \le 1$, is known as the *correlation coefficient*.

The corresponding *joint distribution function* is defined in terms of $p(x,y)$ by

$$P(X \le A, Y \le B) = \int_{-\infty}^{A}\int_{-\infty}^{B} p(x,y)dydx. \qquad (8.2.2)$$

Evidently, we have the following elementary properties as consequences of the definition:

i) $P(X \le -\infty, Y \le B) = P(X \le A, Y \le -\infty) = 0.$
ii) $P(X \le \infty, Y \le \infty) = 1.$
iii) $P(A_1 \le X \le A_2, Y \le B) = P(X \le A_2, Y \le B) - P(X \le A_1, Y \le B)$ and, similarly, $P(X \le A, B_1 \le Y \le B_2) = P(X \le A, Y \le B_2) - P(X \le A, Y \le B_1).$
iv) $p(x,y) = \dfrac{\partial^2 P(X \le x, Y \le y)}{\partial x \partial y}.$

The two quantities $P(X \le \infty, Y \le B)$ and $P(X \le A, Y \le \infty)$ are known as *marginal distributions*. It is seen from the domains of these functions that they cover all the elementary events in the half plane $Y \le B$ and $X \le A$, respectively. It follows that

$$P(X \le \infty, \ Y \le B) = P(Y \le B),$$

$$P(X \le A, \ Y \le \infty) = P(X \le A),$$

as well as

$$p(x) = \int_{-\infty}^{\infty} p(x,y)dy = \frac{\partial P(X \le x, Y \le \infty)}{\partial x},$$

$$p(y) = \int_{-\infty}^{\infty} p(x,y)dx = \frac{\partial P(X \le \infty, Y \le y)}{\partial y},$$

where $p(x)$ and $p(y)$ are known as marginal densities. For example, the marginal densities for the joint Gaussian density function (8.2.1) are

$$p(x) = \frac{1}{\sqrt{2\pi}\sigma_x}e^{-\frac{1}{2}\left(\frac{\xi}{\sigma_x}\right)^2}, \quad p(y) = \frac{1}{\sqrt{2\pi}\sigma_y}e^{-\frac{1}{2}\left(\frac{\eta}{\sigma_y}\right)^2}.$$

The following useful result follows from the definition (8.2.2):

Proposition 27.

$$P(A_1 \leq X \leq A_2, B_1 \leq Y \leq B_2) = P(X \leq A_2, Y \leq B_2)$$
$$-P(X \leq A_1, Y \leq B_2) - P(X \leq A_2, Y \leq B_1) + P(X \leq A_1, Y \leq B_1).$$

A more general version of $P(A_1 \leq X \leq A_2, B_1 \leq Y \leq B_2)$ is

$$P((X,Y)\varepsilon D) = \int \int_D p(x,y)dxdy$$

$$= \text{probability of occurrence of all } (X,Y) \text{ in } D$$

where D is some region in the x, y–plane.

Two random variables X and Y are (*statistically*) *independent* if

$$p(x, y) = p(x)p(y)$$

and correspondingly

$$P(X \leq A, Y \leq B) = P(X \leq A)P(Y \leq B).$$

For example, the random variables X and Y with the joint Gaussian density function (8.2.1) are independent if $r = 0$.

8.2.2   *Expectation, covariance and moments*

As in the single random variable case, we are interested in expected values of functions of several random variables.

Definition 18. Let X and Y be random variables with joint density function $p(x, y)$. Their joint m, n–moment is defined as

$$E[X^m Y^n] = \int_{-\infty}^{\infty} \int_{-\infty}^{\infty} x^m y^n p(x,y)dxdy \equiv \mu_{mn}. \qquad (8.2.3)$$

The special cases of variances and covariance of X and Y are particularly useful in subsequent development. These are given by

$$Var[X] = E[(X - \mu_x)^2] = \int_{-\infty}^{\infty} \int_{-\infty}^{\infty} (x - \mu_x)^2 p(x,y)dxdy \equiv \sigma_{xx},$$

$$Var[Y] = E[(Y - \mu_y)^2] = \int_{-\infty}^{\infty} \int_{-\infty}^{\infty} (y - \mu_y)^2 p(x,y)dxdy \equiv \sigma_{yy},$$

and

$$CoVar[X, Y] = E[(X - \mu_x)(Y - \mu_y)] = \int_{-\infty}^{\infty} \int_{-\infty}^{\infty} (x - \mu_x)(y - \mu_y)p(x,y)dxdy$$

$$= E[XY] - \mu_x \mu_y \equiv \sigma_{xy}.$$

8.2.3   *Multivariate density and distribution*

For a set of random variables $(X_1, X_2, \ldots, X_N)^T \equiv \mathbf{X}$ with a joint density $p(\mathbf{x})$, we may form the variances and covariances σ_{ij} for $i, j = 1, 2, \ldots, N$ and set

$$[\sigma_{ij}] = [E[(X_i - \mu_i)(X_j - \mu_j)]] \equiv S \quad (i, j = 1, 2, \ldots, N).$$

Note that the covariance matrix S is symmetric.

A square matrix A is *positive semi-definite* if $\mathbf{v}^T A \mathbf{v} \geq 0$ for all real-valued non-zero vector $\mathbf{v}$. It is *positive definite* if $\mathbf{v}^T A \mathbf{v} > 0$.

Theorem 20. *The covariance matrix S is positive semi-definite.*

Proof. Let $\mathbf{v} = (v_1, v_2, \ldots, v_N)^T$ and $Z = \sum_{i=1}^{N} v_i(X_i - \mu_i)$. Evidently, we have $Z^2 \geq 0$ and therewith $E[Z^2] \geq 0$. However, we also have

$$E[Z^2] = \sum_{k=1}^{N} \sum_{j=1}^{N} v_j E[(X_j - \mu_j)(X_k - \mu_k)] v_k$$

$$= \sum_{k=1}^{N} \sum_{j=1}^{N} v_j \sigma_{jk} v_k = \mathbf{v}^T S \mathbf{v}.$$

It follows that $\mathbf{v}^T S \mathbf{v} \geq 0$ so that S is positive semi-definite. $\qquad \square$

The (m, n)–moment $E[X^m Y^n]$ (see (8.2.3)) is a special case of a more general expectation $E[g(X, Y)]$ for a function of several random variables. The latter may be defined similarly:

Let $\mathbf{X}$ be an n–vector random variable with joint density function $p(\mathbf{x})$ and the function $\mathbf{y} = g(\mathbf{x})$ is PWS in all component variables. The expectation of $g(\mathbf{X})$ is defined to be

$$E[g(\mathbf{X})] = \int_{-\infty}^{\infty} \cdots \int_{-\infty}^{\infty} g(\mathbf{x}) p(\mathbf{x}) dx_1 \cdots dx_n. \tag{8.2.4}$$

Similar to the scalar case discussed previously, the *characteristic function* of the vector random variable $\mathbf{X}$ is a special case of (8.2.4) with

$$\hat{p}(\mathbf{u}) = E[e^{i\mathbf{u} \cdot \mathbf{X}}] = \int_{-\infty}^{\infty} \cdots \int_{-\infty}^{\infty} e^{i\mathbf{u} \cdot \mathbf{x}} p(\mathbf{x}) dx_1 \cdots dx_n. \tag{8.2.5}$$

Evidently, $\hat{p}(\mathbf{u})$ is the Fourier transform of $p(\mathbf{x})$ with the following inversion formula

$$p(\mathbf{x}) = \frac{1}{(2\pi)^n} \int_{-\infty}^{\infty} \cdots \int_{-\infty}^{\infty} e^{-i\mathbf{u} \cdot \mathbf{x}} \hat{p}(\mathbf{u}) du_1 \cdots du_n, \tag{8.2.6}$$

to the extent that $p(\mathbf{x})$ is absolutely integrable.

The generalization of the Gaussian (normal) distribution (8.1.8) and (8.2.1) to the general multivariate case is

$$p(\mathbf{x}) = \frac{1}{\sqrt{2\pi \det(\Sigma)}} e^{-\frac{1}{2}(\mathbf{x}-\mu)^T \Sigma^{-1}(\mathbf{x}-\mu)} \qquad (8.2.7)$$

where μ is the mean vector of $\mathbf{x}$ and Σ is its covariance matrix. The corresponding characteristic function is

$$\hat{p}(\mathbf{u}) = E[e^{i\mathbf{u}\cdot\mathbf{X}}] = e^{\mu^T \mathbf{u} - \frac{1}{2}\mathbf{u}^T \Sigma \mathbf{u}}. \qquad (8.2.8)$$

The density characterization (8.2.7) assumes the existence of the inverse Σ^{-1}. When it does not, then Σ^{-1} and $\det(\Sigma)$ are to be taken as the generalized inverse Σ^+ and pseudo determinant Det*, respectively. On the other hand, the characteristic function representation of the same normal density function remains valid as it stands. Hence, it is often preferable to work with the latter whenever possible.

8.3 Mean Square Convergence

8.3.1 *Metric space of random variables*

As n increases, the sequence of numbers $\{1, 1/2, 1/3, \ldots, 1/n, \ldots\}$ clearly approaches zero. We say the sequence $\{x_n = 1/n\}$ tends to 0 as n tends to infinity:

$$\lim_{n \to \infty} [x_n] = 0.$$

For all $|x| < \infty$, the sequence of functions $\{f_n(x) = x^n e^{-nx^2}\}$ also tends to the zero function as n tends to infinity:

$$\lim_{n \to \infty} [f_n(x)] = 0.$$

When $\{X_n\}$ is a sequence of random variables, we also would like to ask whether it converges to something, a number or another random variable. For an answer, we need to phrase the question in the form of convergence of sequences of numbers or functions for which we have well developed theories. There are many ways we can do the conversion. Here, we will limit our discussion to one type of convergence known as *convergence in the mean square* (often abbreviated as "mean square convergence" or "limits in the mean (l.i.m.)"). For this purpose, we consider only random variables for which at least their first and second moments exist and are bounded. Such random variables are referred to as *second order random variables*.

The collection of 2nd order random variables forms a vector space since for any real number c and any two 2nd order random variables X and Y, the quantities cX and $X + Y$ are also 2nd order random variables. It can be verified that all other properties of a vector space are also satisfied. Like physical vectors, we also would like to have a way of measuring the magnitude of the random variables, the elements of this new vector space. For convergence in the mean square, we take the magnitude of a scalar random variable X, denoted by $\|X\|$, to be

$$\|X\| = \sqrt{\int_{-\infty}^{\infty} x^2 p(x) dx} = \sqrt{E[X^2]} \geq 0,$$

where $p(x)$ is the probability density function of X. It is also known as the (mean square) **norm** of X. It can be shown that the definition satisfies the requirements of a norm:

- $\|X\| \geq 0$ and $\|X\| = 0$ if and only if (iff) $X = 0$ with probability 1.
- $\|cX\| = |c| \, \|X\|$ for and real number c.
- $\|X + Y\| \leq \|X\| + \|Y\|$.

Verification of the first two properties is straightforward. The proof of the third, known as *triangular inequality*, requires a mean square version of the *Schwarz inequality*:

Lemma 8. *Suppose that X and Y are two second order random variables. Then their joint moment satisfies (the mean square version of) Schwarz's inequality:*

$$E[|XY|]^2 \leq E[X^2] E[Y^2].$$

Proof. $Z = (|X| - c|Y|)^2$ is a second order random variable and is nonnegative for any real number c. In that case, $E[Z]$ is also nonnegative. But

$$E[Z] = E[(|X| - c|Y|)^2] = E[Y^2]c^2 - 2E[|X| \, |Y|]c + E[X^2]$$

$$= (c, 1) \begin{bmatrix} E[Y^2] & -E[|XY|] \\ -E[|XY|] & E[X^2] \end{bmatrix} \begin{pmatrix} c \\ 1 \end{pmatrix} \equiv (c, 1) U \begin{pmatrix} c \\ 1 \end{pmatrix}.$$

The quadratic function of the real constant c corresponding to $E[Z]$ can only be nonnegative *if and only if* (iff) the matrix U is positive semi-definite. Since U is positive semi-definite iff the determinant (and the principal subdeterminants) of U are all non-negative. We have therefore

$$E[|XY|]^2 \leq E[X^2] E[Y^2]$$

which is the inequality we set out to prove. $\qquad\square$

We now use Schwarz's inequality to prove the triangular inequality for the mean square norm. The proof amounts to writing out $\|X + Y\|^2$ and applying the Schwarz inequality to get (with $E[XY] \leq E[|XY|]$)

$$\|X + Y\|^2 = E[(X + Y)^2] = E[X^2] + E[Y^2] + 2E[XY]$$
$$\leq E[X^2] + E[Y^2] + 2\sqrt{E[X^2]E[Y^2]}$$
$$= \left\{ \sqrt{E[X^2]} + \sqrt{E[Y^2]} \right\}^2 = \{\|X\| + \|Y\|\}^2$$

which is just the property claimed.

8.3.2 Limit in the mean square

With a well-defined norm, we can talk about various kinds of convergence:

Definition 19. A sequence of second order random variables $\{X_n\}$ is said to *converge in the mean square* to a real number c iff $\|X_n - c\| \to 0$ as $n \to \infty$, often written as

$$\lim_{n \to \infty} \|X_n - c\| = 0 \quad \text{or} \quad \underset{n \to \infty}{l.i.m.} [X_n] = c.$$

The following example can be found in many text:

Example 16. $X_n = \begin{cases} 1 & \text{with } P(1) = \dfrac{1}{n} \\[2mm] 0 & \text{with } P(0) = 1 - \dfrac{1}{n} \end{cases}$
(with $P(X_n) = 0$ for $X_n \neq 0, 1$).

Since $P(1) \to 0$ as $n \to \infty$, we expect that the sequence $\{X_n\}$ to tend to $c = 0$ (at least in the mean square sense as defined above). To see that this is in fact the case, we apply the definition of convergence in the mean square to get (with $c = 0$)

$$\lim_{n \to \infty} \|X_n - c\| = \lim_{n \to \infty} \|X_n\| = \left[\lim_{n \to \infty} \int_{-\infty}^{\infty} x_n^2 \, p(x) dx \right]^{1/2}$$
$$= \lim_{n \to \infty} \sqrt{1 \cdot \frac{1}{n} + 0 \cdot (1 - \frac{1}{n})} = \lim_{n \to \infty} \sqrt{\frac{1}{n}} = 0$$

as we wanted to show. Note that X_n may be 1 however large n may be. Thus, the sequence $\{X_n\}$ does not converge in the ordinary sense of convergence.

Mean square convergence of a sequence $\{X_n\}$ may also be to another random variable X instead of a number as illustrated by the following example:

Example 17. Let $\{c_n\}$ be a sequence of real numbers converging to a number c, i.e., $\lim_{n \to \infty} [c_n] = c$. Suppose X is a random variable with density function $p(x)$ and $\{X_n\} = \{c_n X\}$ is a sequence of identically distributed random variables. Then the sequence $\{X_n\} \to cX$ in the mean square as $n \to \infty$.

This is seen from

$$\lim_{n \to \infty} \|X_n - cX\| = \lim_{n \to \infty} E[(c_n X - cX)^2\} = \lim_{n \to \infty} E[(c_n - c)^2 X^2\}$$
$$= \lim_{n \to \infty} (c_n - c)^2 E[X^2] = 0.$$

In general, there are some additional requirements or restrictions in order for mean square convergence to another random variable to make sense. In the example above, the limiting random variable X has the same density function as the X_n's, i.e., all the random variables involved are i.i.d. (independent and identically distributed). If this is not so for another set of random variables, then we would need to know the joint density function before we can compute $\|X_n - X\|^2 = E[(X_n - X)^2]$. Assuming these requirements and/or restrictions are met, we have the following more general definition of mean square convergence:

Definition 20. A sequence of second order random variables $\{X_n\}$ is said to *converge in the mean square* to another random variable X iff $\|X_n - X\| \to 0$ as $n \to \infty$, often abbreviated as

$$\lim_{n \to \infty} \|X_n - X\| = 0 \quad \text{or} \quad \underset{n \to \infty}{l.i.m.} [X_n] = X,$$

where

$$\|X_n - X\|^2 = \int_{-\infty}^{\infty} \int_{-\infty}^{\infty} (x_n - x)^2 p(x_n, x) dx_n dx$$

with

$$\|X_n - c\|^2 = \int_{-\infty}^{\infty} (x_n - c)^2 p(x_n) dx_n$$

if $X = c$ is a constant.

The notation $\underset{n \to \infty}{l.i.m.}$ for mean square convergence is also known as *limit in the mean (square)*. Note that if $\{X_n\}$ and X are i.i.d. with density $p(\cdot)$, then

$$\|X_n - X\|^2 = \int_{-\infty}^{\infty} \int_{-\infty}^{\infty} (x_n - x)^2 p(x_n) p(x) dx_n dx.$$

Proposition 28. *If* $\underset{n \to \infty}{l.i.m.} \|X_n - X\| = 0$ *and* $\underset{n \to \infty}{l.i.m.} \|Y_n - Y\| = 0$, *then, for any constants a and b,*

$$\lim_{n \to \infty} \|(aX_n + bY_n) - (aX + bY)\| = 0.$$

Proof. (exercise) □

Proposition 29. *If a sequence of random variables is mean square convergent, then its mean square limit is unique.*

Proof. (exercise) □

8.3.3   *Mean square Cauchy sequences*

Often and especially for convergence to a random variable, we do not know the limiting random variable (or constant). To decide on the mean square convergence of such sequences, we have the equivalent of Cauchy sequence for random variables:

Definition 21. A sequence of second order random variables $\{X_n\}$ is a Cauchy sequence in the mean square (often abbreviated as mean square Cauchy) if $\|X_k - X_m\| \to 0$ as $k, m \to \infty$ in any manner whatsoever.

Theorem 21. *A sequence of random variables $\{X_n\}$ converges to X if and only if the sequence is mean square Cauchy.*

Proof. (similar to the proof in [44] for numerical sequences)   □

The following example can be found in many text on continuous probability.

Example 18. $X_n = \begin{cases} n & \text{with } P(n) = \dfrac{1}{n^2} \\ 0 & \text{with } P(0) = 1 - \dfrac{1}{n^2} \end{cases}$

Since we do not have a good idea about a limiting random variable or a constant (if there is one at all), we apply Proposition 21 by forming

$$\|X_m - X_k\|^2 = E[(X_m - X_k)^2] = E[X_m^2 - 2X_mX_k + X_k^2]$$
$$= \left[\frac{m^2}{m^2} + 0\right] - 2\left[\frac{m}{m^2}\frac{k}{k^2}\right] + \left[\frac{k^2}{k^2} + 0\right]$$
$$= 2\left(1 - \frac{1}{mk}\right) \to 2 \quad \text{as } m \to \infty \text{ and } k \to \infty.$$

Hence, the sequence does not converge in the mean square.

Example 19. $X_n = \begin{cases} n & \text{with } P(n) = \dfrac{1}{n^3} \\ 0 & \text{with } P(0) = 1 - \dfrac{1}{n^3} \end{cases}$

The situation is quite different here. As long as we are not sure whether the sequence of second order r.v. would converge (in the mean square), we should check to see if the sequence is mean square Cauchy. Calculations similar to the previous example gives

$$\|X_m - X_k\|^2 = E[(X_m - X_k)^2] = E[X_m^2 - 2X_m X_k + X_k^2]$$

$$= \left[\frac{m^2}{m^3} + 0\right] - 2\left[\frac{m}{m^3}\frac{k}{k^3}\right] + \left[\frac{k^2}{k^3} + 0\right]$$

with

$$\lim_{k,m \to \infty} \|X_m - X_k\|^2 = \lim_{k,m \to \infty}\left(\frac{1}{m} + \frac{1}{k} - \frac{2}{m^2 k^2}\right) = 0.$$

Hence, the sequence is mean square Cauchy and converges in the mean square by Theorem 21. With mean square convergence established, we can guess at the limit point of the sequence to be 0 and apply the definition of mean square convergence to validate our guess:

$$l.i.m._{n \to \infty}\{X_m\} = \lim_{m \to \infty} \|X_m - 0\| = \lim_{m \to \infty} \sqrt{\frac{1}{m}} = 0.$$

Example 20. Suppose we try to repeatedly measure and record some available information (such as repeatedly sampling the depolarization voltage of a nerve axon in a refractory stage) which is actually μ. Because of the noisy environment, what is recorded through each sampling is $X_n = \mu + \eta_n$ where the (second order) random noise variables $\{\eta_n\}$ are *i.i.d.* with *zero* mean and variance σ^2 (uncorrelated from sample to sample). Experimentalists typical would average their samples to get rid of the effects of the noise by letting

$$Y_n = \frac{1}{n}\sum_{k=1}^{n} X_k.$$

Theoretically, we have $E[Y_n] = \mu$ and $Var[Y_n] = \sigma^2/n$ (see Proposition 31 in the next section); thus Y_n does appear to average out the noise. The question is: Does the random sequence $\{Y_n\}$ converges in the mean square?

Since we do not know the limiting random variable, the question can only be answered with the possibility of a Cauchy sequence by looking at the sequence $\|Y_m - Y_k\|^2$ with

$$\|Y_m - Y_k\|^2 = E[(Y_m - Y_k)^2] = E[\{(Y_m - \mu) - (Y_k - \mu)\}^2]$$

$$= E[(Y_m - \mu)^2 - 2(Y_m - \mu)(Y_k - \mu) + (Y_k - \mu)^2]$$

$$= \frac{\sigma^2}{m} - 2E[(Y_m - \mu)(Y_k - \mu)] + \frac{\sigma^2}{k}.$$

For $m > k$, we have

$$E[(Y_m - \mu)(Y_k - \mu)] = \int_{-\infty}^{\infty}\int_{-\infty}^{\infty}(y_m - \mu)(y_k - \mu)p(x_m, x_k)dx_m dx_k = 0$$

given the different sampling noises are *i.i.d.* so that $p(x_m, x_k) = p(x_m)p(x_k)$. This leaves

$$\|Y_m - Y_k\|^2 = \frac{\sigma^2}{m} + \frac{\sigma^2}{k} \to 0 \quad \text{as } m \text{ and } k \to \infty.$$

Thus, the sequence $\{Y_n\}$ is mean square Cauchy and therefore converges in the mean square to some random variable Y which is what we want to know.

Exercise 30. Let $\{X_1, X_2, \ldots\}$ be independent random variables with $P[X_i = 1] = P[X_i = -1] = 0.5$. Compute the characteristic function of the random variables X_1, $S_n = X_1 + \cdots + X_n$ and $V_n = S_n/\sqrt{n}$ and decide whether the sequences $\{S_n\}$ and $\{V_n\}$ converge in the mean square.

In the study of stochastic differential equations, one of the most often invoked properties is the commutativity between l.i.m. and another operation. The proposition below establishes the first such commutativity relation: that l.i.m. commutes with the operation of calculating expectation.

Proposition 30. *If a sequence of second order random variables $\{X_n\}$ converges to a second order random variable X, then*

$$\lim_{n \to \infty} E[X_n] = E[X].$$

Proof. By Schwarz's inequality (with $X = X_n$ and $Y = 1$), we have

$$E[X_n] \leq E[|X_n|] \leq \sqrt{E[X_n^2]} = \|X_n\|.$$

Similarly, we have also by Schwarz's inequality

$$|E[X_n] - E[X]| = \left| \int_{-\infty}^{\infty} \int_{-\infty}^{\infty} (X_n - X) \, p(x, x_n) dx dx_n \right|$$

$$\leq \int_{-\infty}^{\infty} \int_{-\infty}^{\infty} |X_n - X| \, p(x, x_n) dx dx_n = E[|X_n - X|]$$

$$\leq \|X_n - X\|.$$

The proposition follows from the mean square convergence of $\{X_n\}$ to X (so that $\|X_n - X\| \to 0$ as $n \to \infty$). $\qquad \square$

8.4 Chebyshev Inequality and Sample Size

One topic related to the idea of convergence of random variables is the Chebyshev inequality. Suppose that $\{X_1, X_2, \ldots, X_n\}$ is a sequence n *i.i.d.* random variables with common mean μ and common variance σ^2. Let

$$Y_n = \frac{1}{n} \sum_{k=1}^{n} X_k.$$

Evidently, Y_n is the average of the n random variables $\{X_i\}$ and is itself a random variable. Furthermore, as n increases, we have another new sequence of random variables $\{Y_1, Y_2, \ldots, Y_k \ldots\}$. We can of course investigate the mean square convergence of either sequence. But for a finite n (corresponding to a finite sample of the data from a population), we note the following relations between $\{X_i\}$ and $\{Y_k\}$:

Proposition 31. *Suppose that $\{X_1, X_2, \ldots, X_n\}$ is a sequence n i.i.d. random variables with common mean μ and common variance σ^2 and Y_n is the average of the n X's. Then*

$$\mu_n = E[Y_n] = \mu \quad and \quad \sigma_n^2 = Var[Y_n] = \frac{\sigma^2}{n}.$$

Proof. (left as an exercise keeping in mind that the i.i.d. random variables $\{X_k\}$ are uncorrelated) $\qquad\square$

A particular realization of Y_n, calculated from a particular sample of $\{X_1, X_2, \ldots, X_n\}$, will generally not have $\mu_n = \mu$. We want to know the difference between Y_n and μ. In particular, we may want to know what is the probability of $|Y_n - \mu| < t$ (using t for tolerance instead of the usual ubiquitous ϵ) for a prescribed t value. This question is answered by the Chebyshev inequality:

Theorem 22. *Suppose Y_n is random variable with mean μ_n and variance σ_n^2. Then*

$$P\{|Y_n - \mu_n| \geq t\} \leq \frac{\sigma_n^2}{t^2}, \quad P\{|Y_n - \mu_n| < t\} > 1 - \frac{\sigma_n^2}{t^2}.$$

Proof. Observe that Y_n is also *i.i.d.* with the same density function as X_k so that

$$\sigma_n^2 = \int_{-\infty}^{\infty} (y_n - \mu_n)^2 p(y_n) dy_n$$

$$= \int_{-\infty}^{\mu_n - t} \Delta_n^2 p(y_n) dy_n + \int_{\mu_n + t}^{\infty} \Delta_n^2 p(y_n) dy_n + \int_{\mu_n - t}^{\mu_n + t} \Delta_n^2 p(y_n) dy_n$$

$$= \int_{|y_n - \mu_n| \geq t} (y_n - \mu_n)^2 p(y_n) dy_n + A^2 \qquad (8.4.1)$$

with $\Delta_n = y_n - \mu_n = y_n - \mu$ and

$$A^2 = \int_{\mu_n - t}^{\mu_n + t} (y_n - \mu)^2 p(y_n) dy_n \geq 0,$$

given the integrand is non-negative. From (8.4.1) we obtain

$$\sigma^2 \geq \int_{|y_n - \mu| \geq t} (y_n - \mu)^2 p(y_n) dy_n \geq t^2 \int_{|y_n - \mu| \geq t} p(y_n) dy_n = t^2 P\{|Y_n - \mu|\}$$

and the first inequality follows (while the second is an simple consequence of the first). $\qquad\qquad\qquad\qquad\qquad\qquad\qquad\qquad\qquad\qquad\qquad\qquad\qquad\qquad\qquad\Box$

Example 21. Suppose $\mu_n = \sigma_n^2 = 9$. Use Chebyshev inequality to determine $P\{|Y_n - \mu_n| < 5)$. Compare the result with the corresponding probability if the X's are Poisson distributed with $\lambda \ (= \mu_n = \sigma_n^2) = 9$.

It is straightforward to apply the Chebyshev inequality to get

$$P\{|Y_n - \mu_n| < t) = P\{|Y_n - 9| < 5) = P\{4 < Y_n < 14) \geq 1 - \frac{\sigma_n^2}{t^2} = 1 - \frac{9}{25} = 0.64.$$

The corresponding probability for $P\{4 < Y_n < 14)$ from a Poisson distribution is

$$P\{4 < Y_n < 14) = e^{-9} \sum_{k=5}^{14} \frac{9^k}{k!} = 0.937307\ldots.$$

Evidently the Probability obtained from Chebyshev inequality is quite conservative. On the otherhand, it does not require any knowledge of the underlying density or distribution of the random variables involved.

Example 22. Suppose Y_n is the average of the n i.i.d. random variables $\{X_i\}$ of mean μ and variance σ^2 (such as n samples of a population). Determine the size n of the sample in order for $P\{|Y_n - \mu| < 5) \geq 0.95$.

Given $\mu_n = \mu$ and $\sigma_n^2 = \sigma^2/n$ from Proposition 31, we have

$$P(|Y_n - \mu| < t) = P(|Y_n - \mu| < 5) > 1 - \frac{\sigma_n^2}{5^2} = 1 - \frac{\sigma^2}{25n}.$$

We do not know μ and σ^2 (the mean and variance of the population of interest), but may have some estimate of the variance from data available for the population.

For example, if we know the range of the X_i's (which are *i.i.d.*), say $\alpha \leq X \leq \beta$, we can take the average of the two extreme values to get $\sigma^2 \leq (\beta - \alpha)^2/4$. In that case, we have for $t = 5$

$$P(|Y_n - \mu| < t) \geq 1 - \frac{\sigma_n^2}{t^2} = 1 - \frac{\sigma^2}{nt^2} > 1 - \frac{(\beta - \alpha)^2}{100n}.$$

If we want this probability to be > 0.95, then we take

$$\frac{(\beta - \alpha)^2}{100n} < 0.05 \quad \text{or} \quad n > \frac{(\beta - \alpha)^2}{5}$$

or $n > 80$ if $\beta - \alpha = 20$. In other words, the sample size should be 80 or larger for a better than 95% chance that a sample mean Y_n to be within 5 of the unknown population mean μ.

For another example, if $X_i's$ are *i.i.d.* Bernoulli trial with the proportion of successes equal to p and failures $1 - p$. Then we have $\mu_n = E[Y_n] = p$ and $\sigma_n^2 = p(1 - p)/n$ (given $\sigma_n^2 = (1 - \mu_n)^2 p + (0 - \mu_n)^2 (1 - p) = p(1 - p)$ for the each X_i). In that case,

$$P(|Y_n - p| < t) > 1 - \frac{\sigma_n^2}{t^2} = 1 - \frac{p(1 - p)}{nt^2} \geq 1 - \frac{1}{4nt^2}.$$

For $t = 0.1$ and the desired probability $P(|Y_n - p| < t)$ to be > 0.95, we should take

$$\frac{1}{4nt^2} < 0.05 \quad \text{or} \quad n > \frac{1}{4 \cdot 0.05 \cdot 0.1^2} = 500.$$

8.5 Characteristic Functions and Central Limit Theorem

The characteristic function of a multivariate random variable can also be defined analogous to the single random variable case.

Definition 22. The characteristic function of two random variables X and Y with joint probability density function $p(x, y)$ is

$$\hat{p}(u, v) = E[e^{iuX + ivY}] = \int_{-\infty}^{\infty} \int_{-\infty}^{\infty} e^{i(ux + vy)} p(x, y) dx dy. \tag{8.5.1}$$

Evidently, $\hat{p}(u, v)$ is just the Fourier transform of $p(x, y)$. To the extent that $p(x, y)$ is absolutely integrable, we have the usual inversion formula for the transform which will not be listed here. Instead, we note the important fact that the characteristic function and hence the joint probability density function can be determined from the collection of all joint moments the random variables:

Proposition 32. *Suppose X and Y are two random variables with joint probability density function $p(x, y)$ and the corresponding characteristic function $\hat{p}(u, v)$. Then,*

$$\frac{\partial^{m+k} \hat{p}(u, v)}{\partial u^m \partial v^k} \bigg|_{u=v=0} = i^{m+k} E[X^m Y^k]$$

where $i = \sqrt{-1}$ is the imaginary unit.

Proof. (exercise). $\square$

For more random variables, we use the vector notation to write $(X_1, X_2, \ldots, X_N)^T$ as $\mathbf{X}$ and

$$\hat{p}(\mathbf{u}) = E[e^{i\mathbf{u} \cdot \mathbf{x}}] = \int_{-\infty}^{\infty} \cdots \int_{-\infty}^{\infty} e^{i\mathbf{u} \cdot \mathbf{x}} p(\mathbf{x}) dx_1 \cdots dx_N.$$

The relation between the joint moments and the characteristic function now reads

$$\frac{\partial^m \hat{p}(\mathbf{u})}{\partial u_1^{k_1} \cdots \partial u_N^{k_N}}\bigg|_{\mathbf{u}=0} = i^m E[X_1^{k_1} \cdots X_N^{k_N}], \quad (m = k_1 + \cdots + k_N).$$

Suppose $\{X_1, X_2, \ldots, X_n\}$ is a sequence of mutually *independent and identically distributed (i.i.d.)* random variables with means μ and variances σ^2. Let $X = \sum_{k=1}^n X_k$ and Y be the normalized random variable:

$$Y = \frac{1}{\sqrt{n\sigma^2}}(X - n\mu) = \frac{1}{\sqrt{n\sigma^2}} \sum_{k=1}^n (X_k - \mu).$$

Theorem 23. *The pdf of Y converges to the zero mean, unit variance Gaussian pdf as $n \to \infty$.*

Proof. The characteristic function of Y is given by

$$\hat{p}_Y(u) = E[e^{iuY}] = \left[e^{-iu\mu/\sqrt{n\sigma^2}} \hat{p}(u/\sqrt{n\sigma^2}) \right]^n$$

$$= \left[e^{-iu\mu/\sqrt{n\sigma^2}} \left\{ 1 + \frac{iu\mu}{\sqrt{n\sigma^2}} - \frac{\mu_2}{2!} \left(\frac{u}{\sqrt{n\sigma^2}} \right)^2 + \cdots \right\} \right]^n$$

$$= \left[1 - \frac{u^2}{2n} + o\left(\frac{u^2}{n} \right) \right]^n \quad (\text{as } n \to \infty).$$

where we have recalled the relation $\mu_2 = \sigma^2 + \mu^2$ and where

$$\lim_{z \to 0} \frac{o(z)}{z} = 0.$$

With the $o(u^2/n)$ negligible compared the first two terms in the expansion for $\hat{p}_Y(u)$, we have

$$\hat{p}_Y(u) \sim \left[1 - \frac{u^2}{2n} \right]^n \to e^{-u^2/2} \quad \text{as } n \to \infty,$$

and $e^{-u^2/2}$ is the characteristic function of a zero mean and unit variance Gaussian pdf. □

Before moving on to applications and extensions of the basic continuous probability theory summarized above, we wish to note explicitly our adopted notation of using capital letters for random variables, though not all capitalized quantities are random variables. A matrix is obviously not a random variable while it is often denoted by M. Whether or not a capitalized quantity is a random variable will usually be obvious from the context of the discussion; it will be explicitly noted otherwise.

Chapter 9

Transformations and Stochastic ODE

9.1 Introduction

For a continuous random variable X with a pdf $p_X(x)$, we previously defined and calculated its expected value, variance, higher moments and characteristic function. All of these are special cases of the more general expected value,

$$E[g(X)] = \int_{-\infty}^{\infty} g(x)p_X(x)dx, \qquad (9.1.1)$$

of a function $g(X)$ of the random variable X. As $g(X)$ is also a random variable, the expression (9.1.1) may be viewed in a different way by setting $Y = g(X)$ so that we have $E[g(X)] = E[Y]$ which is just the expectation of the random variable Y. Now $E[Y]$ is known to be given by

$$E[Y] = \int_{-\infty}^{\infty} yp_Y(y)dy \qquad (9.1.2)$$

where $p_Y(y)$ is the pdf of the random variable Y. While it is not given, it should exist for a well-defined $g(X)$ and must be related to $p_X(x)$ in such a way that the value on the right side of (9.1.2) is the same as $E[g(X)]$. To ensure $E[Y] = E[g(X)]$ for any well-defined $g(X)$, we should have the correct relation between $p_Y(y)$ and $p_X(x)$ for $x = g^{-1}(y)$. This determination of the pdf and the corresponding probability distribution for the transformed (scalar or vector) random variable Y is a principal goal of this chapter.

There are other equally compelling reasons why we should be interested in finding $p_Y(y)$ and the corresponding probability distribution $P_Y(y)$ given $Y = g(X)$ and $p_X(x)$ including the more general case where both X and Y are vector random variables. When viewed as a random output Y generated by a processor (machine, factory, etc.) $g(\cdot)$ receiving a random input X, we are interested here mainly in the probabilistic information of the random output, such as its pdf (or the joint pdf of the components in the case of a vector random output). It is expected that such probabilistic information on the output can be derived from the known probabilistic information of the random input. Listed below are three examples for phenomena governed by an ODE system that will be used to illustrate how the transformation theory for $Y = g(X)$ may be useful for learning about output of ODE systems in an uncertain environment.

Example 23. $y' = Cy,$ $y(0) = X,$ $(y' = dy/dt)$

The exact solution of this IVP is

$$y(t) = Xe^{Ct}.$$

When X and/or C are known only probabilistically, $y(t)$ for a fixed t is a random variable Y_t,

$$Y_t = Xe^{Ct} = g(X, C; t),$$

that is a function of two random variables X and C with t as a parameter. Given the joint pdf of the two random variable X and C, we would like to know the pdf of Y_t (for that fixed t).

Example 24. $y' = Cy^2,$ $y(0) = X,$ $(y' = dy/dt)$

The exact solution of this IVP is

$$y(t) = \frac{X}{1 - XCt}.$$

Note that the solution becomes unbounded as $t \uparrow 1/(CX)$. When X and C are known only probabilistically, $y(t)$ for a fixed t is a random variable Y_t with

$$Y_t = \frac{X}{1 - XCt} = g(X, C; t).$$

Example 25. $u' = v,$ $v' = -u,$ with $u(0) = X,$ $v(0) = Y.$

The exact solution of this IVP for simple harmonic motion is

$$u(t) = X\cos(t) + Y\sin(t), \quad v(t) = -X\sin(t) + Y\cos(t).$$

When X and Y are random variables with known statistics, the quantities $u(t)$ and $v(t)$ for a fixed t are two random variables U_t and V_t with

$$U_t = X\cos(t) + Y\sin(t), \quad V_t = -X\sin(t) + Y\cos(t).$$

Ideally, we know the input random variables through their joint pdf or distribution and we would like to obtain the corresponding probabilistic information on the output random variables. We develop in this chapter a method for doing this for multiple input random variables and multiple output random variables (that evolves with time). More often, this ambitious goal cannot be met for a number of reasons. One obvious reason is when we do not have an exact explicit solution for the IVP. Another would be that we do not have the joint pdf of the input random variables, only limited statistical information such as means and variances. As we shall see later, having the joint pdf for the output random variables with t as a parameter is still insufficient for a complete probabilistic characterization of an evolving phenomenon and we need to obtain other probabilistic information (such as joint distribution for output at different times). We will deal with these more complicated requirements in later chapters of this volume.

9.2 Functions of a Random Variable

9.2.1 *Basic relation on the probability distributions*

We begin the development of a transformation theory by focusing on the case of a function of one random variable. Suppose X is a random variable with a known (probability) density $p_X(x)$ or the corresponding (probability) distribution $P(X \leq x)$, which will often be denoted also by $P_X(x)$ for brevity. Let $y = g(x)$ be a real-valued (typically PWS) function of the real variable x. Consider the corresponding transformation

$$Y = g(X) \tag{9.2.1}$$

of the random variable X into another random variable Y. Since X is real-valued, so is Y. We are interested in the density function $p_Y(y)$ and the distribution $P(Y \leq y) \equiv P_Y(y)$. The solution to this problem is both subtle and technical. Before presenting the basic idea that leads to the solution for our problem, we need to be cognizant of the following items:

- How the range of g is related to its domain, i.e., the range of x. Often, we speak of the former as the *image* set and the latter as the *pre-image*.
- For every pair A and B, the set $\{A \leq g(X) \leq B\}$ must be an event; otherwise the probability distribution $P(A \leq Y \leq B) = P_Y(B) - P_Y(A) = 0$.
- The events $\{Y = g(X) = \pm\infty\}$ must be assigned a probability of 0 and 1, respectively:

$$P_Y(-\infty) = 0, \quad P_Y(\infty) = 1$$

since $P_Y(\infty) = \text{Prob}\{Y \leq \infty\}$ is a certain event.

The *indexed set* I_y for the transformation $g(\cdot)$ is defined to be

$$I_y = \{x : g(x) \leq y\}. \tag{9.2.2}$$

Evidently, the composition of I_y changes with y. For the linear function $y = ax + b$ with $a > 0$, the indexed set for a given value of y is the half line $x \leq (y-b)/a$ that expands to the entire real x-line as $y \to \infty$. For the quadratic function $y = x^2$, the indexed set for $y > 0$ is the interval $-\sqrt{y} \leq x \leq \sqrt{y}$ which extends to the entire real

x-line as $(0 <) \, y \to \infty$. On the other hand, for $y < 0$, the indexed set I_y (over the reals) is empty; there is no value x for which $x^2 < 0$. Evidently, **the probability of $Y \le y$ must be the same as the probability of X spanning the indexed set I_y**:

$$P_Y(y) = P[Y \le y] = P[Y = g(X) \le y] = P_X[x \in I_y] = [P_X(x)]_{x \in I_y}. \qquad (9.2.3)$$

This observation provides the basis and a practical method for computing the distribution function and the pdf of Y. It reduces the problem to finding the indexed set I_y for the particular transformation.

Exercise 31. For $Y = \sin(X)$, find I_y for i) $0 < y < 1$; ii) $y > 1$; and iii) $y = 0$.

9.2.2 Two examples

In this section, we illustrate the application of the basic idea on the probability distribution of the transformed variable by a couple of examples and extract from them the main results for the general case when the pdf $p_X(x)$ is known.

Example 26. $y = g(x) = ax + b$, where $a > 0$ and b are constants and X is a random variable with density function $p_X(x)$. Determine the pdf $p_Y(y)$ and distribution $P_Y(y)$ of the random variable $Y = g(X)$ in terms of $p_X(x)$.

For the density function p_Y of Y, we have from the transformation of $y = g(x)$ the inverted relation $x = (y - b)/a$ for the entire real line so that we may write $p_X(x) = p_X((y - b)/a)$. For a given value $\bar{x}$ (instead of x to avoid confusion with the dummy variable of integration), we have $\bar{y} = a\bar{x} + b$ or $\bar{x} = (\bar{y} - b)/a$. The desired $p_Y(y)$ is related to $p_X(x)$ by the equality of the two versions of the same probability distribution (9.2.3). For $a > 0$, it requires

$$\int_{-\infty}^{\bar{x}} p_X(x) dx = \int_{-\infty}^{\bar{x}} p_X\left(\frac{y-b}{a}\right) dx = \int_{-\infty}^{\bar{y}} p_X\left(\frac{y-b}{a}\right) \frac{dx}{dy} dy = \int_{-\infty}^{\bar{y}} p_Y(y) dy.$$

With $dx/dy = (dg/dx)^{-1} = 1/a$, this leads to

$$p_Y(\bar{y}) = \frac{1}{a} p_X\left(\frac{\bar{y} - b}{a}\right). \qquad (9.2.4)$$

For the distribution of Y, we let

$$I_y = \{x \colon g(x) = ax + b \le y\} = \{x : x \le (y - b)/a\}$$

and note the relation that gives P_Y in terms of p_X to get:

$$P_Y(\bar{y}) = \int_{-\infty}^{\bar{y}} p_Y(y) dy = \int_{-\infty}^{\bar{y}} \frac{1}{a} p_X\left(\frac{y-b}{a}\right) dy$$

or

$$P_Y(\bar{y}) = \int_{-\infty}^{\bar{y}} p_Y(y) dy = \int_{-\infty}^{(\bar{y}-b)/a} \frac{1}{a} p_X(x) \frac{dy}{dx} dx$$

$$= P_X\left(\frac{\bar{y}-b}{a}\right) = [P_X(x)]_{x \in I_{\bar{y}}}.$$

The last part of the result above is important if we know $P_X(x)$ instead of $p_X(x)$. In that case, we obtain the distribution $P_Y(y)$ from $[P_X(x)]_{x \in I_y}$, with $I_y = \{x \le (y-b)/a\}$ for this example. From $P_Y(y)$, we obtain $p_Y(y)$ by differentiation.

Of course, if we know $P_X(x)$ for the random variable X, we also get immediately $p_X(x)$ by differentiating $P_X(x)$ and use (9.2.4) to get $p_Y(y)$. But for nonlinear or more complex $g(\cdot)$, it may well be simpler to calculate $p_Y(y)$ from $p_X(x)$ by way of the general analogues of the relation (9.2.4), to be established as Theorem 24 and Corollary 12 later. For this reason, our emphasis in this chapter will be for input random variables with known pdf.

Exercise 32. Given $p_X(x)$, determine $P_Y(y)$ and $p_Y(y)$ for $Y = aX + b$ with $a < 0$.

Unlike the previous example, the inverse relation for other functions $g(X)$ is not always so straightforward. The next simple example illustrates the possible complications and points to the general results.

Example 27. Let $Y = g(X) = X^2$ for the random variable X with a pdf $p_X(x)$.

To find $p_Y(y)$ and $P_Y(y)$, we note the following features of the pre-image of $Y \le \bar{y}$:

- For the pdf $p_Y(y)$, we note that $y = x^2$ has no real solution if $y < 0$. In that case, we have $p_Y(y) = 0$ for $y < 0$.
- If $y > 0$, then $y = x^2$ has two solutions $x_1 = \sqrt{y}$ and $x_2 = -\sqrt{y}$ with $-\sqrt{\bar{y}} \le x \le \sqrt{\bar{y}}$ being the pre-image of $0 \le y \le \bar{y}$. It follows that

$$P(Y \le \bar{y}) = \begin{cases} 0 & (\bar{y} < 0) \\ P(X_1 \le \sqrt{\bar{y}}) - P(X_2 \le -\sqrt{\bar{y}}) & (\bar{y} > 0) \end{cases} \qquad (9.2.5)$$

where X_1 and X_2 denote the relevant branch of the random variable X. In arriving at the result for the random variable Y for $\bar{y} > 0$, we noted that the range of the corresponding random variable X is necessarily restricted to $-\sqrt{\bar{y}} \le X \le \sqrt{\bar{y}}$, for otherwise we would have $Y = X^2 > \bar{y}$.

To obtain the corresponding density function, we work with the expression for $P_Y(Y \le \bar{y})$ for $Y > 0$:

$$P(Y \le \bar{y}) = \int_{-\infty}^{\bar{y}} p_Y(y)dy = \int_{0}^{\bar{y}} p_Y(y)dy = \int_{-\sqrt{\bar{y}}}^{\sqrt{\bar{y}}} p_X(x)dx$$

where the last equality is by the basic equality (9.2.3) relating the probability for the transformed random variable Y to that of its pre-image set in X. After a change of variable of integration from x to y in the last integral, we get from the relation above for $\bar{y} > 0$

$$P_Y(\bar{y}) = \int_{\bar{y}}^{0} p_X(x_2(y))\frac{dx_2}{dy}dy + \int_{0}^{\bar{y}} p_X(x_1(y))\frac{dx_1}{dy}dy$$

$$= \int_{\bar{y}}^{0} p_X(-\sqrt{y})\frac{-1}{2\sqrt{y}}dy + \int_{0}^{\bar{y}} p_X(\sqrt{y})\frac{1}{2\sqrt{y}}dy$$

$$= \frac{1}{2}\int_{0}^{\bar{y}} \frac{p_X(\sqrt{y}) + p_X(-\sqrt{y})}{\sqrt{y}}dy.$$

From this follows

$$p_Y(y) = \begin{cases} 0 & (y < 0) \\ \dfrac{1}{2\sqrt{y}}\left[p_X(\sqrt{y}) + p_X(-\sqrt{y})\right] & (y > 0) \end{cases} \qquad (9.2.6)$$

and

$$P_Y(\bar{y}) = \left\{P_X\left(\sqrt{\bar{y}}\right) - P_X\left(-\sqrt{\bar{y}}\right)\right\} H(\bar{y}).$$

The latter is consistent with (9.2.5) while the former may be written as

$$p_Y(y) = \begin{cases} 0 & (y < 0) \\ p_X(x_1(y))\left|\dfrac{dx_1}{dy}\right| + p_X(x_2(y))\left|\dfrac{dx_2}{dy}\right| & (y > 0) \end{cases} \qquad (9.2.7)$$

for the purpose of generalization to a general function $g(x)$.

9.2.2.1 The general case

In general, if $y = g(x)$ has a unique inverse, then we can solve for a unique x in terms of y to get $x = g^{-1}(y)$. We did this in the first example above. When the inverse is multi-valued, we will have to work with all the different branches of the inverse. That is, we obtain the different solutions of $y = g(x)$, denoted by $\{x_1(y), x_2(y), \ldots, x_n(y)\}$ (such that $y = g(x_k(y))$, $k = 1, 2, 3, \ldots, n$) are distinctly different pre-images $\{x_k\}$ of y on the real x-line as in second example above. The range of each $x_k(y)$ covers a different part of the x-line; the union of the non-overlapping ranges of $x_k(y)$, $0 \le k \le n$, cover the entire x-line. It is possible that the domains of all the $x_k(\cdot)$ do not span the entire $y-$line (e.g., $y < 0$ is not in the range of $g(x)$ for the second example). For any such unreachable interval of the $y-$line, we have $p_Y(y) = 0$. These observations lead to the following theorem for the general one variable transformation $y = g(x)$. Its proof will help to elucidate the relations between the image set of $g(\cdot)$ and the pre-image sets that are the ranges of the inverses $\{x_k(\cdot)\}$.

Theorem 24. *Let X be a (scalar) random variable with density function $p_X(x)$ and $g(\cdot)$ is a PWS function on the real line so that $Y = g(X)$ is another random variable. Then the pdf for Y is given by*

$$p_Y(y) = \sum_{k=1}^{n} p_X(x_k(y)) \left| \frac{dx_k}{dy} \right| \tag{9.2.8}$$

with $p_Y(y) = 0$ if y is not in the range of $g(\cdot)$, i.e., not reachable by the transformation.

Proof. It suffices to prove the theorem for a cubic like transformation $y = g(x)$ as shown in Fig. 9.1. For the generic y value chosen for establishing the theorem, we have three pre-images $x_k(y) = g^{-1}(y)$, $k = 1, 2$ and 3, with $x_1(y) < x_2(y) < x_3(y)$. For the incrementally larger value $y + dy > y$, the corresponding pre-images are $x_1 + dx_1, x_2 + dx_2$ and $x_3 + dx_3$ with dy chosen sufficiently small so that $x_1 + dx_1 < x_2 + dx_2 < x_3 + dx_3$. While dx_1 and dx_3 are positive, dx_2 is negative. For dy sufficiently small, we have from (9.2.3)

$$P(y \le Y \le y + dy) = p_Y(y)dy$$

$$= p_X(x_1)dx_1 + p_X(x_2) |dx_2| + p_X(x_3)dx_3 \tag{9.2.9}$$

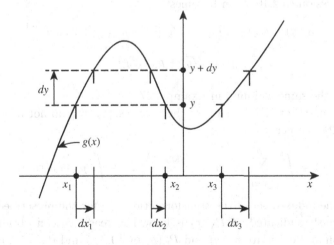

Fig. 9.1.   A general (cubic-like) transformation $y = g(x)$.

except for terms of higher order in dy and dx_k. Note that the absolute value $|dx_2|$ in the second term is required since probability must be non-negative. It follows from (9.2.9) that

$$p_Y(y) = p_X(x_1)\left|\frac{dx_1}{dy}\right| + p_X(x_2)\left|\frac{dx_2}{dy}\right| + p_X(x_3)\left|\frac{dx_3}{dy}\right|.$$

More generally, we have, for a typical value y on the real y-line, the more general result (9.2.8) where n is the number of pre-image of y for the transformation $g(\cdot)$.   □

It is not always possible to find an explicit expression for $x_k = \left[g^{-1}(y)\right]_k$ let alone the corresponding dx_k/dy. A more practical approach is to differentiate $y = g(x)$ and take advantage of the relation

$$\frac{dx_n}{dy} = \frac{1}{dy/dx_n} = \left[\frac{1}{g'(x)}\right]_{x=x_n(y)}$$

where $x_n(y)$ may be found numerically. For that reason, we often work with the following corollary of the result above:

Corollary 12.

$$p_Y(y) = \sum_{k=1}^{n} p_X(\mathbf{x}_k(\mathbf{y}))\left|\left(\frac{dg}{dx}\right)_{x=x_k(y)}\right|^{-1}. \tag{9.2.10}$$

For the example $g(x) = x^2$ with $x_k(y) = \pm\sqrt{y}$, application of (9.2.10) leads to

$$\left(\frac{dg}{dx}\right)_{x=x_k(y)} = 2x_k(y) = \begin{cases} 2\left(-\sqrt{y}\right) & (k=1) \\ 2\left(\sqrt{y}\right) & (k=2) \end{cases}$$

and the expression (9.2.10) then becomes

$$p_Y(y) = p_X(-\sqrt{y})\left|2\left(-\sqrt{y}\right)\right|^{-1} + p_X(\sqrt{y})\left|2\left(\sqrt{y}\right)\right|^{-1}$$

$$= \frac{1}{2\sqrt{y}}\left[p_X(-\sqrt{y}) + p_X(\sqrt{y})\right].$$

The result is the same as found in Example 27.

To the extent that the domains of the inverses $\{x_k(y)\}$ do not overlap, we can integrate (9.2.8) to get

$$P_Y(y) = \int_{-\infty}^{y} \sum_{k=1}^{n} p_X(x_k(z))\left|\frac{dx_k(z)}{dz}\right| dz = \int_{I_y} p_X(x)dx, \tag{9.2.11}$$

where I_y is the indexed set of the transformation $g(\cdot)$ for the prescribed y value.

As previously indicated, if $P_X(x)$ is known instead, we can obtain $p_X(x)$ by differentiation and calculate $p_Y(y)$ and $P_Y(y)$ by (9.2.8) and (9.2.11), respectively. Nevertheless, it is instructive to see how $P_Y(y)$ may be obtained directly from

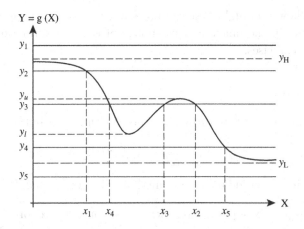

Fig. 9.2. $P(Y \leq y)$ in terms of $P(X \leq x)$ given $Y = g(X)$.

$P_X(x)$ for a general transformation $g(\cdot)$. To illustrate the solution process, consider the transformation $g(x)$ as sketched in Fig. 9.2 (with $y_L \leq g(x) \leq y_H$).

i) $y = y_1 > y_H$: We have $g(x) < y_1$ for all $-\infty \leq x \leq \infty$. In that case, we have

$$P_Y(y_1) = P(-\infty < Y \leq y_1) = P(-\infty < X \leq \infty) = 1.$$

ii) $y = y_2$ $(y_u < y_2 < y_H)$: We have $g(x) \leq y_2$ for all $x_1 \leq x \leq \infty$. In that case, we have

$$P_Y(y_2) = P(-\infty < Y \leq y_2) = P(x_1 \leq X \leq \infty) = 1 - P_X(x_1).$$

iii) $y = y_3$ $(y_\ell < y_3 < y_u)$: We have $g(x) \leq y_3$ for all $(x_4 \leq x \leq x_3)$ $\cup$ $(x_2 \leq x \leq \infty)$. In that case, we have

$$P_Y(y_3) = P(-\infty < Y \leq y_3) = P(x_4 \leq X \leq x_3) + P(x_2 \leq X \leq \infty)$$
$$= 1 - P_X(x_4) + P_X(x_3) - P_X(x_2).$$

iv) $y = y_4$ $(y_L < y_4 < y_\ell)$: We have $g(x) \leq y_4$ for all $(x_5 \leq x \leq \infty)$. In that case, we have

$$P_Y(y_4) = P(-\infty < Y \leq y_4) = P(x_5 \leq X \leq \infty) = 1 - P_X(x_5).$$

v) $y = y_5$ $(< y_L)$: We have $g(x) > y_5$ for all $(-\infty \leq x \leq \infty)$. In that case, we have

$$P_Y(y_5) = P(-\infty < Y \leq y_5) = P(X = -\infty) = 0.$$

As indicated at the start of this section, the key to obtaining the probability density function and distribution of the transformed random variable $Y = g(X)$ is to determine the indexed set for the relevant transformation. From the few examples worked out above, this important task is seen to be far from straightforward. Other intricacies not shown through these examples include possible jump discontinuities and flat segments in the image and pre-image of the transformation to be found among assigned exercises.

9.2.3 Applications

Example 28. Let $Y = X^2$ and $p_X(x)$ be the Rayleigh distribution

$$p_X(x) = \frac{x}{\alpha^2} e^{-(x/\alpha)^2/2} H(x)$$

where $H(x)$ is the Heaviside unit step function. Find $p_Y(y)$ and $P_Y(\bar{y})$.

From (9.2.6) of Example 27 (effectively a consequence of Theorem 24 or Corollary 12), we get the following exponential density function for Y:

$$p_Y(y) = \frac{1}{2\sqrt{y}} [p_X(\sqrt{y}) + p_X(-\sqrt{y})] H(y)$$

$$= \frac{1}{2\sqrt{y}} \frac{1}{\alpha^2} \left[\sqrt{y} e^{-y/2\alpha^2} H(\sqrt{y}) + (-\sqrt{y}) e^{-y/2\alpha^2} H(-\sqrt{y}) \right] H(y)$$

$$= \frac{1}{2\alpha^2} e^{-y/2\alpha^2} H(y).$$

Upon integration, we get

$$P_Y(\bar{y}) = \int_{-\infty}^{\bar{y}} \frac{1}{2\alpha^2} e^{-y/2\alpha^2} H(y) dy = \begin{cases} 0 & (\bar{y} < 0) \\ \int_0^{\bar{y}} \frac{1}{2\alpha^2} e^{-y/2\alpha^2} dy & (\bar{y} > 0) \end{cases}$$

$$= \left\{ 1 - e^{-\bar{y}/2\alpha^2} \right\} H(\bar{y}).$$

Example 29. $Y = X^{-2}$

Again $x^2 = 1/y$ has no solution for $y < 0$ and therewith $P_Y(y) = 0$, $(y < 0)$. For $y > 0$, $x^2 = 1/y$ has two solutions:

$$x_1(y) = -\frac{1}{\sqrt{y}}, \quad x_2(y) = \frac{1}{\sqrt{y}}.$$

Also, $X^{-2} \leq y$ requires $X^2 \geq 1/y$ or $X \geq x_2(y) = 1/\sqrt{y}$ and $X \leq x_1(y) = -1/\sqrt{y}$. It follows that

$$P_Y(y) = P(Y \leq y) = P(Y = X^{-2} \leq y) = P(X \leq x_1(y)) + P(X \geq x_2(y))$$

$$= P_X(x_1(y)) + [1 - P_X(x_2(y))]$$

$$= P_X\left(-\frac{1}{\sqrt{y}}\right) + \left[1 - P_X\left(\frac{1}{\sqrt{y}}\right)\right] \quad (y > 0).$$

To obtain the corresponding pdf $p_Y(y)$ in terms of the pdf $p_X(x)$, we again differentiate $P_Y(y)$ to get

$$p_Y(y) = \frac{dP_Y}{dy} = \frac{d}{dy}\left[P_X\left(-\frac{1}{\sqrt{y}}\right) + \left\{1 - P_X\left(\frac{1}{\sqrt{y}}\right)\right\}\right]$$

$$= \frac{1}{2y^{3/2}}p_X\left(-\frac{1}{\sqrt{y}}\right) - \left(\frac{-1}{2y^{3/2}}\right)p_X\left(\frac{1}{\sqrt{y}}\right)$$

$$= \frac{1}{g'(x_1(y))}p_X(x_1(y)) - \left(\frac{1}{g'(x_2(y))}\right)p_X(x_2(y)) \quad (y > 0)$$

with

$$g'(x_1) = -\frac{2}{[x_1(y)]^3} = 2y^{3/2} > 0 \quad g'(x_2) = -\frac{2}{[x_2(y)]^3} = -2y^{3/2} < 0.$$

The expression for pdf $p_Y(y)$ may again be written as

$$p_Y(y) = \frac{p_X(x_1(y))}{|g'(x_1(y))|} + \frac{p_X(x_2(y))}{|g'(x_2(y))|}.$$

Exercise 33. Let X be a random variable with $p_X(x) \sim N(0, \sigma^2)$ (meaning X is normally distributed with mean zero and variance σ^2) and $Y = X^2$. Show (by way of $E[e^{iuY}]$ or otherwise) that

$$p_Y(y) = \frac{1}{\sqrt{2\pi\sigma^2 y}}e^{-y/2\sigma^2}H(y).$$

Example 30. $Y = X(X^2 - 1)$

For this example, we obtain the pdf $p_Y(y)$ directly without first obtaining $P_Y(y)$. For this purpose, we consider two neighboring values $y = 1/5$ and $y = 1/5 + dy$. The horizontal line $y = 1/5$ intersects the curve $y = x^3 - x$ at three locations: $x_1 < -2/3^{3/2}$, $-2/3^{3/2} < x_2 < 0$ and $x_3 > 2/3^{3/2}$. It follows that

$$P(y < Y \leq y + dy) = P_1 + P_2 + P_3$$

where

$$P_k = P(x_k < X \leq x_k + dx_k) \quad (k = 1, 3),$$

$$P_2 = (x_2 + dx_2 < X \leq x_2) \quad (dx_2 < 0).$$

For sufficiently small $|dx_k|$, we have except for small terms of higher orders in dx_k

$$p_Y(y)dy = p_X(x_1)dx_1 - p_X(x_2)dx_2 + p_X(x_3)dx_3$$

or, with $dx_2 < 0$,

$$p_Y(y) = p_X(x_1)\frac{dx_1}{dy} + p_X(x_2)\left|\frac{dx_2}{dy}\right| + p_X(x_3)\frac{dx_3}{dy}.$$

Since $g'(x_1) > 0$ and $g'(x_3) > 0$, we may re-write $p_Y(y)$ as

$$p_Y(y) = \sum_{k=1} \frac{p_X(x_k(y))}{|g'(x_k(y))|} = \sum_{k=1} \frac{p_X(x_k(y))}{|3x_k^2(y) - 1|}. \tag{9.2.12}$$

9.2.4 *Expected value of a transformed random variable*

For completeness, we return to our original observation on the appropriateness of the definition of $E[g(X)]$ in terms of the pdf $p_X(x)$ of X. More specifically, the definition should be consistent with setting $g(X)$ as a random variable Y with its own pdf $p_Y(y)$. Now that we have establish $p_Y(y)$ in terms of $p_X(x)$ (see Theorem 24 and Corollary 12), we are in a position to state and trivially prove the following consistency theorem:

Theorem 25. *Let X be a random variable with probability density function $p_X(x)$ and $Y = g(X)$ where $g(\cdot)$ is continuously differentiable (or PWS). Then*

$$E[Y] = \int_{-\infty}^{\infty} g(x)p_X(x)dx.$$

Proof. The proof starts with the same argument that proves Theorem 24. For the same configuration sketched out in Fig. 9.1 we have again $p_Y(y)dy = p_X(x_1)dx_1 + p_X(x_2)|dx_2| + p_X(x_3)dx_3$ so that

$$yp_Y(y)dy = g(x_1)p_X(x_1)dx_1 + g(x_2)p_X(x_2)|dx_2| + g(x_3)p_X(x_3)dx_3$$

where $\{x_k(y)\}$ are the three distinct pre-images and $dx_2 < 0$ for our choice of y and the particular graph of $g(x)$. As the image variable y ranges over the entire y-axis, the indexed set I_y (with non-overlapping intervals) would span the entire x-axis (possibly with some part of the y-axis not reached by $g(\cdot)$ so that we have $p_Y(y) = 0$ for these y values). It follows that

$$\int_{-\infty}^{\infty} yp_Y(y)dy = \int_{I_1} g(x_1)p_X(x_1)dx_1 + \cdots + \int_{I_3} g(x_3)p_X(x_3)dx_3$$

so that

$$E[Y] = \int_{-\infty}^{\infty} g(x)p_X(x)dx = E[g(x)]. \qquad \square$$

Exercise 34. Let X be a random variable with $p_X(x) \sim N(0, \sigma^2)$ and $Y = |X|^n$. Find $E[Y]$ for $n = 2m$ (an even integer) and $n = 2m + 1$ (an odd integer).

9.3 A Function of Two Random Variables

Sometimes a random variable is the output generated by two other random variables in the form of

$$Z = g(X, Y)$$

where g is at least piecewise differentiable in both arguments. Suppose we have the joint pdf or joint distribution of X and Y. We expect to be able to get the pdf and distribution of the output random variable Z, $p_Z(z)$ and $P_Z(z)$. Unlike the case of $y = g(x)$ studied in the previous section, it is not possible to solve the one relation $z = g(x, y)$ for the two input variables x and y in terms of the output z. To get some ideas on how to work with this new class of problems, we again start with a few examples for which we can obtain concrete results for the pdf and distribution of Z. In all cases, we limit discussion to problems with known joint pdf $p_{XY}(x, y)$ of X and Y. If the joint distribution $P_{XY}(x, y)$ is given instead, we can obtain the corresponding joint density by differentiation.

9.3.1 *Linear transformations*

9.3.1.1 *The simplest case*

For the simplest linear transformation $Z = X + Y$, we have the region $X + Y \leq z$ in the x, y–plane as the pre-image D_z for the range $Z \leq z$ (see Fig. 9.3). In that case, the probability $P(Z \leq z)$ is given by

$$P_Z(z) = P(Z = X + Y \leq z) = \int\int_{D_z} p_{XY}(x, y) dx dy$$

$$= \int_{-\infty}^{\infty} \int_{-\infty}^{z-y} p_{XY}(x, y) dx dy = \int_{-\infty}^{\infty} \int_{-\infty}^{z-x} p_{XY}(x, y) dy dx.$$

The corresponding probability density function for transformed variable is obtained by differentiation with respect to z:

$$p_Z(z) = \frac{dP_Z(z)}{dz} = \int_{-\infty}^{\infty} p_{XY}(z - y, y) dy = \int_{-\infty}^{\infty} p_{XY}(x, z - x) dx.$$

9.3.1.2 *Independent input variables*

When X and Y are independent, the relevant integral for our problem simplifies somewhat to the convolutional form

$$p_Z(z) = \int_{-\infty}^{\infty} p_X(z - y) p_Y(y) dy = \int_{-\infty}^{\infty} p_X(x) p_Y(z - x) dx. \tag{9.3.1}$$

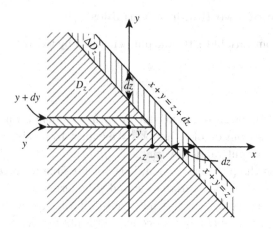

Fig. 9.3. The pre-image set D_z of the linear transformation $Z = X + Y$.

The more general linear transformation $Z = aX + bY$ has sufficiently wide applications to warrant highlighting the corresponding expression for $p_Z(z)$:

Proposition 33. *Suppose the two r.v. X and Y are independent with pdf $p_X(x)$ and $p_Y(y)$, respectively. If $Z = aX + bY$, then we have the following generalization of (9.3.1) for $p_Z(z)$:*

$$p_Z(z) = \frac{1}{|ab|} \int_{-\infty}^{\infty} p_X\left(\frac{z - u}{a}\right) p_Y\left(\frac{u}{b}\right) du$$

$$= \frac{1}{|ab|} \int_{-\infty}^{\infty} p_X\left(\frac{v}{a}\right) p_Y\left(\frac{z - v}{b}\right) dv.$$

9.3.1.3 Applications

The following specific applications of (9.3.1) for the transformation $Z = X + Y$ when the two random variables are *independent*:

a) For $p_X(x) = \alpha e^{-\alpha x} H(x)$ and $p_Y(y) = \beta e^{-\beta y} H(y)$ where $H(t)$ is the unit step function, we have from (9.3.1)

$$p_Z(z) = \int_{-\infty}^{\infty} \alpha e^{-\alpha(z-y)} H(z - y) \beta e^{-\beta y} H(y) dy$$

$$= H(z)\alpha\beta \int_0^z e^{-\alpha(z-y)} e^{-\beta y} dy = \alpha\beta e^{-\alpha z} \int_0^z e^{-(\beta-\alpha)y} dy$$

$$= H(z) \begin{cases} \dfrac{\alpha\beta}{\beta - \alpha} \left[e^{-\alpha z} - e^{-\beta z}\right] & (\alpha \neq \beta) \\ \alpha\beta z e^{-\alpha z} & (\alpha = \beta). \end{cases}$$

The corresponding distribution function $P_Z(z)$ can be obtained from $p_Z(z)$ by integration.

b) If X and Y are both uniformly distributed over the interval (x_1, x_2) and (y_1, y_2), respectively, we may write them as

$$p_X(x) = \frac{1}{\Delta x} \{H(x - x_1) - H(x - x_2)\},$$

$$p_X(x) = \frac{1}{\Delta y} \{H(y - y_1) - H(y - y_2)\}$$

where $\Delta x = x_2 - x_1$ and $\Delta y = y_2 - y_1$. In that case, (9.3.1) becomes

$$p_Z(z) = \frac{1}{\Delta x \Delta y} \int_{-\infty}^{\infty} h(y; z) \{H(y - y_1) - H(y - y_2)\} dy$$

$$= \frac{1}{\Delta x \Delta y} \int_{y_1}^{y_2} h(y; z) dy$$

where

$$h(y; z) = H(z - y - x_1) - H(z - y - x_2)$$

since $H(y-y_1) - H(y-y_2) = 0-0 = 0$ for $y < y_1$ and $H(y-y_1) - H(y-y_2) = 1-1 = 0$ for $y > y_2$. It is also not difficult to see that

$$p_Z(z) = 0 \quad (z < x_1 + y_1, \quad z > x_2 + y_2).$$

For $z < x_1+y_1$, we have $z-y-x_2 < z-y-x_1 < y_1-y \le 0$ so that $h(y; z) = 0-0 = 0$.
For $z > x_2+y_2$, we have $z-y-x_1 > z-y-x_2 > y_2-y > 0$ so that $h(y; z) = 1-1 = 0$.

Exercise 35. For $\Delta x = \Delta y$, show that the graph of $p_Z(z)$ is an isosceles triangle in the interval $x_1 + y_1 < z < x_2 + y_2$ of height $1/\Delta x = 1/\Delta y$:

$$p_Z(z) = \frac{1}{\Delta x \Delta y} \begin{cases} (z - x_1 - y_1) H(z - y_1 - x_1) & (z \le \frac{1}{2}(\Delta x + \Delta y)) \\[2mm] (x_2 + y_2 - z) H(y_2 + x_2 - z) & (z \ge \frac{1}{2}(\Delta x + \Delta y)). \end{cases}$$

Exercise 36. For $\Delta x \ne \Delta y$, show by graphical method that the graph of $p_Z(z)$ is linear upwards adjacent to $z = x_1 + y_1$, linear downward adjacent to $z = x_2 + y_2$ and flat in an interval centered at $z = \frac{1}{2}(\Delta x + \Delta y)$ connecting the two linear segments.

9.3.2 The general case

It is evident from the preceding examples that we need to work with the pre-image set $D_z = \{(x, y) \; g(X, Y) \le z\}$ to determine the distribution $P_Z(z)$ and the corresponding pdf $p_Z(z)$ from the basic relation

$$P_Z(z) = P(Z = g(X, Y) \le z) = \int \int_{D_z} p_{XY}(x, y) dx dy, \tag{9.3.2}$$

analogous to (9.2.3) for the single input variable case. The corresponding pdf $p_Z(z)$ is then obtained by differentiating $P_Z(z)$ with respect to z (which appears as a parameter on the right hand side of (9.3.2)) to get

$$p_Z(z) = \frac{dP_Z(z)}{dz}. \tag{9.3.3}$$

For a sufficiently small increment dz, the two simple (non-self-intersecting) closed curves $g(x, y) = z$ and $g(x, y) = z + dz$ in the x, y–plane are non-intersecting and concentric (see Fig. 9.4). Let dD_z be the region between these two curves. Note that dD_z may not be simply connected and may consist of several non-overlapping regions so that $dD_z = dD_z^{(1)} \cup dD_z^{(2)} \cup \cdots \cup dD_z^{(m)}$. In that case, we have

$$p_Z(z)dz = P(z \le Z \le z + dz) = P(Z \le z + dz) - P(Z \le z) \tag{9.3.4}$$

$$= \sum_{k=1}^{m} p_{XY}(x^{(k)}, y^{(k)})dD_z^{(k)} = p_{XY}(x, y)dD_z.$$

As z ranges over the interval $(-\infty, z)$, we sum up all such increments to get $P_Z(z) = P(Z = g(X, Y) \le z)$. The result is the same as $P_{XY}(x, y) \, \epsilon D_z$ where D_z is the pre-image set of $-\infty \le Z \le z$, possibly consisting of several non-overlapping regions:

$$\int_{-\infty}^{z} p_Z(z)dz = P_Z(z) = \iint_{D_Z} p_{XY}(x, y)dxdy. \tag{9.3.5}$$

Note that the pre-image set D_z is the analogue of the indexed set for the one variable transformation case discussed in the previous section. While the determination of the pdf and distribution for the transformed variable Z is reduced to finding D_z, it is not always simple to execute the known task. Here it is necessary however to limit further discussion on this topic to a few illustrating examples in the next subsection and among the assigned exercises in order to get to the heart of

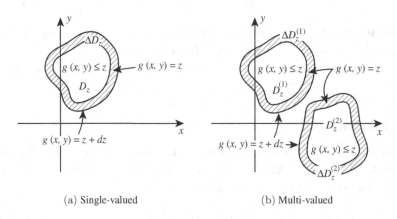

(a) Single-valued   (b) Multi-valued

Fig. 9.4.   The pre-images set D_z of a general transformation $Z = g(X, Y)$.

the material on stochastic differential equations. The brief discussion of this section however suffices for three main purposes:

(1) To introduce reader to an approach for determining the density and distribution of a scalar output random variable Z given the density of the two input variable X and Y.
(2) To sensitize readers to the basic idea as well as the difficult aspects of the tasks involved in such activities.
(3) To alert readers to the need to investigate alternative approaches for obtaining the needed probabilistic information.

9.3.3 *Some nonlinear transformations*

A number of nonlinear transformation $Z = g(X, Y)$ will be investigated in the assigned exercises at the end of this volume. We illustrate by two examples below the process for determining the pdf and distribution of the transformed variable Z of this type of transformations.

9.3.3.1 $Z = X/Y$

For this nonlinear transformation, we note the pre-image set D_z for $Z \le z$ varies with the sign of y. For $z > 0$, we have

$$D_z = \begin{cases} x \le yz & (y > 0) \\ x \ge yz & (y < 0) \end{cases} \tag{9.3.6}$$

(see Fig. 9.5). The pre-image set with the same two disjoint regions can also be given in terms of the sign of x but will not be done here (as it is evident from a

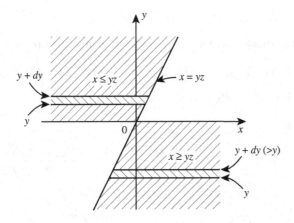

Fig. 9.5.   The pre-image set D_z of the nonlinear transformation $Z = X/Y$.

sketch of D_z). In terms of D_z as characterized by (9.3.6), we have

$$P_Z(z) = P\left((x,y)\epsilon D_z\right) = \int\int_{D_z} p_{XY}(x,y)dxdy$$

$$= \int_0^\infty \int_{-\infty}^{yz} p_{XY}(x,y)dxdy + \int_{-\infty}^0 \int_{yz}^\infty p_{XY}(x,y)dxdy.$$

Correspondingly, the pdf $p_Z(z)$ is obtained by differentiating $P_Z(z)$ to get

$$p_Z(z) = \frac{dP_Z}{dz} = \int_0^\infty yp_{XY}(yz,y)dy - \int_{-\infty}^0 yp_{XY}(yz,y)dy$$

$$= \int_0^\infty yp_{XY}(yz,y)dy + \int_{-\infty}^0 |y|p_{XY}(yz,y)dy$$

$$= \int_{-\infty}^\infty |y|p_{XY}(yz,y)dy.$$

If $p_{XY}(x,y)$ is even in its argument so that $p_{XY}(-x,-y) = p_{XY}(x,y)$, the density and distribution function simplify to

$$p_Z(z) = 2\int_0^\infty yp_{XY}(yz,y)dy$$

$$P_Z(z) = 2\int_0^\infty \int_{-\infty}^{yz} p_{XY}(x,y)dxdy.$$

Exercise 37. Determine $p_Z(z)$ and $P_Z(z)$ if the two input variables are jointly normal with

$$p_{XY}(x,y) = \frac{1}{2\pi\sigma_x\sigma_y\sqrt{1-r^2}} \exp\left\{-\frac{1}{2(1-r^2)}\left(\frac{x^2}{\sigma_x^2} - \frac{2rxy}{\sigma_x\sigma_y} + \frac{y^2}{\sigma_y^2}\right)\right\}.$$

9.3.3.2 $Z = \min(X,Y)$

For this nonlinear transformation, we have from elementary probability

$$P_Z(z) = P((X \le z) \cup (Y \le z)) = P(X \le z) + P(Y \le z) - P((X \le z) \cap (Y \le z))$$

$$= P_X(z) + P_Y(z) - P_{XY}(z,z).$$

From this, we get

$$p_Z(z) = p_X(z) + p_Y(z) - \left(\frac{\partial P_{XY}(x,y)}{\partial x} + \frac{\partial P_{XY}(x,y)}{\partial y}\right)_{x=y=z}$$

where

$$p_X(z) = \int_{-\infty}^\infty p_{XY}(z,y)dy, \quad p_Y(z) = \int_{-\infty}^\infty p_{XY}(x,z)dx.$$

The expression for pdf $p_Z(z)$ simplifies if X and Y are *independent* with $p_{XY}(x,y) = p_X(x)p_Y(y)$. In that case, we have

$$p_Z(z) = p_X(z) + p_Y(z) - (p_X(z)P_Y(z) + p_Y(z)P_X(z))$$

and

$$P_Z(z) = P_X(z) + P_Y(z) - P_X(z)P_Y(z).$$

9.3.4 Expected value

The expectation of $g(X,Y)$ is given by (8.2.4) of the previous chapter. However, as in the single variable case, it may be viewed in a different way by setting $Z = g(X,Y)$ so that $E[g(X,Y)] = E[Z]$. The expectation of Z is known to be given by

$$E[Z] = \int_{-\infty}^{\infty} z p_Z(z) dz \qquad (9.3.7)$$

with the pdf $p_Z(z)$ for the random variable Z to be obtained from $p(x,y)$. For consistency between (8.2.4) and (9.3.7), we have the following result:

Theorem 26. *Let X and Y be random variables with joint density $p(x,y)$ and $Z = g(X,Y)$ for some continuous piecewise differentiable function $g(\cdot,\cdot)$. Then the expectation $E[Z]$ is given by*

$$E[Z] = E[g(X,Y)] = \int_{-\infty}^{\infty} \int_{-\infty}^{\infty} g(x,y)p(x,y)dxdy. \qquad (9.3.8)$$

Proof. The relation (9.3.4) enables us to write

$$z p_Z(z) dz = \sum_{k=1}^{m} g(x^{(k)}, y^{(k)}) p_{XY}(x^{(k)}, y^{(k)}) dD_z^{(k)} = g(x,y)p_{XY}(x,y)dD_z.$$

The theorem follows upon after summing all such increments. □

9.4 Several Functions of Several Random Variables

9.4.1 Two functions of two random variables

The theory for two transformations of two random variables,

$$U = f(X,Y), \quad V = g(X,Y), \qquad (9.4.1)$$

is more typical and illustrative of the more general case of the number of transformations equal to the number of input random variables. Given the joint pdf $p_{XY}(x,y)$, we can determine the joint distribution of the output variables $P_{UV}(u,v)$ by noting that the restrictions $U = f(X,Y) \leq u$ and $V = g(X,Y) \leq v$ specify a pre-image set

D_{uv} in the x,y–plane consisting of all points in region(s) in that plane satisfying the two relations $f(x,y) \le u$ and $g(x,y) \le v$. It follows that

$$P_{UV}(u,v) = P(U \le u, V \le v) = P(f(X,Y) \le u, g(X,Y) \le v)$$

$$= \int\int_{D_{uv}} p_{XY}(x,y)dxdy.$$

The corresponding pdf $p_{UV}(u,v)$ is obtained by differentiating the distribution function $P_{UV}(u,v)$

$$p_{UV}(u,v) = \frac{\partial^2 P_{UV}}{\partial u \partial v}.$$

An alternative approach is to work with the incremental region ΔD_{uv} in the x,y–plane consisting of all points (x,y) specified by

$$u \le U \le u + du, \quad v \le V \le v + dv,$$

where the quantities (U,V) are related to (X,Y) by (9.4.1). Consider the simple transformation

$$U = \alpha X + \beta Y, \quad V = \gamma X + \delta y$$

and let $(x_k(u,v), y_k(u,v))$ be the unique solution of

$$f(x,y) = \alpha x + \beta y = u, \quad g(x,y) = \gamma x + \delta y = v,$$

or

$$x_k(u,v) = \frac{1}{\Delta}(\delta u - \beta v), \quad x_k(u,v) = \frac{1}{\Delta}(\alpha v - \gamma u),$$

where $\Delta = \alpha \delta - \beta \gamma$ (assuming $\Delta \ne 0$). Since $(x_k(u,v), y_k(u,v))$ is the only solution, $p_{XY}(x_k, y_k)dxdy$ constitutes the same incremental probability in the x,y–space for

$$p_{UV}(u,v)dudv \simeq P(u \le U(X,Y) \le u + du, v \le V \le v + dv)$$

$$= P((x_k(u,v) \le X \le x_k(u,v) + dx_k, \; y_k(u,v) \le Y \le y_k(u,v) + dy_k)$$

$$\simeq p_{XY}(x_k(u,v), y_k(u,v))dx_k dy_k.$$

But elementary calculus enables us to re-write the left side in terms $dxdy$ so that

$$p_{UV}(u,v) |J_k(u,v)| \, dx_k dy_k \simeq p_{XY}(x_k(u,v), y_k(u,v))dx_k dy_k$$

where $J_k(u,v)$ is the determinant of the Jacobian matrix $J(x,y)$ evaluated at the unique solution (x_k, y_k) of (9.4.1) for a given pair of (u,v),

$$J_k(u,v) = \det \begin{bmatrix} f_{,x}(x,y) & f_{,y}(x,y) \\ g_{,x}(x,y) & g_{,y}(x,y) \end{bmatrix}_{x=x_k(u,v), y=y_k(u,v)}, \qquad (9.4.2)$$

and $|J_k(u, v)|$ is the absolute value of $J_k(u, v)$. Since $(x_k(u, v), y_k(u, v))$ is the only pre-image of (u, v), we have in the limit of $(du, dv) \to (0, 0)$ (and hence $(dx_k, dy_k) \to (0, 0)$)

$$p_{UV}(u, v) = \frac{p_{XY}(x_k(u, v), y_k(u, v))}{|J_k(u, v)|}.$$

It should be noted that the notation $J(x, y)$ is used in some texts and journal articles to denote the Jacobian of n functions of n independent variables, e.g.,

$$J(x, y) = \begin{bmatrix} f_{,x}(x, y) & f_{,y}(x, y) \\ g_{,x}(x, y) & g_{,y}(x, y) \end{bmatrix} \tag{9.4.3}$$

for the two variable case, and not its determinant and certainly not its determinant evaluated at a pre-image of the transformed variables, $(x_k(u, v), y_k(u, v))$. In the transformation theory developed in this chapter, we only need the determinant of the Jacobian matrix and not the Jacobian matrix itself. That we need only the absolute value of the determinant evaluated at a pre-image point justifies our choice of notation. Had we let J be the Jacobian matrix, then we would have $|J|$ or $\det[J]$ for its determinant and then $|\det[J]|$ or $\|J\|$ for its absolute value. The former seems cumbersome and the latter may be taken for some norm of matrices or functions; hence, the notation (9.4.2) is preferred and adopted.

For the more general case of (9.4.1) where f and g are general functions, there may be more than one solutions for

$$f(x, y) = u, \quad g(x, y) = v, \tag{9.4.4}$$

denoted by $\{(x_k(u, v), y_k(u, v))\}$, $k = 1, 2, \ldots, m$, $p_{UV}(u, v)$ would be the sum of the contribution from all the different pre-image regions summarized in the theorem below:

Theorem 27. *For the transformation of two input random variables into two output random variables in the form of (9.4.1), the joint pdf of the output random variables is given in terms of the joint pdf of the input random variables by*

$$p_{UV}(u, v) = \sum_{k=1}^{m} \frac{p_{XY}(x_k(u, v), y_k(u, v))}{|J_k(u, v)|},$$

where $(x_k(u, v), y_k(u, v))$ is the kth root of (9.4.4) and $J_k(u, v)$ is the determinant of the Jacobian given by (9.4.2).

Exercise 38. Suppose X and Y are independent random variables with $p_X(x) = e^{-x} H(x)$ and $p_Y(y) = e^{-y} H(y)$. Find the joint pdf of $U = X + Y$ and $V = X/(X + Y)$.

9.4.2 More input than output variables

Transformation involving fewer output random variables $(Z_1, Z_2, \ldots, Z_m)$ than input r.v.'s $(X_1, X_2, \ldots, X_n)$ may be converted into one with equal number of input and output variables by adding some *auxiliary variables*. With $m < n$, we may add $\{Z_{m+1} = X_{m+1}, Z_{m+2} = X_{m+2}, \ldots, Z_n = X_n\}$. For the example of a single transformed variable Z of X and Y,

$$Z = XY,$$

we may add the auxiliary variable $W = X$. For a prescribed pair (z, w), we can solve for X and Y to get the unique inversion

$$X = w, \quad Y = z/w$$

with

$$J = J_k(u, v) = \det \begin{bmatrix} y & x \\ 1 & 0 \end{bmatrix} = x(z, w) = w.$$

We can then apply the theorem above to get

$$p_{ZW}(z, w) = \frac{p_{XY}(w, z/w)}{|w|}.$$

To get $p_Z(z)$ for the original problem, we integrate $p_{ZW}(z, w)$ to get

$$p_Z(z) = \int_{-\infty}^{\infty} \frac{1}{|w|} p_{XY}\left(w, \frac{z}{w}\right) dw.$$

Note that other choices of the additional output random variables are also possible and may be chosen to suit a specific purpose.

9.4.3 General vector random variables

With the addition of auxiliary variables as shown in the previous subsection, it suffices to consider only the case of the vector transformation $\overrightarrow{Z} = \overrightarrow{h}(\overrightarrow{X})$ of an n vector $\overrightarrow{X}$ where $h(\cdot)$ is an n-vector function. Given the joint pdf $p(\overrightarrow{X})$ and $\overrightarrow{Z} = \overrightarrow{h}(\overrightarrow{X})$ having a unique inverse $\overrightarrow{X} = \overrightarrow{h}^{-1}(\overrightarrow{Z})$. Then we have

$$P(Z \le A) = \int_{R_Z} p_Z(\overrightarrow{z}) d\overrightarrow{z} = \int_{R_X} p(\overrightarrow{x}) d\overrightarrow{x} = \int_{R_Z} p(\overrightarrow{h}^{-1}(\overrightarrow{z})) |J| d\overrightarrow{z} \quad (9.4.5)$$

where J is the determinant of the Jacobian matrix of $\overrightarrow{h}^{-1}(\overrightarrow{Z})$.

Proposition 34. *Under the hypothesis leading to (9.4.5), we have (as an alternative outcome in the development for the two variables case above)*

$$p_Z(\vec{z}) = p(\vec{h}^{-1}(\vec{z})) \, |J| \qquad (9.4.6)$$

where

$$J = \det\left[\nabla_z\left(\mathbf{h}^{-1}(\mathbf{z})\right)\right] = \det\begin{bmatrix} \frac{\partial x_1}{\partial z_1} & \frac{\partial x_2}{\partial z_1} & \cdots & \frac{\partial x_n}{\partial z_1} \\ \frac{\partial x_1}{\partial z_2} & \frac{\partial x_2}{\partial z_2} & & \cdot \\ \cdot & & & \cdot \\ \cdot & & & \cdot \\ \frac{\partial x_1}{\partial z_n} & \frac{\partial x_2}{\partial z_n} & \cdots & \frac{\partial x_n}{\partial z_n} \end{bmatrix}$$

with $\nabla_z(\mathbf{w}(\mathbf{z}))$ being the gradient of $\mathbf{w}(\mathbf{z})$ in $\mathbf{z} = (z_1, z_2, \ldots, z_n)^T$ space.

Remark 6. The expression for $p_Z(\mathbf{z})$ in the proposition above is consistent with the known mathematical fact

$$J_k(\mathbf{z}) = J^{-1}(\mathbf{x}(\mathbf{z})),$$

i.e., the determinant version of the inverse function theorem (that the matrix inverse of the Jacobian matrix of an invertible function is the Jacobian matrix of the inverse function).

9.4.4 *More output than input variables*

Several random variables $(Z_1, Z_2, \ldots, Z_m)$ may be outputs of a fewer number of input random variables $(X_1, X_2, \ldots, X_n)$ so that $n < m$. An example may be the location $L(t)$ and velocity $V(t)$

$$L(t) = A\cos(t) + B\sin(t), \quad V(t) = -A\sin(t) + B\cos(t) \qquad (9.4.7)$$

of a particle as functions of time starting with some initial position-velocity pair (A, B). Normally, we know the initial position of the particle with high degree of certainty, e.g., $A = 1$, but measurements of the initial velocity B are of considerable variability. Suppose we think of B as a random variable with pdf $p_B(b)$. Then for a fixed t, the quantities $L_t = L(t)$ and $V_t = V(t)$ are random variables with some joint pdf $p(x_t, v_t)$. How $p(\ell_t, v_t)$ is determined in terms of $p_B(b)$ is generally not addressed in texts on elementary probability.

Similar to what we did for the more input than output variables $(m < n)$ case in a previous subsection, our approach to transformations when $m > n$ is to augment the input variables by adding $\{X_{n+1}, X_{n+2}, \ldots, X_m\}$ with an attendant joint density function $p_X(\mathbf{x}) = p_X(x_1, x_2, \ldots, x_m)$ depending on our choice of the augmented variables. In our example of the two random variable L_t and V_t in which A is known to be 1 so that we really have

$$L(t) = \cos(t + B\sin(t)), \quad V(t) = -\sin(t) + B\cos(t), \qquad (9.4.8)$$

we may choose to augment the single random input variable B by working with (9.4.7) with A as the second random variable. By Proposition 34, we have for the joint density of the general case

$$p_Z(\mathbf{z}) = p_X(\mathbf{x}(\mathbf{z}))\left|J^{-1}(\mathbf{x}(\mathbf{z}))\right| \qquad (9.4.9)$$

where $p_X(\mathbf{x}(\mathbf{z}))$ is a suitably chosen joint density for the augmented set of input variables.

For our specific example, the absolute value of the relevant determinant of the Jacobian matrix is 1 and the joint pdf for the two new random variables should be $p_{AB}(a, b) = p_B(b)\delta(a - 1)$. By (9.4.6) or (9.4.9), we have for the output random variables

$$p(\ell_t, v_t) = p_0(a(\ell_t, v_t), b(\ell_t, v_t)) \cdot 1 = p_B(b(\ell_t, v_t))\delta(a(\ell_t, v_t) - 1).$$

To make use of this result, we need the inverse of (9.4.7):

$$A = L\cos(t) - V\sin(t), \quad B = L\sin(t) + V\cos(t). \qquad (9.4.10)$$

With (9.4.10), we can obtain useful information about the output r.v. such as the marginal distribution:

$$P(V_t \le v) = \int_{-\infty}^{v}\int_{-\infty}^{\infty} p_B(b(\ell_t, v_t))\delta(a(\ell_t, v_t) - 1)d\ell_t dv_t$$

$$= \int_{-\infty}^{v}\int_{-\infty}^{\infty} p_B(b(\ell_t, v_t))\delta(\ell_t\cos(t) - v_t\sin(t) - 1)d\ell_t dv_t.$$

(1) For $t \ne k\pi$ and $t \ne (k + \frac{1}{2}\pi)$, $k = 0, \pm 1, \pm 2, \ldots$, the relation above simplifies to

$$P(V_t \le v) = \int_{-\infty}^{v} [p_B(b(\ell_t, v_t))]_{\ell_t = \{v_t\sin(t)+1\}/\cos(t)}\, dv_t$$

$$= \int_{-\infty}^{v} p_B(\{v_t + \sin(t)\}/\cos(t))dv_t = \cos(t)P_B\left(B \le \frac{v + \sin(t)}{\cos(t)}\right)$$

where

$$P_B(B \le v) = \int_{-\infty}^{v} p_B(b)db.$$

(2) For $t = k\pi$, we have instead

$$P(V_t \leq v) = \int_{-\infty}^{v} \int_{-\infty}^{\infty} p_B(b(\ell_t, v_t))\delta((-)^k\ell_t - 1)d\ell_t dv_t$$

$$= \int_{-\infty}^{v} p_B((-)^k v_t)dv_t = (-)^k \int_{-(-)^k\infty}^{(-)^k v} p_B(y)dy$$

$$= \begin{cases} P_B(B \leq v) & (k \text{ even}) \\ 1 - P_B(B \leq -v) & (k \text{ odd}) \end{cases}$$

(3) If $t = (k + \frac{1}{2}\pi)$ instead, we would have

$$P(V_t \leq v) = \int_{-\infty}^{v} \int_{-\infty}^{\infty} p_B((-)^k\ell_t))\delta((-)^k v_t - 1)d\ell_t dv_t$$

$$= \int_{-\infty}^{v} \delta((-)^k v_t - 1)dv_t$$

$$= \begin{cases} H(v - 1) & (k \text{ even}) \\ H(1 - v) & (k \text{ odd}) \end{cases}$$

where $H(x)$ is the Heaviside unit step function.

Exercise 39. Obtain the marginal distribution $P(L_t \leq \ell)$ for the two random variable L_t and V_t in (9.4.8).

9.5   Applications to Stochastic ODE

As an important class of applications of the transformation theory developed above, we consider here the IVP for the ODE system,

$$X'_j = f_j(X_1, \ldots, X_n, t; C_1, \ldots, C_m), \quad X_j(t_0) = A_j, \quad (j = 1, 2, \ldots, n), \quad (9.5.1)$$

where the initial data $\mathbf{A} = (A_1, A_2, \ldots, A_n)$ and the m (time-invariant) coefficients $\mathbf{C} = (C_1, C_2, \ldots, C_m)$ may be *random variables* with known statistics. (Note that we could have also written the scalar ODE system as a vector ODE for $\mathbf{X}(t) = (X_1(t), X_2(t), \ldots, X_n(t))$.) In principle, the solution of the corresponding deterministic IVP is a vector function of time t with $\{A_j\}$ and $\{C_k\}$ as parameters, denoted by $\mathbf{X}(t; \mathbf{A}, \mathbf{C})$. When $\mathbf{C}$ and $\mathbf{A}$ are vector random variables, $\mathbf{X}_\tau = \mathbf{X}(\tau; \mathbf{A}, \mathbf{C})$, for any fixed instant of time τ, is a vector random variable whose statistics depend on the known statistics of $\mathbf{C}$ and $\mathbf{A}$. For example, the joint probability density function of the components of $\mathbf{X}_\tau$, $p_{\mathbf{X}}(x_1, \ldots, x_n; \tau)$, depends on the corresponding input joint probability density functions of $\mathbf{C}$ and $\mathbf{A}$, $p(c_1, \ldots, c_m, a_1, \ldots, a_n) = p(\mathbf{c}, \mathbf{a})$. The determination of the former in terms of the latter is seen to be a possible application of the transformation theory.

9.5.1   *Initial data*

Example 31. $X' = 2X, \qquad X(0) = A$

The exact solution of this simple IVP is $X(t; A) = Ae^{2t}$. We are interested in the case where A is a random variable with probability density function $p_A(a)$ and would like to know the corresponding pdf, $p_X(x; t)$ of X_t and the probability distribution $P_X(x; t) = P(X_t \leq x)$. With

$$A = Xe^{-2t}, \quad \frac{da}{dx} = e^{-2t},$$

we have from Theorem 24

$$p_X(x; t) = p_A(xe^{-2t})e^{-2t} \quad (t > 0). \tag{9.5.2}$$

The corresponding probability distribution is

$$P(X_t \leq x; t) = \int_{-\infty}^{x} p_A(ze^{-2t})e^{-2t}dz \quad (t > 0). \tag{9.5.3}$$

Alternatively, the same result can also be obtained directly from $p_A(a)$ by

$$P(X_t \leq x; t) = \int_{-\infty}^{x} p_X(x; t)dx = \int_{-\infty}^{a(x)} p_A(a)da = \int_{-\infty}^{xe^{-2t}} p_A(a)da. \tag{9.5.4}$$

For the process $X(t)$ to be initiated by the initial data A at $t = 0$ (and does not exist prior to that initial time) that is known to have a pdf $p_A(a) = e^{-\lambda a}H(a)$ with $\lambda > 0$ and $H(z)$ being the Heaviside unit step function, we have from (9.5.3)

$$P(X_t \leq x; t) = \int_{-\infty}^{x} e^{-\lambda z e^{-2t}} H(ze^{-2t})e^{-2t}dz$$

$$= \int_{0}^{x} e^{-\lambda z e^{-2t}} e^{-2t}dz = \frac{1}{\lambda}\left(1 - e^{-\lambda x e^{-2t}}\right)$$

and the same result from (9.5.4).

Having (9.5.2), we can compute the statistics of the response $X(t)$ for any fixed positive t. For example, the expected value of $X(t)$ is given by

$$E[X(t)] = \int_{-\infty}^{\infty} xp_X(x; t)dx = \int_{-\infty}^{\infty} xp_A(xe^{-2t})e^{-2t}dx$$

$$= e^{2t} \int_{-\infty}^{\infty} zp_A(z)dz.$$

From this result, we see immediately that

(1) $E[X(t)] = 0$ if $p_A(z)$ is even.
(2) $E[X(t)] = 2e^{2t} \int_{0}^{\infty} zp_A(z)dz$ if $p_A(z)$ is odd.

Example 32. $X' = -X^2$, $X(0) = A$.

The exact solution for this IVP is

$$X(t; A) = \frac{A}{1 + At}$$

with

$$A = \frac{X_t}{1 - X_t t}, \quad 1 + At = \frac{1}{1 - X_t t}$$

$$\frac{dX_t}{dA} = \frac{1}{(1 + At)^2} = (1 - X_t t)^2.$$

It follows from Corollary 12 that

$$p_X(x; t) = \frac{1}{(1 - xt)^2} p_A \left(\frac{x}{1 - xt} \right).$$

Exercise 40. Determine $P_X(x)$ in terms of $p_A(a)$.

Example 33. $X' = -\frac{1}{2}X^3$, $X(0) = A$.

The exact solution of the IVP is

$$X(t; A) = \frac{A}{\sqrt{1 + A^2 t}}.$$

For a fixed t, the inverse of the relation above is multi-valued given by

$$A^{(1)}(X_t) = \frac{X_t}{\sqrt{1 - tX_t^2}}, \quad A^{(2)}(X_t) = \frac{-X_t}{\sqrt{1 - tX_t^2}}$$

with

$$\frac{dA^{(1)}}{dX_t} = \frac{1}{(1 - tX_t^2)^{3/2}}, \quad \frac{dA^{(2)}}{dX_t} = \frac{-1}{(1 - tX_t^2)^{3/2}}.$$

It follows from Theorem 24 that

$$p_X(x; t) = \frac{1}{(1 - x^2 t)^{3/2}} \left[p_A \left(a^{(1)}(x) \right) + p_A \left(a^{(2)}(x) \right) \right].$$

Example 34. $\mathbf{X}' = \begin{bmatrix} 0 & 1 \\ -1 & 0 \end{bmatrix} \mathbf{X}$, $\mathbf{X}(0) = (A_1, A_2)^T = \mathbf{A}$.

The exact solution of this IVP is well known to be

$$X_1(t) = A_1 \cos(t) + A_2 \sin(t), \quad X_2(t) = -A_1 \sin(t) + A_2 \cos(t)$$

with the following determinant of its Jacobian matrix

$$|J_X| = \begin{vmatrix} \cos(t) & \sin(t) \\ -\sin(t) & \cos(t) \end{vmatrix} = 1$$

and a unique inverse

$$A_{1t} = X_{1t} \cos(t) - X_{2t} \sin(t), \quad A_{2t} = X_{1t} \sin(t) + X_{2t} \cos(t).$$

By Proposition 34, we have the following pdf for $\mathbf{X}_t$

$$p_X(\mathbf{x};t) = p_A(\mathbf{a}(\mathbf{x};t)) |J_A| = \frac{1}{|J_X|} p_A(\mathbf{a}(\mathbf{x};t)) = p_A(\mathbf{a}(\mathbf{x};t)). \qquad (9.5.5)$$

As in an earlier example of the trajectory of a moving particle, the initial location of the particle is usually know accurately but the initial velocity is more difficult to know accurately. With $A_1 = 1$ and $A_2 = B$ a random variable (with some estimated mean), we have again a case of more output random variables (position X_1 and velocity X_2) than the single input random variable B. For a fixed instant in time, the joint pdf for the two output has already been determined in the previous section (see subsection 9.4.4). Here we would like to obtain the expected location of the particle at some fixed time t given generally by

$$E[X_1(t)] = \int_{-\infty}^{\infty} \int_{-\infty}^{\infty} x_1 p_X(x_1, x_2, t) dx_1 dx_2.$$

Instead of making use of the transformation (9.5.5) for our example, we work with

$$E[X_1(t)] = \int_{-\infty}^{\infty} \int_{-\infty}^{\infty} x_1(a_1, a_2, t) p_A(a_1, a_2, t) da_1 da_2$$

$$= \int_{-\infty}^{\infty} \int_{-\infty}^{\infty} \{a_1 \cos(t) + a_2 \sin(t)\} p_B(a_2) \delta(a_1 - 1) da_1 da_2$$

$$= \int_{-\infty}^{\infty} \{\cos(t) + a_2 \sin(t)\} p_B(a_2) da_2$$

$$= \cos(t) + \sin(t) \int_{-\infty}^{\infty} a_2 \sin(t) p_B(a_2) da_2 = \cos(t) + \sin(t) E[B].$$

The result above for our simple problem could have been obtained more simply by taking the expectation of both sides of the formal solution (for $A_1 = 1$)

$$X_1(t) = \cos(t) + A_2 \sin(t),$$

to get

$$E[X_1(t)] = \cos(t) + E[A_2]\sin(t).$$

Our more circuitous calculation demonstrates the consistency of the more general approach of adding more input random variables so that we can appeal to Proposition 34.

9.5.2 *Random forcing*

Example 35. $X' = c_1 X + C_2, \qquad X(0) = 3.$

For this example, the additive term C_2 on the right does not involve the unknown $X(t)$ and is known as a forcing term or a source term in deterministic ODE problems. The special case of $c_1 = 0$ has as the exact solution $X(t) = C_2 t + 3$ for the IVP. When C_2 is a random variable with pdf $p_C(c_2)$, the pdf for the response X_t of $X(t) = C_2 t + 3$ at the instant t is given by Theorem (24) to be

$$p_X(x; t) = \frac{1}{t} p_C((x - 3)/t).$$

When c_1 is a known constant and not equal to 0, the exact solution of the IVP is

$$X(t) = \left(3 + \frac{C_2}{c_1}\right) e^{c_1 t} - \frac{C_2}{c_1}$$

with

$$\frac{dX}{dC_2} = \frac{1}{c_1}(e^{c_1 t} - 1)$$

The equation for $X(t)$ may be solved for C_2 in terms of X_t to get

$$C_2 = \frac{c_1 (X_t - 3e^{c_1 t})}{e^{c_1 t} - 1}.$$

The pdf for X_t is then given by

$$p_X(x; t) = \frac{c_1}{e^{c_1 t} - 1} p_C\left(\frac{c_1 (x - 3e^{c_1 t})}{e^{c_1 t} - 1}\right).$$

Example 36. $\mathbf{X}' = \begin{bmatrix} 0 & 1 \\ -1 & 0 \end{bmatrix} \mathbf{X} + \begin{pmatrix} C_1 \\ C_2 \end{pmatrix}, \qquad \mathbf{X}(0) = (0, 0)^T.$

For this example, the forcing terms C_1 and C_2 are random variables. The exact solution of the IVP is

$$X_1 = -C_2 \cos(t) + C_1 \sin(t) + C_2, \quad X_2 = C_2 \sin(t) + C_1 \cos(t) - C_1$$

with the following determinant of its Jacobian matrix

$$\det[J_X] = \det \begin{bmatrix} \frac{\partial X_1}{\partial C_1} & \frac{\partial X_1}{\partial C_2} \\ \frac{\partial X_2}{\partial C_1} & \frac{\partial X_2}{\partial C_2} \end{bmatrix} = \det \begin{bmatrix} \sin(t) & -\cos(t) \\ \cos(t) & \sin(t) \end{bmatrix} = 1$$

and a unique inverse

$$C_1 = \frac{1}{2} \left\{ \frac{\sin(t)}{1 - \cos(t)} X_{1t} - X_{2t} \right\}, \quad C_2 = \frac{1}{2} \left\{ X_{1t} + \frac{\sin(t)}{1 - \cos(t)} X_{2t} \right\}.$$

By Proposition 34, we have the following pdf for $\overrightarrow{X}_t$

$$p_X(\mathbf{x}; t) = p_A(\mathbf{a}(\mathbf{x}; t)) |J_A| = \frac{1}{|\det[J_X]|} p_C(\mathbf{c}(\mathbf{x}; t)) = p_C(\mathbf{c}(\mathbf{x}; t)).$$

Example 37. $X' + X = CX^\beta, \quad X(0) = 1 \quad (0 < \beta < 1).$

The ODE is a Bernoulli equation and may be transformed into a linear ODE by a change of variable $Y = X^{1-\beta}$. Upon divide both sides by X^β, we may re-arrange the resulting equation into

$$Y' + (1 - \beta)Y = (1 - \beta)C, \quad Y(0) = 1,$$

which is a linear ODE with random forcing when C is a random variable. The exact solution for the new IVP is

$$Y(t) = C \left\{ 1 - e^{-(1-\beta)t} \right\} = X_t^{1-\beta}$$

with

$$C = \frac{Y_t}{1 - e^{-(1-\beta)t}}, \quad \frac{dC}{dX_t} = \frac{(1 - \beta) X_t^{-\beta}}{1 - e^{-(1-\beta)t}}.$$

It follows from Theorem (24) that the pdf of X_t is given by

$$p_X(x; t) = \frac{1 - \beta}{x^\beta \left\{ 1 - e^{-(1-\beta)t} \right\}} p_C \left(\frac{x^{1-\beta}}{1 - e^{-(1-\beta)t}} \right).$$

9.5.3   Random coefficients

The solutions for problems worked out in this and the next subsection are straightforward applications of the transformation theory developed in this chapter. They are presented here to provide validation for a more general method of solution to be discussed in the next chapter for problems for which there are no explicit solutions (so that the transformation theory is not useful).

Example 38. $X' = CX$, $X(0) = 1$.

The exact solution for this simple problem is $X(t) = e^{Ct}$ for $t > 0$. Evidently $X(t) > 0$ for all t and C. As such, $x \leq 0$ is unreachable and hence $p_X(x) = 0$ for all $x \leq 0$. From this solution, we get

$$\frac{dX_t}{dC} = te^{Ct}, \quad C = \frac{1}{t}\ln(X_t).$$

The pdf for X_t is given by Theorem 24 (and the fact that $x \leq 0$ is unreachable)

$$p_X(x;t) = \frac{e^{-Ct}}{t}p_C\left(\frac{\ln(x)}{t}\right)H(x) = \frac{1}{tx}p_C\left(\frac{\ln(x)}{t}\right)H(x) \quad (t > 0).$$

Example 39. $X' = CX^2$, $X(0) = 1$.

The exact solution is

$$X(t) = \frac{1}{1 - Ct}$$

with

$$\frac{dX}{dC} = \frac{t}{(1 - Ct)^2} = tX^2, \quad C = \frac{X - 1}{tX}.$$

Note that unlike the previous example, there is no unreachable region for the pre-image set. By Theorem 24, the pdf of X_t is given by

$$p_X(x;t) = \frac{1}{tx^2}p_C\left(\frac{x - 1}{tx}\right). \tag{9.5.6}$$

Exercise 41. (a review of ODE)

$$x' = -x + y + Cx^2, \quad y' = y(1 - 2Cx), \quad x(0) = 1, \quad y(0) = 0$$

Show that the only solution of the IVP above requires $y(t) \equiv 0$ *and*

$$x' = -x + Cx^2, \quad x(0) = 1. \tag{9.5.7}$$

<u>Hint</u>: Show that the corresponding first order ODE

$$\frac{dy}{dx} = \frac{y(1 - 2Cx)}{-x + y + Cx^2}$$

is exact whose solution may be taken in the form

$$xy - \frac{1}{2}y^2 - Cx^2y = C_0$$

where C_0 is a constant of integration to be determined by the prescribed initial conditions.

Exercise 42. (a review of ODE) Show that the exact solution of the IVP (9.5.7) is

$$x(t) = \frac{1}{C + (1 - C)e^t}.$$

Exercise 43. Suppose C is a random variable with pdf $p_C(c)$. For a fixed t, determine the pdf of the random variable X_t.

9.5.4 Multi-input random variables

Example 40. $X' = C, \qquad X(0) = A.$

The exact solution of this simple IVP is $X(t) = Ct + A$. Suppose that both C and A are random variables with a joint pdf $p_{AC}(a, c)$. To determine the pdf of X_t for a fixed t, we begin with the probability distribution

$$P_X(x) = P(X_t \le x) = \int\int_{D_x} p_{AC}(a, c)\,da\,dc$$

$$= \int_{-\infty}^{\infty} \int_{-\infty}^{x - ct} p_{AC}(a, c)\,da\,dc \quad (t > 0).$$

We then we differentiate $P_X(x)$ to get

$$p_X(x; t) = \int_{-\infty}^{\infty} p_{AC}(x - ct, c)\,dc.$$

Example 41. $X' = CX, \qquad X(0) = A.$

C and A are random variables with joint pdf $p_{CA}(c, a)$. With the exact solution $X(t) = Ae^{Ct}$, we have for the probability distribution for X_t

$$P_X(x) = P(X_t \le x) = \int_{-\infty}^{\infty} \int_{-\infty}^{xe^{-ct}} p_{AC}(a, c)\,da\,dc. \tag{9.5.8}$$

Upon differentiating with respect to x, we get

$$p_X(x; t) = \int_{-\infty}^{\infty} p_{AC}(xe^{-ct}, c)e^{-ct}\,dc. \tag{9.5.9}$$

Exercise 44. Obtain $P_X(x) = P(X_t \le x)$ and $p_X(x; t)$ but now first integrate with respect to c to get

$$p_X(x; t) = \frac{1}{tx} \int_{-\infty}^{\infty} p_{AC}\left(a, \frac{1}{t}\ln\left(\frac{x}{a}\right)\right) da. \tag{9.5.10}$$

Example 42. $X' = Ct + B, \quad X(0) = A.$

For the three random variables A, B, and C with joint density function $p_{CBA}(c, b, a)$, we have as the probability distribution for the random variable X_t

$$P_X(x) = P(X_t \leq x) = \int \int \int_{D_x} p_{CBA}(c, b, a) da\, db\, dc.$$

With $X(t) = A + Bt + \frac{1}{2}Ct^2$ as the exact solution of the IVP, the region D_x is $A + Bt + \frac{1}{2}Ct^2 \leq x$. We may then write the triple integral as

$$P_X(x) = \int_{-\infty}^{\infty} \int_{-\infty}^{\infty} \int_{-\infty}^{x - bt - \frac{1}{2}ct^2} p_{CBA}(c, b, a) da\, db\, dc.$$

From this we obtain

$$p_X(x; t) = \frac{dP_X(x)}{dx} = \int_{-\infty}^{\infty} \int_{-\infty}^{\infty} p_{CBA}(c, b, x - bt - \frac{1}{2}ct^2) db\, dc.$$

Example 43. $X' = -X + C, \quad X(0) = A.$

With $X(t) = (A - C)e^{-t} + C$, we may take

$$P_X(x) = P(X_t \leq x) = \int_{-\infty}^{\infty} \int_{-\infty}^{(x-c)e^t + c} p_0(c, a) da\, dc.$$

It follows that

$$p_X(x; t) = \frac{dP_X(x)}{dx} = e^t \int_{-\infty}^{\infty} p_0(c, (x - c)e^t + c) dc.$$

Exercise 45. Determine the pdf of X_t in terms of the joint pdf $p_0(c, b, a)$ for the IVP $X' = CX + B, \ X(0) = A$ where A, B and C are random variables with a joint pdf $p_{CBA}(c, b, a)$.

9.6 Liouville Equation for Random Initial Data

9.6.1 The Liouville equation

When the solution of the general IVP

$$\mathbf{x}'(t) = \mathbf{f}(\mathbf{x}(t), t), \quad \mathbf{x}(t_0) = \mathbf{x}^o,$$

for the vector function $\mathbf{x} = (x_1, x_2, \ldots, x_n)$ is known to exist, it is not always possible to obtain an explicit expression of the exact solution of the problem in terms of known functions. In this section, we develop a general method for obtaining probabilistic information about the solution for this problem when the only randomness of the problem is in the initial data of the unknown. When the initial data is a vector random variable $\mathbf{A}$ with a prescribed pdf $p_A(\mathbf{a})$, the corresponding solution at a particular instant of time is a random variable $\mathbf{X}_t$ with a pdf that depends on the pdf of the initial data. This more general approach requires that we solve a partial differential equations (PDE) known as the *Liouville equation* for the (joint) pdf $p(\mathbf{x},t)$.

To simplify notations somewhat, we often write $\{p(\mathbf{x},t), p_0(\mathbf{a})\}$ for $\{p_{\mathbf{X}}(\mathbf{x},t), p_A(\mathbf{a})\}$ in the subsequent development unless indicated otherwise.

Theorem 28. (*Liouville*) *Suppose the solution of the nonlinear stochastic IVP*

$$\mathbf{X}'(t) = \mathbf{f}(\mathbf{X}(t),t), \quad \mathbf{X}(t_0) = \mathbf{A} \tag{9.6.1}$$

exists. Then the probability density function $p(\mathbf{x},t) = p_X(x_1,\ldots,x_n;t)$ *is a solution of the first order PDE (known as the Liouville equation)*

$$\frac{\partial p}{\partial t} + \sum_{k=1}^{n} \frac{\partial (pf_j)}{\partial x_j} = 0 \tag{9.6.2}$$

where $\mathbf{f}(\mathbf{x},t) = (f_1(\mathbf{x},t),\ldots,f_n(\mathbf{x},t))^T$ *is the right hand side of the stochastic DE* (9.6.1).

Proof. Since taking expectation does not involve operating on the parameter t, we may interchange the order of that operation and differentiation with respect to t so that

$$E\left[\frac{\partial}{\partial t}\left(e^{i\mathbf{u}\cdot\mathbf{X}(t)}\right)\right] = \frac{\partial}{\partial t}\left(E\left[e^{i\mathbf{u}\cdot\mathbf{X}(t)}\right]\right) = \frac{\partial \left[\hat{p}(\mathbf{u};t)\right]}{\partial t}, \tag{9.6.3}$$

where $\hat{p}(\mathbf{u};t)$ is the joint characteristic function of the joint pdf $p(\mathbf{x},t)$ given by

$$\hat{p}(\mathbf{u};t) = E[e^{i\mathbf{u}\mathbf{X}}] = \int_{-\infty}^{\infty}\cdots\int_{-\infty}^{\infty} e^{i\mathbf{u}\cdot\mathbf{x}}p(\mathbf{x},t)dx_1\cdots dx_n.$$

Note that $\hat{p}(\mathbf{u};t)$ is also the multi-variate Fourier transform of the pdf $p(\mathbf{x},t)$ in $\mathbf{x}$. With

$$\frac{\partial}{\partial t}\left(e^{i\mathbf{u}\cdot\mathbf{X}(t)}\right) = i\mathbf{u}\cdot\frac{d\mathbf{X}(t)}{dt}e^{i\mathbf{u}\cdot\mathbf{X}(t)} = i\mathbf{u}\cdot\mathbf{f}(\mathbf{X}(t),t)e^{i\mathbf{u}\cdot\mathbf{X}(t)}, \tag{9.6.4}$$

the relation (9.6.3) may be written as

$$\frac{\partial \hat{p}(\mathbf{u};t)}{\partial t} = E[i\sum_{k=1}^{n} u_k f_k(\mathbf{X}(t),t)e^{i\mathbf{u}\cdot\mathbf{x}(t)}]$$

$$= \int_{-\infty}^{\infty}\cdots\int_{-\infty}^{\infty}\left\{i\left[\sum_{k=1}^{n} u_k f_k(\mathbf{X}(t),t)\right]e^{i\mathbf{u}\cdot\mathbf{x}(t)}p(\mathbf{x},t)\right\}dx_1\cdots dx_n.$$

$$\tag{9.6.5}$$

The right hand side is known to be the multi-variate Fourier transform of

$$-\sum_{k=1}^{n} \frac{\partial}{\partial x_k} [f_k(\mathbf{X}(t), t)p(\mathbf{x}, t)] = -\nabla \cdot (p(\mathbf{x}, t)\mathbf{f}(\mathbf{x}, t)).$$

Taking the inverse Fourier transform of both sides of (9.6.5), we obtain

$$\frac{\partial p(\mathbf{x};t)}{\partial t} + \nabla \cdot (p(\mathbf{x}, t)\mathbf{f}(\mathbf{x}, t)) = 0.$$

$\square$

With t as a parameter, this theorem enables us to reduce the problem of determining the (first order) pdf $p(\mathbf{x}, t)$ of the random variable $\mathbf{X}_t$ to solving a single first order linear (deterministic) partial differential equation for $p(\mathbf{x}, t)$ augmented by the initial condition

$$p(\mathbf{x}, 0) = p_0(\mathbf{a}).$$

Shown below are the Liouville equations for a few familiar IVP for SDE.

Example 44. $X'(t) = cX(t)$, $X(0) = A$ where c is a constant and A is a second order random variable with a prescribed density function $p_0(a)$.

The Liouville equation for this problem is

$$\frac{\partial p}{\partial t} + cx\frac{\partial p}{\partial x} + cp = 0$$

supplemented by the initial condition

$$p(x, 0) = p_0(a).$$

Example 45. $\qquad X'' + \omega^2 X = 0, \quad X(0) = X^o, \quad X'(0) = V^o.$

Let $X_1 = X(t)$ and $X_2 = X'(t)$ to write the ODE as

$$\mathbf{X}' = \begin{bmatrix} 0 & 1 \\ -\omega^2 & 0 \end{bmatrix} \mathbf{X}, \qquad \mathbf{X}(0) = \mathbf{A} = \begin{pmatrix} X^o \\ V^o \end{pmatrix}.$$

For this problem, we have $\partial f_k/\partial x_k = 0$, $k = 1, 2$. In that case, the PDE (9.6.2) becomes

$$\frac{\partial p}{\partial t} + x_2\frac{\partial p}{\partial x_1} - \omega^2 x_1\frac{\partial p}{\partial x_2} = 0.$$

The reduced equation is a first order linear PDE in three independent variables. It can be solved by the method of characteristics with the initial condition

$$p(\mathbf{x}, 0) = p_0(\mathbf{a}) = p_0(x^o, v^o).$$

It remains to solve the IVP for the Liouville equation to obtain the same solution as found by transformation theory.

9.6.2 *Solution by method of characteristics*

The Liouville equation for the pdf of the solution of the stochastic IVP (9.6.1) may be re-written as

$$\frac{\partial p}{\partial t} + \sum_{k=1}^{n} f_j \frac{\partial p}{\partial x_j} + p\boldsymbol{\nabla} \cdot \mathbf{f} = 0 \qquad (9.6.6)$$

where

$$\boldsymbol{\nabla} \cdot \mathbf{f} = \sum_{k=1}^{n} \frac{\partial f_j}{\partial x_j}.$$

Equation (9.6.6) is a first order linear PDE for the pdf $p(\mathbf{x}, t)$ of the independent variables $(t, x_1, x_2, \ldots, x_n)$. For this equation, the characteristic equations are

$$\frac{dt}{d\xi} = 1, \quad \frac{dp}{d\xi} = -p\boldsymbol{\nabla} \cdot \mathbf{f}, \quad \frac{dx_j}{d\xi} = f_j \quad (j = 1, 2, \ldots, n).$$

The characteristic system being autonomous, the first two equation can be combined as a single equation

$$\frac{dp}{dt} = -p\boldsymbol{\nabla} \cdot \mathbf{f}(\mathbf{x}, t)$$

or

$$p(\mathbf{x}, t) = p(\mathbf{x}(0), 0) \exp\left\{ -\int^{t} \boldsymbol{\nabla} \cdot \mathbf{f}(\mathbf{x}(\tau), \tau) d\tau \right\}$$

$$= p_0(\mathbf{a}) \exp\left\{ -\int^{t} \boldsymbol{\nabla} \cdot \mathbf{f}(\mathbf{h}(\mathbf{a}, \tau), \tau) d\tau \right\}_{\mathbf{x}^\circ = \mathbf{h}^{-1}(\mathbf{x}, t)}.$$

The remaining characteristic equations may be combined to give

$$\frac{dx_j}{dt} = f_j(x_1, \ldots, x_n, t), \quad x_j(0) = a_j \quad (j = 1, 2, \ldots, n)$$

which merely reproduce the original IVP.

9.6.3 *Some illustrating examples*

Example 46. $X'(t) = cX(t)$, $X(0) = A$ where c is a constant and A is a second order random variable with a prescribed density function $p_0(a)$.

The Liouville equation for this problem is

$$\frac{\partial p}{\partial t} + cx\frac{\partial p}{\partial x} + cp = 0,$$

with $p(x(0), 0) = p_0(a)$. The characteristic equations for this problem is

$$\frac{dt}{ds} = 1, \quad \frac{dx}{ds} = cx, \quad \frac{dp}{ds} = -cp.$$

The first two equations can be combined to give a deterministic version of the original stochastic DE with the explicit solution

$$x(t) = ae^{ct} \quad \text{or} \quad a = xe^{-ct}$$

while the first and third give

$$p(x, t) = p_0(a)e^{-ct} = p_0(xe^{-ct})e^{-ct}$$

which is the same result as that obtained through transformation theory.

Example 47. $\quad X' = cX^2, \quad X(0) = A.$

Here c is a known constant and A is a second order random variable with a prescribed density function $p_0(a)$.

The Liouville equation for this SDE is

$$\frac{\partial p}{\partial t} + cx^2\frac{\partial p}{\partial x} + 2cxp = 0,$$

with $p(x(0), 0) = p_0(a)$. The characteristic equations for this PDE are

$$\frac{dx}{dt} = cx^2, \quad \frac{dp}{dt} = -2cxp.$$

They are augmented by the initial conditions

$$x(0) = a, \quad p(0) = p_0(a).$$

The solution of the IVP for these characteristic equation is

$$x(t) = \frac{a}{1 - act}, \quad p(t) = p_0(a)\left(1 - act\right)^2$$

with

$$a = \frac{x}{1 + xct}.$$

Upon using the inverse relation above to eliminate a from $p(t)$, we get

$$p(x, t) = p_0\left(\frac{x}{1 + xct}\right)\frac{1}{(1 + xct)^2}.$$

When $c = -1$, the solution above is the same as that obtained by the transformation theory in the last chapter.

Example 48. The mass-spring problem (previously recast) in the form

$$\mathbf{X}' = \begin{bmatrix} 0 & 1 \\ -\omega^2 & 0 \end{bmatrix} \mathbf{X}, \qquad \mathbf{X}(0) = \mathbf{A}.$$

For this problem, we have

$$\boldsymbol{\nabla} \cdot \mathbf{f}(\mathbf{x}(\tau),\tau) = 0.$$

The corresponding Liouville equation (9.6.2) is

$$\frac{\partial p}{\partial t} + x_2 \frac{\partial p}{\partial x_1} - \omega^2 x_1 \frac{\partial p}{\partial x_2} = 0,$$

augmented by the initial condition

$$p(\mathbf{x}(0),0) = p_0(\mathbf{a}).$$

The characteristic IVP consists of

$$\frac{dx_1}{dt} = x_2, \quad \frac{dx_2}{dt} = -\omega^2 x_1, \quad \frac{dp}{dt} = 0,$$
$$x_1(0) = a_1, \quad x_2(0) = a_2, \quad p(0) = p_0(a_1, a_2).$$

The part of the solution of this IVP to be singled out is $p(t)$ being a constant so that

$$p(x_1, x_2; t) = p_0(a_1, a_2).$$

The rest of the characteristic curve characterization is given by

$$x_1 = a_1 \cos \omega t + \frac{1}{\omega} a_2 \sin \omega t, \quad x_2 = -\omega a_1 \sin \omega t + a_2 \cos \omega t.$$

Upon solving for a_1 and a_2, we obtain

$$a_1 = x_1 \cos \omega t - \frac{1}{\omega} x_2 \sin \omega t, \quad a_2 = \omega x_1 \sin \omega t + x_2 \cos \omega t.$$

The *pdf* for stochastic process $\mathbf{X}(t)$ is therefore

$$p(x_1, x_2; t) = p_0(a_1, a_2)$$
$$= p_0(x_1 \cos \omega t - \frac{1}{\omega} x_2 \sin \omega t, \ \omega x_1 \sin \omega t + x_2 \cos \omega t).$$

Example 49. (Dynamics of glucose and insulin) Glucose is a key metabolic substrate in many physiological processes; abnormally low or high level of glucose would result in serious health issues. The control and regulation of blood glucose is by pancreatic hormone insulin. The normal balance of glucose and insulin in an healthy individual can be disrupted by an increased level of glucose in blood (after eating for example) while the mechanism for generating the needed level of counteracting insulin may fail to perform its role of upregulating insulin (caused by pancreatic abnormality for example). For the purpose of medical intervention, we first investigate the tolerance level of a normal individual's constitution.

For a normal person, there is usually a balanced steady state level (corresponding to a fixed point of a dynamical system model) of glucose and insulin in the blood stream. Let the deviations from this steady state level be denoted by G and I, respectively. The excess of glucose concentration (initiated by a small one time uptake) generally decreases due to a (slow) natural degradation of (what is left of the excess) and its neutralization by the insulin presence. This change is taken to be in the form of the first differential equation below. Similarly, there are also adjustments of insulin concentration inherent in humane physiology from natural degradation on the one hand and additional production of insulin induced by pancreatic response to glucose excess. This leads to the second equation in the following second order dynamical system model for the glucose dynamics:

$$\frac{d}{dt}\begin{pmatrix} G \\ I \end{pmatrix} = \begin{bmatrix} -\delta & -\gamma \\ \beta & -\alpha \end{bmatrix}\begin{pmatrix} G \\ I \end{pmatrix} = A\begin{pmatrix} G \\ I \end{pmatrix}$$

The rate constants δ, γ, β and α are to be estimated from the various glucose and insulin measurements (in canines or humans). Suppose at $t = 0$, there is a one time abnormal uptake of glucose corresponding to the initial conditions

$$G(0) = G^o, \quad I(0) = 0.$$

Of interest is the evolution of the glucose concentration with time. Given the assumptions made, it is not surprising that the change due to the one time uptake would decrease with time. The question is whether it decreases sufficiently (to nearly zero) before the next abnormal uptake, taken here to be a random variable with known first and second order statistics.

For the Liouville equation approach, we need the important factor $\nabla \cdot \mathbf{f}$ which in this case is

$$\nabla \cdot \mathbf{f}(\mathbf{x},t) = \frac{\partial f_1}{\partial g} + \frac{\partial f_2}{\partial i} = -\alpha - \delta.$$

The corresponding Liouville equation (9.6.2) is

$$\frac{\partial p}{\partial t} - (\delta g + \gamma i)\frac{\partial p}{\partial g} - (\alpha g - \beta i)\frac{\partial p}{\partial i} - (\alpha + \delta)p = 0,$$

to be augmented by the initial conditions

$$p(g, i; 0) = p_0(a_1, a_2),$$

with the two random variables a_1 and a_2 corresponding to G^o and I^o.

Exercise 46. Determine by the method of characteristics the solution for $p(g, i; t)$.

Exercise 47. The production of red blood cells (RBC) in a human body has been modeled by the following system of ODE

$$\frac{dR}{dt} = -\alpha R + \beta M, \quad \frac{dM}{dt} = \gamma(R_c - R) - \varepsilon M. \tag{9.6.7}$$

Each day, RBC (in concentration R) are lost at the rate $-\alpha R$ but replenish by new ones produced by available bone marrow (in concentration M) at the rate βM. Bone marrow concentration naturally decreases with age at a very slow rate $-\varepsilon M$ while a concentration of RBC below the normal concentration R_c stimulates some production of bone marrow to help with additional RBC production. The evolution of $R(t)$ and $M(t)$ according to the prescribed dynamics leads to the dynamical system model (9.6.7). Through these reactions, human bodies maintain an adequate amount of RBC for a healthy existence. The question of interest is whether a perturbation from this ongoing steady state characterized by a set of random perturbed initial conditions would trigger an irreversible deterioration.

Exercise 48. For a mature (but not aging) adult, there is negligible loss of bone marrow over a long period of time. For these individuals, we may take $\varepsilon = 0$. Suppose $R(0) = A < R_c$ with a pdf $p_0(a)H(R_c - a)$. Obtain the joint pdf $p(r, m; t)$ by i) the method of Liouville equation, and ii) transformation theory.

9.7 ODE with Random Coefficients

When an evolving phenomenon with unpredictable outcomes is modeled by a stochastic ODE where the stochasticity is only in the form of randomness in the system properties (such as the rate constants in a chemical reaction), we have a problem for stochastic ODE with random coefficients. For the evolution of a single evolving process, the general problem takes the form

$$X'(t) = f(X(t), t; C), \quad X(t_0) = A$$

where C is a random variable with a known density function, possibly as a joint density with A if the latter is also a random variable. If there is an explicit solution for ODE system for the problem, we showed earlier how the transformation theory developed in this chapter may be applied to determine the pdf of the unknown $X(t)$ for different t of interest. If an explicit solution of the ODE is not available, another method would be needed for obtaining information about the response process.

One approach suggested in the literature (see [50] for example) is to transform the problem with random coefficients into one with initial data. For the scalar problem above, such an approach would introduce a stochastic process $Z(t)$ as the solution of the following IVP for an new SDE with random initial condition:

$$Z' = 0, \quad Z(t_0) = C. \tag{9.7.1}$$

The new IVP is then combined with the original problem to form the following IVP for a vector SDE for $\mathbf{X}(t) = (X(t), Z(t))^T$ where stochasticity is only in the random initial conditions:

$$\mathbf{X}' = \begin{pmatrix} f(\mathbf{X}, t) \\ 0 \end{pmatrix}, \quad \mathbf{X}(t_0) = \begin{pmatrix} A \\ C \end{pmatrix} \equiv \begin{pmatrix} A_1 \\ A_2 \end{pmatrix}. \tag{9.7.2}$$

The extended IVP (9.7.2) involves a vector differential equation with stochasticity appearing only in the initial conditions. The following examples suggest the approach should be used with some care.

Example 50. A frequently encountered example in life science literature is

$$X'(t) = CX, \quad X(0) = A$$

where C is a random variable with a known density function, possibly as a joint density with A, if the latter is also a random variable.

For the general approach described above, we introduce a new stochastic process $Z(t)$ as the solution of the IVP (9.7.1). Together with the original problem, we have the following IVP for the vector SDE

$$\frac{d}{dt} \begin{pmatrix} X \\ Z \end{pmatrix} = \begin{pmatrix} ZX \\ 0 \end{pmatrix}, \quad \begin{pmatrix} X(0) \\ Z(0) \end{pmatrix} = \begin{pmatrix} A \\ C \end{pmatrix}.$$

The problem may be written in vector form for $\mathbf{X}(t) = (X_1, X_2)^T = (X, Z)^T$ with

$$\mathbf{X}' = \mathbf{f}(\mathbf{X}, t) = \begin{pmatrix} X_1 X_2 \\ 0 \end{pmatrix} \equiv \begin{pmatrix} f_1(\mathbf{X}, t) \\ f_2(\mathbf{X}, t) \end{pmatrix}, \quad \mathbf{X}(0) = \mathbf{A} = \begin{pmatrix} A \\ C \end{pmatrix} \equiv \begin{pmatrix} A_1 \\ A_2 \end{pmatrix}.$$

For this new system, the method of Liouville equation applies with the joint pdf $p(\mathbf{x}; t)$ for the component output processes $(X_1, X_2) = (X(t), Z(t))$ determined by

$$\frac{\partial p}{\partial t} + zx \frac{\partial p}{\partial x} + (0) \frac{\partial p}{\partial z} + (z + 0)p = 0, \quad p(\mathbf{x}; 0) = p_0(\mathbf{a}). \tag{9.7.3}$$

The corresponding characteristic IVP is

$$x' = xz, \quad z' = 0, \quad p' = -zp, \tag{9.7.4}$$

$$x(0) = a_1, \quad z(0) = a_2, \quad p(0) = p_0(a_1, a_2).$$

The solution for this IVP problem is

$$z(t) = a_2, \quad x(t) = a_1 e^{a_2 t}, \quad p(t) = p_0(a_1, a_2)e^{-a_2 t}. \tag{9.7.5}$$

Upon using the first two relations to eliminate a_1 and a_2 from the third, we obtain

$$p(\mathbf{x};t) = p_0(xe^{-zt}, z)e^{-zt} \qquad (9.7.6)$$

Here, we treat the case where A is a known constant a, corresponding to $p_0(a_1, a_2) = p_C(a_2)\delta(a_1 - a)$ so that

$$p(\mathbf{x};t) = p_C(z)\delta(xe^{-zt} - a)e^{-zt}. \qquad (9.7.7)$$

which is the solution for the Liouville equation problem. To compare it with the previous solution $p(x,t)$ by transformation theory, we need to integrate $p(\mathbf{x};t) = p(x,z;t)$ over z to get the marginal pdf

$$p(x,t) = \int_{-\infty}^{\infty} p_0(z)\delta(xe^{-zt} - a)e^{-zt}dz = \frac{a}{tx}p_0\left(\frac{1}{t}\ln(x/a)\right)H\left(\frac{x}{a}\right). \qquad (9.7.8)$$

Evidently, the interval $x \leq 0$ is an unreachable region in the pre-image set if $a > 0$ (and for $x > 0$ if $a < 0$). For $a = 1$, the final solution (9.7.8) is the same as the exact solution previously obtained by transformation theory.

Example 51. Continue with the same example but now both A and C are random variables with a joint pdf $p_0(a_1, a_2)$.

For this more complicated problem, the IVP for the Liouville equation for the joint pdf $p(\mathbf{x};t)$ remains unchanged as given by (9.7.3) with the solution for $p(\mathbf{x};t)$ given by (9.7.6). To compare with the solution previously obtained by transformation theory, we need the marginal pdf $p_x(x;t)$ again obtained by integrating over the entire range of z:

$$p_X(x;t) = \int_{-\infty}^{\infty} p_0(xe^{-zt}, z)e^{-zt}dz \qquad (9.7.9)$$

The solution (9.7.9) is the same as the exact solution (9.5.9). No further simplication without addition information on the initial joint pdf $p_0(a_1, a_2)$.

Example 52. A second example involving a nonlinear ODE is

$$X'(t) = CX^2, \quad X(0) = A$$

where C is a random variable with a known density function, possibly as a joint density with A, if the latter is also a random variable.

For the general approach described above, we introduce an augmented stochastic process $Z(t)$ as the solution of the IVP (9.7.1). Together with the original problem, we have the following IVP for the vector SDE

$$\frac{d}{dt}\begin{pmatrix} X \\ Z \end{pmatrix} = \begin{pmatrix} ZX^2 \\ 0 \end{pmatrix}, \quad \begin{pmatrix} X(0) \\ Z(0) \end{pmatrix} = \begin{pmatrix} A \\ C \end{pmatrix}.$$

For this new system, the method of Liouville equation applies with the joint pdf $p(\mathbf{x}; t)$ for the component output processes $\mathbf{X} = (X_1, X_2) = (X(t), Z(t))$ is determined by

$$\frac{\partial p}{\partial t} + zx^2 \frac{\partial p}{\partial x} + (0) \frac{\partial p}{\partial z} + (2xz + 0)p = 0, \quad p(\mathbf{x}; 0) = p_0(\mathbf{a}).$$

The corresponding characteristic IVP is

$$x' = zx^2, \quad z' = 0, \quad p' = -2zxp,$$

$$x(0) = a_1, \quad z(0) = a_2, \quad p(0) = p_0(a_1, a_2).$$

The solution for this problem is

$$z(t) = a_2, \quad x(t) = \frac{a_1}{1 - a_1 a_2 t}, \quad p(t) = p_0(a_1, a_2)(1 - a_1 a_2 t)^2.$$

Upon using the first two relations to eliminate a_1 and a_2 from the third, we obtain

$$a_2 = z, \quad a_1 = \frac{x}{1 + ztx}, \quad p(\mathbf{x}; t) = p_0\left(\frac{x}{1 + ztx}, z\right)\left(\frac{1}{1 + ztx}\right)^2.$$

It remains to integrate the expression $p(\mathbf{x}; t) = p(x, z; t)$ with respect to z over $(-\infty, \infty)$ to obtain the marginal pdf $p_X(x, t)$.

For the case A is a known constant a, we have $p_0(a_1, a_2) = p_C(a_2)\delta(a_1 - a)$ and

$$p_x(x; t) = \int_{-\infty}^{\infty} p_C(z)\,\delta\left(\frac{x}{1 + ztx} - a\right)\frac{dz}{(1 + ztx)^2}$$

$$= \frac{1}{t}\int_{-\infty}^{\infty} p_C\left(\frac{y}{t}\right)\delta\left(\frac{x}{1 + yx} - a\right)\frac{dy}{(1 + yx)^2}$$

$$= \left[\frac{1}{t}p_C\left(\frac{y}{t}\right)\left(\frac{1}{1 + yx}\right)^2\right]_{y=(x-a)/ax} = \frac{a^2}{x^2 t}p_C\left(\frac{x - a}{xat}\right)$$

For $a = 1$, the solution (9.7.8) is the same as the exact solution (9.5.6) obtained by transformation theory in a previous section.

Chapter 10

Continuous Stochastic Processes

10.1 Random Variables with Continuous Indexing

10.1.1 *Continuous stochastic processes*

By now, we have encountered stochastic processes (s.p.) $X(t)$ with a discrete or continuous sample space for each (discrete) stage or (continuous) instant of the indexing parameter t. In all cases, we have limited our discussion to the pdf or probability distribution of the random variable $X_\tau = X(\tau)$ of the s.p. at a specific "time" τ (whether it is discrete or continuous indexing). In reality, many evolving s.p., particularly those associated with the solution of stochastic differential equations, are driven by some underlying dynamics. A well known example is the Cartesian coordinates of the position of a particle in Brownian motion (movement of a dust particle in space in the presence of other particles). Each coordinate is a s.p. with the whole real line as its sample space at any instant as it evolves continuously with (continuous) time. For such s.p., the random variable $X_m = X(t_m)$ is generally not statistically independent of the random variable $X_k = X(t_k)$ at a different index $t_k \neq t_m$. For a more complete understanding of this type of s.p., we need to know not only the probabilistic information about the random variable $\{X_j\}$ for different t_j, but also additional probabilistic information on the relations among them, such as their joint pdf of various orders.

For the simple example of a population increasing at a linear growth rate, the evolving biomass of the population $x(t)$ may be taken as a continuous function of time modeled by the IVP

$$x' = cx, \quad x(0) = x_0.$$

If measurements of the initial data includes some random errors with some known probabilistic measure (its probability distribution or the probability density function for example), then the initial data may be taken to be $x_0 = \mu + A$ where μ is the actual (or statistical mean) population size while A is a random variable with a continuous sample space of $(-\infty, \infty)$. In that case, the solution of the ODE at a particular time t_k is a random variable X_k related to the random variable A by

$$X_k = (\mu + A)\, e^{ct_k}. \tag{10.1.1}$$

Through the transformation theory developed in the last chapter, the relation (10.1.1) enables us to determine the pdf of X_k in terms of the pdf $p_A(a)$ of the random variable, A. All these observations have essentially been made in the previous chapters. Not dwelled upon in the previous development was the fact that t is not fixed and X_k changes with t_k even in the deterministic case. When A is random, we are interested not only in the random variable X_k for a fixed t_k, but also the probabilistic information on the collection of $\{X_m\}$ as t changes. Determination of information such as their joint pdf is intrinsically more complex and will be the main topic of discussion for the remaining chapters of this volume.

10.1.2 *The Kolmogorov compatibility conditions*

As t changes continuously over a range of values (t_0, τ), we have a stochastic process indexed by the continuous parameter t (which is time in most cases) to be denoted by X_t or $X(t)$. Now for any set of n time instants $\{t_1, t_2, \ldots, t_n\}$, the corresponding $\{X_1, X_2, \ldots, X_n\}$ is a set of random variables that generally are not independent of each other. As such, we need to know about their joint probability distributions or density functions.

To describe formally what we call a continuous stochastic process $X(t)$ for t in a time interval (t_0, τ) (which may span the entire real line), we have implicitly suggested in the discussion up to now that we learn about the stochastic process by working with $\{X_1 = X(t_1), \ldots, X_n = X(t_n)\}$ for any collection of time instants $\{t_1, t_2, \ldots, t_n\}$ as a collection of random variables with a joint probability density function among them (which now varies with the indexing sequence $\{t_k\}$ as well). Since there are many possible sets of time instances in (t_0, τ), there are many such collections of random variables associated with different combinations of $\{t_k\}$. Here we give another characterization of this aspect of continuous stochastic processes to offer a different perspective:

Definition 23. A **stochastic process** $X(t, \varsigma)$ is a function of two variables, say t and ς. The domain of ς is an index of the sample set (the six possible faces of a die) and the domain of t is in general the real line $(-\infty, \infty)$. For a specific value of $\varsigma = \varsigma_i$, the quantity $X(t, \varsigma_i)$ is an ordinary function corresponding to an elementary event that varies with t or a *sample function*. For a specific $t = t_k$, $X(t_k, \varsigma)$ is a random variable which ranges over the sample space as ς varies over its domain. Finally, for any pair (t_k, ς_i), $X(t_k, \varsigma_i)$ is a mere number.

It is customary to omit the appearance of the stochastic parameter ς and simply write $X(t)$ for a stochastic process. This is consistent with the omission of the sample index parameter ς for random variables in the previous chapter. As such, $X(t)$ represents four different things: *i*) A family of functions (when both t and ς allowed to vary); *ii*) a single function of t (with an assigned value for ς); *iii*) random variable (when t is fixed), and *iv*) a single number (when both t and ς are fixed).

We stipulate that two s.p. are *equal* when the corresponding sample functions are identical for any outcome parameter value ς_i. Mathematical operations, such as sum, product, differentiation, etc., on one or more s.p. processes operate on their sample functions.

Stochastic processes are generally complicated. One example of a rather complex s.p. is the *Brownian motion* or its mathematical characterization known as the *Wiener process*. Originally from Robert Brown studying the zigzag movements of particles in a fluid resulting from collisons with other fluid molecules, the sample paths (functions) of the particle for different ς_i cannot be described by a formula and knowing the past does not help to predict the future direction of the path.

Sample paths of a stochastic process can also be very regular as illustrated by the following coin tossing experiment: A coin is tossed. If a head turns up then $X(t) = sin(t)$. But if a tail turns up, we have $X(t) = t$. Even if the sample paths are simple regular curves, we still have a stochastic process.

To the extent that $X(t)$ is a random variable for a fixed value of t, it is endowed with a probability density function $p(x,t)$ which may vary with the value of the time-indexing parameter t. As $X(t_1)$ and $X(t_2)$ are generally two related random variables, there would be a joint probability density function $p_2(x_1, x_2; t_1, t_2)$ for them with

(1) the joint density function invariant under a permutation of the order of the arguments;

$$p_2(x_1, x_2; t_1, t_2) = p_2(x_2, x_1; t_2, t_1); \qquad (10.1.2)$$

(2) the marginal density functions given by

$$p_1(x_k; t_k) = \int_{-\infty}^{\infty} p_2(x_1, x_2; t_1, t_2)dx_j \equiv p(x_k; t_k), \qquad (10.1.3)$$

for $j = 1$ or 2 and $k \neq j$.

The two requirements (10.1.2) and (10.1.3) are known as the *Kolmogorov compatibility conditions* for $p_2(x_1, x_2; t_1, t_2)$. Note that we now use the notation $p_2(x_1, x_2; t_1, t_2)$, instead of $p_X(x_1, x_2; t_1, t_2)$, for the second order joint density function when we wish to convey its specific order.

It is not obvious that $p_1(x_k; t_k) \equiv p(x_k; t_k)$ should not depend on t_j explicitly after integration with respect to x_j in (10.1.3). To appreciate this requirement, we note that the pdf of $X(t)$ for the simple example in (10.1.1) with a pdf $p_A(a)$ for the random initial data A is form exactly by transformation theory to be

$$p_1(x_1, t_1) = p_A(x_1 e^{-ct_1}) e^{-ct_1}$$

For a later time t_2, the corresponding pdf is $p_1(x_2, t_2) = p_A(x_2 e^{-ct_2}) e^{-ct_2}$. By the relation

$$p_2(x_1, x_2; t_1, t_2) = p(x_1, t_1 | x_2, t_2) p_1(x_2, t_2)$$

with $p(x_1, t_1 | x_2, t_2)$ being the pdf of $X(t_1)$ conditioned on the occurrence of $X(t_2)$. If we assume that $X(t_1)$ and $X(t_2)$ are statistically independent, we would have $p(x_1, t_1 | x_2, t_2) = p_1(x_1, t_1)$ so that

$$p_2(x_1, x_2; t_1, t_2) = p_A(x_1 e^{-ct_1}) e^{-ct_1} p_A(x_2 e^{-ct_2}) e^{-ct_2}.$$

It follows that

$$p_1(x_1; t_1) = \int_{-\infty}^{\infty} p_2(x_1, x_2; t_1, t_2) dx_2$$

$$= p_A(x_1 e^{-ct_1}) e^{-ct_1} \int_{-\infty}^{\infty} p_A(x_2 e^{-ct_2}) e^{-ct_2} dx_2$$

$$= p_A(x_1 e^{-ct_1}) e^{-ct_1}$$

satisfying the Kolmogorov first compatibility condition.

We can introduce higher order joint density functions and the corresponding probability distributions by looking at the s.p. at more instants in time $\{t_i, t_2, \ldots\}$ to get a better description of a continuous stochastic process. In theory, a continuous stochastic process $X(t)$ is effectively seen through its various joint density functions (or distributions): For every finite set of $\{t_i, t_2, \ldots, t_n\}$, there corresponds for a s.p. $X(t)$ a collection of n random variables $\{X_1 = X(t_i), \ldots, X_n = X(t_n)\}$ with a joint probability density function $p_n(x_1, \ldots, x_n; t_i, \ldots, t_n)$ and the corresponding probability distribution

$$P(X(t_1) \le A_1, \ldots, X(t_n) \le A_n) = \int_{-\infty}^{A_n} \cdots \int_{-\infty}^{A_1} p_n(x_1, \ldots, x_n; t_i, \ldots, t_n) dx_1 \cdots dx_n,$$

subject to the two *Kolmogorov compatibility conditions*:

$$p_2(x_1, \ldots, x_n; t_1, \ldots, t_n) = p_2(x_{i_1}, \ldots, x_{i_n}; t_{i_1}, \ldots, t_{i_n}), \tag{10.1.4}$$

where $(i_1, \ldots, i_n)$ is any permutation of $(1, 2, \ldots, n)$, and

$$p_m(x_{i_1}, \ldots, x_{i_m}, t_{i_1}, \ldots, t_{i_m}) = \int_{-\infty}^{\infty} \cdots \int_{-\infty}^{\infty} p_n(x_1, \ldots, x_n; t_1, \ldots, t_n) dx_{i_{m+1}} \cdots dx_{i_n}$$

$$\tag{10.1.5}$$

for $m < n$.

The definitions above and their implications can be extended to vector stochastic processes such as $\mathbf{X}(t) = (X_1(t), \ldots, X_\ell(t))^T$. For either scalar or vector s.p., their joint density functions and probability distributions are either specified, estimated from available data, or determined from other stochastic processes. We are particularly interested in the third possibility where stochastic processes of interest are outputs of functional transformations, ODE and PDE with random input (random forcing, random system properties, random initial data or all of the above). If the growth rate constant c in (10.1.1) is also a random variable, then the output $X(t)$ would be a function of two random variables A and C with its density function determined by the known joint density function $p_{AC}(a, c)$.

10.2 Moments and Characteristic Functions

As in the case of random variables, we are interested in various joint moments and characteristic functions of stochastic processes. The main difference is now a dependence of these quantities on the "time" variable t. For example, we have for the nth moment of a s.p. $X(t)$

$$E[X^n(t)] = \int_{-\infty}^{\infty} x^n p(x, t) dx = \mu_n(t)$$

with $\mu_1(t) = \mu(t)$ being the *mean* or *expectation* of $X(t)$.

For the joint moments, we have

$$E[X^n(t_1)X^m(t_2)] = \int_{-\infty}^{\infty} x_1^n x_2^m p_2(x_1, x_2; t_1, t_2) dx_1 dx_2 = \mu_{nm}(t_1, t_2)$$

with the central moments defined to be

$$\mu_{nm}^o(t_1, t_2) = E[\{X(t_1) - \mu(t_1)\}^n \{X(t_2) - \mu(t_2)\}^m].$$

The central moment $\mu_{11}^o(t_1, t_1)$ is the variance of $X(t)$:

$$\mu_{11}^o(t_1, t_1) = E\left[\{X(t_1) - \mu(t_1)\}^2\right] = \sigma_X^2(t_1).$$

For many problems of interest, it suffices to work with the central moments as a change of variable from $X(t)$ to $Z(t) = X(t) - \mu(t)$ results in a related s.p. which may be simpler to process. Unless clearly indicated otherwise, we take $X(t)$ to be of zero mean for all t in the domain T henceforth.

Cross-moments of two stochastic processes $X(t)$ and $Y(t)$ such as

$$E[X^n(t_1)Y^m(t_2)] = \int_{-\infty}^{\infty} x_1^n y_2^m p_2(x_1, x_2; t_1, t_2) dx_1 dy_2 = \eta_{nm}(t_1, t_2)$$

also arise in applications with the corresponding central cross-moments

$$\eta_{nm}^o(t_1, t_2) = E[\{X(t_1) - \mu_X(t_1)\}^n \{Y(t_2) - \mu_Y(t_2)\}^m].$$

Among these moments, most often encountered are the following second order moments of stochastic processes:

$$C_{XX}(t_1, t_2) = E[X(t_1)X(t_2)] = \int_{-\infty}^{\infty} \int_{-\infty}^{\infty} x_1 x_2 p_2(x_1, x_2; t_1, t_2) dx_1 dx_2 = \mu_{11}(t_1, t_2)$$

$$C_{XY}(t_1, t_2) = E[X(t_1)Y(t_2)] = \int_{-\infty}^{\infty} \int_{-\infty}^{\infty} x_1 y_2 p_2(x_1, y_2; t_1, t_2) dx_1 dy_2 = \eta_{11}(t_1, t_2).$$

They are also known as the *autocorrelation function* and *cross-correlation function*, respectively. The corresponding central moments $\mu_{11}^o(t_1, t_2)$ and $\eta_{11}^o(t_1, t_2)$ are known as *autocovariance function and cross-covariance function*, respectively. Below are some properties of these functions:

Proposition 35. (i) $C_{XX}(t_1, t_2) = C_{XX}(t_2, t_1)$. (ii) $C_{XX}(t, t) \geq 0$.
  (iii) $C_{XY}(t_1, t_2) = C_{YX}(t_2, t_1)$. (iv) $C_{XX}^2(t_1, t_2) \leq C_{XX}(t_1, t_1)C_{XX}(t_2, t_2)$.
  (v) $C_{XY}^2(t_1, t_2) \leq C_{XX}(t_1, t_1)C_{YY}(t_2, t_2)$.

Proof. (exercise) □

Proposition 36. *The matrix* $[C_{XX}(t_i, t_j)] = [C_{ij}]$ *is positive semi-definite.*

Proof. (exercise) □

Example 54. i) For $X(t) = At$ with A being a random variable of zero mean and variance σ^2, show that $C_{XX}(t, s) = \sigma^2 ts$ (exercise).
  ii) For $Y(t) = Ae^{-ct}$ with c being a known constant, A being a random variable of zero mean and variance σ^2, find $C_{YY}(t, s)$ (exercise).

As in the case of random variables, we can define joint characteristic functions for the joint probability density functions of stochastic processes:

$$\hat{p}_n(\mathbf{u}) = \int_{-\infty}^{\infty} \cdots \int_{-\infty}^{\infty} e^{i\mathbf{u}\cdot\mathbf{x}} p_n(\mathbf{x}; t_1, t_2, \ldots, t_n) dx_1 \cdots dx_n$$

where $\mathbf{x} = (x_1, x_2, \ldots, x_n)^T$ and $\mathbf{u} = (u_1, u_2, \ldots, u_n)^T$ (with bold face indicating a vector quantity). For stochastic processes however, we have the additional possibility of taking the Fourier transform in time of some of their statistics as we will do in the next section.

10.3 Stationary Stochastic Processes

A stochastic process $X(t)$ may be independent of the reference time with its probabilistic and statistical properties not affected by a shift of time. That is, the probabilistic and statistical properties of $X(t)$ and $X(t+\tau)$ are the same. Such a s.p. is said to be (strictly) stationary. Two stochastic processes $X(t)$ and $Y(t)$ are jointly (strictly) stationary if the joint statistics of $\{X(t), Y(t)\}$ are the same as those of $\{X(t+\tau), Y(t+\tau)\}$. Note that stochastic processes may be individually stationary but not jointly stationary. For a single stationary s.p. $X(t)$, it follows from the above characterization of (strict) stationarity that its nth order joint density function must have the property

$$p_n(x_1, x_2, \ldots, x_n; t_1, t_2, \ldots, t_n) = p_n(x_1, x_2, \ldots, x_n; t_1 + \tau, t_2 + \tau, \ldots, t_n + \tau).$$

As an immediate consequence, we have the following useful simplifications:

Lemma 9. *For a stationary process $X(t)$, its density function is independent of t, i.e., $p(x; t) = p(x)$ and consequently $E[X(t)] = \mu$ is also independent of t.*

Proof. For any $\epsilon > 0$, we have $p(x; t+\epsilon) = p(x; t)$ for a stationary process. As this must be true for every ϵ, $p(x; t)$ must be independent of t. With

$$\mu(t) = E[X(t)] = \int_{-\infty}^{\infty} xp(x; t)dx = \int_{-\infty}^{\infty} xp(x)dx,$$

it follows that $\mu(t) = E[X(t)]$ is a constant. $\qquad\square$

Lemma 10. *For a stationary process $X(t)$, its second order joint density function $p_2(x_1, x_2; t_1, t_2)$ depends only on the time increment between the two instants of time involved, namely $\tau = t_2 - t_1$, i.e., $p_2(x_1, x_2; t_1, t_2) = p_2(x_1, x_2; t_2 - t_1)$. Correspondingly, the autocorrelation function $C_{XX}(t_1, t_2)$ also depends only on τ and not on t_1 and t_2, individually.*

Proof. (exercise) $\qquad\square$

We may assume $t_2 > t_1$ so that τ is positive only on the basis of the following observation:

Lemma 11. $\qquad\qquad C_{XX}(-\tau) = C_{XX}(\tau)$

Proof. (exercise) $\qquad\square$

Definition 24. A second order stochastic process $X(t)$ is a **wide-sense (or weakly) stationary** s.p. if
(i) $E[X(t)]$ and $Var[X(t)]$ are constants (independent of t), and
(ii) $E[X(t+\tau)X(t)] = C_{XX}(\tau)$.

Note that the definition of wide (or weak) sense stationarity says nothing about higher order probabilistic and statistical properties being invariant under a time translation. More importantly, even the joint density $p_2(x_1, x_2; t_1, t_2)$ may not be invariant under a time shift. The two s.p. in the Example 54 (with $X(t) = At$ and $Y(t) = Ae^{-at}$) are clearly not stationary, not even wide sense stationary. However, the s.p. below is.

Example 55. For $X(t) = a\cos(\omega t + \phi)$ where the only random variable ϕ is uniformly distributed in the interval $(0, 2\pi)$, show that $E[X(t)] = 0$ and $C_{XX}(t+\tau, t) = \frac{1}{2}a^2\cos(\omega\tau)$. (exercise)

While the s.p. above is wide sense stationary, it can be verified that $p(x; t)$ is not independent of t.

Example 56. Show that the random telegraph transmission processes of Problems 8 and 9 of Assignment VI are at least wide sense stationary.

Example 57. Let Y be a random variable with a uniform density function on $[0, 2\pi]$ and $X(t)$ be defined in terms of Y by

$$X(t) = \cos(t + Y)$$

for all t in $(-\infty < t < \infty)$. Then $X(t)$ is wide sense stationary.

In practice, we often limit discussion to first and second moments for second order s.p. Such processes can be shown to be independent of time when they are stationary.

With $C_{XX}(t_1, t_2) = C_{XX}(t_2 - t_1) = C_{XX}(\tau)$ for a wide sense stationary process, we can consider its Fourier transform with respect to τ.

Definition 25. When the autocorrelation function of a wide sense stationary process is absolutely integrable, the **power spectrum** or **spectral density** $S(\omega)$ of the s.p. is defined to be

$$S(\omega) = \int_{-\infty}^{\infty} e^{i\omega\tau} C_{XX}(\tau) d\tau.$$

For real-valued s.p. (and our discussion is limited to such processes), we have the following consequence of Proposition 11:

Proposition 37. *The spectral density of a stationary s.p. is an even function in its argument $S(-\omega) = S(\omega)$.*

Theorem 29. *Given the power spectral density $S(\omega)$, the autocorrelation function of a wide sense stationary stochastic process is given by the inversion formula:*

$$C_{XX}(\tau) = \frac{1}{2\pi} \int_{-\infty}^{\infty} e^{-i\omega\tau} S(\omega)d\omega$$

with equality taken to be the average value at point of discontinuity of $C_{XX}(\tau)$.

Proof. The result follows from the Fourier inversion formula. □

The pair of formulas for $\{C_{XX}(\tau), S(\omega)\}$ are known as the Wiener–Khinchin relations in stochastic processes. In some field of applications, the sign of the exponentials are reversed.

Corollary 13.

$$\frac{1}{2\pi} \int_{-\infty}^{\infty} S(\omega)d\omega \geq 0$$

Proof. The result follows from the inversion formula and the definition of the autocorrelation function,

$$\frac{1}{2\pi} \int_{-\infty}^{\infty} S(\omega)d\omega = C_{XX}(0) = E[X^2(t)] = \sigma^2,$$

with the second moment being a nonnegative constant for a wide sense stationary process. □

10.4 Random Walk and the Wiener Process

10.4.1 *Random walk revisited*

Starting at time $t = 0$ and with the position of a marker at $X(0) = 0$, a coin is tossed every T seconds. If a head turns up, a step of length s is taken to the right. If a tail turns up instead, a step of the same length is taken to the left. After t seconds, the marker position $X(t)$ depends on the coin tossing outcomes (which are random) and hence a stochastic process known as random walk (in one dimension). The sample function is an up and down staircase with equal step length s over each T time interval and with jump discontinuities at $t_n = nT$. It is a phenomenon we have previously encountered in an earlier chapter. Here we are concerned with the statistical aspects of the phenomenon.

Suppose that for the first n tosses of the coin, k heads turn up. Then at $t = nT$, the marker has moved k steps to the right and $n - k$ steps to the left and therewith

$$X(nT) = ks - (n - k)s = (2k - n)s \equiv rs$$

where $r = 2k - n$. Since k varies from sample to sample, $X(nT)$ is a random variable taking the value rs with r possibly assuming one of the elementary events in the sample space $\{-n, -n+1, \ldots -1, 0, 1, 2, \ldots, n-1, n\}$. In other words, $\{X(nT) = rs\}$ is the event $\{k$ heads in n tosses$\}$ where $k = (n + r)/2$. We know from binomial distribution that

$$p(x; t) = \binom{n}{k} p^k (1 - p)^{n-k} \delta(x - k), \quad (t = nT)$$

or, for a fair coin,

$$p(x; t) = \frac{1}{2^n} \binom{n}{\frac{n+r}{2}} \delta(x - k), \quad (t = nT)$$

$$P(k \text{ heads}) = P\left(X(nT) = rs\right) = \frac{1}{2^n} \binom{n}{\frac{n+r}{2}}.$$

Evidently, the random variables $\{X(jT) = X_j\}$ are i.i.d. so that

$$p_n(x_1, \ldots x_n; t_1, \ldots, t_n) = p(x_1; t_1) \cdots p(x_n; t_n).$$

It follows that

$$E[X(nT)] = 0, \quad E[X^2(nT)] = ns^2, \tag{10.4.1}$$

since

$$E[X(nT)] = E\left[\sum_{j=1}^{n} X(jT)\right] = \sum_{j=1}^{n} E[X(jT)] = n \cdot 0 = 0$$

$$E[X^2(nT)] = E\left[\sum_{j=1}^{n} X^2(jT)\right] = \sum_{j=1}^{n} E[X^2(jT)] = n\left[ps^2 + qs^2\right] = ns^2.$$

For large n and $r = O(\sqrt{n})$, it can be shown (see Sec. VII.2 of [11]) the following asymptotic relation holds

$$P\left\{X(nT) = rs\right\} \sim \sqrt{\frac{2}{n\pi}} e^{-r^2/2n},$$

with

$$P\left\{X(nT) \leq rs\right\} \sim \frac{1}{2} + \mathrm{erf}\left(\frac{r}{\sqrt{n}}\right), \quad \mathrm{erf}(x) = \frac{1}{\sqrt{2\pi}} \int_0^x e^{-u^2/2} du.$$

10.4.2　The Wiener process

By setting $t = nT$, we re-write the results of the previous section as

$$E[X(t)] = 0, \quad E[X^2(t)] = \frac{t}{T} s^2.$$

Suppose we keep t fixed and allow s and T to tend to zero. The variance $E[X^2(t)]$ approaches a limit only if $s^2 = DT$ where D is a constant. We denote the limiting process by $B(t)$ in honor of Robert Brown for his pioneering work on Brownian motion but call it the Wiener process for Wiener's work relating Brownian motion to stochastic process. With the relation $s^2 = DT$, we get from (10.4.1) for the random walk problem

$$E[B(t)] = 0, \quad E[B^2(t)] = Dt. \tag{10.4.2}$$

Moreover, we can prove the following related deeper result:

Theorem 30. *In the limit of $s^2 = DT \to 0$ while $t = nT$ remaining fixed, the limiting stochastic process $B(t)$ is normally distributed with mean zero and variance Dt.*

Proof. For the finite T and s case, we have with $t = nT$ and $rs = x$,

$$P\{X(t) \le x\} \sim \frac{1}{2} + \mathrm{erf}\left(\frac{r}{\sqrt{n}}\right) = \frac{1}{2} + \mathrm{erf}\left(\frac{x/s}{\sqrt{t/T}}\right) = \frac{1}{2} + \mathrm{erf}\left(\frac{x}{\sqrt{Dt}}\right).$$

In the limit as $s, T \to 0$ with $s^2/T \to D$, we have $X(t) \to B(t)$ with the probability distribution

$$P\{B(t) \le x\} = \frac{1}{2} + \mathrm{erf}\left(\frac{x}{\sqrt{Dt}}\right) = \frac{1}{2} + \frac{1}{\sqrt{2\pi}} \int_0^{x/\sqrt{Dt}} e^{-z^2/2} dz$$

and the corresponding density function

$$p(x; t) = \frac{1}{\sqrt{2\pi Dt}} e^{-x^2/2Dt}.$$

$\square$

Proposition 38. $\qquad C_{BB}(t_1, t_2) = \min[Dt_1, Dt_2].$

Proof. (exercise) $\qquad\qquad\qquad\qquad\qquad\qquad\qquad\qquad\qquad\qquad$ $\square$

The Wiener (or Wiener–Lévy) process $B(t)$ has been related to the interesting and complex physical phenomenon of Brownian motion mentioned at the beginning of this chapter. In 1923, MIT mathematician Norbert Wiener established the existence of a Gaussian process with the properties expected of Brownian movements including

(1) continuity of its sample functions;
(2) $E[B(t) - B(s)] = 0$;
(3) $E[\{B(t_4) - B(t_3)\}\{B(t_2) - B(t_1)\}] = 0$ if $t_1 < t_2 \le t_3 < t_4$, and
(4) $E[\{B(t) - B(s)\}^2] = D|t - s|$ for $t, s > 0$.

These properties of the Wiener process will be used extensively in subsequent developments.

10.5   Mean Square Continuity

By knowing a specific s.p., we mean having knowledge of joint pdf or joint moments of all orders for the process at different instants in time. In practice, the probabilistic information about the process accessible is most often limited to first and second order pdf or joint moments of the process. With the phenomenon of interest typically modeled by ODE or PDE when there is no uncertainty, it is natural to make use of the same mathematical model to study the phenomenon when there is randomness in the phenomenon. This practice is illustrated by the simple population growth model (10.1.1) in the first section of this chapter, some of the examples in last three sections of Chapter 9 and the many *stochastic differential equation* (SDE) models for phenomena in the neuroscience, epidemiology and developmental and cell biology to be formulated and analyzed in the later chapters.

Since an SDE involves derivatives of the stochastic processes $X(t)$ such as $dX(t)/dt$ in (10.1.1) and $X(t, \zeta_i)$ is a different random variable for an instant $\bar{t}$ and a different instant $\bar{t} + \Delta t$. The output $X(\bar{t} + \Delta t, \zeta_j)$ of the latter random variable may be an event in its sample space quite different from the output $X(\bar{t}, \zeta_i)$ of the random variable of the earlier instant. Questions necessarily arise about the meaning of the derivative of a s.p. (since $X(\bar{t} + \Delta t, \zeta_j)$ and $X(\bar{t}, \zeta_i)$ are generally not close to each other). In fact similar questions already arise about the meaning of a s.p. being continuous before that. In short, we need to develop a calculus for s.p. before we begin to derive the desired probabilistic information about these processes.

There are a good number of approaches to a calculus for s.p. discussed in texts such as [50]. It is our goal here to focus on one such calculus that are generally acceptable and more or less equivalent to other approaches (see [50]). The approach we adopt that leads to a calculus to be applied thereafter is based on the *mean square norm* for measuring the magnitude of random variables. We are to extend and apply it to stochastic processes to help us formulate continuity, differentiability and integrability of stochastic processes in the context of limit in the mean square sense. With this approach, it is necessary to limit ourselves to stochastic processes that are at least second order, i.e., all relevant associated random variables $\{X(t_k) = X_k\}$ have finite first and second moment. To the extent that, for a fixed t, $Y(t) = X(t) - E[X(t)]$ is also a second order random variable with zero mean, we will (unless specifically indicated otherwise) consider here stochastic processes with zero mean and finite variance whenever appropriate, with the simplification of $Var[X(t)] = E[X^2(t)]$. We also abbreviate "mean square" as "m.s." for brevity.

Definition 26. A second order s.p. $X(t)$ for t in the interval $(0, T)$, at times abbreviated by T, is continuous in the mean square (or m.s. continuous) at a fixed t if

$$l.i.m._{\tau \to 0}[X(t + \tau)] = X(t), \quad (t + \tau \text{ in } T)$$

with the notation "l.i.m." being an abbreviation for "limit in the mean square", i.e.,

$$\lim_{\tau \to 0} \|X(t + \tau) - X(t)\|^2 = \lim_{\tau \to 0} E[\{X(t + \tau) - X(t)\}^2] = 0.$$

Remark 7. We could have used the criterion

$$\lim_{t \to t_0} E[\{X(t) - X(t_0)\}^2] = 0.$$

But the choice of writing $t = t_0 + \tau$ would simplify complications in subsequent developments.

Definition 27. If a second order s.p. is m.s. continuous at every t in $(t_1, t_2) \varepsilon T$, then $X(t)$ is m.s. continuous on the interval (t_1, t_2).

Theorem 31. *A second order s.p. $X(t)$ is m.s. continuous on an interval $[t_1, t_2]$ in T iff $C(t, s)$ is continuous at (t, t) for every t in (t_1, t_2).*

Proof. (omitted, see [50] if interested) □

Example 58. Recall the solution $y(t) = y_0 e^{a(t - t_0)}$ of the simple exponential growth problem. Consider the case that y_0 is a second order random variable with density $p(y_0)$. To see whether the resulting s.p. is $Y(t)$ is continuous, we note that

$$C_{YY}(t, s) = E[Y(t)Y(s)] = \int_{-\infty}^{\infty} y_0^2 p(y_0) dy_0 e^{a(t+s)} = E[y_0^2] e^{a(t+s)}$$

is continuous at $t = s$. Theorem 31 assures us that $Y(t)$ is m.s. continuous.

10.6 Mean Square Differentiation

With m.s. continuity, we can now discuss the m.s. differentiability of stochastic processes.

Definition 28. A second order s.p. $X(t)$, for all t in T has a m.s. derivative $X'(t)$ at t if

$$l.i.m._{\tau \to 0} \frac{X(t + \tau) - X(t)}{\tau} = X'(t). \tag{10.6.1}$$

Higher derivatives are defined analogously.

In discussing continuity of stochastic processes, we know the function being considered and hence what the mean square limit, if it exists, would be in many cases. In deciding on whether a stochastic process is differentiable, we often do not know its limit m.s. derivative $X'(t)$. Hence, it is not possible to investigate differentiability by the definition (10.6.1) or its actual requirement:

$$\lim_{\tau \to 0} E\left[\left\{\frac{X(t + \tau) - X(t)}{\tau} - X'(t)\right\}^2\right] = 0.$$

Fortunately, we have previously shown that we really do not need to know $X'(t)$ or work with the definition (10.6.1) (that requires it to be known). We only need to establish the corresponding *mean square Cauchy sequence* for the convergence in the mean square of $(X(t+\tau) - X(t))/\tau$. This can be achieved by setting

$$Z_n(t) = \frac{X(t+n) - X(t)}{n}$$

with $n \downarrow 0$ (corresponding to $\tau = n$) or

$$Y_n(t) = \frac{X(t+n^{-1}) - X(t)}{n^{-1}}$$

with $n \uparrow \infty$. With such a device, we have the following theorem to convert the definition of differentiability into one working with ordinary functions:

Theorem 32. *With $C(t,s) \equiv C_{XX}(t,s)$ as its autocorrelation function, a second order s.p. $X(t)$, for all t in T is m.s. differentiable iff the second ("generalized") derivative, $\lim_{\tau,\tau' \to 0} [\Delta\Delta C(t,s)/\tau\tau']$, exists at (t,t) and is finite with*

$$\Delta\Delta C(t,s) = C(t+\tau, s+\tau') - C(t+\tau, s) - C(t, s+\tau') + C(t,s). \qquad (10.6.2)$$

Proof. We prove the "if" part of the theorem by showing the sequence $\{Y_n(t)\}$ is a Cauchy sequence in the mean square if the mixed derivative (10.6.2) exists and is finite at the point (t,t). Form

$$\|Y_n(t) - Y_m(t)\|^2 = E\left[\{Y_n(t) - Y_m(t)\}^2\right] = E[Y_n^2(t)] + E[Y_m^2(t)] - 2E[Y_n(t)Y_m(t)]$$

with

$$E[Y_n(t)Y_m(t)] = \frac{1}{\tau s} E\left[\{X_n(t+\tau) - X(t)\}\{X_m(t+\tau') - X(t)\}\right]$$

$$= \frac{1}{\tau s}\left[C_{nm}(t+\tau, t+\tau') - C_{0m}(t, t+\tau') - C_{n0}(t+\tau, t) + C_{00}(t,t)\right].$$

Letting n and $m \to \infty$ (in any way), we get

$$E[Y_n(t)Y_m(t)] \to \frac{1}{\tau s}\left[C(t+\tau, t+\tau') - C(t, t+\tau') - C(t+\tau, t) + C(t,t)\right].$$

Similarly, the other two terms also lead to the same expression in the limit as n and $m \to \infty$:

$$E[Y_n^2(t)] \to E[Y_m^2(t)] \to \frac{1}{\tau s}\left[C(t+\tau, t+\tau') - C(t, t+\tau') - C(t+\tau, t) + C(t,t)\right].$$

It follows that

$$\|Y_n(t) - Y_m(t)\|^2 \to 0 \quad \text{as} \quad n, m \to \infty \text{ in any way whatsoever.}$$

For the "only if" part, we assume that $X(t)$ is mean square differentiable so that

$$\lim_{n,m\to\infty} E[Y_n(t)Y_m(t)] = E[\{X'(t)\}^2]$$

exists for some $X'(t)$. But the existence of the limit on the left hand side is also the same as the existence of the autocorrelation function.

It can be shown that the existence of the mixed (generalized) derivative of $C(t_1, t_2)$ at the point (t, t) in $T \times T$ implies the existence of $C(t_1, t_2)$ for all (t_1, t_2) in $T \times T$ [50]. $\qquad\square$

Remark 8. For problems in most applications, the generalized derivative in (10.6.2) is the same as the ordinary mixed derivative $\partial^2 C(t, s)/\partial t \partial s$. There are pathological examples for which the generalized derivative of (10.6.2) does not exist (or is unbounded) while the ordinary mixed derivative does (and is bounded).

Example 59. Suppose $X(t) = At$ for all t in T with A being a second order r.v. with mean zero and variance σ^2. Then $C(t, s) = \sigma^2 ts$ with

$$\lim_{\tau,\tau'\to 0} \frac{\sigma^2}{\tau\tau'} \{(t + \tau)(s + \tau') - (t + \tau)s) - t(s + \tau') + ts\} = \sigma^2.$$

Hence, $X(t)$ is m.s. differentiable at every time t.

Definition 29. If a second order s.p. $X(t)$ is m.s. differentiable at every t in (t_1, t_2) in T, then $X(t)$ is m.s. differentiable on the interval $[t_1, t_2]$.

Recall that a stochastic process $X(t)$ is second order wide-sense stationary s.p. if i) $E[X(t)]$ and $E[X^2(t)]$ exist; ii) both are finite constants, and iii) $E[X(t)X(s)] = C(t, s) = C(t - s)$. For the differentiability of such processes, we have

Corollary 14. *A wide-sense stationary second order s.p. $X(t)$ is m.s. differentiable if, and only if, the first and second order derivatives of $C(\tau)$ exist and are finite at $\tau = 0$.*

10.7   Mean Square Integration

Suppose $x(t)$ is an ordinary function defined on $[a, b]$ and (uniformly) continuous in (a, b). We introduce (for simplicity) an equally space mesh on $[a, b]$ by setting $t_0 = a$, $\Delta t = (b - a)/n$, $t_k = a + k\Delta t$ so that $t_n = b$. Now form an upper *Riemann sum*

$$y_n = \sum_{k=1}^{n} x(t_k)\Delta t.$$

The (uniformly continuous) function $x(t)$ is said to be *integrable* if the limit of y_n exists (and equal to y) as $n \to \infty$ and $\Delta t \to 0$, so that

$$\lim_{n\to\infty} y_n = \lim_{n\to\infty} \sum_{k=1}^{n} x(t_k)\Delta t = y$$

It is customary to relate y to $x(t)$ by writing

$$\lim_{n\to\infty} \sum_{k=1}^{n} x(t_k)\Delta t = \int_a^b x(t)dt.$$

The right hand side is called the *integral of* $x(t)$ over the interval $[a, b]$.

There are other ways of forming a Riemann sum. For example, we may form a lower Riemann sum and take its limit to get

$$\lim_{n\to\infty} [y_n] = \lim_{n\to\infty} \left[\sum_{k=1}^{n} x(t_{k-1})\Delta t\right]. \tag{10.7.1}$$

The new limit would still be y for a uniformly continuous $x(t)$. Had an uneven mesh $\{\Delta_k\}$ been used instead of the same increment Δt, we should still get the same limit for the resulting Riemann sum as long the maximum mesh size $|\Delta_k|_{\max} \to 0$ as $n \to \infty$. This may not be the case if $x(t)$ is not uniformly continuous. In that case, the integral of $x(t)$ exists only if the limit is independent of our choice of Riemann sum.

Suppose $X(t)$ is a second order stochastic process defined over $[a, b]$ and we form similarly

$$Y_n = \sum_{k=1}^{n} X(t_k)\Delta_k. \tag{10.7.2}$$

Here $\{X(t_k)\}$ and Y_n are random variables. As n increases (and $|\Delta_k|_{\max}$ decreases), we have a longer and longer sequence of random variables $\{Y_n\}$. It is natural to define the convergence of the sequence to a limit Y in the mean square sense by the following definition:

Definition 30. Suppose $X(t)$ is the second order s.p. on the interval $[a, b]$ with the associated random variables $\{Y_n\}$ as defined above. $X(t)$ is said to be mean square Riemann integrable if the sequence $\{Y_n\}$ converges in the mean square to a limit Y as $n \to \infty$ and $|\Delta_k|_{\max} \to 0$,

$$\underset{\substack{n\to\infty \\ [\Delta_k]_{\max}\to 0}}{l.i.m.} [Y_n] = Y.$$

As in the case of an ordinary function $x(t)$, it is customary to write the limit, if it exists, as an integral as well:

$$\underset{\substack{n\to\infty \\ [\Delta_k]_{\max}\to 0}}{l.i.m.} \left[\sum_{k=1}^{n} X(t_{k-1})\Delta_k\right] \equiv \int_a^b X(t)dt.$$

We note again the notation $\underset{\{n\to\infty,\,|\Delta_k|_{\max}\to 0\}}{l.i.m.}[Y_n] = Y$ is merely a short hand for

$$\|Y_n - Y\|^2 = E[(Y_n - Y)^2] \to 0$$

as $n \to \infty$ and $|\Delta_k|_{\max} \to 0$ which is what we have to check about integrability of the stochastic process $X(t)$.

Similar to the process of checking mean square differentiability of a s.p., we also do not know the limiting (Riemann) integral Y when we try to apply the definition of integrability of $X(t)$. Fortunately, we can again work with the alternative test of a Cauchy sequence in the mean square which does not require the knowledge of the limit sought. The use of this alternative test then allows us to reduce the verification of the integrability of $X(t)$ to an investigation of the integrability of its autocorrelation function $C(t, s)$ which is an ordinary function.

Theorem 33. *Suppose $X(t)$ is a second order s.p. over $[a, b]$ and Y_n is as defined above. Then a finite limit Y of the sequence of random variables $\{Y_n\}$ exists as $n \to \infty$ and $|\Delta_k|_{\max} \to 0$ if the following ordinary double integral of the continuous autocorrelation function $C(t, s)$ of $X(t)$ exists and is finite:*

$$\int_a^b \int_a^b C(t, s)dtds = \text{ a finite constant.}$$

(Note that continuity of $C(t, s)$ ensures the m.s. continuity of $X(t)$.)

Proof. We start with Y_n as defined in (10.7.2) and form

$$\|Y_n - Y_m\|^2 = E[(Y_n - Y_m)^2] = E[Y_n^2] + E[Y_m^2] - 2E[Y_nY_m]$$

$$= \sum_{k=1}^{n}\sum_{\ell=1}^{n} C_{X_nX_n}(t_k, t_j)\Delta_n^2 + \sum_{i=1}^{m}\sum_{j=1}^{m} C_{X_mX_m}(t_i, t_j)\Delta_m^2$$

$$- 2\sum_{p=1}^{n}\sum_{q=1}^{m} C_{X_nX_m}(t_p, t_q)\Delta_n\Delta_m.$$

Since $X(t)$ is a second order s.p., $C_{X_nX_m}(t_\alpha, t_\beta)$ exists and tends to $C(t, s)$ as $n, m \to \infty$. If the double integral of $C(t, s)$ exists and is bounded, we have in the limit

$$\lim_{n,m\to\infty} \|Y_n - Y_m\|^2 = 0.$$

$\square$

As in integrals of ordinary functions, we have the following useful bound of an integral of a stochastic process.

Lemma 12. *If $X(t)$ is mean square continuous on $[a, b]$, then*

$$\left\| \int_a^b X(t)dt \right\| \leq M(b - a),$$

where $M = \max_{a \leq t \leq b} \|X(t)\|$.

Proof. Mean square continuity of $X(t)$ assures the existence of the mean square (Riemann) integral of $X(t)$. Given

$$\left\| \int_a^b X(t)dt \right\| \leq \lim_{n \to \infty} \sum_{k=1}^n \|X(t'_k)\| (t_k - t_{k-1}) = \int_a^b \|X(t)\| dt, \qquad (10.7.3)$$

the lemma follows from the definition of M. Note that the last integral exists because $\|X(t)\|$ is a continuous ordinary real-valued function. □

For later applications, we extend the above results somewhat by considering the stochastic integral of $f(t, \xi)X(t)$ over the interval $[a, b]$ where $X(t)$ is again a second order stochastic process with a continuous $C(t, s)$ and $f(t, \xi)$ is an ordinary continuous function of two variables defined for t in $[a, b]$ and ξ in another interval $[c, d]$. Consider the sequence of random variables

$$Z_n(\xi) = \sum_{k=1}^n f(t_k, \xi)X(t_k)\Delta_k \qquad (10.7.4)$$

relevant to the integrability of $f(t, \xi)X(t)$. The following theorem is analogous to Theorem 33 for the existence of the more general m.s. limit integral

$$l.i.m._{\substack{n \to \infty \\ |\Delta_k|_{\max} \to 0}} Z_n(\xi) = Z(\xi) \equiv \int_a^b f(t, \xi)X(t)dt.$$

Theorem 34. *The mean square limit of $Z(\xi)$ exists (so that $f(t, \xi)X(t)$ is m.s. Riemann integrable) if*

$$\int_a^b \int_a^b f(t, \xi)f(s, \eta)C_{XX}(t, s)dtds = F(\xi, \eta)$$

for some well defined bounded function $F(\xi, \eta)$ of ξ and η.

Proof. (omitted) □

An analogous formulation of a m.s. *Riemann–Stieltjes integral* has also been worked out and will be used when needed in subsequent development. Hence, the (m.s.) integrability of a stochastic process may be either in the sense of *Riemann* or *Riemann–Stieltjes*, whichever is appropriate.

10.8   Additional Tools in Mean Square Calculus

Proposition 30 assures us that, for a mean square convergent sequence of random variables $\{X_n\}$, the operation of taking the expectation of X_n commutes with the operation of taking the limit (in the mean square) of the X_n. Before we describe the important consequences of this proposition, we note at this opportune time that the expected value of the random variable X_n is also known as the *ensemble average* of X_n denoted by the notation $\langle X_n \rangle$ with

$$\mu = E[X] = \int_{-\infty}^{\infty} xp(x)dx \equiv \langle X \rangle$$

$$\sigma^2 = E[(X - \mu)^2] = \int_{-\infty}^{\infty} (x - \mu)^2 p(x)dx \equiv \langle (X - \mu)^2 \rangle,$$

etc. The notation $\langle X \rangle$ is most often used in the analysis of stochastic differential equations. For that reason, the same notation is extended for stochastic processes as well with

$$\mu(t) = E[X(t)] = \int_{-\infty}^{\infty} xp(x;t)dx \equiv \langle X(t) \rangle$$

$$\sigma^2(t) = E[(X(t) - \mu)^2] = \int_{-\infty}^{\infty} (x - \mu(t))^2 p(x;t)dx \equiv \langle (X(t) - \mu(t))^2 \rangle,$$

etc.

   Applications of Proposition 30 gives some useful tools in dealing with stochastic DE. A few of these are listed below while others will be mentioned when needed.

Proposition 39. *If $f(t, \xi)$ is a function of two real variables and continuous for t in (a, b), and $X(t)$ is m. s. integrable in $[a, b]$ with*

$$Y(\xi) = \int_a^b f(t, \xi)X(t)dt,$$

then

$$\langle Y(\xi) \rangle = \int_a^b f(t, \xi)\langle X(t) \rangle dt.$$

Proof. (omitted)   □

Proposition 40. *If $f(t, \xi)$ is a function of two real variables and differentiable in t in some interval (a, b), and $X(t)$ is m. s. differentiable in (a, b) with*

$$Y(t, \xi) = \frac{d}{dt}[f(t, \xi)X(t)],$$

then

$$\langle Y(t, \xi) \rangle = \frac{d}{dt}\{f(t, \xi)\langle X(t) \rangle\}.$$

Proof. (omitted)   □

Proposition 41. (*Leibniz Rule*) *Under the same hypotheses as Proposition 40,*
suppose

$$Y(s) = \int_{a(s)}^{b(s)} f(s,t)X(t)dt.$$

Then

$$\frac{dY}{ds} = \int_{a(s)}^{b(s)} \frac{\partial f(s,t)}{\partial s}X(t)dt. + [f(s,t))X(t)]_{t=b(s)}\frac{db}{ds} - [f(s,t))X(t)]_{t=a(s)}\frac{da}{ds}.$$

Proof. (omitted)   □

There are many others such tools that are analogous to those encountered in
(ordinary) calculus to be mentioned as they arise in subsequent developments.

10.9 White Noise

Here seems to be a good place to introduce an important stochastic process in
applications, known as the **white noise** process, with the following definition:

Definition 31. A stochastic process $W(t)$ is said to be a *white noise* process if it
has zero mean and is temporally uncorrelated so that

$$E[W(t)] = 0, \quad C_{WW}(t,s) = D\delta(t-s)$$

where D is a positive constant and $\delta(\cdot)$ is the *Dirac delta function.*

(We will work with the delta function in the usual way [4, 70], fully recognizing that
it is not an ordinary function.)

Since $W(t)$ is stationary, we can compute its power spectral density to get

$$S_W(\omega) = D.$$

The value of $S_W(\omega)$ for a white noise process is therefore the same at all frequencies.
For that reason, the stochastic process is called *white* to draw an analogy with the
spectrum of white light. As a conventional integral the inverse Fourier transform of
$S_W(\omega)$,

$$\frac{1}{2\pi}\int_{-\infty}^{\infty} e^{-i\omega\tau}S_W(\omega)d\omega = \frac{D}{2\pi i\tau}\lim_{\Omega\to\infty}\left[e^{-i\Omega\tau} - e^{i\Omega\tau}\right] = \frac{D}{\pi\tau}\lim_{\Omega\to\infty}\left[\sin\left(\Omega\tau\right)\right],$$

does not exist. If we work in the context of distribution theory (or generalized
function theory) as is done in this area, the integral is known to be $D\delta(\tau)$ consistent
with $C_{WW}(\tau)$ that gave rise to $S_W(\omega)$.

The complication however is that the white noise process is not a second order stochastic process, since the second moments of $W(t)$ does not exist. Yet we do want to make use of such a stochastic process in a mean square theory, given that white noise is a very good approximation and idealization of many stochastic processes that arise in applications and enables us to perform many kinds of analyses that would not be possible or very cumbersome otherwise. As such, a great deal of effort has been made to relate white noise to a mean square process. This is accomplished by the observation that $W(t)$ may be considered as the *derivative* of the *Wiener process*; the latter is a perfectly legitimate second order process given Theorems 30 and 38.

With Theorem 38, we have

$$\left\langle \frac{dB(t)}{dt} \frac{dB(s)}{ds} \right\rangle = \frac{\partial^2}{\partial t \partial s} \langle B(t)B(s) \rangle = \frac{\partial^2 C_{BB}(t,s)}{\partial s \partial t}$$

$$= \frac{\partial^2}{\partial s \partial t} \{D \min(t,s)\} = \frac{\partial^2}{\partial t \partial s} \begin{cases} Dt & (t<s) \\ Ds & (t>s) \end{cases}.$$

Upon taking the partial derivative of $C_{BB}(t,s)$ first with respect to s, we get

$$\frac{\partial^2}{\partial t \partial s} \{D \min(t,s)\} = D \frac{\partial}{\partial t} H(t-s).$$

It follows that

$$\left\langle \frac{dB(t)}{dt} \frac{dB(s)}{ds} \right\rangle = D \frac{\partial}{\partial t} H(t-s) = D\delta(t-s).$$

Proposition 42. *Both the white noise process $W(t)$ and the derivative of the Wiener process, $B'(t)$, are of zero mean and temporally uncorrelated so that*

$$\left\langle \frac{dB(t)}{dt} \frac{dB(s)}{ds} \right\rangle = D\delta(t-s) = \langle W(t)W(s) \rangle.$$

With the white noise process sharing the same (zero) first moment and temporally uncorrelated autocorrelation function with the first derivative of the Wiener process $B'(t)$, it is a general practice to think of white noise as the derivative a Wiener process whenever it suits our purpose. The main reason for this practice is the reality that we often do not know much about the stochastic input in our ODE beyond the first and second order statistics. In subsequent discussions of stochastic differential equations, it will become clear that replacing white noise by (the derivative of) a Wiener process does not necessarily remove completely the difficulty caused by the unacceptable mathematical features of white noise in certain application.

Part 4
Stochastic Ordinary Differential Equations

Chapter 11

Linear ODE with Random Forcing

11.1 Existence and Uniqueness of a Mean Square Solution

11.1.1 *Mean square solution*

Up to now, the discussion of evolving phenomena governed by differential equations in the presence of uncertainties, or *stochastic differential equations* (SDE) for brevity, has been confined to issues that can be addressed by focusing on an instant in time. For a fixed t, an activity in the mathematical model for the evolving phenomenon is a random variable and the probabilistic tools developed in earlier chapters can be applied to extract useful information about the random variable. Now that we have formulated the concept of *stochastic process* for evolving phenomena with uncertainty and developed additional mathematical machinery for such process, we are in a position to learn about their evolving features as time changes, features not accessible without the mean square (m.s.) calculus of the previous chapter.

We confine our discussion in this chapter to *linear* ordinary SDE which may be written in vector form as

$$\mathbf{X}' = C(t)\mathbf{X} + \mathbf{f}(t), \quad \mathbf{X}(t_0) = \mathbf{A}, \tag{11.1.1}$$

with $(\ )' = d(\ )dt$. In (11.1.1), C is a known $n \times n$ matrix whose elements are continuous (and possibly constant) functions of time t in some interval $[t_0, t_T]$ (or simply the interval T), $\mathbf{f}(t)$ is an n-component *second order, m.s. continuous* vector stochastic process in that interval and $\mathbf{A}$ is a second order vector random variable with a prescribed pdf $p_A(\mathbf{a})$. With the norm $\|Y_j\|$ for the jth component Y_j of an n-component vector random variable $\mathbf{Y}$ already defined to be the positive square root of its second moment $E[Y_j^2]$, we will take the norm for the vector random variable $\mathbf{Y}$ to be

$$\|\mathbf{Y}\|_n = \max_{1 \le j \le n} \ [\|Y_j\|].$$

Definition 32. Suppose $C(t)$, $\mathbf{A}$ and $\mathbf{f}$ and the norms for the relevant second order random variables are as stipulated above, the vector stochastic process $\mathbf{X}(t)$ is said to be a mean square (m.s.) solution of the IVP (11.1.1) in the interval T if

i) $\mathbf{X}(t)$ is mean square continuous on T;

ii) $\mathbf{X}(t_0) = \mathbf{A}$, and

iii) $C(t)\mathbf{X}(t) + \mathbf{f}(t)$ is the mean square derivative of $\mathbf{X}(t)$ on the interval T.

Our goal here is to show that a mean square solution of the stochastic IVP (11.1.1) exists and is unique. We do this in two steps, $\mathbf{f}(t) = \mathbf{0}$ first and $\mathbf{f}(t) \neq \mathbf{0}$ in a later subsection.

11.1.2 Random initial data only

Theorem 35. With $\mathbf{f}(t) = \mathbf{0}$ (with probability one), there exist a unique m.s. solution for the IVP (11.1.1) given by

$$\mathbf{X}(t) = \Phi(t, t_0)\mathbf{A} \qquad (11.1.2)$$

in the interval T, where $\Phi(t, \tau)$ is the **principal matrix solution** for the given linear system, i.e., a fundamental matrix solution with $\Phi(\tau, \tau) = I$ (the $n \times n$ identity matrix).

Proof. By definition, the principal matrix solution satisfies the IVP $\Phi'(t, t_0) = C(t)\Phi(t, t_0)$, $\Phi(t_0, t_0) = I$ and therewith

$$\Phi(t, t_0) = I + \int_{t_0}^{t} C(s)\Phi(s, t_0)ds.$$

Upon post-multiplying both sides by A to get

$$\Phi(t, t_0)\mathbf{A} = I\mathbf{A} + \int_{t_0}^{t} C(s)\Phi(s, t_0)ds\mathbf{A}$$

$$= \mathbf{A} + \int_{t_0}^{t} C(s)\Phi(s, t_0)\mathbf{A}ds,$$

we see that $\mathbf{X}(t) = \Phi(t, t_0)\mathbf{A}$ satisfies the IVP (11.1.1) for $\mathbf{f}(t) = \mathbf{0}$. Furthermore, $\mathbf{X}(t) = \Phi(t, t_0)\mathbf{A}$ is a second order stochastic process (given that $\mathbf{A}$ is a second order random variable by hypothesis) and m.s. continuous (which follow from the continuity of $\Phi(t, t_0)$ in t). Hence, $\mathbf{X}(t) = \Phi(t, t_0)\mathbf{A}$ is m.s. differentiable and a m.s. solution for our IVP.

To prove uniqueness, suppose there are two m.s. solutions, then the difference $\mathbf{Y}(t)$ between them satisfies the ODE $\mathbf{Y}' = C(t)\mathbf{Y}$ and the initial condition

$\mathbf{Y}(t_0) = \mathbf{0}$ with probability one, or

$$\mathbf{Y}(t) = \int_{t_0}^{t} C(s)\mathbf{Y}(s)ds \equiv L\left[\mathbf{Y}(t)\right] \tag{11.1.3}$$

where the linear operator $L\left[\cdot\right]$ multiplies its vector argument by the matrix C and integrate as shown in the middle term of the above relation. An upper bound of $L\left[\mathbf{Y}(t)\right]$ is given by

$$\|L\left[\mathbf{Y}(t)\right]\| \leq \int_{t_0}^{t} \|C(s)\| \, \|\mathbf{Y}(s)\| \, ds$$

$$\leq cy(t - t_0)$$

with

$$c = \max_{s \, \epsilon \, T} \, \|C(s)\|, \quad y = \max_{s \, \epsilon \, T} \, \|\mathbf{Y}(s)\|$$

where $\|C(s)\|$ is the matrix norm of $C(s)$. A second application of (11.1.3) on the right side of that relation results in

$$\mathbf{Y}(t) = L[\mathbf{Y}(t)] = L\left[L[\mathbf{Y}(t)]\right] = \int_{t_0}^{t} C(\tau) \left\{ \int_{t_0}^{\tau} C(s)\mathbf{Y}(s)ds \right\} d\tau$$

so that

$$\|\mathbf{Y}(t)\| \leq \frac{1}{2}c^2 y(t - t_0)^2.$$

After m such applications, we have

$$\|\mathbf{Y}(t)\| \leq \frac{1}{m!}c^m y(t - t_0)^m \sim \frac{y}{\sqrt{2\pi m}} \left(\frac{ec(t - t_0)}{m} \right)^m$$

where the Stirling asymptotic formula has been used to obtain the last term. The right hand side tends to zero as $m \to \infty$ and thereby uniqueness. $\qquad \square$

11.1.3 *Probability density function absent of forcing*

11.1.3.1 *The principal matrix solution*

For linear systems of ODE, we saw from the development above that the principal matrix solution $\Phi(t, z)$ plays an important role in the existence and uniqueness of a m.s. solution for the stochastic IVP problem with or without random forcing (similar to its important role in deterministic theory of linear systems). Here we show how $\Phi(t, z)$ is constructed from a collection of n linearly independent complementary solution of the given vector linear ODE. This is done by lining up these complementary solution vectors as different columns of a matrix $S(t)$,

$$S(t) = [\mathbf{v}_1(t), \mathbf{v}_2(t), \dots, \mathbf{v}_n(t)], \tag{11.1.4}$$

with

$$S'(t) = [\mathbf{v}_1'(t), \mathbf{v}_2'(t), \dots, \mathbf{v}_n'(t)] = C(t)S(t).$$

Let

$$\Phi(t, z) = S(t)S^{-1}(z). \tag{11.1.5}$$

Then $\Phi(t, z)$ satisfies both the given ODE and the corresponding stochastic integral equation (SIE) with

$$\Phi'(t, t_0) = S'(t)S^{-1}(t_0) = C(t)S(t)S^{-1}(t_0) = C(t)\Phi(t, t_0), \tag{11.1.6}$$

$$\Phi(z, z) = I. \tag{11.1.7}$$

It follows from (11.1.6) and (11.1.7)

$$\Phi(t, t_0) = I + \int_{t_0}^{t} C(s)\Phi(s, t_0)ds. \tag{11.1.8}$$

Exercise 48. Show

$$\Phi(t, z)\Phi(z, s) = \Phi(t, s). \tag{11.1.9}$$

11.1.3.2   *Probability density function*

To obtain the density function for $\mathbf{X}(t)$, we note that the relation $\mathbf{X}(t) = \Phi(t, t_0)\mathbf{A}$ is invertible because $\Phi(t, t_0)$ is non-singular for the range of t of interest:

$$\mathbf{A} = \Phi^{-1}(t, t_0)\mathbf{X}(t).$$

Transformation theory can then be used in conjunction with the joint density function $p_A(\mathbf{a})$ to determine $p_X(\mathbf{x}, t)$ to be

$$p_X(\mathbf{x}, t) = p_A(\Phi^{-1}(t, t_0)\mathbf{x}(t)) \, |J|$$

where

$$J = \det \left[\frac{\partial A_k}{\partial X_j} \right] = \det \left[\left[\Phi^{-1}(t, t_0) \right]_{ij} \right]$$

is the determinant of the relevant Jacobian.

11.1.4   *Random forcing*

Theorem 36. *For the IVP (11.1.1) with a prescribed non-zero second order m.s. continuous vector stochastic process* $\mathbf{f}(t)$, *there exists is a unique m.s. solution, given by*

$$\mathbf{X}(t) = \Phi(t,t_0)\mathbf{A} + \int_{t_0}^{t} \Phi(t,s)\mathbf{f}(s)ds, \tag{11.1.10}$$

in the interval T.

Proof. Differentiate both sides of (11.1.10) with respect to t to get

$$\mathbf{X}'(t) = \left[\frac{\partial}{\partial t}\Phi(t,t_0)\right]\mathbf{A} + \frac{d}{dt}\int_{t_0}^{t}\Phi(t,s)\mathbf{f}(s)ds.$$

Given that differentiating the first term involves no stochastic process), it follows from the definition of the principal matrix solution that

$$\mathbf{X}'(t) = [C(t)\Phi(t,t_0)]\mathbf{A} + \frac{d}{dt}\int_{t_0}^{t}\Phi(t,s)\mathbf{f}(s)ds.$$

The integral on the right side is a m.s. integral since $\mathbf{f}(s)$ is a second order stochastic process and $\Phi(t,s)$ is continuous. But $\mathbf{f}(s)$ is also m.s. continuous (by hypothesis), hence the m.s. integral is also m.s. differentiable with

$$\frac{d}{dt}\int_{t_0}^{t}\Phi(t,s)\mathbf{f}(s)ds = \int_{t_0}^{t}\frac{\partial\Phi(t,s)}{\partial t}\mathbf{f}(s)ds + \Phi(t,t)\mathbf{f}(t)$$

$$= \int_{t_0}^{t}C(t)\Phi(t,s)\mathbf{f}(s)ds + \mathbf{f}(t)$$

by the m.s. Leibniz rule so that

$$\mathbf{X}'(t) = [C(t)\Phi(t,t_0)]\mathbf{A} + \int_{t_0}^{t}C(t)\Phi(t,s)\mathbf{f}(s)ds + \mathbf{f}(t)$$

$$= C(t)\left\{\Phi(t,t_0)\mathbf{A} + \int_{t_0}^{t}\Phi(t,s)\mathbf{f}(s)d\right\} + \mathbf{f}(t)$$

$$= C(t)\mathbf{X}(t) + \mathbf{f}(t).$$

This proves existence of a m.s. solution. Uniqueness is proved exactly as in the preceding theorem. □

With the theorem above ensuring the existence and uniqueness of its m.s. solution, we can focus our attention on effective and efficient methods of solution for the stochastic forced IVP. For simplicity, we limit our discussion to the case $\mathbf{A} = 0$. In practice, initial data are generally statistically independent of the random forcing.

The solution for the more general case involves no new conceptual complications, only a higher level of complexity in the details of the solution process.

11.2   Integrate and Fire Models of Motoneuron

It should be evident from specific applications in the previous chapters that ODE models are prevalent in the life sciences. One life science area with such models not discussed so far is *neuroscience*. Since it is a major source of mathematical models involving both linear and nonlinear SDE, we briefly summarize in this section the basic elements of neurophysiology and some linear models for motoneurons in the central nervous system. Nonlinear models of neurons will be discussed in lster chapters.

11.2.1   *Motoneurons*

Nerve cell activities are physiologically complex and their mathematical characterization challenging. For example, there are different kinds of nerve cells (*neurons*) distinguished by their location (such as brain, spinal cord), shape and functions. They are interconnected to form neural networks and interact with other neurons through electrical signals. In this section, we limit discussion mainly to a few simple models for subthreshold voltage depolarization of *motoneurons* that are located in the spinal cord. (Other types of neurons include *pyramidal* cells located in the

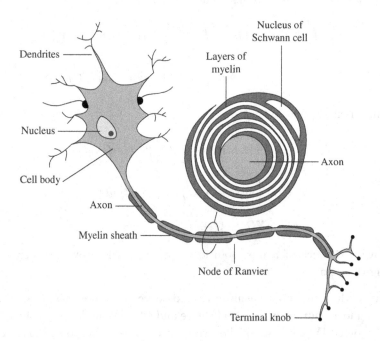

Fig. 11.1.   Schematic representation of a nerve cell.

cerebral cortex and Purkinje cells in the cerebellum.) These models characterize the evolution of the electrical voltage across the cellular membrane of a single motoneuron idealized as a point centered at its main cell body (*soma*) and hence known as point model. (The idealization is similar to treating planets in the solar system as mass points in the study of their motion.)

In reality, motoneurons are typically with a long *axon* and multiple *dendrites*. Through the branches of their *dendritic trees*, motoneurons make contact with axons and dendrites of other cells. By their own synapses and synapses of other axons and dendrites of other neurons, they send and receive activating electrical signals. As signals may reach a nerve cell at any synapse of its dendritic branches, it is not possible for any point model of a neuron to replicate or account for the spatial effects of all neuronal interactions.

Like other cells, neurons are surrounded by a membrane composed of a lipid bilayer with proteins embedded in it. The membrane serves both as an insulator and as a barrier to the movement of ions. Ions differential across the membrane due to a synaptic electrical signal establishes a concentration gradient and creates a voltage between the two sides of the membrane that changes when the ion channels allow ions to move across the membrane. The evolution of this *membrane voltage*

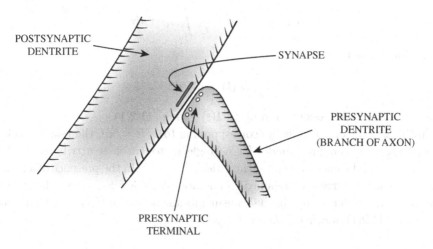

Fig. 11.2.   Axons, dentrites and synaptic contacts.

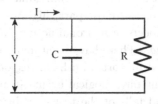

Fig. 11.3.   RC circuit point model for subthreshold voltage neuron activity.

(also known as *transmembrane potential*) is a key component of the physiology of the central nervous system. An important aspect of neuroscience is to understand how the change in transmembrane potential of a motoneuron relates to the generation and transmission of electrical signals that ultimately direct the activities of biological organisms including cognitive and muscular activities.

11.2.2   *Leaky integrator and the Lapicque model*

In the simplest model, the voltage change across the cellular membrane is modeled as a parallel RC circuit with mathematical characterization

$$C\frac{dV}{dt} + \frac{V}{R} = I \tag{11.2.1}$$

where C is the capacitance of the capacitor and R is the resistance of resistor. The capacitor and resistor in parallel are driven by an input current $I(t)$ which splits into two components: $I(t) = I_R(t) + I_C(t)$. The component $I_R(t)$ passes through the linear resistor and is related to the voltage $V(t)$ across the resistor by Ohm's law (to be equal to V/R). The other current component $I_C(t)$ charges the capacitor to a level $V(t)$ with

$$V(t) = V(t_0) + \frac{1}{C}\int_{t_0}^{t} I_C(\tau)d\tau$$

or, in differential form,

$$\frac{1}{C}I_C(t) = dV/dt.$$

The two current components add up to $I(t)$ giving (11.2.1).

In the absence of the resistor (corresponding to $1/R = \infty$), the relation (11.2.1) as an "integrate and fire" model for a single motoneuron is attributed to Louis Lapicque [31] and sometimes called the *Lapicque model*. In the presence of a resistor, some of the input current is siphoned off and not available for charging the capacitor resulting in a smaller change in the transmembrane potential $V(t)$. For that reason, the model (11.2.1) is called a *Leaky Integrator*.

11.2.3   *Action potential*

It has been observed that the activity of individual neurons in the mammalian central nervous system is, for the most part, highly irregular and unpredictable. Of particular interest are aspects of neuronal activity associated with the onset of an *action potential* or "*firing.*" There have been many experimental studies of the electrical spike trains emitted by cortical cells in response to natural and artificial stimulation. These important physiological events in motoneurons are the occurrence of such rapid rises and falls of the transmembrane voltage magnitude. For actual neurons, action potentials play a central role in cell-to-cell communication by the propagation of signals along the neuron's axon towards synaptic boutons (the presynaptic axon terminal) situated at the ends of an axon (or a branch of the dentritic tree of it); these signals can then reach other neurons at synapses, or to motor cells or glands. (In other types of cells, their main function is to activate intracellular processes. In muscle cells, for example, an action potential is the first step in the chain of events leading to contraction. In beta cells of the pancreas, they provoke release of insulin.) Action potentials in neurons are also known as "nerve impulses" or "spikes", and the temporal sequence of action potentials generated by a neuron is called a "spike train".

A point model such as the leakage integrator has idealized the neuron in a way that only captures the evolution of its transmembrane potential prior to an action potential. It does not include any cellular process responsible for initiating an action potential and consequently incapable of inducing firing. As already pointed out earlier, it is also not capable of delineating effects of spatial locations of signaling on neuronal physiology. Because of its simplicity, a leaky integrator type model was nevertheless used to study the onset of action potentials by setting a prescribed threshold voltage for firing. For this purpose, we complete the formulation of the Leaky Integrator point model by imposing the condition that an action potential is generated when $V(t_i) = \theta_T$ (30 mV). To be consistent with neurophysiology, there should be a period of quiescence of duration τ_q after the firing of a nerve impulse before the neuron can began to increase the depolarization voltage toward the firing threshold again. The period of quiescence is known as a *refractory period* with a short period of absolute quiescence of about one millisecond during which action potential is not possible. It is followed by a *relative refractory period* with a threshold significantly higher than θ_T. This second refractory period varies in duration depending on the type of neuron.

For the purpose of mathematical analysis and numerical computation, the threshold for an action potential in point models is typically taken to be $\theta(t)$ with

$$\theta(t) = \begin{cases} \infty & (t_i < t < t_i + \tau_q) \\ \theta_T & (t_i + \tau_q \leq t < t_{i+1}) \end{cases}$$

for the period (t_i, t_{i+1}), $i = 1, 2, \ldots$ with an action potential fired at t_i $i = 1, 2, 3, \ldots$. For simplicity, the higher threshold condition for the relative refractory period is relaxed and reduced to the normal threshold θ_T. Since there is no reset mechanism for a refractory period in the leaky integrator model, we would have to impose also the artificial reset of

$$V(t) = \begin{cases} V_r & (t_i < t \leq t_i + \tau_q) \\ V_i(t) & (t_i + \tau_q \leq t < t_{i+1}) \end{cases}$$

where V_r is resting potential (typically about -60 mV) and $V_i(t)$ is the solution of the leaky integrator for the period (t_i, t_{i+1}).

11.2.4   *Time constant and estimation of parameters*

For the Lapicque model with a time-invariant capacitance and a constant input current, the evolution of the transmembrane voltage is given by the exact solution

$$V(t) = V_0 + \frac{I}{C}t$$

of the IVP

$$C\frac{dV}{dt} = I, \quad V(0) = V_0$$

where V_0 is the initial transmembrane voltage normally taken to be V_r (<0). With the transmembrane voltage monotone increasing, it would exceed any prescribed threshold θ_T should the input current persist. Even if the input should be a non-negative rectangular wave train with a substantial gap of no current, the response would still continue to climb between interval of constant voltage resulting in a staircase of slanted escalation between flat platforms.

The situation is quite different for a leaky integrator of finite resistance characterized by (11.2.1). The exact solution for a constant current input is

$$V(t) = V_r + IR(1 - e^{-t/RC})$$

where V_0 is typically the resting potential V_r. While the output depolarized voltage is monotone increasing, it tends to a maximum steady state value of

$$\lim_{t \to \infty} [V(t)] = V_r + IR$$

assuming the constant current persists. The maximum transmembrane voltage $V_r + IR < IR$ may or may not be greater than the threshold θ_T.

In the case of a non-negative rectangular wave train input with maximum magnitude I_0, we work with

$$I(t) = I_0 \sum_{k=0}^{N} [H(t_k) - H(t_{k+1})], \quad t_j = j\Delta,$$

where $H(\cdot)$ is now the Heaviside unit step function

$$H(t) = \begin{cases} 0 & (t < 0) \\ 1 & (t > 0) \end{cases} \tag{11.2.2}$$

(and not to be confused with the notation $H(u)$ for the *Hamiltonian* in the analysis of optimal control models for the spread of Chlamydia). For this input current, we have over the first block of the rectangular wave train input

$$V(t) = V_1(t) = V_r + I_0 R(1 - e^{-t/RC}) \quad (0 \le t \le t_1).$$

To simplify book-keeping, we let $V(t) = U(t) + V_0$ so that $U(0) = 0$ and

$$U(t) = U_1(t) = I_0 R(1 - e^{-t/RC}) \quad (0 \le t \le t_1).$$

For $t > t_1$, the solution for (11.2.1) subject to the continuity condition at t_1 is

$$U(t) = U_2(t) = U_1(t_1)e^{-t/RC}$$
$$= I_0 R(1 - e^{-t_1/RC})e^{-(t-t_1)/RC} \quad (t_1 \le t \le t_2).$$

Suppose $U_1(t_1)$ is below the threshold $(\theta_T - V_r)$. For the special case $N = 0$ so that $U_2(t)$ applies for $t_1 \le t < \infty$, let T_2 be defined by

$$U_2(T_2) = U_1(t_1)e^{-(T_2-t_1)/RC}.$$

The product $RC \equiv \tau$ known as the *time constant* for the cell membrane (or growth rate constant for general first order growth process) can be determined once we know (from measuring $U_2(t)$) the time when $U_2(t)$ drops to e^{-1} of its value at t_1. A second output measurement such as the measured transmembrane voltage $U_1(t_1)$ at t_1 enables us to determine R and C separately.

The solution for the more general problem with $N > 0$ may be obtained by the same solution process.

11.2.5 *The Ornstein–Uhlenbeck process*

While more general problems with time-varying capacitance and resistance may be solved exactly with the help of an integrating factor (see [72]), we are interested in the more realistic problem that the input current is not prescribed or known explicitly. Rather, it is usually the result of stimuli generated by other neurons in contact with the cell of interest with the activities of these other neurons in turn dependent ultimately on its very complex environment. As such, the input current to our particular cell is more likely to be a random process (or processes). It is the goal of this chapter and next to develop methods of solution for the corresponding stochastic voltage response for such current input and other phenomena modeled by SDE.

Many different kinds of stochastic point neuron models other than the Lapique model have been formulated in the neuroscience literature. Among those where stochasticity is introduced through random external stimuli, we mention here the Ornstein–Uhlenbeck model as it is the stochastic counterpart of the leaky integrator. For a nerve cell receiving Poisson type random synaptic input (excitation and inhibition), the subthreshold random depolarization $X(t)$ $(< \theta_T)$ in the diffusion approximation satisfies the stochastic ODE

$$dX = (-X + a)\, dt + b dW$$

where $W(t)$ is a Wiener process with zero mean and variance t (with time measured in unit of the membrane time constant). The subthreshold response voltage in this case is known as an Ornstein–Uhlenbeck process.

11.2.6 *Interacting motoneurons*

The physiology of the spinal cord nervous system shows that nerve cells are interconnected to form a neural network. In a point model of motoneurons, a general network of n neurons interacting linearly may be modeled mathematically by the linear ODE system

$$C\frac{d\mathbf{V}}{dt} + K\mathbf{V} = G\mathbf{I}(t) \tag{11.2.3}$$

for the transmembrane voltages $(V_1, \ldots, V_n)$ of the n neurons constituting the n components of the voltage vector $\mathbf{V}(t)$. In this linear model, the $n \times n$ matrix C is a diagonal matrix with the kth element being the capacitance of the kth leaky integrator model neuron. The jth diagonal element of the matrix K is the conductance ($= 1/$resistance) of the jth neuron. Connectivity among neurons typically transmits the output of one neuron and convert it into an input to another neuron. An off-diagonal element K_{ij} of the matrix K assigns a weight to the effects of input from the neuron j on the transmembrane potential of the neuron i. The elements of the m-vector $I(t)$ are input current generated by different external stimuli impacting directly some (or all) of the different neurons without passing through other neurons in the neural net. The elements of the $n \times m$ matrix G correspond to the weight assigned to the effects of each external input current on the different neurons.

If the input to any neuron in the network is solely through input currents generated by external sources, then the K matrix is diagonal since there would be no influence from other neurons. At the other extreme, if an exciting source should stimulate an input current for only one neuron, then the matrix G would have only a single non-zero element. Other than for the purpose of estimation of parameters of the linear system (11.2.3) as described in a previous subsection, we will be interested mainly in the case of $\mathbf{I}(t)$ being a vector stochastic process. To develop the solution process for such linear systems, we first focus in this chapter on the exact solutions for the moments of the response process made possible by the linearity of the voltage-current dynamics. Examples of nonlinear stochastic ODE models including those for motoneurons and methods of solution for such models will be discussed in the next chapter.

11.3 The Scalar Linear Problem

We begin with the scalar problem,

$$X'(t) = A(t)X(t) + F(t), \quad X(t_0) = 0, \tag{11.3.1}$$

where we work with more generic quantities $F(t)$ and $X(t)$ (instead of $I(t)$ and $V(t)$) for input and output processes and where the scalar "time constant" $A(t)$ is more generally allowed to be a known function of time. With any random initial data typically independent of the random forcing, we avoid unnecessary bookkeeping details by taking the response process to be initially quiescent. In theory, we should obtain the joint probability densities of $X(t)$ of all orders for various comination of different time instants if we are to know all about the response process. In most applications, we often need (or settle for) only the first and second moments of $X(t)$ (particularly its mean, variance and correlation). For this limited goal, we first explore the application of Theorem 36 for the determination of the desired statistics of the m.s. solution for the problem. (There will be some discussion on the joint density functions and characteristic functions of the response process $X(t)$ in the next chapter for those interested in these more general results.)

11.3.1 *Evaluation of integral representations*

By taking the relevant expectations, we get from (11.1.10)

$$\mu_x(t) = \langle X(t) \rangle = E[X(t)] = \int_{t_0}^{t} \Phi(t, z) \langle F(z) \rangle dz$$

$$C_{XF}(s, z) = \langle X(s)F(z) \rangle = E[X(s)F(z)]$$

$$= \int_{t_0}^{t} \Phi(s, \xi) \langle F(\xi)F(z) \rangle d\xi = C_{FX}(z.s)$$

$$C_{XX}(t, s) = \langle X(t)X(s) \rangle$$

$$= \int_{t_0}^{t} \Phi(t, z) \langle F(z)X(s) \rangle dz = \int_{t_0}^{t} \Phi(t, z) C_{FX}(z.s) dz$$

with $\langle \cdots \rangle = E[\cdots]$ and

$$C_{XX}(t, t) = \int_{t_0}^{t} \Phi(t, z) \langle F(z)X(t) \rangle ds = \int_{t_0}^{t} \Phi(t, z) C_{XF}(t, z) ds.$$

From these, we can calculate the variance of $X(t)$ usually important in most applications. In the sequence of results above, a statistical measure of the unknown response process is given in term of the statistics of the known stochastic forcing. As such, the problem is reduced to evaluating a number of ordinary integrals, numerically if necessary. In principle, we can continue the process to determine higher moments by evaluating more integrals. As such the problem is solved.

In practice, the fundamental solution $\Phi(t, s)$ (also known as *impulse response* in engineering literature) may not be determined exactly and explicitly in terms of known functions (even for the scalar case) and some numerical methods would be needed for its determination before performing the various required integration. The approach outlined above would require that we store several two dimensional arrays of numerical data (and higher dimensional arrays for vector unknowns). For each pair of (t, s), we need $\Phi(s, \xi)$ and $C_{FF}(\xi, z)$ for a range of ξ and z to calculate $C_{XF}(s, z)$ (for a needed range of z). The results are then used together with $\Phi(t, z)$ (for a range of z) to calculate the one value $C_{XX}(t, s)$. We have to do this for a two dimensional array of t and s for a numerical solution of the problem. This may not be much of a problem given today's computing capacity. However, the computing requirements escalate geometrically as the number of components in an unknown vector response process $\mathbf{X}(t)$ increases. It seems desirable to have an alternative method that would reduce these requirements, the most serious of which (in today's computing environment) seems to be the storage requirements.

11.3.2 *IVP for response statistics*

One way to avoid storing a double array of $\Phi(t_i, s_j)$ is to calculate the mean and correlation function by solving the IVP for this quantities in the t variable for different fixed values of s. For the mean $\mu_X(t)$, we take the expectation of both side of the ODE for $X(t)$ and get (with the help of commutativity of l.i.m. and expectation)

$$\frac{d\mu_X}{dt} = A(t)\mu_X + \mu_F(t), \quad \mu_X(t_0) = 0. \tag{11.3.2}$$

Since the mean $\mu_F(t)$ of the forcing term is known, we have in (11.3.2) a deterministic forced IVP in ODE for the mean $\mu_X(t)$ of the response process $X(t)$ that can be solved in the usual way, analytically or numerically. Note that the mean is only a function of one variable t, we have avoided storing a double array of $\Phi(t_i, s_j)$ by this new approach.

If we now set $F(t) = F^*(t) + \mu_F(t)$ and $X(t) = X^*(t) + \mu_X(t)$, we get by the linearity of the IVP (11.3.1) that $X^*(t)$ is the solution of the IVP

$$\frac{dX^*(t)}{dt} = A(t)X^*(t) + F^*(t), \quad X^*(t_0) = 0 \tag{11.3.3}$$

with $\langle F^*(t) \rangle = 0$ and therewith $\mu_{X^*}(t) = 0$. Since the problem (11.3.2) for the mean response is straightforward with no theoretical complication or unacceptable computational burden, we will henceforth assume that we have separated out the problem for the mean response and focus on the problem (11.3.3) or (11.3.1) where the forcing term $F(t)$ is of zero mean.

Next, we multiply the given ODE and initial condition by $X(s)$ and *ensemble average* (take the expectation of)

$$\frac{\partial C(t,s)}{\partial t} = A(t)C(t,s) + C_{FX}(t,s), \quad C(t_0,s) = 0$$

where we have written $C(t,s)$ for $C_{XX}(t,s)$ to simplify notations (as there is no possible ambiguity) and $C_{FX}(t,s) = \langle F(t)X(s) \rangle = C_{XF}(s,t)$. In arriving at the ODE for $C(t,s)$, we have made use of the commutativity of m.s. differentiation and expectation.

Since $C_{FX}(t,s)$ involves the unknown solution process $X(t)$, it is itself an unknown. To determine $C_{FX}(t,s)$, we form $\langle F(t)\{dX(s)/ds\} \rangle$ to get

$$\frac{\partial C_{FX}(t,s)}{\partial s} = A(s)C_{FX}(t,s) + C_{FF}(t,s), \quad C_{FX}(t,t_0) = 0.$$

For a fixed value of t, this last IVP in the s variable is straightforward by any available computing software such as MatLab, Mathematica and Maple. However, for the purpose of determining the correlation function $C(t,s)$ for fixed s, we need $C_{FX}(t,s)$ for the entire array of $\{t_i\}$ for a numerical solution for that particular s. These can of course be obtained by repeating the solution of the IVP for $C_{FX}(t,s)$ for all mesh points $\{t_i\}$ in the array for the same s value. The computing requirement involved is very modest and, more importantly, preferrable to storing the required double array of $\Phi(t,z)$ for the application of (11.1.10). As such, the new approach to obtain $C(t,s)$ for a fixed s requires an acceptable amount of computation (by today's standards) and significantly less storage.

If we want to obtain $C(t,s)$ for the same range of t and a group of s values (instead of just one fixed s), we would have to repeat the solution process outlined above for each additional s_j. The computing requirement would escalate if we do not wish to store the double array of $C_{FX}(t,s)$ values on the rectangular mesh. It is still preferable to pay the price of more computing for an order of magnitude reduction of storage requirement even if it is no longer as attractive. However, the real merit of the new approach is that it leads to a third approach that involves much more attractive reduction of both storage and computational requirements. This new third approach and its benefits will be described in the next few sections.

11.4 White Noise Excitation

11.4.1 *Response variance*

In this section, we formulate a third method for calculating the second order statistics of the response process $X(t)$ given that we are working with forcing process of zero mean. Instead of trying to determine the correlation function $C(t,s)$, this new method first seeks the second moment $\langle X^2(t) \rangle = C(t,t) \equiv V(t)$ which is also the variance of $X(t)$ (denoted by $V(t)$ henceforth), since the response process is of zero mean.

To obtain an ODE for the variance $V(t)$, we differentiate $\langle X^2(t) \rangle$ to get

$$V'(t) = 2\langle X(t)X'(t) \rangle = 2A(t)V(t) + 2\langle X(t)F(t) \rangle$$

where the unknown $V_{FX}(t) = \langle X(t)F(t) \rangle$ is given by

$$V_{FX}(t) = \langle X(t)F(t) \rangle = \int_{t_0}^{t} \Phi(t, \xi)\langle F(\xi)F(t) \rangle d\xi. \qquad (11.4.1)$$

The right hand side of (11.4.1) is a known quantity since $\langle F(\xi)F(t) \rangle$ is known. While we can evaluate $V_{FX}(t)$ numerically (and possibly analytically in a few cases) to be used in the equation for $V(t)$, we look instead at a special case when $V_{FX}(t)$ is completely known without any calculation. This is the case when the random forcing is temporally uncorrelated so that

$$\langle F(\xi)F(t) \rangle = D\delta(\xi - t),$$

where D is a known constant. A particular class of process with such an autocorrelation function is the white noise process. For such process, the "integral" for $V_{FX}(t)$ simplifies to

$$V_{FX}(t) = \int_{t_0}^{t} \Phi(t, \xi)D\delta(\xi - t)d\xi$$

$$= \int_{-\infty}^{t} \Phi(t, \xi)D\delta(\xi - t)d\xi = \frac{1}{2}\Phi(t, t)D = \frac{1}{2}D \qquad (11.4.2)$$

where we made use of the properties $\Phi(t, t) = 1$ of the fundamental solution as well as the convention

$$\int_{-\infty}^{t} \delta(t - y)dy = \frac{1}{2}. \qquad (11.4.3)$$

Upon substituting this result into the ODE for $V(t)$, we obtain

$$V'(t) = 2A(t)V(t) + D, \quad V(t_0) = 0. \qquad (11.4.4)$$

This is a conventional IVP for a deterministic ODE; it determines the variance $V(t)$ of the (zero mean) response process independent of any other response statistics.

It should be noted that white noise or temporarily uncorrelated s.p. are not second order processes. In principle, we should, instead of (11.3.1), work with Ito type stochastic integral form of the problem:

$$X(t) = \int_{t_0}^{t} \{A(s)X(s)ds + dB(s)\}$$

or its differential abbreviation

$$dX = A(t)X(t)dt + dB(t)$$

where $B(t)$ is the Wiener process with $\langle[B(t) - B(s)]^2\rangle = D(t - s)$ and $dB(t) = B(t + dt) - B(t)$. We do this in the following chapter for nonlinear problems. The linear problems treated in this chapter are sufficiently straightforward that we can stay with the white noise formalism without reverting to the more proper approach in terms of the Wiener process.

11.4.2 *The autocorrelation function*

With $V(t)$ determined, we can develop an efficient method for calculating the autocorrelation $C(t, s) = \langle X(t)X(s)\rangle$ of the m.s. response process $X(t)$. This is accomplished by forming $\langle X(s)X'(t)\rangle$ to get

$$\frac{\partial C(t, s)}{\partial t} = A(t)C(t, s) + \langle X(s)F(t)\rangle.$$

Next, we observe that

$$C_{FX}(t, s) = \langle X(s)F(t)\rangle = \int_{t_0}^{s} \Phi(s, z)\langle F(z)F(t)\rangle dz$$

and, for $\langle F(z)F(t)\rangle = D\delta(t - z)$,

$$C_{FX}(t, s) = \int_{t_0}^{s} \Phi(s, z)D\delta(t - z)dz = 0 \quad (t > s)$$

It follows that for $t > s$

$$\frac{\partial C(t, s)}{\partial t} = A(t)C(t, s), \quad C(s, s) = V(s). \tag{11.4.5}$$

With $V(s)$ known from the solution of the IVP (11.4.4) at the end of the last section, the IVP (11.4.5) determines $C(t, s)$ for all $t > s$ without doing a separate calculation to get $C_{FX}(t, s)$. Since we already have found $V(t)$ for all $t > t_0$ (in the previous subsection), we can obtain $C(t, s)$, $t \geq s$, for all $s \geq t_0$. Note that we also have the symmetry condition $C(t, s) = C(s, t)$ for $t < s$.

11.5 Correlated Forcing

The efficient method for finding the second order statistics of the solution process developed above depends critically on the assumption that the input process is temporally uncorrelated (or *delta-correlated*) with $\langle F(\xi)F(t)\rangle = D\delta(\xi - t)$. If the random forcing $F(t)$ is *temporally correlated*, we would need to modify the development above to handle the more general random forcing process. The needed modification is based on the observation that correlated noise processes are usually the output of passing white noise through a filter. Mathematically, this corresponds to the solution of some dynamical system (ODE) with a delta-correlated random forcing. For example, the (temporally correlated) Ornstein–Uhlenbeck process $U(t)$ may be considered as the steady state solution of the stochastic IVP

$$\frac{dU}{dt} + \alpha U = DW(t), \quad U(t_0) = 0 \tag{11.5.1}$$

where $W(t)$ is the white noise process with $\langle W(t)W(s)\rangle = \delta(t-s)$[65]. The integral representation of the stochastic response process is

$$U(s) = \int_{t_0}^{s} e^{-\alpha(s-\zeta)} DW(\zeta)d\zeta.$$

Correspondingly, we have for $z > t_0$

$$C_{UW}(s,z) = E[U(s)W(z)] = \int_{t_0}^{s} e^{-\alpha(s-\zeta)} D\langle W(\zeta)W(z)\rangle d\zeta$$

$$= \int_{t_0}^{s} e^{-\alpha(s-\zeta)} D\delta(\zeta - z)d\zeta = De^{-\alpha(s-z)} H(s-z)H(z-t_0)$$

$$= \begin{cases} 0 & (z < t_0) \\ De^{-\alpha(s-z)} & (t_0 < z < s) \\ 0 & (z > s), \end{cases}$$

where $H(\cdot)$ is the Heaviside unit step function, and therewith

$$C_{UU}(t,s) = D\int_{t_0}^{t} e^{-\alpha(t-z)}\langle W(z)U(s)\rangle dz$$

$$= D\int_{t_0}^{t} e^{-\alpha(t-z)} e^{-\alpha(s-z)} DH(s-z)H(z-t_0)dz$$

$$= D^2 \begin{cases} \frac{1}{2\alpha}\left[e^{-\alpha(t-s)} - e^{-\alpha(s+t-2t_0)}\right] & (t_0 < s < t) \\ \frac{1}{2\alpha}\left[e^{-\alpha(s-t)} - e^{-\alpha(s+t-2t_0)}\right] & (t_0 < t < s) \end{cases}$$

with

$$C_{UU}(t,s) \to \frac{D^2}{2\alpha} e^{-\alpha|t-s|} \quad (t \gg t_0, s \gg t_0).$$

If $F(t)$ in the original stochastic IVP (11.3.1) is an Ornstein–Uhlenbeck process $U(t)$, we can convert the IVP with temporally correlated forcing to one with uncorrelated forcing by appending to the original IVP the additional IVP (11.5.1), resulting in an IVP for the stochastic vector process $\mathbf{Y}(t) = (X(t), U(t))^T$:

$$\mathbf{Y}' = M(t)\mathbf{Y}(t) + \mathbf{F}(t), \quad \mathbf{Y}(t_0) = \mathbf{0}, \qquad (11.5.2)$$

where

$$M(t) = \begin{bmatrix} A & 1 \\ 0 & -\alpha \end{bmatrix}, \quad \mathbf{F}(t) = \begin{pmatrix} 0 \\ DW(t) \end{pmatrix}. \qquad (11.5.3)$$

A vector version of the method of the previous subsection for calculating the corresponding covariance matrix and then the correlation matrix function will have to be developed for this linear IVP with a delta-crrelated forcing. We do this in the next section.

11.6 Linear Vector IVP with Random Forcing

11.6.1 *The covariance matrix*

Suppose now the stochastic IVP is for an n dimensional vector unknown $\mathbf{Y}(t)$ that is the solution of the vector ODE (11.5.2) where $M(t)$ in that ODE is an $n \times n$ matrix whose elements are continuous functions of t in an appropriate range of t and where $\mathbf{F}(t)$ is a second order n vector stochastic process. The ODE is augmented by an initial condition which we take to be $\mathbf{Y}(t_0) = \mathbf{0}$ (where we have taken the initial data $\mathbf{Y}^o = \mathbf{0}$ to simplify the presentation). Guided by the development of the scalar case, we are interested in formulating a conventional IVP for the covariance matrix $V(t) = \langle \mathbf{Y}(t)\mathbf{Y}^T(t) \rangle$ of the output process $\mathbf{Y}(t)$. Note that $V(t)$ is an $n \times n$ matrix function (distinctly different from the scalar function $\langle \mathbf{Y}^T(t)\mathbf{Y}(t) \rangle$).

With the only randomness in the problem provided by the second order s.p. $\mathbf{F}(t)$, the existence and uniqueness of an m.s. solution is assured by Theorem 36. We can therefore concentrate on obtaining the m.s. solution for the problem. Analogous to the scalar case, we assume the forcing process is zero mean so that the variance is the same as the second moment of the solution. With

$$V'(t) = \langle \mathbf{Y}'(t)\mathbf{Y}^T(t) \rangle + \langle \mathbf{Y}(t)\{\mathbf{Y}^T(t)\}' \rangle,$$

we use the ODE for $\mathbf{Y}'(t)$ write

$$V'(t) = M(t)V(t) + V(t)M^T(t) + V_C(t), \quad V(t_0) = 0 \qquad (11.6.1)$$

where

$$V_C(t) = \langle \mathbf{F}(t)\mathbf{Y}^T(t) \rangle + \langle \mathbf{Y}(t)\mathbf{F}^T(t) \rangle.$$

For the unknown $V_{YF}(t) = \langle \mathbf{Y}(t)\mathbf{F}^T(t) \rangle (\neq \langle \mathbf{F}(t)\mathbf{Y}^T(t) \rangle = V_{FY}(t))$, we make use of the integral representation (11.1.10) for $\mathbf{Y}(t)$ to obtain

$$V_{YF}(t) = \int_{t_0}^{t} \Phi(t, \xi) \langle \mathbf{F}(\xi)\mathbf{F}^T(t) \rangle d\xi.$$

While we can evaluate $V_{YF}(t)$ numerically (and possibly analytically in a few cases) to be used in the equation for $V(t)$, our experience from the scalar case suggests that it suffices to consider only the special case

$$\langle F_i(\xi)F_j(t)\rangle = D_{ij}\delta(\xi - t), \qquad (11.6.2)$$

where $\{D_{ij}\}$ are known constants. A particular class of stochastic processes with such an autocorrelation function are the white noise proceses. For such processes, the integral for $V_{FX}(t)$ simplifies to

$$V_{FX}(t) = \int_{t_0}^t \Phi(t,\xi)[D_{ij}]\,\delta(\xi-t)d\xi = \frac{1}{2}\Phi(t,t)[D_{ij}] = \frac{1}{2}[D_{ij}]$$

where we have made use of (11.4.3) and the property $\Phi(t,t) = I$ of the principal matrix solution. Upon substituting this result into the ODE for $V(t)$, we obtain the following matrix differential equation for the covariance matrix $V(t)$:

Theorem 37. *If the random forcing vector process is delta-correlated so that (11.6.2) holds and the matrix $M(t)$ is non-singular, then its covariance matrix $V(t)$ is determined by the IVP*

$$V'(t) = M(t)V(t) + V(t)M^T(t) + [D_{ij}], \quad V(t_0) = 0. \qquad (11.6.3)$$

The matrix equation (11.6.3) constitutes a conventional IVP for a deterministic ODE system. It completely determines the elements of the symmetric covariance matrix $V(t)$ of the m.s. vector response process $\mathbf{Y}(t)$.

11.6.2 *Autocorrelation*

Having found the covariance matrix $V(t)$, we can now develop an efficient method for calculating the autocorrelation of the solution process. This is accomplished by forming $\langle \mathbf{Y}'(t)\mathbf{Y}^T(s)\rangle$ to get

$$\frac{\partial C(t,s)}{\partial t} = M(t)C(t,s) + \langle \mathbf{F}(t)\mathbf{Y}^T(s)\rangle.$$

Next, we form

$$C_{YF}(s,t) = \langle \mathbf{Y}(s)\mathbf{F}^T(t)\rangle = \int_{t_0}^s \Phi(s,z)\langle \mathbf{F}(z)\mathbf{F}^T(t)\rangle dz$$

and observe $\langle \mathbf{F}(z)\mathbf{F}^T(t)\rangle = [D_{ij}]\,\delta(t-z)$ to get

$$C_{YF}(s,t) = \int_{t_0}^s \Phi(s,z)[D_{ij}]\delta(t-z)dz = 0 \quad (t > s).$$

We have then the following result:

Theorem 38. *Under the same hypotheses as Theorem 37, the correlation matrix function $C(t, s) = \langle \mathbf{Y}(t)\mathbf{Y}^T(s) \rangle$ for the vector response process $\mathbf{Y}(t)$ of (11.5.2) is determined by the IVP*

$$\frac{\partial C(t, s)}{\partial t} = M(t)C(t, s), \quad (t > s), \quad C(s, s) = V(s). \tag{11.6.4}$$

for $t > s$.

With $V(s)$ known from the solution of the IVP (11.6.3) at the end of the last section, the IVP (11.6.4) determines $C(t, s)$ for all $t > s$ without doing a separate calculation to get $C_{YF}(s, t) = C_{FY}(t, s)$. The relation $C(t, s) = C^T(s, g)$ enables us to obtain $C(t, s)$ for $t < s$ when needed.

11.7 Time-Invariant Systems

11.7.1 *Particular solution for a non-defective M*

For the special case of a constant matrix, $M(t) = M$, a time-independent particular solution $V^{(p)}$ of the IVP (11.6.3) is an $n \times n$ constant symmetric matrix determined by the matrix equation

$$MV^{(p)} + V^{(p)}M^T = -[D_{ij}]. \tag{11.7.1}$$

Given the broad range of applications and implications of the matrix equation above, a few words should be said about how we may solve it for $V^{(p)}$ without transforming it into an equation for the $n(n+1)/2$ dimensional vector consisting of all the distinct entries of the symmetric covariance matrix $V^{(p)}$.

When the constant matrix M is *nondefective* (see [72]) so that it has a full set of (linearly independent) eigenvectors with eigen-pairs $\{\lambda_i, \mathbf{p}^{(i)}\}$, $i = 1, 2, \ldots, n$, M^T also has a full set of eigenvectors with eigen-pairs $\{\lambda_i, \mathbf{q}^{(i)}\}$. Form the modal matrices

$$P = [\mathbf{p}^{(1)}\mathbf{p}^{(2)} \cdots \mathbf{p}^{(n)}], \quad Q = [\mathbf{q}^{(1)}\mathbf{q}^{(2)} \cdots \mathbf{q}^{(n)}]. \tag{11.7.2}$$

Then pre-multiply the matrix equation (11.7.1) by P^{-1} and post-multiply by Q to get

$$\Lambda U + U\Lambda = [(\lambda_i + \lambda_j)U_{ij}] = -\bar{D} \tag{11.7.3}$$

with

$$U = P^{-1}V^{(p)}Q, \quad \bar{D} = [\bar{D}_{ij}] = P^{-1}[D_{ij}]Q, \tag{11.7.4}$$

$$\Lambda = P^{-1}BP = Q^{-1}B^TQ = [\lambda_i\delta_{ij}] \tag{11.7.5}$$

where δ_{ij} is the Kronecker delta. The solution of (11.7.3) is

$$U = [U_{ij}] = -\left[\frac{\bar{D}_{ij}}{\lambda_i + \lambda_j}\right].$$

We have then the following proposition:

Proposition 44. *If the constant matrix M is non-singular and nondefective, the solution of the matrix equation* (11.7.1) *is given by*

$$V^{(p)} = -P\left[\frac{\bar{D}_{ij}}{\lambda_i + \lambda_j}\right]Q \tag{11.7.6}$$

with the modal matrices P and Q defined in (11.7.2) *and $\bar{D} = P^{-1}[D_{ij}]Q$.*

Exercise 49. Show that $V^{(p)}$ is a synmetric matrix.

11.7.2 The complete solution

For the IVP (11.6.3) itself when $M(t)$ is a constant matrix M, we may make use of the relevant superposition principle and write $V(t) = V^{(c)}(t) + V^{(p)}$ where $V^{(c)}(t)$ is the complementary solution, the solution of

$$\frac{dV^{(c)}}{dt} = MV^{(c)} + V^{(c)}M^T. \tag{11.7.7}$$

The initial condition $V(t_0) = 0$ requires

$$V^{(c)}(t_0) = -V^{(p)}. \tag{11.7.8}$$

In determining the complementary matrix solution $V^{(c)}(t)$, it is more efficient and elegant to avoid converting the matrix ODE (11.7.7) into a vector ODE $\mathbf{v}' = A\mathbf{v}$ for the vector $\mathbf{v}(t)$ whose components are the $n(n+1)/2$ unknown components of $V^{(c)}(t)$. Instead, we make use of the same matrix diagonalization method applied to the matrix M to reduce both the storage requirements and the amount of computation involved. With (11.7.2) and (11.7.4), we again can re-write the ODE (11.7.7) as

$$\frac{dV^{(c)}}{dt} = P\left(\Lambda U^{(c)} + U^{(c)}\Lambda\right)Q^{-1} = P\left[(\lambda_i + \lambda_j)U_{ij}^{(c)}\right]Q^{-1}$$

so that

$$\frac{dU^{(c)}}{dt} = \left[(\lambda_i + \lambda_j)U_{ij}^{(c)}\right] \quad \text{or} \quad \frac{dU_{ij}^{(c)}}{dt} = (\lambda_i + \lambda_j)U_{ij}^{(c)}.$$

To satisfy the initial condition (11.7.8), we take the solution in the form

$$U_{ij}^{(c)}(t,t_0) = e^{(\lambda_i + \lambda_j)(t-t_0)} c_{ij}, \quad U_{ij}^{(c)}(t_0,t_0) = c_{ij}.$$

Altogether we found

$$U^{(c)} = \left[e^{(\lambda_i + \lambda_j)(t-t_0)} c_{ij} \right], \quad V^{(c)}(t) = P \left[e^{(\lambda_i + \lambda_j)(t-t_0)} c_{ij} \right] Q^{-1}.$$

The initial condition (11.7.8) is satisfied by choosing the n^2 constants of integration $\{c_{ij}\}$ as the solution of the matrix equation

$$PU^{(c)}(t_0,t_0)Q^{-1} = -V^{(p)} = P \left[\frac{\bar{D}_{ij}}{\lambda_i + \lambda_j} \right] Q^{-1}$$

where we have made use of (11.7.6) for the last equality. It follows that

$$[c_{ij}] = \left[\frac{\bar{D}_{ij}}{\lambda_i + \lambda_j} \right].$$

Exercise 50. Show that $V^{(c)}(t)$ is a symmetric matrix.

11.7.3   Defective coefficient matrix

If the time-invariant coefficient matrix M does not have a full set of eigenvectors, we know from linear algebra that we can still find matrices P and Q (whose columns are the generalized eigenvectors of M and M^T, respectively). The same operations applied to (11.7.1) then lead to

$$JU + UJ = [(\lambda_i + \lambda_j)U_{ij} + U_{i+1,j} + U_{i,j-1}] = -\bar{D} \qquad (11.7.9)$$

where J is the Jordan matrix for M. Our experience with reduction of a linear system of first order ODE (see [72]) suggests that Jordan form effectively de-couples the linear system so that we only have to solve single first order linear ODE starting with the last (or the first if the *Jordan* matrix has its 1's in the sub-diagonal positions). If we now look at the $\bar{D}_{nn}$ elements of the $\bar{D}$ matrix, the matrix equation (11.7.9) gives

$$2\lambda_n U_{nn} + U_{n,n-1} = -\bar{D}_{nn}$$

which involves two unknowns. If we look at the equation for the element $\bar{D}_{11}$, then (11.7.9) gives

$$2\lambda_1 U_{11} + U_{21} = -\bar{D}_{11}.$$

It is again coupled to other equations. On the other hand, the component equation of (11.7.9) with the element $\bar{D}_{n1}$, namely

$$(\lambda_n + \lambda_1)U_{n1} = -\bar{D}_{n1} \qquad (11.7.10)$$

is uncoupled from the others and can be solved for the only unknown U_{n1}.

To see how we may continue to obtain de-coupled equations for more unknown elements, we examine equations corresponding to the $\bar{D}_{k1}$ elements for $k = n - 1, n - 2, n - 3, \ldots, 1$ with

$$(\lambda_k + \lambda_1)U_{k1} + U_{k+1,1} = -\bar{D}_{k1}.$$

For $k = n - 1$ in which the term $U_{k+1,1} = U_{n1}$ is known from (11.7.10) and we can solve that equation for

$$U_{k1} = -\frac{\bar{D}_{k1} - U_{k+1,1}}{\lambda_k + \lambda_1} \tag{11.7.11}$$

for $k = n - 1$. Working backwards starting from the unknown located at the lower left hand corner of the matrix, the relation (11.7.11) deterinines U_{k1} sequentially for $k = n, n - 1, n - 2, \ldots, 2, 1$.

With the first column of the matrix U now known, the next group of equations containing only one unknown each is seen to come from

$$U_{jk} = -\frac{\bar{D}_{jk} - U_{j+1,k} - U_{j,k-1}}{\lambda_j + \lambda_k} \quad (j = n, (n-1), \ldots, 2, 1). \tag{11.7.12}$$

For $k = 2$, the last term in the numerator on the right hand side is known for all j. Hence, the situation is reduced to the $k = 1$ case so that (11.7.12) determines U_{j2} for all $j = n, n - 1, \ldots, 2, 1$. It should be clear that we may continue the same solution process by apply (11.7.12) sequentially for $k = 3, 4, \ldots, n$ to get the solution of the matrix equation (11.7.1) as shown in the following proposition:

Proposition 45. *The exact solution of the matrix Riccati equation (11.7.1) is given in terms of*

$$U_{jk} = -\frac{\bar{D}_{jk} - U_{j+1,k} - U_{j,k-1}}{\lambda_j + \lambda_k} \quad (\{j = n, (n-1), \ldots, 1\}, \ k = 1, 2, \ldots, n)$$

with $U_{jk} = 0$ for $j < 1$, $j > n$, $k < 1$ or $k > n$. The (particular) solution $V^{(p)}$ is then obtained from (11.7.4) as

$$V^{(p)} = PUQ^{-1}.$$

While not explicitly stated, the result above was more or less suggested by Bartels and Stewart [3] in the description of their method of solution for applications in control theory.

11.7.4 *Solution of IVP*

An alternative way to find the complete solution of the IVP for $V(y)$ is by direct substitution into the matrix DE the following matrix representation for $V(t)$ in terms of the matrix exponential $e^{B(t-t_0)}$:

$$V^{(c)}(t) = e^{(t-t_0)M} C e^{(t-t_0)M^T}$$

where C is a constant matrix of integration, keeping in mind that $e^{Mt}M = Me^{Mt}$. By the relevant superposition principle, the general solution for the matrix DE in (11.6.3) is

$$V(t) = e^{(t-t_0)M} C e^{(t-t_0)M^T} + V^{(p)}.$$

The initial condition $V(t_0) = 0$ requires $C = -V^{(p)}$ so that

$$V(t) = V^{(p)} - e^{(t-t_0)M} V^{(p)} e^{(t-t_0)M^T}.$$

Note the solution applies whether or not M is defective and whether or not M has only eigenvalues with negative real parts so that a steady state exists.

11.8 Storage Reduction for Separable PDE

While the matrix equation (11.7.1) arises naturally in our approach to determining the variance of a stochastic process, the equation finds applications in many other different areas of science and engineering, particularly areas involving control theory. In connection with subsequent development herein on linear PDE with random forcing, we mention here a particular use of the matrix equation approach in solving conventional partial differential equations. This is illustrated below with the simplest boundary value problem for elliptic PDE, namely the Dirichlet problem for Poisson's equation on a unit square R with its boundary denoted by ∂R:

$$\nabla^2 v = -f(\mathbf{x}) \quad (\mathbf{x} \text{ in } R), \quad v(\mathbf{x}) = \mathbf{0} \quad (\mathbf{x} \text{ in } \partial R).$$

It will be evident that the same benefits of a Riccati equation approach extends to other types of problems in PDE with a "separable" structure.

Suppose we wish to obtain solution of the Dirichlet problem above by some numerical method. Let R be covered by an equally spaced set of mesh points, $\{(x_i, y_j)\}$, $i, j = 0, 1, \ldots, n+1$ with the same spacing Δ between neighboring mesh points in both x and y directions. Given that v vanishes on the boundary, we have altogether n^2 unknowns $\{V_{ij} = v(x_i, y_j)\}$ for $i, j = 1, 2, \ldots, n$. The PDE is then written as a 2-dimensional finite difference equation:

$$\{V_{i+1,j} - 2V_{ij} + V_{i-1,j}\} + \{V_{i,j+1}, -2V_{ij} + V_{i,j-1}\} = -F_{ij}, \quad F_{ij} = \Delta^2 F(x_i, y_j).$$

The conventional approach would be to line up the unknowns and the corresponding forcing terms as $n^2 \times 1$ vectors:

$$\mathbf{v} = (V_{11}, V_{21}, \ldots, V_{n1}, V_{12}, V_{22}, \ldots, V_{n2}, \ldots, V_{1n}, \ldots, V_{nn})^T$$

and

$$\mathbf{f} = (F_{11}, \ldots, F_{n1}, F_{12}, \ldots, F_{nn})^T$$

and write the linear system as $K\mathbf{v} = \mathbf{f}$ to be solved by any one of the available methods of solution. However, the coeficient matrix K for the new system is $n^2 \times n^2$. For $n = 100$, K would be a $10^4 \times 10^4$ matrix.

If we store the unknowns $\{V_{ij}\}$ (rather naturally) as a matrix $[V_{ij}]$, we can write the linear system for V_{ij} as the matrix equation,

$$AV + VA = -F, \tag{11.8.1}$$

where A, V and $F = [F_{ij}]$ are all $n \times n$ matrices with A being tridiagonal for the simplest case of a three-point central difference scheme. Hence, the storage requirement is reduced by an order of magnitude. For $n = 100$, the reduction is from one $10^4 \times 10^4$ coefficient matrix to one 100×100 coefficient matrix, though we may make use of the band structure of the coefficient matrix K for a more efficient solution scheme to be accompanied by some storage reduction.

Even if the spacings are not the same in the x and y direction, we would still have only one $n \times n$ and one $m \times m$ coefficient matrix in the matrix equation instead of one much larger $nm \times nm$ coefficient matrix for a vector equation. Moreover, the available methods for solving the matrix equation involves only $n \times n$ (or $n \times m$) matrices along the way to the final solution [3]. Evidently, the reduction in storage requirements would be even more dramatic for conventional PDE problem in higher spatial dimensions (see [38]) as long as the PDE system is of the separable type. We will see similar benefits in stochastic PDE problems in the next chapter.

11.9   The Nonautonomous Case

When the coefficient matrix $M(t)$ varies with t, we may still obtain the solution of the matrix ODE (11.6.3) without converting it to a vector ODE. Since the matrix ODE is linear, we may look for an integrating factor similar to a single ODE for one scalar unknown. Here, we consider a matrix integrating factor $E(t)$ and form the composite matrix function $E(t)V(t)E^T(t)$. Upon differentiating the composite function, we get

$$\left[E(t)V(t)E^T(t)\right]' = E'(t)V(t)E^T(t) + E(t)V'(t)E^T(t) + E(t)V(t)\left[E^T(t)\right]'. \tag{11.9.1}$$

With the identity

$$E(t)V'(t)E^T(t) = E(t)M(t)VE^T(t) + E(t)VM^T(t)E^T(t) + E(t)DE^T(t)$$

for the middle term of (11.9.1), we may simplify that relation to

$$W'(t) = E(t)DE^T(t) \tag{11.9.2}$$

for $W(t) = E(t)V(t)E^T(t)$ by taking

$$E'(t) + E(t)M(t) = 0, \tag{11.9.3}$$

with $M^T(t)E^T(t) + \left[E^T(t)\right]' = \left[E(t)M(t) + E'(t)\right]^T = 0$. The first order ODE system for the matrix unknown $E(t)$ is linear and can be solved numerically with the initial condition

$$E(t_0) = I. \tag{11.9.4}$$

Given

$$W(t_0) = [E(t)V(t)E^T(t)]_{t=t_0} = \{E(t_0)V(t_0)E^T(t_0)]_{t=t_0},$$

we have as the initial condition for the ODE (11.9.2)

$$W(t_0) = V(t_0) = 0. \tag{11.9.5}$$

The IVP defined by (11.9.2) and (11.9.5) is also linear and can be solved numerically for $W(t)$ to give

$$V(t) = \left[E^{-1}(t)\right] W(t) \left[E^T(t)\right]^{-1} = \left[E^{-1}(t)\right] W(t) \left[E^{-1}(t)\right]^T.$$

Chapter 12

General Nonlinear ODE Systems

12.1 Nonlinear Stochastic ODE Models

There are many nonlinear stochastic ODE models in the quantitative studies of life science phenomena. For illustrative purpose, we describe below a few such models for neurons, AIDS, and cancer.

12.1.1 *The quadratic model of a single motoneuron*

An example of nonlinear point model for a single motoneuron is the quadratic model [9, 17, 32]:

$$\tau X' = (X - x_r)(x_c - X) + RF(t) \tag{12.1.1}$$

where τ, x_r, x_c and R are known system parameters that can be estimated from empirical data. As a point model for a motoneuron, R corresponds to the resistance in the leaky integrator, τ is the time constant and the two constant x_r and $x_c(> x_r)$ limit the transmembrane voltage to the range (x_r, x_c) as long as $X(0) < x_c$ and $F(t) = 0$. If $X(0) > x_c$, $X(t)$ increases without bound (and exceeds the threshold θ_T at some time T_1) with or without a nonnegative input current $F(t)$) so that an action potential is triggered. The parameter x_c can therefore be interpreted as the critical voltage for spike initiation by some very short pulse current $F(t) = I_0\delta(t)$ of sufficient strength with $\delta(t)$ being the Dirac delta function.

In reality, the input current is not known exactly and usually resulting from random stimuli of the surrounding environment so that $F(t)$ is usually a stochastic process with some known statistics. If $F(t)$ is white noise or a filtered white noise, the scalar stochastic ODE for $X(t)$ (or the corresponding vector version for filtered white noise) is in the form of an *Ito SDE* and may be analyzed for response statistics (by method described later in this chapter).

12.1.2 *Lefever and Garay tumor growth model*

Among the early scalar stochastic model for tumor growth under immune surveillance and chemotherapy is the Lefever and Garay model featuring the following stochastic differential equation for the evolution of the tumor (cell population) size:

$$dX = \left\{ r_0 X \left(1 - \frac{X}{x_c} \right) - \frac{\beta X^2}{1 + X^2} + A_0 \cos(\omega t) X \left(1 - \frac{X}{x_c} \right) \right\} dt + X \left(1 - \frac{X}{x_c} \right) dB.$$

Here, $X(t)$ is the biomass of the tumor cell population evolving at a logistic natural growth rate in the absence of external intervention with r_0 being the linear per capita birth rate of cancer cells and x_c the carrying capacity of the environment. The factor $A_0 \cos(\omega t)$ characterizes the influence of a periodic chemotherapy treatment, β the influence of the immune system (that include macrophages, cytotoxic T lymphocite (CTL) cells and natural killer cells), and the term involving the standard Brownian noise (Wiener process) $B(t)$ capturing the randomness of the input. The scalar nonlinear stochastic ODE is in the form of an Ito SDE whose noise term has an *envelop function* nonlinear in the state variable.

12.1.3 *The stochastic FitzHugh–Nagumo (FHN) model*

The FitzHugh-Nagumo (FHN) model is for a prototype excitable neuron capable of generating action potentials (spikes) as well as repetitive activity (periodic solutions) in certain ranges of stimuli. The ODE system of this model consists of the two couple ODE

$$V' = V - \frac{1}{3} V^3 - U + I, \quad \tau U' = V + a - bU$$

for the membrane voltage V and a recovery variable U. The system exhibits a characteristic excursion in phase space if the external stimulus I exceeds a certain threshold value. It then relaxes back to its rest state with the variables V and U returning to their resting state values (possibly after some undershooting). This behavior is typical for spike generations (a short, nonlinear elevation of transmembrane voltage V, diminished over time by a slower, linear recovery variable U) in a neuron after stimulation by an external input current. The FHN model is a simplified version of the Hodgkin–Huxley (HH) model, a PDE model that captures in a detailed manner activation and deactivation dynamics of a spiking neuron [64] but is considered too complex for mathematical analysis to require simplifications. Considerably more will be said about the HH model and its simplifications in a later chapter.

More relevant to this section is the more general stochastic version of the FHN model in which the voltage process $X = X(t)$ and the recovery process $Y = Y(t)$

evolve according to the following SDE:

$$dX = [f(X) - Y + I]dt + \sigma dB, \quad dY = \frac{1}{\tau}(X - bY)dt.$$

In these equations, $I = I(t)$ is a deterministic input current (stimulus) which may be constant or time-varying; and the function f is a nonlinear natural growth rate for the membrane voltage often taken to be the cubic function

$$f(x) = kx(x - a)(1 - x), \quad (0 < a < 1).$$

The new element in the stochastic problem is a noisy component of the input current $W(t)$ being the derivative of a standard Wiener process $B(t)$ (of zero mean and variance t) with a constant σ characterizing the overall variability of the input noise. This SDE system is in the form of a nonlinear vector Ito SDE with an envelop independent of the state variables.

12.1.4   Nonlinear immunogenic tumor model

Another model on tumor growth was formulated and analyzed in [27]. Unlike the previous model, this immunogenic tumor model explicitly tracks the dynamics of the antipathogen cells in the immune system, collectively characterized by the population size variable $x(t)$. They are generated in response to the tumor cells of population size $y(t)$. The stochastic differential equations governed the evolution of these two population are taken to be:

$$dX = \{a_1 - a_2 X + a_3 XY\} dt + \{b_{11}(X - x_e) + b_{12}(Y - y_e)\} dB_1,$$

$$dY = \{b_1 Y (1 - b_2 Y) - XY\} dt + \{b_{21}(X - x_e) + b_{22}(Y - y_e)\} dB_2.$$

In these equations, $B_1(t)$ and $B_2(t)$ are standard Wiener processes and the parameters x_e an y_e are the stochastic equilibrium points constituting thresholds for generating more cells of each type. Briefly, the model assumes that the amount of noise increases with the distance to the equilibrium point. We have again a nonlinear Ito SDE system with a noise term linear in the state variables.

12.1.5   Early dynamics model for HIV

The analysis of cocktail drug treatment for AIDS in [59] considers a four component viral dynamic model subject to two different types of drug treatment (see also [72] and references therein). The viral dynamics aspects of that model is related to earlier models as described in [59] and may be simplified considerably for earlier

stages of the HIV virus growth. This was accomplished in [59] by the following
assumptions:

- The large amount of CD4+T cells (normalized to 1) essentially remains
  unchanged after losses due to viral infection.
- The conversion to HIV infected cells through chance encounters between normal
  (healthy) cells and HIV virions is probabilistically modeled by a term $\sqrt{1 \cdot V} dW$
  in the growth rate equation for the infected cell population $Y(t)$.
- Upon a chance encounter and interaction between a virion and a normal immune
  cell, the virion is lost and a new infected cell created.

Consequently, the four-component model may be simplified to the following two
stochastic ODE:

$$dY = (V - aY)dt + \sqrt{V}dB,$$
$$dV = (cY - \gamma V)dt - \sqrt{kV}dB.$$

Note that the coefficient of dB in both equations involves both the virion population $V(t)$ and the population of normal cells which has been normalized to 1.
The system is in the form of the vector Ito SDE

$$d\mathbf{X} = \mathbf{f}(\mathbf{X}, t)dt + G(\mathbf{X}, t)d\mathbf{B}, \quad \mathbf{X(t)} = (Y(t), V(t))^T$$

that can be analyzed for the response statistics.

12.1.6 *Shortest time to cancer*

Another example of nonlinear SDE system is the problem of carcinogenesis induced
by genetic instability. For this problem, let $X_1(t)$ be the population size of normal
(harmless) cells at time t normalized by its initial population size N and $X_2(t)$ be the
population size of cancerous mutant cells normalized by its final target population
size M. The evolution of these two populations are governed by the two following
two ODE (see [24])

$$X_1' = \ell(U)X_1 - [1 - m(U)]X_1^2 \equiv g_1(X_1, X_2; U) \tag{12.1.2}$$

$$X_2' = \frac{1}{\sigma}(\mu + U)X_1 + [1 - m(U)](a - X_1)X_2 \equiv g_2(X_1, X_2; U) \tag{12.1.3}$$

where $(\ )' = d(\ )/dt$, $\sigma = M/N \gg 1$ and

$$\ell(U) = -(\mu + U) + 1 - m(U) = -(\mu + U) + (1 - U)^\alpha. \tag{12.1.4}$$

Among the other parameters in these equations, $a \geq 2$ is the natural growth rate of
the faster proliferating cancerous cells, $0 < \mu \ll 1$ is basic mutation rate constant

of the normal cells in the homeostatic state, and U is the (normalized) per cell mutation rate with

$$0 \le U \le 1. \tag{12.1.5}$$

Typically, we have $\sigma \ge 10$, $a = 2$, and $\mu = 10^{-1} \ll 1$. For simplicity, we have focused our discussion herein to the death rate of the form:

$$m(U) = 1 - (1 - U)^\alpha \quad (\alpha > 0)$$

for some real parameter $\alpha > 0$. An in-depth discussion for the more general death rate $m(U)$, characterized only by its convexity type, can be found in [24],[73].

The two ODE (12.1.2) and (12.1.3) are subject to the following three auxiliary conditions:

$$X_1(0) = 1, \quad X_2(0) = 0, \quad X_2(T) = 1$$

where T is the (unknown) terminal time when the cancerous mutant cell population reaches the target size. The evolution of the normalized mutation rate $U(t)$ is constrained to be among the class of piecewise continuous functions on the interval $[0, T]$, denoted by Ω and by the inequality (12.1.5): Needless to say, the two normalized cell populations must be non-negative.

Here we are interested in the effects of a noisy environment causing the mutation rate U to deviate from the normal (optimal) rate for reaching the target cancerous population in the shortest time. In the final chapter of this volume, we analyze the simpler problem of a deviation from a time-invariant normal rate u_{op} so that $U = u_{op} + V$ for some random variable V. The resulting SDE system is one with random coefficients. Of interest is the mean time to target population to be determined from the prescribed statistics of V.

12.2   Existence and Uniqueness

Before analyzing the more complex nonlinear models in the life sciences listed in the previous section, we first consider the simplest nonlinear model in stochastic DE, one with known system properties and only uncertain initial data. This problem takes the mathematical form of

$$\mathbf{X}'(t) = \mathbf{f}(\mathbf{X}(t), t), \quad \mathbf{X}(t_0) = \mathbf{A} \tag{12.2.1}$$

where $\mathbf{X}(t)$ is an n vector (short for n dimensional vector) s.p. (with each component $X_i(t)$ being a stochastic process), $\mathbf{A}$ is an n vector second order random variable (with each vector component being a second order random variable) and $\mathbf{f}$ is a continuous vector function of the n component stochastic processes $\{X_i(t)\}$ and the independent variable t. In component form, (12.2.1) is equivalent to

$$X'_j = f_j(X_1, \ldots, X_n, t), \quad X_j(t_0) = A_j, \quad (j = 1, 2, \ldots, n). \tag{12.2.2}$$

Definition 33. In addition to the hypotheses $\mathbf{X}(t)$, $\mathbf{A}$ and $\mathbf{f}$ stipulated above, suppose the components of $\mathbf{f}$, $\{f_j\}$, are continuous in all of its arguments on the time interval $T = [t_0, t_T]$ and all components $\{A_j\}$ of the initial data $\mathbf{A}$ are second order random variables. Then $\mathbf{X}(t)$ is said to be a mean square (m.s.) solution of the IVP (12.2.1) in the interval T if

  i) $\mathbf{X}(t)$ is mean square continuous on T;
 ii) $\mathbf{X}(t_0) = \mathbf{A}$, and
iii) $\mathbf{f}(\mathbf{X}(t), t)$ is the mean square derivative of $\mathbf{X}(t)$ on the interval T.

If instead of random initial data, the quantities $\{A_j\}$ are known with certainty and $\mathbf{X}(t)$ is any solution of the IVP (12.2.1), we may re-write it as

$$\mathbf{X}(t) = \mathbf{A} + \int_{t_0}^{t} \mathbf{f}(\mathbf{X}(s), s)ds. \tag{12.2.3}$$

This equivalent integral equation form of the IVP has a number of advantages. The most important of these is its facilitation of a Picard iterations proof of existence and uniqueness of the solution of (12.2.1). When $\mathbf{A}$ is a random variable, the corresponding integral equation form is meaningful as a stochastic integral equation only if the integral on the right hand side exists as a m.s. integral for the function $f(\cdot, \cdot)$. More on this equivalence can be found in [50]. We limit our discussion here to a sufficiency condition for the existence and uniqueness of a m.s. solution of the stochastic IVP (12.2.1) for a prescribed second order random initial data $\mathbf{A}$.

Definition 34. $\mathbf{f}(\mathbf{X}, t)$ is mean square Lipschitz continuous in $\mathbf{X}$ if

$$\|\mathbf{f}(\mathbf{X}, t) - \mathbf{f}(\mathbf{Y}, t)\| \le L(t)\|\mathbf{X} - \mathbf{Y}\|$$

for some ordinary function $L(t)$ (often called the Lipschitz function for $\mathbf{f}(\mathbf{X}, t)$) with

$$\int_{t_0}^{t_T} L(t)dt < \infty.$$

Theorem 39. *If $\mathbf{f}(\mathbf{X}, t)$ is m.s. Lipschitz continuous in $\mathbf{X}$, there is a unique m.s. solution $\mathbf{X}(t)$ for any initial condition $\mathbf{A}$.*

Proof. The proof is essentially analogous to the corresponding proof for the deterministic version of the problem by Picard's iterations and will be omitted here. $\square$

This fundamentally important result is still rather restrictive. There are several reasons for the limitation. One of them is not allowing for randomness in other part of the model in addition to the initial data. Another limitation is the absence of a method for determining the probabilistic/statistical properties of the solution. In the rest of this chapter, we will direct our attention to methods of solution for broader classes of problems involving more general stochastic differential equations. For this purpose, we need the notion of conditional probability for continuous stochastic processes.

12.3 Kinetic Equation for a Stochastic Process

12.3.1 *Conditional probability*

The methods described in the previous chapter on the evolution of a stochastic phenomenon are limited to obtaining first and second order statistics of linear differential equations in which stochasticity appears only through the external forcing, i.e., the inhomogeneous term of the linear (scalar or vector) equation that does not involve the unknown response process. It is certainly desirable to develop also method for determining the pdf and joint pdf of the response process as we did in the chapter on transformation theory. Furthermore, we also need to be able to do the same for nonlinear differential equations. The goal of this chapter is to work toward a very general method for this purpose. We start on this task by formulating the *kinetic equation* for the probability density function or the joint density function of the output of a general first order SDE. The derivation will be for a scalar equation though it may be extended to vector equations as well.

The method of derivation is based on the concept of *conditional probability*. For the simple problem of rolling a fair die, the sample space of the corresponding random variable X is known to consist of six elementary events: $S_X = \{1, 2, 3, 4, 5, 6\}$. For a fair die, the probability of turning up a 2 is $1/6$ for any roll, $P(X = 2) = 1/6$. However, the probability of getting a 2 would be greater if it is known (or assumed) that the outcome is even. With the new sample space consisting only of three elementary events, the probability of getting a 2 (given that the outcome is even) is increased to $P(X = 2 \mid X = even) = 1/3$. The notation $P(X = 2 \mid X = even)$ means the *probability of $X = 2$ given (or "conditioning on") X being even*. We would like to extend this concept and its consequences to problems with continuous random variables and to stochastic processes.

To accomplish this, we recall from Chapter 1 that for the case of a discrete sample space

$$P(B \mid A) = \frac{P(B \cap A)}{P(A)},$$
(12.3.1)

where B and A may be from the same sample space or different discrete sample spaces. The counterpart of (12.3.1) for continuous random variables is

$$P(x_1 \leq X \leq x_2 \mid y_1 \leq Y \leq y_2) = \frac{P(x_1 \leq X \leq x_2, y_1 \leq Y \leq y_2)}{P(y_1 \leq Y \leq y_2)}$$

which may be expressed in terms of the relevant density functions as

$$P(x_1 \leq X \leq x_2 \mid y_1 \leq Y \leq y_2) = \frac{\int_{x_1}^{x_2} \int_{y_1}^{y_2} p(u,v)\,dvdu}{\int_{-\infty}^{\infty} \int_{y_1}^{y_2} p(u,v)\,dvdu}. \tag{12.3.2}$$

The following two properties are evident:

$$P(-\infty \leq X \leq -\infty \mid y_1 \leq Y \leq y_2) = 0,$$

$$P(-\infty \leq X \leq \infty \mid y_1 \leq Y \leq y_2) = 1.$$

We now specialize (12.3.2) by setting $x_1 = -\infty$, $x_2 = x$, $y_1 = y$ and $y_2 = y + \Delta y$ to get

$$P(-\infty \leq X \leq x_2 \mid y \leq Y \leq y + \Delta y) = \frac{\int_{-\infty}^{x} \int_{y}^{y+\Delta y} p(u,v)\,dvdu}{\int_{y}^{y+\Delta y} p_Y(v)\,dv} \tag{12.3.3}$$

where

$$p_Y(v) = \int_{-\infty}^{\infty} p(u,v)\,du. \tag{12.3.4}$$

As $\Delta y \to 0$, (12.3.3) tends to

$$P(-\infty \leq X \leq x \mid Y = y) = \frac{\int_{-\infty}^{x} p(u,y)\,dvdu}{p_Y(y)} \tag{12.3.5}$$

giving the probability of $X \leq x$ conditioning on $Y = y$.

The conditional density function is obtained by differentiating both sides of (12.3.5) with respect to x to get

$$p(x|y) = \frac{p(x,y)}{p_Y(y)}. \tag{12.3.6}$$

Definition 35. The random variables X and Y are statistically independent if $p_{XY}(x|y) = p_X(x)$.

As a consequence, we have as the analogue of what we know from the discrete sample space case the following relation for the joint probability density function of two statistically independent random variables:

$$p_{XY}(x,y) = p_X(x)p_Y(y).$$

12.3.2   Kinetic equation for the density function

For stochastic DE, we are also interested in conditional probability and conditional density functions of two random variables of its response process. The solution of a stochastic DE is a stochastic process $X(t)$. For fixed t and $t+\Delta t$, the quantities $X(t)$ and $X(t + \Delta t)$ are two different random variables (similar to the random variable

X and Y in the die rolling experiment). From the development of the previous section, we have

$$p(x, t + \Delta t; y; t) = p(x, t + \Delta t | X(t) = y)p(y, t),$$

where we have omitted the subscripts of the density functions (especially when Y and X are just X at different time), and therewith

$$p(x, t + \Delta t) = \int_{-\infty}^{\infty} p(x, t + \Delta t | y, t)p(y, t)dy. \tag{12.3.7}$$

The corresponding characteristic function for the conditional density $p(x, t + \Delta t | y, t)$ is

$$\phi(u, t + \Delta t | y, t) = E[e^{iu\Delta X} | y, t] = \int_{-\infty}^{\infty} e^{iu\Delta x} p(x, t + \Delta t | y, t)dx \tag{12.3.8}$$

where $\Delta X = X(t + \Delta t) - X(t)$ and Δx is an abbreviation for $x - y$. The corresponding inverse Fourier transform is

$$p(x, t + \Delta t | y, t) = \frac{1}{2\pi} \int_{-\infty}^{\infty} e^{-iu\Delta x} \phi(u, t + \Delta t | y, t)du.$$

We now expand $\phi(u, t + \Delta t \mid y, t)$ as a Taylor series in u to get

$$p(x, t + \Delta t | y, t) = \sum_{n=0}^{\infty} \frac{1}{n!} \left. \frac{\partial^n \phi(u, t + \Delta t | y, t)}{\partial u^n} \right|_{u=0} \frac{1}{2\pi} \int_{-\infty}^{\infty} u^n e^{-iu\Delta x} du$$

$$= \sum_{n=0}^{\infty} \frac{\phi^{(n)}(0.t + \Delta t | y, t)}{n!} \frac{1}{2\pi} \int_{-\infty}^{\infty} \frac{1}{(-i)^n} \frac{\partial^n}{\partial x^n} \left\{ e^{-iu(x-y)} \right\} du$$

$$= \sum_{n=0}^{\infty} \frac{(-)^n}{n!} \sigma_n(y, t) \frac{\partial^n}{\partial x^n} \frac{1}{2\pi} \int_{-\infty}^{\infty} \left\{ e^{-iu(x-y)} \right\} du$$

where

$$\Delta x = x - y; \quad \phi^{(n)}(0.t + \Delta t | y, t) = \left. \frac{\partial^n \phi(u, t + \Delta t | y, t)}{\partial u^n} \right|_{u=0},$$

and

$$\sigma_n(y, t) = \frac{1}{(i)^n} \phi^{(n)}(0.t + \Delta t | X(t) = y).$$

In view of (12.3.8), we may rewrite $\sigma_n(y, t)$ as

$$\sigma_n(y, t) = \frac{1}{(i)^n} \int_{-\infty}^{\infty} (i\Delta x)^n p(x, t + \Delta t | y, t)dx = E[(\Delta X)^n | X(t) = y]$$

$$= E[\{X(t + \Delta t) - X(t)\}^n | X(t) = y]. \tag{12.3.9}$$

Since the inverse Fourier transform of 1 is the Dirac delta function,

$$\frac{1}{2\pi} \int_{-\infty}^{\infty} e^{-iu\Delta x} du = \frac{1}{2\pi} \int_{-\infty}^{\infty} e^{-iu(x-y)} du = \delta(x-y),$$

the expression for $p(x, t+\Delta t \mid y, t)$ is simplified to

$$p(x, t+\Delta t | y, t) = \sum_{n=0}^{\infty} \frac{(-)^n}{n!} \sigma_n(y, t) \frac{\partial^n}{\partial x^n} \left\{\delta(x-y)\right\}.$$

We now substitute the above expression for $p(x, t+\Delta t | y, t)$ into (12.3.7) and obtain after repeated integration by parts

$$p(x, t+\Delta t) = \sum_{n=0}^{\infty} \frac{(-)^n}{n!} \frac{\partial^n}{\partial x^n} \int_{-\infty}^{\infty} \sigma_n(y, t) p(y, t) \delta(x-y) dy$$

$$= \sum_{n=0}^{\infty} \frac{(-)^n}{n!} \frac{\partial^n}{\partial x^n} \left[\sigma_n(x, t) p(x, t)\right].$$

This is essentially the result sought except for some rearrangement of the terms. Upon moving the first term on the right side to the left side and dividing through by Δt, we get in the limit as $\Delta t \to 0$

$$\lim_{\Delta t \to 0} \frac{p(x, t+\Delta t) - p(x, t)}{\Delta t} = \frac{\partial p(x, t)}{\partial t} = \sum_{n=1}^{\infty} \frac{(-)^n}{n!} \frac{\partial^n}{\partial x^n} \left[b_n(x, t) p(x, t)\right] \quad (12.3.10)$$

where

$$b_n(x, t) = \lim_{\Delta t \to 0} \frac{\sigma_n(x, t)}{\Delta t} = \lim_{\Delta t \to 0} \frac{1}{\Delta t} E[(\Delta X)^n | X(t) = x]$$

$$= \lim_{\Delta t \to 0} \frac{1}{\Delta t} \left\{\phi^{(n)}(0, t+\Delta t | X(t) = x)\right\}. \quad (12.3.11)$$

The relation (12.3.10) with coefficients $\{b_n(x, t)\}$ defined in (12.3.11) is known as the **kinetic equation** for a general (scalar) stochastic process $X(t)$ and the coefficients themselves are sometimes known as *derivate moments* of $X(t)$.

12.3.3   *Possible vanishing of higher order derivate moments*

In order to determine the collection of derivate moments, we need to specify the kind of $X(t)$ we are considering. It will be shown in the next section that for the Ornstein–Uhlenbeck process (that satisfied the Langevin's equation), the *derivate moments* of order higher than 3 vanish. This reduces the kinetic equation (12.3.10) to a second order linear PDE of the parabolic type and thereby makes it solvable by known methods, numerically if necessary. Naturally, we would like to know when does something similar happen for solutions of other SDE, perhaps not vanishing for all derivate monments of order higher than 3 but for all orders higher than some finite integer k. In that case, the infinite series of derivatives terminates after a finite number of terms making the kinetic equation useful for the determination of $p(x, t)$. While we would like to have a set of conditions for this termination in terms of the prescribed quantities such as the structure of the SDE (12.6.1), we first give here a result that simplifies the search for the termination of the series.

Theorem 40. *Suppose the derivate moments $\{b_j(x, t)\}$ exist for all $j = 1, 2, 3, \ldots$ and one vanishes for some even $j = 2k$, $k > 1$, i.e., $b_{2k}(x, t) = 0$ for some integer $k > 1$. Then $b_n(x, t) = 0$ for all $n \geq 3$.*

Proof. If $n = 2m + 1 \geq 3$ is an odd integer, we write the derivate moment $b_n(x, t)$ as

$$b_n(x, t) = \lim_{\Delta t \to 0} \frac{1}{\Delta t} E[(\Delta X)^n | X(t) = x]$$

$$= \lim_{\Delta t \to 0} \frac{1}{\Delta t} E[(\Delta X)^{(n-1)/2}(\Delta X)^{(n+1)/2} | X(t) = x].$$

By Schwarz's inequality, we have

$$b_n^2(x, t) \leq \lim_{\Delta t \to 0} \frac{1}{(\Delta t)^2} E[(\Delta X)^{n-1} | X(t) = x] E[(\Delta X)^{n+1} | X(t) = x]$$

$$= b_{n-1}(t)b_{n+1}(t) \quad (n \text{ odd}, \ n \geq 3). \tag{12.3.12}$$

Similarly, we have for even $n = 2m \geq 4$

$$b_n^2(x, t) \leq b_{n-2}(t)b_{n+2}(t) \quad (n \text{ even}, \ n \geq 4). \tag{12.3.13}$$

By hypothesis, we have $b_{2k}(x, t) = 0$ for some $k \geq 2$. By setting $m = k - 1$ and $m = k$ in (12.3.12), we get

$$b_{2k-1}^2(x, t) \leq b_{2k-2}(t)b_{2k}(t) \quad (k \geq 2), \tag{12.3.14}$$

$$b_{2k+1}^2(x, t) \leq b_{2k+2}(t)b_{2k}(t). \quad (k \geq 1). \tag{12.3.15}$$

Similarly, by setting $n = 2k - 2$ and $2k + 2$ in (12.3.13), we get

$$b_{2k-2}^2(x, t) \leq b_{2k-4}(t)b_{2k}(t) \quad (k \geq 3), \tag{12.3.16}$$

$$b_{2k+2}^2(x, t) \leq b_{2k+4}(t)b_{2k}(t) \quad (k \geq 1). \tag{12.3.17}$$

Since $b_{2k}(x,t) = 0$ (and $b_n(x,t)$ exists for all n), the inequalities (12.3.16) and (12.3.17) iteratively and sequentially require

$$b_{2k}(x,t) = 0 \quad (k \geq 2) \tag{12.3.18}$$

and the inequalities (12.3.14) and (12.3.15) imply

$$b_{2k+1}(x,t) = 0 \quad (k \geq 1)). \tag{12.3.19}$$

Together, (12.3.18) and (12.3.19) mean $b_n(x,t) = 0$ for all $n \geq 3$. $\qquad\square$

12.4 Langevin's Equation

12.4.1 *The kinetic equation for Langevin equation*

There are two issues related to the use of the kinetic equation for the solution of a specific problem. One obvious issue is whether the infinite series of "spatial" derivative terms terminates after a finite number of terms. (We would not have a conventional PDE for p if it does not terminate.) The other is the evaluation of $b_n(x,t)$ in the limit as $\Delta t \to 0$. In one way, the two issues are related. If we can evaluate $b_n(x,t)$ for all n, we would know whether or not the series terminates. But it is still desirable to know about the series terminating after a finite umber of terms even if we cannot determine $b_n(x,t)$ explicitly. In this section, we illustrate how to evaluate the derivate moments. Our analysis shows that the series (12.3.10) terminates for this problem and gives some impetus to an effort toward more general results on the termination issue.

Suppose we want to find the density function $p(x,t)$ for the s.p. $X(t)$ that is the m.s. solution of the stochastic IVP

$$dX(t) = -cX(t)dt + dB(t), \quad X(0) = A \tag{12.4.1}$$

where c is a known constant and $B(t)$ is a Wiener process with $\Delta B = B(t+\Delta t) - B(t)$ being Gaussian with mean zero and variance $D\Delta t$. The simple SDE is known as Langevin's equation which gives rise to the Ornstein–Uhlenbeck process. In the context of a model for a motoneuron, it is the Leaky Integrator with white noise input current. Here we approximate it by the finite difference analogue:

$$\Delta X(t) = -cX(t)\Delta t + \Delta B(t) \tag{12.4.2}$$

neglecting terms small of higher order in Δt, i.e. the difference between $\Delta X(t)$ and $dX(t)$ is $o(\Delta t)$ with

$$\lim_{\Delta t \to 0} \frac{o(\Delta t)}{\Delta t} = 0.$$

To apply the kinetic equation, we need to evaluate the expression

$$b_n(x,t) = \lim_{\Delta t \to 0} \frac{\sigma_n(x,t)}{\Delta t} = \lim_{\Delta t \to 0} \frac{1}{\Delta t} E[(\Delta X)^n | X(t) = x] \qquad (12.4.3)$$

in (12.3.11). With $\Delta X(t)$ given by (12.4.2), we have

$$b_1(x,t) = \lim_{\Delta t \to 0} \frac{1}{\Delta t} E[\Delta X | x, t] = \lim_{\Delta t \to 0} \frac{1}{\Delta t} E[-cX(t)\Delta t + \Delta B(t) | x, t]$$

$$= \lim_{\Delta t \to 0} \frac{1}{\Delta t} E[-cX(t)\Delta t | X(t) = x] + \lim_{\Delta t \to 0} \frac{1}{\Delta t} E[\Delta B(t) | X(t) = x].$$

The last term on the right vanishes because $\Delta B(t)$ is zero mean and the first term is $-cx\Delta t$ since the expectation is conditioned on $X(t) = x$. We have then

$$b_1(x,t) = -cx.$$

For the second order moment $b_2(x,t) = E[\{\Delta X\}^2 | X(t) = x]/\Delta t$, we note that

$$b_2(x,t) = \lim_{\Delta t \to 0} \frac{1}{\Delta t} E[\{\Delta X\}^2 | X(t) = x]$$

$$= \lim_{\Delta t \to 0} \frac{1}{\Delta t} E[\{-cX(t)\Delta t + \Delta B\}^2 | X(t) = x]$$

$$= \lim_{\Delta t \to 0} \frac{1}{\Delta t} E\left[\left\{c^2 X^2(t)(\Delta t)^2 - 2cX(t)\Delta t\Delta B + (\Delta B)^2\right\} | X(t) = x\right]$$

$$= \lim_{\Delta t \to 0} \left\{c^2 x^2 \Delta t\right\} - 2cx E[\Delta B] + \lim_{\Delta t \to 0} \frac{1}{\Delta t} E[\{\Delta B\}^2]$$

$$= \lim_{\Delta t \to 0} \frac{1}{\Delta t} E[\{\Delta B\}^2] = D$$

keeping in mind that $\Delta B(t) = B(t + \Delta t) - B(t)$ is of mean zero and variance $D\Delta t$.

Proposition 46. $b_n(x,t) = 0$ *for* $n \geq 3$.

Proof. It is simpler to prove this by way of the characteristic function (12.3.8)

$$\phi(u, t + \Delta t | y, t) = E[e^{iu\Delta X} | y, t] = E[e^{iu\{-cy\Delta t + \Delta B\}} | y, t]$$

$$= e^{iu(-cy\Delta t)} E[e^{iu\Delta B} | y, t]$$

since $X(t) = y$ and $e^{-iucy\Delta t}$ does not involve x and is independent of ΔB. Since ΔB is Gaussian of zero mean and variance $D\Delta t$, we have altogether (see Proposition 24)

$$\phi(u, t + \Delta t | y, t) = e^{iu\{-cy\Delta t\} - \frac{1}{2} u^2 D\Delta t}.$$

By Proposition 21, we have

$$b_n(x,t) = \lim_{\Delta t \to 0} \frac{1}{\Delta t} E[(\Delta X)^n | X(t) = y] = \lim_{\Delta t \to 0} \frac{1}{\Delta t} \frac{1}{i^n} \left[\frac{\partial^k \phi}{\partial u^k} \right]_{u=0}.$$

It is a straightforward calculation to verify that $b_1(x,t)$ and $b_2(x,t)$ are as found earlier and $b_n(x,t) = O\left([\Delta t]^{n-2}\right)$ and hence $\to 0$ for $n \geq 3$ as $\Delta t \to 0$. □

It follows that the kinetic equation reduces to

$$\frac{\partial}{\partial t}[p(x,t)] = \frac{\partial}{\partial x}[cxp(x,t)] + \frac{D}{2}\frac{\partial^2}{\partial x^2}[p(x,t)] \qquad (12.4.4)$$

in this case. As a PDE second order in space and first order in time, we expect one initial condition and two boundary conditions. For x spanning the reals, we have for the latter

$$p(-\infty, t) = p(\infty, t) = 0 \qquad (12.4.5)$$

while an initial condition may be a prescribed density function

$$p(x,0) = p_0(x). \qquad (12.4.6)$$

The linear PDE (12.4.4) and the auxiliary conditions (12.4.5)–(12.4.6) define a conventional IBVP. For $c = 0$, the kinetic equation simplifies further to

$$\frac{\partial}{\partial t}[p(x,t)] = \frac{D}{2}\frac{\partial^2}{\partial x^2}[p(x,t)]$$

which was deduced in an earlier chapter as the diffusive limit of the unbiased random walk problem. Its fundamental solution and solutions on a bounded interval have been discussed in some detail there.

In general, the stochastic problem is considered solved (to first order) once we have reduced the determination of the pdf $p(x,t)$ of the stochastic response process $X(t)$ to solving the IBVP (12.4.4)–(12.4.6) as the solution for such deterministic IBVP is routine by known analytical and numerical methods for PDE. Before we move on to discuss the derivate moments for more complicated SDE and the determination of second and higher order joint density functions, $p_2(x_1, x_2; t_1, t_2)$, $p_3(x_1, x_2, x_3; t_1, t_2, t_3)$, etc., we sketch the solution for $c \neq 0$ case in the next subsection.

12.4.2 Eigenfunction expansions

For $c \neq 0$, the IBVP (12.4.4)–(12.4.6) can be solved by the method of separation of variables. Though not necessary to do so, we transform the PDE into one conforming to a standard ODE after separation of variables. For this purpose, we set

$$\tau = ct, \quad y = ix\sqrt{\frac{c}{D}}, \quad f(y, \tau) = p(x, t)$$

and re-write the PDE (12.4.4) as

$$\frac{\partial f}{\partial \tau} = -\frac{\partial^2 f}{\partial y^2} + y\frac{\partial f}{\partial y} + f. \tag{12.4.7}$$

This linear PDE (12.4.7) can be solved by the method of separation of variables and eigenfunction expansions. For this approach, we seek solution of the form $f(y, \tau) = S(y)T(\tau)$. For this type of solution, equation (12.4.7) may be rearranged to read

$$\frac{T^{\cdot}(\tau)}{T(\tau)} = \frac{[yS(y)]' - [S''(y)]}{S(y)} = \lambda^2$$

for some constant λ^2. (The sign of the separation constant is chosen anticipating real eigenvalues for the spatial part of the problem.) From the τ dependent part, we get

$$T_\lambda(\tau) = A_\lambda e^{\lambda^2 \tau}.$$

The corresponding y dependent part $S_\lambda(y)$ is determined by the ODE

$$S''_\lambda - yS'_\lambda + (\lambda^2 - 1)S_\lambda = 0.$$

It may be taken in the standard Sturm–Liouville form:

$$\left(e^{-y^2/2}S'_\lambda\right)' + (\lambda^2 - 1)e^{-y^2/2}S_\lambda = 0$$

to show the weight function $e^{-y^2/2}$ for orthogonality of the eigenfunction solutions $\{\phi_n(y)\}$:

$$\int_{-\infty}^{\infty} \phi_n(y)\phi_m(y)e^{-y^2/2}dy = 0 \quad (m \neq n). \tag{12.4.8}$$

For the solution of the ODE for S_λ, we note that $y = 0$ is an ordinary point of the ODE. By Fuchs' theorem, we expect two linearly independent complementary solutions in the form of Taylor series about the origin convergent for all $|y| < \infty$.

The usual method of undetermined coefficients gives

$$S_\lambda(y) = c_1 s_\lambda^{(1)}(y) + c_2 s_\lambda^{(2)}(y),$$

with

$$s_\lambda^{(1)}(y) = \sum_{k=0}^{\infty} \alpha_{2k}(\lambda)y^{2k}, \quad s_\lambda^{(2)}(y) = \sum_{k=0}^{\infty} \beta_{2k+1}(\lambda)y^{2k+1}$$

where $\alpha_0 = 1$, $\beta_1 = 1$ and

$$\alpha_{2(k+1)} = \frac{2k - (\lambda^2 - 1)}{(2(k+1))(2k)}a_{2k}, \quad \beta_{2(k+1)+1} = \frac{(2k+1) - (\lambda^2 - 1)}{(2(k+1)+1)(2(k+1))}\beta_{2k=1}$$

$k = 0, 1, 2, \ldots$. The first series solution $s_\lambda^{(1)}(y)$ reduces to a polynomial of degree $2m$ when $\lambda^2 - 1 = 2m$ or $\lambda^2 = 2m + 1$ for some integer $m \geq 0$. The second series solution $s_\lambda^{(2)}(y)$ reduces to a polynomial of degree $2m + 1$ when $\lambda^2 - 1 = 2m + 1$ or $\lambda^2 = 2m + 2$ for some $m \geq 0$. These polynomials, denoted by $H_j(y)$, $j = 0, 1, 2, \ldots$, are the well-known Hermite polynomials of degree j. They constitute the complete set of eigenfunctions for the problem with eigenvalues $\{\lambda_j^2 = j + 1\}$ and satisfy the orthogonality relations (12.4.8) with the polynomials ensuring its integrability.

We may take the solution $f(y, \tau)$ in the form of an eigenfunction expansion

$$f(y, \tau) = \sum_{n=0}^{\infty} A_n e^{(n+1)\tau} H_n(y) \tag{12.4.9}$$

with the Fourier coefficients $\{A_n\}$ determined by the initial density function $p(x, 0) = p_0(x_0)$ with the help of the orthogonality relation (12.4.8):

$$\gamma_n A_n = \int_{-\infty}^{\infty} \phi_n(z)f_0(z)e^{-z^2/2}dz$$

where

$$f_0(y_0) = f(y, 0) = p(x, 0) = p_0(x_0), \quad \gamma_n = \int_{-\infty}^{\infty} \phi_n^2(y)e^{-y^2/2}dy$$

Evidently, our choice of polynomial eigenfunctions for the problem ensure integrability of the various singular integrals involved in the method of solution above.

For the special case where the initial data for $X(t)$ is a known constant $\bar{a}$ with $p_0(a) = \delta(x_0 - \bar{a})$, the Fourier–Hermite series solution for the density function simplifies to a Gaussian form (see [50] for a derivation of this result from the eigenfunction expansion solution (12.4.9)).

12.4.3 On linear systems with constant coefficients

In trying to extend the method of solution to a general system of linear ODE system with only random forcing, we need a vector version of (12.3.10) and (12.3.11). However, solving an nth order linear system with (known) constant coefficients,

$$d\mathbf{X}(t) = -C\mathbf{X}(t)dt + d\mathbf{B}(t), \quad \mathbf{X}(0) = \mathbf{A},$$

by the method of separation of variables would be rather tedious if not intractable. To see the complexity of the solution process, suppose it is possible to de-couple the system by diagonalizing the matrix C to get

$$d\mathbf{Y}(t) = -\Lambda\mathbf{Y}(t)dt + d\bar{\mathbf{B}}(t),$$

with

$$\mathbf{Y} = Q^{-1}\mathbf{X}, \quad \bar{\mathbf{B}} = Q^{-1}\mathbf{B}$$

where Q is the modal matrix of C and Λ is the diagonal matrix with the eigenvalues of C along its diagonal. By calculations similar to the scalar case, we find that all derivate moments of order higher the 2 vanish so that the corresponding PDE for $p(\mathbf{y}, t)$ takes the form

$$\frac{\partial p}{\partial t} = \sum_{k=1}^{n} \frac{\partial}{\partial y_k} [\lambda_k y_k p] + \sum_{k,j=1}^{n} \frac{1}{2!} \frac{\partial^2}{\partial y_k \partial y_j} [\bar{D}_{kj} p] \qquad (12.4.10)$$

where

$$\bar{D}_{kj} = [Q^{-1}DQ]_{kj}, \quad D = [D_{ij}], \quad E[\Delta B_i(t)\Delta B_j(t)] = D_{ij}\Delta t$$

with D_{ij} being the covariance of the Wiener increments $\Delta B_i(t)$ and $\Delta B_j(t)$. The PDE for $p(\mathbf{y}, t)$ is augmented by the limiting conditions as $|y_k| \to \infty$

$$\lim_{y_k \to \pm\infty} [p(\mathbf{y}, t)] = 0, \quad (k = 1, 2, \ldots, n\}, \qquad (12.4.11)$$

and the initial condition

$$p(\mathbf{y}, t_0) = p(\mathbf{a}, t_0). \qquad (12.4.12)$$

For the case where the vector $\mathbf{A}$ is a known constant (instead of a random variable), the initial condition takes the form

$$p(\mathbf{y}, t_0) = \Pi_{j=1}^{n}\delta(y_j - \bar{a}_j) \qquad (12.4.13)$$

where $\bar{\mathbf{A}} = (\bar{a}_1, \ldots, \bar{a}_n)^T = Q^{-1}\mathbf{A}$.

An exact solution of this IBVP by the method of separation of variables is possible, at least in principle, but rather tedious and challenging. A more tractable method for such linear stochastic system is by way of the characteristic function.

12.4.4   *Solution by characteristic functions*

Upon taking the Fourier transform of the PDE for $p(\mathbf{y}, t)$, we obtain after integration by parts

$$\frac{\partial \hat{p}}{\partial t} = \sum_{k=1}^{n} \lambda_k u_k \frac{\partial \hat{p}}{\partial u_k} - \frac{1}{2!} \sum_{k,j=1}^{n} \bar{D}_{kj} \left[u_k u_j \hat{p} \right] \qquad (12.4.14)$$

where

$$\hat{p}(\mathbf{u}, t) = E[e^{i\mathbf{u}\cdot\mathbf{y}} p(\mathbf{y}, t)] = \int_{-\infty}^{\infty} \cdots \int_{-\infty}^{\infty} e^{i\mathbf{u}\cdot\mathbf{y}} p(\mathbf{y}, t) dy_1 \cdots dy_n$$

$$\equiv F\left[p(\mathbf{y}, t | \bar{\mathbf{a}}, t_0) \right]. \qquad (12.4.15)$$

is the characteristic function of $p(\mathbf{y}, t)$. The PDE for $\hat{p}(\mathbf{u}, t)$ is a first order PDE with the initial condition

$$\hat{p}(\mathbf{u}, t_0) = F\left[p(\mathbf{y}, t_0) \right] = \int_{-\infty}^{\infty} \cdots \int_{-\infty}^{\infty} e^{i\mathbf{u}\cdot\mathbf{y}} p(\mathbf{y}, t_0) dy_1 \cdots dy_n. \qquad (12.4.16)$$

For the case where $\mathbf{A}$ is a prescribed constant vector $\mathbf{a}$ so that (12.4.13) applies, (12.4.16) becomes

$$\hat{p}(\mathbf{u}, t_0) = e^{i\mathbf{u}\cdot\bar{\mathbf{a}}}. \qquad (12.4.17)$$

The deterministic IVP (12.4.14)–(12.4.17) for a first order linear PDE for $\hat{p}(\mathbf{u}, t)$ may be solved more simply (than the IBVP (8.1.2)–(12.4.12)) by the method of characteristics for a single first order PDE.

Example 60. Solve the IVP (12.4.1) by first obtaining the characteristic function $\hat{p}(u, t)$ of $p(x, t)$.

The IVP for the characteristic function $\hat{p}(u, t)$ is

$$\frac{\partial \hat{p}}{\partial t} + cu \frac{\partial \hat{p}}{\partial u} = -\frac{D}{2!} u^2 \hat{p}, \quad \hat{p}(u, 0) = F\left[p(x, 0) \right] = F\left[p_0(a) \right].$$

The corresponding characteristic equation for the method of characteristics are

$$\frac{d\hat{p}}{dt} = cu, \quad \frac{d\hat{p}}{dt} = -\frac{D}{2} u^2 \hat{p}.$$

The initial conditions for these equations are

$$u(0) = u_0, \quad \hat{p}(0) = F\left[p_0(a) \right] = \int_{-\infty}^{\infty} e^{iu_0 a} p_0(a) da = \hat{p}_0(u_0)$$

so that the solution of the characteristic IVP is

$$u = u_0 e^{ct}, \quad \hat{p}(t) = \hat{p}_0(u_0) e^{\psi(t)}$$

where

$$\psi(t) = \frac{D}{4c} u_0^2 \left(1 - e^{2ct}\right) = -\frac{D}{4c} u^2 \left(1 - e^{-2ct}\right)$$

Upon solving for u_0 in terms of u to get $u_0 = u e^{ct}$ and using the result to eliminate u_0 from $\psi(t)$ and $\hat{p}(t)$, we obtain

$$\psi(t) = -\frac{D}{4c} u^2 \left(1 - e^{-2ct}\right) \equiv \psi(u,t), \quad \hat{p}(t) = \hat{p}_0(u e^{-ct}) e^{\psi(u,t)}. \tag{12.4.18}$$

For $A = \bar{a}$ so that $p_0(a) = \delta(a - \bar{a})$ and $F[p_0(a)] = e^{iu_0\bar{a}} = e^{iu\bar{a}e^{-ct}}$, the solution for $\hat{p}$ becomes

$$\hat{p}(u,t) = e^{iu\bar{a}e^{-ct}} e^{\psi(u,t)}.$$

Inverse transforming this expression for $\hat{p}(u,t)$, we obtain

$$p(x,t) = \frac{1}{2\pi} \int_{-\infty}^{\infty} e^{-iu\left(x - \bar{a}e^{-ct}\right)} e^{\psi(u,t)} du$$

$$= \frac{1}{2\pi} e^{D\left(1 - e^{-2ct}\right)\eta^2/4c} \int_{-\infty}^{\infty} e^{-D\left(1 - e^{-2ct}\right)(u-\eta)^2/4c} du$$

$$= \frac{1}{\sqrt{2\pi}} \frac{e^{D\left(1 - e^{-2ct}\right)\eta^2/4c}}{\sqrt{D\left(1 - e^{-2ct}\right)/2c}}$$

where

$$\eta = \frac{2ic(x - \bar{a}e^{-ct})}{D\left(1 - e^{-2ct}\right)}.$$

Upon expressing η^2 in terms of $\bar{a}$, the expression for $p(x,t)$ becomes

$$p(x,t) = \frac{1}{\sqrt{2\pi}} \frac{e^{-c\left(x - \bar{a}e^{-ct}\right)^2/D\left(1 - e^{-2ct}\right)}}{\sqrt{D\left(1 - e^{-2ct}\right)/2c}}$$

showing that

- $X(t)$ is normally distributed with mean $\bar{a}e^{-ct}$ and variance $D\left(1 - e^{-2ct}\right)/2c$;
- $p(x,t)$ decays rapidly to zero as $|x| \to \infty$ (for $c > 0$), and
- $p(x,t)$ tends to a stationary solution that is Gaussian of mean zero and variance $D/2c$ as $t \to \infty$.

Exercise 51. Obtain the exact solution for the IVP (12.4.14)–(12.4.17) for $\hat{p}(u,t)$ of the Integrate & Fire model of motoneuron with white noise current excitation.

Exercise 52. Inverse transform $\hat{p}(u,t)$ to recover $p(y,t)$.

12.5   Diffusion Processes

12.5.1   *Markov and diffusive processes*

When their kinetic equation terminates after a finite number of derivatives with respective to x, the IBVP for PDE (12.3.10) allows us to determine the pdf $p(x,t)$ of the response process. Strictly speaking, the pdf is a conditional one since it is conditioned on the initial condition $X(t_0) = A$ with A either prescribed to be a known constant a or a second order random variable with prescribed statistics (such as a pdf $p_0(a, t_0)$). In fact, it is more appropriately written as $p(x, t \,|\, a, t_0)$ which is seen to be determined by the kinetic equation (at least for the case of Langevin's equation) and the prescribed pdf $p_0(a, t_0)$ which furnishes the initial condition for the kinetic equation.

Definition 36. A Markov process is a stochastic process that is completely specified by its second order density function.

If $p_0(a, t_0)$ is prescribed, then knowledge of the conditional density function $p(x, t \,|\, a, t_0)$ completely specifies the Markov process $X(t)$ since

$$p(y, \tau \,|\, a, t_0) = \int_{-\infty}^{\infty} p(y, \tau \,|\, x, t) p(x, t \,|\, a, t_0) dy$$

and

$$p(y, x; \tau, t) = p(y, \tau \,|\, x, t) p(x, t).$$

Some stochastic processes, including solutions of some SDE, are Markov process. For example, normal (Gaussian) processes are completely characterized by its mean and variance. Other solutions of SDE are often assumed to be Markovian for there is usually not enough information for the higher density functions otherwise. Even with this assumption, there still the problem of the kinetic equation not terminating after a finite number of terms. Specific cases (such as Langevin's equation and Theorem 40) suggest that we often have $b_n = 0$ for $n \geq 3$. For a tractable solution process and not inconsistent with the nature of the biological process of interest, it is customary to take

$$b_n = 0, \quad n \geq 3. \tag{12.5.1}$$

for the response process when the SDE is not of the Ito type (see also Section 6 of this chapter).

Definition 37. Let $X(t)$ be a Markov process characterized by the conditional pdf $p(x, t \,|\, a, t_0)$ with continuous first partial derivative in t and t_0 and continuous first and second partial derivatives in x and a, $X(t)$ is said to be a *diffusive process* if its derivate moments also satisfy the requirements (12.5.1).

12.5.2 The Fokker–Planck equation for diffusive processes

For a diffusive process, the kinetic equation for a scalar process is simplified to

$$\frac{\partial p(x,t\,|\,a,t_0)}{\partial t} = -\frac{\partial}{\partial x}\left[b_1(x,t)p(x,t\,|\,a,t_0)\right] + \frac{1}{2}\frac{\partial^2}{\partial x^2}\left[b_2(x,t)p(x,t\,|\,a,t_0)\right] \quad (12.5.2)$$

with the two nonvanishing derivate moments determined by the SDE by way of the definition (12.4.3) of these moments. The PDE (12.5.2) is known as the *Fokker–Planck equation* and also as the *Kolmogorov forward equation*. Its adjoint equation is known as the Kolmogorov backward equation. They are routinely used for determining the transition density function of the solution of SDE.

We have previously shown that the output of the IVP for Langevin's equation is effectively a diffusive process and obtained the corresponding Fokker–Planck equation. Consider now the IVP for the more complex SDE

$$dX(t) = f(X(t))dt + G(X(t))dB(t), \quad X(t_0) = A \quad (12.5.3)$$

where $B(t)$ is the usual Wiener process (and A is a second order random variable with prescribed statistics). For this equation, we have except for terms small compared to Δt

$$b_1(x,t) = \lim_{\Delta t \to 0}\frac{1}{\Delta t}E[\Delta X \mid x,t] = \lim_{\Delta t \to 0}\frac{1}{\Delta t}E[f(X(t))\Delta t + G(X(t))\Delta B(t) \mid x,t]$$

$$= f(x) + \lim_{\Delta t \to 0}\frac{1}{\Delta t}E[G(X)\Delta B(t) \mid X(t) = x].$$

$$= f(x) + \lim_{\Delta t \to 0}\frac{G(x)}{\Delta t}E[\Delta B(t)] = f(x)$$

given $\Delta B(t)$ is zero mean. Except for terms small compared to Δt, we have also

$$b_2(x,t) = \lim_{\Delta t \to 0}\frac{1}{\Delta t}E[\{\Delta X\}^2 \mid X(t) = x]$$

$$= \lim_{\Delta t \to 0}\frac{1}{\Delta t}E[\{f(X(t))\Delta t + G(X(t))\Delta B(t)\}^2 \mid X(t) = x]$$

$$= \lim_{\Delta t \to 0}\left\{[f(x)]^2\,\Delta t + 2f(x)G(x)E\left[\Delta B(t)\right]\right\} + \lim_{\Delta t \to 0}\frac{[G(x)]^2}{\Delta t}E[\{\Delta B(t)\}^2]$$

$$= \lim_{\Delta t \to 0}\frac{[G(x)]^2}{\Delta t}E[\{\Delta B(t)\}^2] = D[G(x)]^2$$

since

$$E[\{\Delta B(t)\}^2] = D\Delta t.$$

In that case, the Fokker–Planck equation for (12.5.3) takes the form

$$\frac{\partial p}{\partial t} = -\frac{\partial}{\partial x}[f(x)p] + \frac{D}{2}\frac{\partial^2}{\partial x^2}\left[\{G(x)\}^2\, p(x,t)\right].$$

The PDE for $p(x,t)$ is supplemented by the initial condition at $t = t_0$

$$p(x, t_0\,|\,a, t_0) = p_0(a, t_0)$$

(with $p_0(a, t_0) = \delta(x - a)$ if A is a known constant a), the homogeneous Dirichlet condition at the two ends:

$$\lim_{z \to \pm\infty} p(x, t\,|\,a, t_0) = 0;$$

and, if necessary,

$$\lim_{z \to \pm\infty} \frac{\partial p(x, t\,|\,a, t_0)}{\partial x} = 0.$$

12.5.3 *Vector diffusive processes*

For a system of n first order stochastic ODE, if the vector response process $\mathbf{X}(t)$ is a diffusive process, the kinetic equation is now for the joint pdf $p(\mathbf{x}, t) = p(\mathbf{x}, t\,|\,\mathbf{a}, t_0) = p(x_1, x_2, \ldots, x_n; t)$. The first and second order derivate moments b_1 and b_2 that enter into the Fokker–Planck equation in the scalar case are now replaced by the vector $(\mu_1(\mathbf{x}, t), \ldots, \mu_n(\mathbf{x}, t))^T$ and matrix $[\mu_{ij}(\mathbf{x}, t)]$, respectively, with

$$\mu_k(\mathbf{x}, t) = \lim_{\Delta t \to 0}\frac{1}{\Delta t}E[\Delta X_k\,|\,\mathbf{X}(t) = \mathbf{x}],$$

$$\mu_{ij}(\mathbf{x}, t) = \lim_{\Delta t \to 0}\frac{1}{\Delta t}E[\Delta X_i \Delta X_j\,|\,\mathbf{X}(t) = \mathbf{x}].$$

The Fokker–Planck equation itself now takes the form

$$\frac{\partial p}{\partial t} = -\sum_{j=1}^{n}\frac{\partial}{\partial x_j}[\mu_j(\mathbf{x}, t)p] + \frac{1}{2}\sum_{j=1}^{n}\sum_{i=1}^{n}\frac{\partial^2}{\partial x_i \partial x_j}[\mu_{ij}(\mathbf{x}, t)p] \qquad (12.5.4)$$

The single second order linear PDE is to be augmented by suitable boundary conditions and the initial condition

$$p(\mathbf{x}, t\,|\,\mathbf{a}, t_0) = p_0(\mathbf{a}, t_0)$$

where $p_0(\mathbf{a}, t_0)$ is the prescribed joint density function of the initial data $\mathbf{A}$ for the SDE.

12.6 The Ito Type SDE

12.6.1 *Output is a diffusive process*

The principal goal of this section is to establish that the output from an Ito type SDE is a diffusive process. Consequently, its kinetic equation is a Fokker–Planck equation with known first and second derivate moments. For the scalar case, the s.p. $X(t)$ is the solution of

$$dX(t) = f(X(t), t)dt + G(X(t), t)dB(t), \quad X(t_0) = A, \tag{12.6.1}$$

where $B(t)$ is a Wiener process with zero mean and variance Dt. Recall from a previous discussion that the Wiener process satisfies the following properties:

(1) $B(0) = 0$ (with probability one).
(2) For $\Delta t > 0$, $\Delta B = B(t + \Delta t) - B(t)$ is Gaussian with mean zero and variance $E[(\Delta B)\}^2] = D\Delta t$ for some constant D.
(3) For $t > s > u > v$, the increments $B(t) - B(s)$ and $B(u) - B(v)$ are independent.

As $B(t)$ is not mean square differentiable, the IVP (12.6.1) is only a formal differential equation version of the stochastic integral equation

$$X(t) = A + \int_0^t f(X(s), x)ds + \int_0^t G(X(z), z)dB(z) \quad (0 \le t \le T) \tag{12.6.2}$$

with Ito's interpretation for the part in (12.6.2) involving a Wiener increment. For this, we let the interval $(0, t)$ be subdivided into n mesh points with increment $\Delta_k = t_{k+1} - t_k > 0$, $k = 0, 1, \ldots, n-1$, and Δ be the maximum of $\{\Delta_k\}$.

Definition 38. If the limit of the right hand side of

$$\int_0^t G(X(z), z)dB(z) = \lim_{\substack{\ell \to \infty \\ \Delta \to 0}} \sum_{\ell=0}^{n-1} G(X(z_\ell), z_\ell) \{B(z_{\ell+1}) - B(z_\ell)\} \tag{12.6.3}$$

exists, it is called the Ito stochastic integral of $G(X(z), z)$ with respect to $B(z)$.

To maintain an analogue to conventional ODE, the differentiated form of the Ito equation is written for the differential $dX(t)$ instead of the derivative $X'(t)$ to remind ourselves of the meaning of this relation. In either form, the Ito interpretation enables us to determine the deviate moments below.

The vector version of the Ito type SDE is given by the vector SDE

$$d\mathbf{X}(t) = \mathbf{f}(\mathbf{X}(t), t)dt + G(\mathbf{X}(t), t)d\mathbf{B}(t), \quad \mathbf{X}(t_0) = \mathbf{A} \tag{12.6.4}$$

where $\mathbf{X}(t)$ and $\mathbf{f}(\mathbf{X}, t)$ are n-vectors, $G = [G_{ij}]$ is an $n \times m$ matrix, and $\mathbf{W}(t) = d\mathbf{B}(t)/dt$ has been treated as an m-vector white noise process in the previous chapter when it was convenient to do so. Either as a Wiener increment or a white noise process, it is endowed with the covariance matrix $D = [D_{ij}]$. In the actual method of solution and the derivation of the derivate moments, we work with the discretized form of (12.6.4)

$$\Delta \mathbf{X}(t) = \mathbf{f}(\mathbf{X}(t), t)\Delta t + G(\mathbf{X}(t), t)\Delta \mathbf{B}(t), \quad \mathbf{X}(t_0) = \mathbf{A} \qquad (12.6.5)$$

where $\Delta \mathbf{B}(t)$ is a zero mean m-vector Wiener increment with $E[\Delta B_i \Delta B_j] = D_{ij}\Delta t$ prior to taking the limit as $\Delta t \to 0$.

The following theorem and its proof can be found in most texts of SDE (see [50] for example).

Theorem 41. *The solution process* $\mathbf{X}(t)$ *of the IVP* (12.6.4) *is Markovian.*

Proof. A typical proof does not add insight to solutions of SDE and is omitted. □

As a consequence of the theorem above, the stochastic process $\mathbf{X}(t)$ is completely characterized by its first and second order density functions. With $p_0(\mathbf{a}, t_0)$ prescribed by the initial state at t_0, it is only necessary to determine the conditional density function $p(\mathbf{x}, t \,|\, \mathbf{a}, t_0)$. Furthermore, we can also prove the following theorem:

Theorem 42. *The solution* $\mathbf{X}(t)$ *is diffusion process with* $\mu_j = f_j(\mathbf{x}, t)$ *and* $\mu_{ij} = [GDG^T]_{ij}$ *in the corresponding Fokker–Planck equation.*

Proof. We give here a proof for the scalar case by way of the conditional characteristic function of the random variable $\Delta X = X(t + \Delta t) - X(t)$ given $X(t) = y$

$$\phi(u, t + \Delta t | y, t) = E[e^{iu\Delta X} | y, t] = \int_{-\infty}^{\infty} e^{iu\Delta x} p(x, t + \Delta t | y, t) dx$$

with $\Delta x = x - y$, so that

$$\begin{aligned} \phi(u, t + \Delta t | y, t) &= E[e^{iu\{f(X(t),t)\Delta t + G(X(t),t)\Delta B\}} | y, t] \\ &= E[e^{iu\{f(y,t)\Delta t\}}] E[e^{iu\{G(y,t)\Delta B\}}] \\ &= e^{iu\{f(y,t)\Delta t\}} E[e^{iu\{G(y,t)\Delta B\}}] \end{aligned}$$

since $X(t) = y$ is independent of ΔB in Ito's interpretation. Now ΔB is Gaussian of zero mean and variance $D\Delta t$. We can evaluate the last expected value with the

Gaussian pdf to give

$$\phi(u, t + \Delta t | y, t) = e^{iu\{f(y,t)\Delta t\} - \frac{1}{2} u^2 G^2(y,t)D\Delta t}$$

reducing to that for Langevin's equation when $G(y, t) = 1$. By definition and Proposition 21, we have

$$\mu_1(y, t) = \lim_{\Delta t \to 0} \frac{1}{\Delta t} E[(\Delta X) | X(t) = y] = \lim_{\Delta t \to 0} \frac{1}{\Delta t} \frac{1}{i} \left[\frac{\partial \phi}{\partial u} \right]_{u=0} = f(y, t)$$

$$\mu_{11}(y, t) = \lim_{\Delta t \to 0} \frac{1}{\Delta t} E[(\Delta X)^2 | X(t) = y] = \lim_{\Delta t \to 0} \frac{1}{\Delta t} \frac{1}{i^2} \left[\frac{\partial^2 \phi}{\partial u^2} \right]_{u=0} = DG^2(y, t).$$

It is a straightforward calculation to show that all higher order derivate moments vanish so that

$$\frac{\partial p}{\partial t} = -\frac{\partial}{\partial y} [f(y, t)p] + \frac{D}{2} \frac{\partial^2}{\partial y^2} [G^2(y, t)p].$$

The corresponding results for the vector case can be proved by a similar set of calculations.   □

The following corollary giving the Fokker–Planck equation for the vector Ito SDE is an immdediate consequence of Theorem 42 above:

Corollary 15. *The conditional density function $p(\mathbf{x}, t \,|\, \mathbf{a}, t_0)$ for the n-vector solution $\mathbf{X}(t)$ of the Ito IVP (12.6.4) is determined by*

$$\frac{\partial p}{\partial t} = -\sum_{j=1}^{n} \frac{\partial}{\partial x_j} [f_j(\mathbf{x}, t)p] + \frac{1}{2} \sum_{j=1}^{n} \sum_{i=1}^{n} \frac{\partial^2}{\partial x_i \partial x_j} \left[(GDG^T)_{ij} p \right] \tag{12.6.6}$$

with $[p(\mathbf{x}, t \,|\, \mathbf{a}, t_0)]_{t=0, \mathbf{x}=\mathbf{a}} = p_0(\mathbf{a}, t_0)$.

12.6.2   Expectation of a function of the output of Ito's equation

12.6.2.1   Derivative of an expectation of a function of $\mathbf{X}(t)$

Determination of the pdf for $\mathbf{X}(t)$ by obtaining an explicit analytical solution of the IBVP (12.6.6) for the Ito SDE problem (12.6.4) is generally a difficult task. For useful information about $\mathbf{X}(t)$, it is often necessary to settle for getting the first and second order moments of that process as in the last chapter. (For processes that are Gaussian, these would determine its density function.) For Ito type of SDE, the form of the Fokker–Planck equation enables us to establish a useful result for determining these and higher order moments. To derive the general result for this purpose, we start with the expectation of a scalar function $g(\cdot)$ of the output $\mathbf{X}(t)$ of

(12.6.4): Let $g(\mathbf{X}(t))$ be at least twice continuously differentiable in all components of $\mathbf{X}$. Since $\mathbf{X}(t)$ depends on $p_0(\mathbf{a}, t_0)$, the expectation of $g(\mathbf{X}(t))$ is given by

$$E[g(\mathbf{X}(t))] = \int_{-\infty}^{\infty} \int_{-\infty}^{\infty} g(\mathbf{x}) p(\mathbf{x}, t; \mathbf{a}, t_0) d\mathbf{x}\, d\mathbf{a}$$

$$= \int_{-\infty}^{\infty} \int_{-\infty}^{\infty} g(\mathbf{x}) p(\mathbf{x}, t \mid \mathbf{a}, t_0) p_0(\mathbf{a}, t_0) d\mathbf{x} d\mathbf{a} \equiv M(t)$$

Note that we have omitted the dependence of M on t_0 since the initial time is a fixed prescribed quantity for our stochastic IVP. For $g(\mathbf{X}(t)) = X_1^{k_1}(t) X_2^{k_2}(t) \cdots X_n^{k_n}(t)$, the expectation $E[g(\mathbf{X}(t))] = M^{(k)}(t) = M^{(k_1, \ldots, k_n)}(t)$ is a k $(= k_1 + k_2 + \cdots + k_n)$ moment of $\mathbf{X}(t)$.

If we now differentiate $M(t)$ with respect to t, we obtain

$$M'(t) = \int_{-\infty}^{\infty} \int_{-\infty}^{\infty} g(\mathbf{x}) \frac{\partial p(\mathbf{x}, t \mid \mathbf{a}, t_0)}{\partial t} p_0(\mathbf{a}, t_0) d\mathbf{x} d\mathbf{a} \tag{12.6.7}$$

from which we obtain the following proposition:

Proposition 47. *The time derivative of the expectation $E[g(\mathbf{X}(t))]$ of a function $g(\cdot)$ of the solution $\mathbf{X}(t)$ of the stochastic IVP (12.6.4) is given by*

$$M'(t) = \sum_{j=1}^{n} \left\{ \frac{1}{2} \sum_{i=1}^{n} E\left[(GDG^T)_{ij} \frac{\partial^2 g(\mathbf{X})}{\partial X_i \partial X_j} \right] + E\left[f_j(\mathbf{X}, t) \frac{\partial g(\mathbf{X})}{\partial X_j} \right] \right\}. \tag{12.6.8}$$

Proof. Upon replacing $\partial p / \partial t$ in (12.6.7) by the right hand side of the Fokker–Planck equation (12.6.6) for $\mathbf{X}(t)$ of the IVP (12.6.4), we get after repeated integrations by parts

$$M'(t) = \sum_{j=1}^{n} \int_{-\infty}^{\infty} \int_{-\infty}^{\infty} p_0(\mathbf{a}, t_0) p(\mathbf{x}, t \mid \mathbf{a}, t_0) \Gamma(\mathbf{x}, t) d\mathbf{x} d\mathbf{a}$$

with

$$\Gamma(\mathbf{x}, t) = \frac{1}{2} \sum_{i=1}^{n} (GDG^T)_{ij} \frac{\partial^2 g(\mathbf{x})}{\partial x_i \partial x_j} + f_j(\mathbf{x}, t) \frac{\partial g(\mathbf{x})}{\partial x_j}.$$

The expression for $M'(t)$ may be written as (12.6.8). $\qquad \square$

12.6.2.2   Moments of some SDE output

We now apply the general result (12.6.8) to some examples of Ito type equation to illustrate its execution. The outcomes of such applications are typically a system of deterministic ODE for the moments of the (components of the vector) unknown $\mathbf{X}(t)$.

Example 61. $X' = cX + W(t), \quad X(t_0) = A.$

For this example, we have $n = 1$, $m = 1$, $f_1 = cX$, $G_{11} = 1$ and $W(t)$ is a white noise process (of zero mean). For the first moment $M^{(1)}(t) = E[X(t)]$, we have $g(X) = X(t)$ and, given $W(t)$ being of zero mean, we have

$$\frac{dM^{(1)}(t)}{dt} = cM^{(1)}(t).$$

Subject to the initial condition $M^{(1)}(t_0)] = E[X(t_0)] = E[A]$, the corresponding IVP completely determines the expected value of $X(t)$.

For the second moment $M^{(2)}(t) = E[X^2(t)]$, we have

$$\left\{M^{(2)}(t)\right\}' = 2cM^{(2)}(t) + D,$$

with $E[X^2(t_0)] = E[A^2]$. The IVP completely determines the second moment of $X(t)$.

Example 62. $X' = \{c + W(t)\} X(t), \quad X(t_0) = A.$

For this example, we have $n = 1$, $m = 1$, $f_1 = cX$, $G_{11} = X(t)$ and $W(t)$ is again a white noise forcing. The derivative of the first moment $M^{(1)}(t)$, corresponding to $g(X(t)) = X(t)$, is

$$\left\{M^{(1)}(t)\right\}' = cM^{(1)}(t)$$

with $M^{(1)}(t_0) = E[A]$. The derivative of the second moment $M^{(2)}(t)$ corresponding to $g(X(t)) = X^2(t)$ is

$$\left\{M^{(2)}(t)\right\}' = (2c + D)\,M^{(2)}(t)$$

with $M^{(2)}(t_0) = E[A^2]$. The two deterministic IVP completely determine the expected value and second moment of $X(t)$.

Example 63. $X'' = \{c + W(t)\} X(t), \quad X(t_0) = A_1, \quad X'(t_0) = A_2.$

For this or other single equation of higher order, a more systematic treatment would be to convert it to a system of first order equations. For the given equation, we let $X_1 = X$ and $X_2 = X'$ so that we have

$$\mathbf{X}' = \begin{pmatrix} X_1 \\ X_2 \end{pmatrix}' = \begin{pmatrix} X_2 \\ cX_1 \end{pmatrix} + \begin{pmatrix} 0 \\ X_1 \end{pmatrix} W(t) \equiv \mathbf{f}(\mathbf{X}) + GW.$$

For the resulting first order system, we have $n = 2$ and $m = 1$. There are two first moments and four second moments so that $M^{(1)}(t)$ and $M^{(2)}(t)$ are

$$M^{(1)}(t) = \begin{pmatrix} E[X_1^1 X_2^0] \\ E[X_1^0 X_2^1] \end{pmatrix} = \begin{pmatrix} E[X_1] \\ E[X_2] \end{pmatrix} \equiv \begin{pmatrix} m_{10} \\ m_{01} \end{pmatrix},$$

$$M^{(2)}(t) = \begin{bmatrix} E[X_1^2 X_2^0] & E[X_1^1 X_2^1] \\ E[X_1^1 X_2^1] & E[X_1^0 X_2^2] \end{bmatrix} = \begin{bmatrix} m_{20} & m_{11} \\ m_{11} & m_{02} \end{bmatrix}.$$

For $M^{(1)}(t)$, we have $g(\mathbf{X}(t)) = X_1$ for m_{10} and $g(\mathbf{X}) = X_2$ for m_{01}. It follows from the relation (12.6.8) and the prescribed initial conditions

$$\left\{ M^{(1)}(t) \right\}' = \begin{pmatrix} m_{10}' \\ m_{01}' \end{pmatrix} = \begin{pmatrix} m_{01} \\ cm_{10} \end{pmatrix}, \quad M^{(1)}(t_0) = \begin{pmatrix} E[A] \\ E[B] \end{pmatrix}.$$

The IVP for the second order moment matrix $M^{(2)}(t)$ is given by

$$\left\{ M^{(2)}(t) \right\}' = \begin{bmatrix} m_{20}' & m_{11}' \\ m_{11}' & m_{02}' \end{bmatrix} = \begin{bmatrix} 2m_{11} & cm_{20} + m_{02} \\ cm_{20} + m_{02} & 2cm_{11} + Dm_{20} \end{bmatrix},$$

$$M^{(2)}(t_0) = \begin{bmatrix} E[A^2] & E[AB] \\ E[AB] & E[B^2] \end{bmatrix}.$$

These IVP completely determine the first and second order statistics of the vector unknown $\mathbf{X}(t)$.

Exercise 53. Obtain a typical ODE for each type of third order moments for the SDE in the example above.

Example 64. Nonlinear Leaky Integrator with a Voltage Threshold:

$$X' = X(x_c - X) + I_0 + W(t), \quad X(t_0) = A$$

where x_c and I_0 are prescribed constants and $W(t)$ is a zero mean white noise process.

For this nonlinear SDE of the Ito type (corresponding to (12.1.1) with $x_r = 0$ and $RF(t) = I_0 + W(t)$), the IVP for the first moment can be obtained directly by taking the expected value of (aka ensemble averaging) both sides of the SDE and the initial condition to get

$$m_1' = m_2 - x_c m_1 + I_0, \quad m_1(0) = E[A]. \tag{12.6.9}$$

Unlike all previous examples, the IVP for m_1 involves also the second moment m_2 and cannot be solved without one more equation or condition for the same unknowns. An expression for the derivative of $m_2(t)$ obtained from the kinetic equation (12.6.8) is

$$m_2' = -2m_3 + 2x_c m_2 + 2I_0 m_1 + D, \quad m_2(0) = E[A^2]. \tag{12.6.10}$$

For this relative simple SDE, it is instructive to see a direct derivation by ensemble averaging XX' to get

$$\frac{1}{2}m_2' = x_c m_2 - m_3 + I_0 m_1 + E[XW].$$

For the expression $E[XW]$, we make use of the integral equation for $X(t)$ and form

$$E[W(t)X(t)] = E[AW(t)] + \int_{t_0}^{t} \{E[X(x_c - X) + I_0\} dB + \int_{t_0}^{t} E[W(t)W(\tau)]d\tau$$

$$= \int_{t_0}^{t} D\delta(t - \tau)d\tau = \frac{1}{2}D$$

since $W(t)$ is statistically independent of A and the second term vanishes given the Ito interpretation of the stochastic integral. Upon using this last result to eliminate $E[W(t)X(t)]$ from the expression for m_2' in the line before, we obtain the same ODE (12.6.10) for the second moment previously found by specializing (12.6.8). This equation contains three unknowns, the first, second and third moments, leaving us with two equations for three unknowns and in need of another equation.

It should be evident by now that a set of equations for $\{m_k'\}$, $k = 1, 2, \ldots, m$ would contain $m + 1$ moments of $X(t)$. It is a typical dilemma confronting those who use the moments approach to learn about the properties of the process $X(t)$ determined by a nonlinear SDE. In order to have enough equations for the number of unknowns, various closure schemes have been advanced with varying degree of effectiveness. Other approaches include perturbation methods that take advantage of the presence of small or large parameters. For the present example, the parameter x_c may be large compared to $E[A]$ so that we may normalize X by x_c with the normalized problem containing the small parameter $\varepsilon = E[A]/x_c$. Perturbation and asymptotic methods may then be used to obtain simplified IVP that are self-contained and can be solved for approximate solutions for the moments by known techniques for such deterministic problems. In some cases, such approach may be limited to a bounded range of t. Readers interested in closure and other methods for nonlinear SDE are referred to more advanced text such as [13, 50] and references therein.

Example 65. A Special Second Order Linear System

$$X'' = \{c + W_1(t)\}\, X + I_0 + f(t), \quad X(t_0) = A_1, \quad X'(t_0) = A_2$$

where c and I_0 are prescribed constants, $W_1(t)$ is a zero mean white noise process with variance D_{11} and $f(t)$ is a zero mean stochastic process with prescribed second order statistics and independent of $W_1(t)$ and A_1 and A_2 are two independent random variables providing the initial data for the SDE.

To apply the kinetic equation (12.6.8) to this problem, it is more systematic to denote $X(t)$, $X'(t)$ and $f(t)$ by X_1, X_2 and X_3, respectively, and write the second order SDE as a system of first order equation as in an earlier example:

$$X_1' = X_2, \quad X_2' = (c + W_1)\, X_1 + I_0 + X_3, \quad X_1(t_0) = A_1, \quad X_2(t_0) = A_2.$$

We take $X_3 = f(t)$ to be (approximated by) the output of some filtered white noise such as the Ornstein–Uhlenbeck process characterized by

$$X_3' = -\alpha X_3 + W_2(t), \quad X_3(t_0) = A_3,$$

where α is a known constant, $W_2(t)$ is a zero mean white noise process with variance D_{22} assumed to be independent of $W_1(t)$ and all other mutually independent initial random variables $\{A_k\}$. The system may then be written as

$$\mathbf{X}' = \begin{pmatrix} X_2 \\ cX_1 + I_0 + X_3 \\ -\alpha X_3 \end{pmatrix} + \begin{bmatrix} 0 & 0 \\ X_1 & 0 \\ 0 & 1 \end{bmatrix} \begin{pmatrix} W_1 \\ W_2 \end{pmatrix} \equiv \mathbf{f}\,(\mathbf{X}) + \mathbf{GW}.$$

The first order moments of the SDE system above are easily seen to be determined by the linear IVP:

$$m_{100}' = m_{010}, \quad m_{010}' = cm_{100} + m_{001} + I_0, \quad m_{001}' = -\alpha m_{001},$$

$$m_{100}(t_0) = E[A_1], \quad m_{010}(t_0) = E[A_2], \quad m_{001}(t_0) = E[A_3].$$

This linear IVP for a system of three first order deterministic ODE completely determines the three unknown first order moments $\{m_{100}, m_{010}, m_{001}\}$.

For the second order moments $\{m_{200}, m_{020}, m_{002}, m_{110}, m_{101}, m_{011}\}$, we have from specializing $g(\mathbf{X})$ in (12.6.8) for each and all combinations of $\{X_i X_j\}$,

$i, j = 1, 2, 3$, noting in particular

$$GDG^T = \begin{bmatrix} 0 & 0 \\ X_1 & 0 \\ 0 & 1 \end{bmatrix} \begin{bmatrix} D_{11} & 0 \\ 0 & D_{22} \end{bmatrix} \begin{bmatrix} 0 & X_1 & 0 \\ 0 & 0 & 1 \end{bmatrix} = \begin{bmatrix} 0 & 0 & 0 \\ 0 & D_{11}X_1^2 & 0 \\ 0 & 0 & D_{22} \end{bmatrix},$$

where the two white noise processes are independent (and hence not correlated),

$$m'_{200} = 2m_{110}, \quad m'_{002} = -2\alpha m_{002} + D_{22},$$

$$m'_{020} = 2\left(cm_{110} + I_0 m_{010} + m_{011}\right) + D_{11}m_{200},$$

$$m'_{110} = m_{020} + cm_{200} + I_0 m_{100} + m_{101}, \quad m'_{101} = m_{011} - \alpha m_{101},$$

$$m'_{011} = -\alpha m_{011} + cm_{101} + I_0 m_{001} + m_{002}.$$

Upon forming also appropriate moments of the initial conditions, we get

$$m_{200}(t_0) = E[A_1^2], \quad m_{002}(t_0) = E[A_3^2], \quad m_{020}[t_0] = E[A_2^2],$$

$$m_{110}(t_0) = m_{101}(t_0) = m_{011}[t_0] = 0.$$

Together, they constitute an IVP for a system of six first order equations for six unknowns determining all six second moments for the three processes $\{X_1, X_2, X_3\}$.

We may continue to use (12.6.8) to generate equations for determining the third and higher order moments for the unknown $\mathbf{X}(t)$. The following exercise offers an opportunity to practice this use of (12.6.8):

Exercise 54. Obtain the typical third order moments $m_{300}, m_{030}, m_{210}, m_{120}, m_{201}$ and m_{111} for the last example.

12.6.2.3 *A Non-autonomous transformation of* $\mathbf{X}(t)$

The following more general version of the Proposition 47 can be proved similarly:

Proposition 48. *Let* $h(\mathbf{X}(t),t)$ *be at least continuously differentiable in* t *and twice continuously differentiable in all components of* $\mathbf{X}$ *and* $M(t) = E\left[h(\mathbf{X}(t),t)\right]$ *is the expected value of* $h(\mathbf{X}(t),t)$. *Then the following expression holds for* $M'(t)$:

$$M'(t) = \frac{1}{2}\sum_{i=1}^{n}\sum_{j=1}^{n} E\left[(G^T DG)_{ij}\frac{\partial^2 h}{\partial X_i \partial X_j}\right] + \sum_{j=1}^{n} E\left[f_j\frac{\partial h}{\partial X_j}\right] + E\left[\frac{\partial h}{\partial t}\right].$$

12.7 Stochastic Stability

The stability of steady states is an important issue for dynamical systems in the absence of uncertainty. Stability is also an important but much more complex issue for stochastic systems for a number of reasons. As a start, for each kind of deterministic stability (Poisson, Lyapunov, or asymptotic stability), there are different stochastic stability modes associated with the different kinds of stochastic convergence: stability in probability, stability in distribution, almost sure stability, and stability of statistical moments. Consistent with the developments in this volume, we limit our discussion of stochastic stability to stability of the moments of the response process and mainly to second moments (aka mean square stability) after first pointing out some basic features pertaining to stability in the mean.

12.7.1 *Stability in the mean*

Suppose $\mathbf{X}(t) = (X_1(t), \ldots, X_n(t))^T$ is the (vector) response of the nth order system of SDE

$$X'_j = g_j(X_1, \ldots, X_n, t) \equiv g_j(\mathbf{X}, t), \quad X_j(t_0) = A_j, \quad (j = 1, 2, \ldots, n), \quad (12.7.1)$$

where $(g_1(\mathbf{X}, t), \ldots, g_n(\mathbf{X}, t))^T \equiv \mathbf{g}(\mathbf{X}, t)$ is a known vector functions of the vector unknown process $\mathbf{X}(t)$ and some inhomogeneous parts that may be random variables or stochastic processes with known probabilistic properties taken to be statistically independent of the random initial states $\{A_j\}$. An example of such a system is the vector Ito SDE (12.6.5). In our discussion of stochastic stability, we routinely relocate any steady state of the system being investigated to the origin. The following are definitions relevant to stability of the expected value of $\mathbf{X}(t)$:

Definition 39. The vector stochastic system (12.7.1) is said to be *stable in the mean* if $E[\mathbf{X}(t)]$ is bounded and

$$\lim_{t \to \infty} E[\mathbf{X}(t)] < \mathbf{x}_c$$

for some finite constant vector $\mathbf{x}_c$. It is said to be *asymptotically stable in the mean* if

$$\lim_{t \to \infty} E[\mathbf{X}(t)] = \mathbf{0}.$$

The system is *unstable in the mean* if it is neither asymptotically stable nor stable.

The issues and complications in the study of stochastic stability are illustrated by a few examples below.

Example 66. $X' = cX + f(t)$, $X(0) = A$ where c is a known constant while A and $f(t)$ are statistically independent random quantities with known statistics.

When $f(t)$ is white noise or can be approximated by a filtered white noise process, we have an Ito SDE and the results on the moments of $X(t)$ obtained in the last section can be used to get a deterministic IVP for the mean of $X(t)$, $m_1(t) = \langle X(t) \rangle$. For this simple problem, it is possible to deduce the moment stability of $X(t)$ more directly by working with the known exact solution of the given IVP

$$X(t) = e^{ct} \left\{ A + \int_0^t e^{-c\tau} f(\tau) d\tau \right\}.$$ (12.7.2)

Upon taking the expectation of the above relation, we obtain

$$\langle X(t) \rangle = e^{ct} \left\{ a + \int_0^t e^{-c\tau} \phi(\tau) d\tau \right\}$$ (12.7.3)

where $a \equiv \langle A \rangle$ and $\phi(\tau) \equiv \langle f(\tau) \rangle$ are the expected value of A and $f(\tau)$, respectively. It follows from (12.7.3) that stability in the mean of $X(t)$ is determined by the limiting behavior of the deterministic function of time on the right hand side of (12.7.3). For example, if $c > 0$ and $a \neq 0$, then $X(t)$ is unstable in the mean. Stability in the mean for other combinations of a and $\phi(\tau)$ can be determined similarly.

Exercise 55. Analyze the stability in the mean of the example above for the more general case of a known time-varying coefficient $c(t)$.

Example 67. $X' = CX + f(t)$, $X(0) = A$ where C a random variable with C and A being statistically independent (usually the case in practice).

Taking the expectation of the exact solution now gives

$$\langle X(t) \rangle = \left\{ \langle A \rangle \langle e^{Ct} \rangle + \int_0^t \langle e^{-C(\tau-t)} f(\tau) \rangle d\tau \right\}.$$

Given the statistical properties of A, C and $f(\tau)$, the ordinary function $\langle X(t) \rangle$ may be obtained to determine the stability of $X(t)$ in the mean. In particular, if A is of zero mean, then the expression for $\langle X(t) \rangle$ is reduced to

$$\langle X(t) \rangle = \int_0^t \langle e^{-C(\tau-t)} f(\tau) \rangle d\tau.$$

In addition, if C and $f(t)$ are also statistically independent, the expression above is further simplified to

$$\langle X(t) \rangle = \int_0^t \langle e^{-C(\tau-t)} \rangle \langle f(\tau) \rangle d\tau.$$

In all cases, $\langle X(t) \rangle$ is now a known (ordinary) function; its limiting behavior as $t \to \infty$ determines the stability of $X(t)$ in the mean.

Example 68. $X' = W(t)X + f(t)$, $X(t_0) = A$ where $W(t)$ a white noise process.

If $f(t)$ is a known function, the given SDE is of the Ito type and we may obtain an IVP for $m_1(t) = \langle X(t) \rangle$ by the result of the last section which determines the stability of $X(t)$ in the mean. When $f(t)$ is a stochastic process, instead of treating it as a filtered white noise process so that the results of the last section may be applied, we work directly with

$$\Delta X(t) = X(t)\Delta B(t) + f(t)\Delta t.$$

With the Ito interpretation $X(t)\Delta B(t) = X(t)\left[B(t + \Delta t) - B(t)\right]$ and $X(t)$ statistically independent of $[B(t + \Delta t) - B(t)]$, we have

$$\langle X(t)\Delta B(t) \rangle = \langle X(t) \rangle \langle B(t + \Delta t) - B(t) \rangle = \langle X(t) \rangle \cdot 0 = 0$$

so that

$$\langle X'(t) \rangle = \langle X(t) \rangle' = \langle f(t) \rangle$$

or

$$\langle X(t) \rangle = \langle A \rangle + \int_{t_0}^{t} \langle f(\tau) \rangle d\tau.$$

As such, the stability in the mean of the stochastic process $X(t)$ is determined by the growth behavior of the ordinary antiderivative of the expected value of the random forcing $f(t)$. If the antiderivative term remains bounded, the response process is stable (but not asymptotically stable as long as $\langle A \rangle \neq 0$) in the mean and is unstable otherwise.

Example 69. $X' = C(t)X + f(t)$, where $C(t)$ is now a general (non-white) stochastic process.

For this more general case, taking expectation of both side of the SDE gives

$$\langle X'(t) \rangle = \langle C(t)X(t) \rangle + \langle f(t) \rangle \tag{12.7.4}$$

which involves the joint moment $\langle C(t)X(t) \rangle$. When $C(t)$ is not white noise, the joint moment does not vanish so that (12.7.4) is one equation for two unknowns and another equation is needed to determine them. It is not difficult to see that a new equation formulated for $\langle C(t)X(t) \rangle$ would involve more new unknowns that include a third order joint moment $\langle C^2(t)X(t) \rangle$ and others.

A somewhat more helpful approach is to work with the formal exact solution

$$X(t) = \int_0^t e^{\{I(t)-I(\tau)\}} f(\tau)d\tau, \quad I(t) = \int_0^t C(z)dz.$$

For stability in the mean, we take the expectation of both side of the relation above to get

$$\langle X(t)\rangle = \int_0^t \langle e^{I(t,\tau)} f(\tau)\rangle d\tau, \quad I(t,\tau) = \int_\tau^t C(z)dz.$$

The stability in the mean of $X(t)$ is now seen to depend on the ordinary function defined by the integral on the right involving relevant known joint statistics of the prescribed stochastic process $C(t)$ and $f(t)$.

In practice, the forcing term $f(t)$ and the coefficient process $C(t)$ are usually statistically independent. In that case, the integral simplifies somewhat to read

$$\langle X(t)\rangle = \int_0^t \langle e^{I(t,\tau)}\rangle \phi(\tau)d\tau, \quad I(t,\tau) = \int_\tau^t C(z)dz \qquad (12.7.5)$$

where $\phi(\tau) = \langle f(\tau)\rangle$ and $\langle e^{I(t,\tau)}\rangle$ are known or can be calculated from the known statistics of $f(t)$ and $C(t)$. The limiting behavior of the resulting ordinary integral for large t determines the stability in the mean of $X(t)$.

It is worth noting that we can take $C(t)$ as a filtered white noise and convert the problem into one for a vector Ito type SDE of the form

$$\mathbf{Y}' = \mathbf{f}(\mathbf{Y}, t) + d\mathbf{B}(t), \quad \mathbf{Y}(0) = \mathbf{A}^*.$$

However, the vector SDE for $\mathbf{Y}(t)$ would no longer be linear and a sequence of IVP for the moments of $\mathbf{Y}$ would have the usual closure issue.

Example 70. $\mathbf{X}' = C(t)\mathbf{X} + \mathbf{f}(t)$, with $C(t) = [C_{ij} + \delta_{1j}W_1(t)]$ where $[C_{ij}]$ is a known $n \times n$ constant matrix with a full set of eigenvectors $\{\mathbf{q}_i\}$, $[\delta_{ij}]$ is a matrix of Kronecker deltas and $W_1(t)$ is a scalar white noise process.

To determine the stability in the mean of the vector process $\mathbf{X}(t)$, we diagonalize $[C_{ij}]$ by its modal matrix $Q = [\mathbf{q}_1, \ldots, \mathbf{q}_n] = [q_{ij}]$ with $AQ = Q\Lambda$ to get

$$\mathbf{Y}' = [\lambda_k \delta_{kj}] \mathbf{Y} + \begin{bmatrix} \bar{q}_{11}q_{11} & \bar{q}_{11}q_{12} & \cdots & \bar{q}_{11}q_{1n} \\ \bar{q}_{21}q_{11} & \bar{q}_{21}q_{12} & \cdots & \cdot \\ \cdot & \cdot & \cdots & \cdot \\ \cdot & \cdot & \cdots & \cdot \\ \bar{q}_{n1}q_{11} & \bar{q}_{n1}q_{12} & \cdots & \bar{q}_{n1}q_{1n} \end{bmatrix} \mathbf{Y}W(t) + Q^{-1}\mathbf{f},$$

where $\mathbf{Y} = Q^{-1}\mathbf{X} = [\bar{q}_{ij}]\,\mathbf{X}$ and $Q^{-1} = [\bar{q}_{ij}]$. Upon taking expectation of both sides of the relation above, we get with

$$\langle \mathbf{Y}'(t)\rangle = [\lambda_k \delta_{kj}]\,\langle \mathbf{Y}(t)\rangle + Q^{-1}\langle \mathbf{f}(t)\rangle$$

given $\langle \mathbf{Y}(\mathbf{t})W(t)dt\rangle = \langle \mathbf{Y}(t)dB(t)\rangle = \mathbf{0}$. The exact solution of the deterministic ODE above is

$$\langle \mathbf{Y}(t)\rangle = e^{\Lambda t}\mathbf{K} + \mathbf{F}(t)$$

where

$$\mathbf{K} = Q^{-1}\langle \mathbf{A}\rangle, \quad \mathbf{F}(t) = \int_0^t e^{\Lambda(t-\tau)}Q^{-1}\langle \mathbf{f}(\tau)\rangle d\tau$$

Consequently, $\mathbf{Y}(t)$ (and hence $\mathbf{X}(t)$) is unstable in the mean if any of the eigenvalues $\{\lambda_k\}$ has a positive real part. Otherwise, the stability in the mean of the (vector) response process depends on growth property of $\mathbf{F}(t)$.

Example 71. $X' = X - X^2$, $\quad X(t) = A$, $\quad$ where A is second order random variable.

It is straightforward to obtain an IVP for the mean of $X(t)$, $m_1(t) = \langle X(t)\rangle$, by taking the expected value of both sides of the SDE and the initial condition:

$$m_1' = m_1 - m_2, \quad m_1(0) = \langle A\rangle. \tag{12.7.6}$$

We see immediately that the resulting expression for $m_1'(t)$ contains a term involving a higher order moment. Not unexpectedly, we face with the same closure problem for nonlinear SDE encountered in the previous section. Again, we refer readers to more advanced text and the SDE literature (such as [13, 50] for example) for further discussion of the different approaches for resolving the problem posed by more unknowns than equations. We only wish to point out here that in cases where there is an exact solution for the corresponding deterministic equation, the transformation theory in an earlier chapter may be useful for the present purpose. For this problem, we have from the exact solution

$$X(t) = \frac{A}{A + (1 - A)e^{-t}} \tag{12.7.7}$$

the following expression for the mean of $X(t)$:

$$\langle X(t)\rangle = \int_{-\infty}^{\infty} \frac{a}{a + (1 - a)e^{-t}}p(a)da. \tag{12.7.8}$$

Evaluation of the right hand side of (12.7.8) may have to be carried out numerically even if $p(a)$ is prescribed. If only first and second order statistics of A are known and not its pdf), the behavior of $\langle X(t) \rangle$ as $t \to \infty$ (which is all we need for stability in the mean) is easily seen to be

$$\lim_{t \to \infty} \langle X(t) \rangle = \int_{-\infty}^{\infty} p(a)da = 1.$$

Hence, $X(t)$ is stable (but not asymptotically stable) in the mean.

Exercise 56. Determine the stability in the mean of the solution $X(t)$ of $X' = X^2$, $X(0) = A$, where A is second order random variable.

12.7.2 *Stability in the mean square*

Within the framework of a mean square theory for s.p., it is natural (in fact a requisite) to investigate also the stability of the second moments. The following definitions are relevant for this purpose:

Definition 40. The vector stochastic system (12.7.1) is said to be *stable in the mean square* (or *mean square stable*) if $E[\mathbf{X}(t)\mathbf{X}^T(t)]$ is bounded and

$$\lim_{t \to \infty} E[\mathbf{X}(t)\mathbf{X}^T(t)] < X^c$$

for some finite constant square matrix X^c. It is said to be *asymptotically stable in the mean square* (or *mean square asymptotically stable*) if

$$\lim_{t \to \infty} E[\mathbf{X}(t)\mathbf{X}^T(t)] = O,$$

where O is the zero square matrix. The system is (mean square) *unstable* if it is neither asymptotically stable nor stable (in the mean square).

Though not stated explicitly, it would seem reasonable to be concerned with stability in the mean square only when $\mathbf{X}(t)$ is stable in the mean. The issues and complications in the study of stochastic stability in the mean square is even more complex than stability in the mean as seen from the following examples.

Example 72. Investigate stability in the mean square for the IVP $X' = cX + f(t)$, $X(0) = A$ of Example 69.

We again work to reduce the stochastic problem to one for stability of deterministic functions. Given that there is a formal exact solution (12.7.2) for this simple scalar problem, we can use it to form the second moment of $X(t)$ to get

$$\langle X^2(t)\rangle = m_2(t) = e^{2ct} \left\langle \left\{ A + \int_0^t e^{-c\tau} f(\tau)d\tau \right\}^2 \right\rangle$$

$$= e^{2ct} \left\{ \langle A \rangle^2 + 2\langle A \rangle \varphi_1(t) + \varphi_2(t) \right\},$$

where

$$\varphi_1(t) = \int_0^t e^{-c\tau} \langle f(\tau) \rangle d\tau,$$

$$\varphi_2(t) = \int_0^t \int_0^t e^{-c(\tau+\xi)} \langle f(\tau)f(\xi) \rangle d\tau d\xi.$$

The right hand side is an ordinary function of t; its behavior for large t determines the stochastic stability in the mean square of $X(t)$. In particular, $X(t)$ is stochastically unstable in the mean square if $c > 0$ and $\langle A \rangle \neq 0$. If $c \leq 0$, stability of $X(t)$ in the mean square depends on the growth rate of $\langle f(\tau) \rangle$ and $\langle f(\tau)f(\xi) \rangle$.

For example, if $f(t)$ is a (zero mean) white noise process, we have $\langle f(\tau) \rangle = 0$ and $\langle f(\tau)f(\xi) \rangle = D\delta(\tau - \xi)$ so that

$$\langle X^2(t)\rangle = m_2(t) = e^{2ct} \left\{ \langle A \rangle^2 + \frac{D}{2c} \left(1 - e^{-2ct} \right) \right\}. \tag{12.7.9}$$

It follows that $X(t)$ is stable (but not asymptotically stable) in the mean square if $c < 0$ since

$$\lim_{t \to \infty} \langle X^2(t)\rangle = \frac{D}{2|c|}.$$

Another approach that is useful when an exact solution is not available is to derive a system of deterministic IVP for the first and second order moments of $X(t)$. If $f(t)$ is a white noise process, the SDE is of the Ito type and we get from (12.6.8),

$$m_2'(t) = 2cm_2(t) + D. \tag{12.7.10}$$

Supplemented by the initial conditions $m_2(0) = \langle A^2 \rangle$, the resulting $m_2(t)$ is again given by (12.7.9).

When $f(t)$ is not a white noise process and an exact solution such as (12.7.2) is not available, we take $f(t)$ as (or approximated by) the output of passing a white noise through a suitably chosen linear filter. In this way, the relation (12.6.8) applies to the augmented linear stochastic system.

We note in passing that for many linear time-invariant systems such as the present simple scalar problem, it is possible to investigate stochastic stability in the mean and in the mean square by working with the corresponding characteristic functions. To the extent that this alternative frequency domain approach is not effective for time-varying linear systems and nonlinear systems, we continue to work with the time domain approach for our stochastic stability problems. Readers interested in the frequency domain approach are referred to [50] and references therein.

Exercise 57. Obtain the deterministic IVP for the second moment matrix $M^{(2)}(t) = \langle \mathbf{X}(t)\mathbf{X}^T(t) \rangle$ for assessing stability in the mean square for the second order vector process $\mathbf{X}(t)$ evolving according to the vector stochastic IVP

$$\mathbf{X}' = C\mathbf{X} + \mathbf{f}(t), \quad \mathbf{X}(0) = \mathbf{A}$$

where C is a known $n \times n$ constant matrix, $\mathbf{A}$ is a second order vector random variable with known statistics and $\mathbf{f}(t)$ is a second order vector stochastic process with known statistics and independent of $\mathbf{A}$.

Example 73. $X' = \{c + W(t)\} X(t), \; X(0) = A$ where c is a known constant, A is a random variable with known statistics and $W(t)$ is a white noise process with variance D and statistically independent of A.

For this Ito type stochastic equation, the ODE for $m_2(t)$ from (12.6.8) reads

$$m_2' = (2c + D) \, m_2.$$

The IVP defined by this ODE and the initial condition $m_2(0) = \langle A^2 \rangle$ has as its solution

$$m_2(t) = \langle A^2 \rangle e^{(2c+D)t}.$$

The mean square stability of $X(t)$ is now seen to depend on the sign of the constant $2c + D$.

Exercise 58. (Leaky Integrator with noisy resistor and a white noise input current)

$$X' = \{-c + W_1(t)\} X(t) + \{I_0 + W_2(t)\}, \quad X(0) = A_0$$

where A_0 and I_0 are known constants and $W_1(t)$ and $W_2(t)$ are two (zero mean) white noise processes. Determine the mean square stability of $X(t)$.

Exercise 59. (An RLC circuit with noisy resistor)

$$X'' + 2\beta \{1 + W(t)\} X' + \omega^2 X = 0, \quad X(0) = A, \quad X'(0) = 0$$

where A is a second order random variable independent of the (zero mean) white noise process $W(t)$ (and where β and ω are parameters that are combinations of the inductance, resistance and capacitance of the circuit). Determine the mean square stability of the stochastic voltage process $X(t)$ in terms of β and ω.

Exercise 60. (An RLC circuit with noisy capacitance) $X'' + 2\beta X' + \omega^2\{1 + W_1(t)\}X = W_2(t), \; X(0) = 0, \; X'(0) = 0$ where $W_1(t)$ and $W_2(t)$ are (zero mean) white noise processes. Deduce the IVP for first and second order moments of the solution $X(t)$.

Example 74. $X' = X - X^2$, $X(t) = A$, where A is second order random variable. Determine the mean square stability of $X(t)$.

This is still an Ito SDE but with $G(X, t) = 0$. The ODE for $m_2(t)$ from (12.6.8) is

$$m_2' = 2(m_2 - m_3).$$

There is an obvious closure problem for the moment equations. In fact, it has already shown up in the ODE (12.7.6) for the expected value in the previous subsection. But with the exact solution (12.7.7), we have

$$\langle X^2(t) \rangle = \int_{-\infty}^{\infty} \left[\frac{a}{a + (1 - a)e^{-t}} \right]^2 p(a)\, da,$$

where $p(a)$ is the pdf of the random variable A needed to complete the solution for $m_2(t)$. To determine the stability of $X(t)$ in the mean square, we only need the limiting value of $\langle X^2(t) \rangle$ as $t \to \infty$. For the present problem, we have

$$\lim_{t \to \infty} \langle X^2(t) \rangle = \int_{-\infty}^{\infty} p(a)\, da = 1$$

so that $X(t)$ is stable (but not asymptotically stable) in the mean square.

Part 5
Stochastic Partial Differential Equations

Chapter 13

Linear PDE with Random Forcing

13.1 Cable Model Neuron with Random Forcing

13.1.1 *The linear cable model*

Up to now, all our models for neurons idealize them as a point with the electrical activities of an entire nerve cell modeled by a RC (or RLC) circuit. Historically, such models have generated some useful information about, and insight on neuronal behavior. It is also obvious that they do not provide enough information for our understanding of neuronal activities. The linear dimensions of the axon of a motoneuron can reach a considerable fraction of the size of the whole biological organism. For instance, spinal motor neurons of a giraffe reach a length of several meters. With the typical length of an axon of motoneurons in the order of a meter and in contact with multiple synapses along its length, the spatial features of neurons must have a substantial role in neuronal activities and behavior.

In this chapter, we discussed one model for neurons that captures some essential spatial features of their activities. Instead of a single RC circuit characterization, the model to be discussed effectively treats a spatially distributed neuron as a long electrical cable capable of propagating electrical signals. The electrical cable theory was originally developed by William Thompson (Lord Kelvin) for application to long-distance communication by telegraph. The cable model for a neuron is a linear theory with fixed (membrane) properties, particularly conductance, and therefore applicable only for relatively low depolarization voltage, well below the threshold for triggering an action potential. For this reason, the model, known as a passive cable type model, is also said to be suitable for subthreshold activities. For depolarization voltage closer to (or above) threshold, the conductance is known to change significantly from its resting state value and a more complex nonlinear model would be needed to adequately describe the neuronal behavior (see Chapter 16).

The linear cable theory models the neuron as a thin-walled circular cylindrical shell of inner radius a and wall thickness h (Fig. 13.1). The thin wall cylindrical shell itself is to model the cellular protein membrane equipped with pumps, channels and

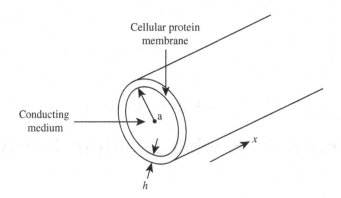

Fig. 13.1.   Cylindrical shell representation of a nerve cylinder membrane.

neurotransmitters. The interior of the membrane is a conducting medium with axi-symmetry along the central axis, the x-axis, of the nerve cylinder. Axi-symmetry enables us to consider a one-dimensional theory where current and voltage quantities are functions of x and t (time). To simplify the development of this sub-threshold linear cable model, we make the (unnecessary) simplifying assumption that the medium exterior to the nerve cell membrane is infinitely conducting with zero resistance and voltage.

With the span of the nerve cylinder divided into incremental segments of length Δx, the mathematical equations for our cable model emerges from the conservation of currents relative to the segment $(x, x + \Delta x)$ of the cylinder. The four electric currents of interest are: $I_i(x, t)$, $I_i(x + \Delta x, t)$, $I_m(x + \Delta x, t)$ and $I_A(x + \Delta x, t)$ as shown in Fig. 13.2:

$$I_i(x + \Delta x, t) = I_i(x, t) + I_m(x + \Delta x, t) + I_A(x + \Delta x, t). \tag{13.1.1}$$

The first two are longitudinal (directed in the positive x-direction) internal to the cylindrical membrane while the last two are cross-membrane currents with $I_A(x + \Delta x, t)$ being an external current injection and $I_m(x + \Delta x, t)$ being the trans-membrane current induced by the RC circuit characterization of the membrane; both directed to the point $x + \Delta x$.

Among these currents, we have the following relation between the current I_m and the trans-membrane voltage similar to the Leaky Integrator:

$$I_m = -\frac{V_i}{R_m} - C_m \frac{\partial V_i}{\partial t}, \quad R_m = r_m/\Delta x, \quad C_m = c_m \Delta x. \tag{13.1.2}$$

The voltage $V_i(x, t)$ is the *depolarization* voltage, the voltage above the resting state potential inside the neuron's cylindrical membrane. We take $V_i = 0$ when the neuron is at the resting state (at about $-70\,\text{mV}$). For the two longitudinal currents,

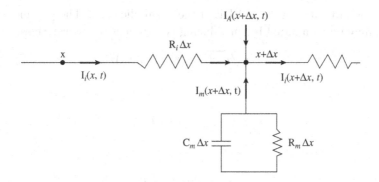

Fig. 13.2. Currents at the point $x + \Delta x$ of the inner surface of nerve cell membrane.

we have

$$V_i(x + \Delta x, t) = V_i(x, t) - R_i I_i(x, t), \quad R_i = r_i \Delta x.$$

As $\Delta x \to 0$, this becomes

$$\frac{\partial V_i}{\partial x} = -r_i I_i(x, t)$$

and after further differentiation

$$\frac{\partial^2 V_i}{\partial x^2} = -r_i \frac{\partial I_i(x, t)}{\partial x}.$$

Upon taking the limit of (13.1.1) as $\Delta x \to 0$, we get

$$c_m r_m \frac{\partial V}{\partial t} = \frac{r_m}{r_i} \frac{\partial^2 V}{\partial x^2} - V + r_m I_A(x, t) \tag{13.1.3}$$

where we have made use of the expression (13.1.2) to eliminate I_m in (13.1.1) and omitted the subscript i in $V_i(x, t)$.

For completeness, we note the following terminology for the various parameters:

$V(x, t) =$ depolarization voltage in volts

$r_i =$ internal resistance per unit length in ohms per centimeter

$r_m =$ membrane resistance $\times$ unit length in ohms-centimeter

$c_m =$ membrane capacitance per unit length in farads per centimeter

$I_A =$ applied current density in amperes per centimeter

Note that $I_A(x, t) dx dt$ is the amount of electrical charges passing through $(x, x + dx)$ during the time interval $(t, t + dt)$.

Equation (13.1.3) is a linear PDE of the parabolic type. The parameters in the equation may be eliminated by introducing two dimensionless variables,

$$x = y\sqrt{r_m/r_i}, \quad t = (r_m c_m)\tau,$$

to give

$$\frac{\partial V}{\partial \tau} = \frac{\partial^2 V}{\partial y^2} - V + I(y, \tau), \tag{13.1.4}$$

where

$$I(y, \tau) = r_m I_A(x.t).$$

A more thorough development of the linear cable theory including the auxiliary conditions to be introduced below can be found in [42, 53] and references therein.

13.1.2 The cable equation with white noise current injection at a point

13.1.2.1 *The initial-boundary value problem*

The cable equation (13.1.4) is known to adequately characterize subthreshold behavior of axon as well as the potential of a dendritic tree when it can be represented by an "equivalent cylinder". At the same time, the findings of many experimental studies suggest that electrophysiological activity of neurons involves a surprising degree of variability in their firing behavior. The time between consecutive action potentials is rarely if ever in regular or predictable temporal patterns. The causes of such irregular activities have yet to be completely identified and too complex to be modeled and analyzed for those that are known (see [65] for a more thorough discussion on this issue). To gain some insight to the irregular neuronal activities, a great deal of effort has been devoted to investigating the effects of appropriate types of noise on neuronal firing in a cable model neuron. Early effort in this direction undertook diffusion approximations to smooth out the rapid and significant changes in synaptic input. To illustrate this type of effort, we consider here the nerve cylinder model formulated above subject to a white noise current with drift injected at the point x_0. Returning to the x, t notation for the scaled independent variables, we have for

$$I(x, t) = \{a + W(t)\}\, \delta(x - x_0) \tag{13.1.5}$$

the following specialization of (13.1.4)

$$\frac{\partial V}{\partial t} = \frac{\partial^2 V}{\partial x^2} - V + \{a + W(t)\}\, \delta(x - x_0) \tag{13.1.6}$$

where a is a *drift parameter* and the zero mean white noise $W(t)$ is of variance D. We note that experiment results for such a setting (with the source active over only a very small patch of a nerve cylinder) has been obtained for a clamped squid axon in [16] (see also [74] for additional electrophysiological aspects of the problem).

Various boundary conditions may be imposed on the stochastic partial differential (13.1.6) corresponding to cases of "sealed end", "killed end" or "lumped soma" at two ends $x = 0$ and $x = L$ of the equivalent nerve cylinder. For brevity, we limit ourselves in this discussion to the sealed end conditions

$$\left[\frac{\partial V}{\partial x}\right]_{x=0} = \left[\frac{\partial V}{\partial x}\right]_{x=L} \tag{13.1.7}$$

so that there is no leakage at either end. We also take the nerve cylinder to be initially in its resting state so that

$$V(x,0) = 0. \tag{13.1.8}$$

It should be evident from the subsequent development that other boundary conditions and/or an inhomogeneous initial condition can also be handled by the same method of solution.

13.1.2.2 *Eigenfunction expansions*

The IBVP for the cable model can be solved by eigenfunction expansions. By the method of separation of variables developed in Chapter 7, it is straightforward to show that the eigenvalues and (normalized) eigenfunctions, $\{\lambda_n, \phi_n(x)\}$ (aka eigenpairs), for the homogeneous problem (when $a = W(t) = 0$) are

$$\lambda_0^2 = 1, \qquad \lambda_n^2 = \left(\frac{n\pi}{L}\right)^2 + 1, \qquad (n = 1, 2, 3, \ldots) \tag{13.1.9}$$

$$\phi_0(x) = \frac{1}{L}, \qquad \phi_n(x) = \sqrt{\frac{2}{L}}\cos\left(\frac{n\pi x}{L}\right), \qquad (n = 1, 2, 3, \ldots). \tag{13.1.10}$$

The eigenfunctions $\{\phi_n(x)\}$ satisfy the orthogonality relations

$$\int_0^L \phi_n(x)\phi_m(x)dx = \begin{cases} 0 & (m \neq n) \\ 1 & (m = n) \end{cases} \tag{13.1.11}$$

can be verified by direct evaluation of the relevant integrals.

The set of eigenfunctions is also known to be complete so that we may write $V(x,t)$ in the form of its Fourier series

$$V(x,t) = \sum_{n=0}^{\infty} V_n(t)\,\phi_n(x), \tag{13.1.12}$$

for some unknown coefficients $\{V_n(t)\}$ to be determined presently. Upon substituting the Fourier series (13.1.12) into the PDE for $V(x,t)$, we rearrange the result to get

$$\sum_{n=0}^{\infty} \left(\frac{dV_n}{dt} + \lambda_n^2 V_n \right) \phi_n(x) = \{a + W(t)\} \delta(x - x_0). \qquad (13.1.13)$$

Orthogonality can then be used to obtain from (13.1.13)

$$\frac{dV_n}{dt} + \lambda_n^2 V_n = \int_0^L \phi_n(x) \{a + W(t)\} \delta(x - x_0) dx = \{a + W(t)\} \phi_n(x_0).$$
$$(13.1.14)$$

Correspondingly, the initial condition (13.1.8) becomes

$$V_n(0) = 0. \qquad (13.1.15)$$

The solutions of the IVP for the coefficients $\{V_n(t)\}$ are

$$V_n(t) = a v_n(t) + u_n(t)$$

where

$$v_n(t) = \frac{\phi_n(x_0)}{\lambda_n^2} \left(1 - e^{-\lambda_n^2 t} \right), \quad u_n(t) = \phi_n(x_0) \int_0^t e^{-\lambda_n^2(t-\tau)} W(\tau) d\tau. \qquad (13.1.16)$$

Re-assembling all the pieces back into the expression (13.1.12), we obtain

$$V(x, t) = a v(x, t; x_0) + u(x, t; x_0) \qquad (13.1.17)$$

where

$$v(x, t; x_0) = \sum_{n=0}^{\infty} \frac{1 - e^{-\lambda_n^2 t}}{\lambda_n^2} \phi_n(x_0) \phi_n(x) \qquad (13.1.18)$$

$$u(x, t; x_0) = \int_0^t G(x, x_0, t - \tau) W(\tau) d\tau \qquad (13.1.19)$$

with

$$G(x, x_0; t - \tau) = \sum_{n=0}^{\infty} e^{-\lambda_n^2(t-\tau)} \phi_n(x) \phi_n(x_0) \qquad (13.1.20)$$

known as the Green's function for the given IBVP. The Green's function, $G(x, c; t - \tau)$, is the cable response to a point source of unit strength applied at the point c and the instant τ. Mathematically, $G(x, c; t - \tau)$ satisfies the PDE

$$\frac{\partial G}{\partial t} = \frac{\partial^2 G}{\partial x^2} - G + \delta(x - c)\delta(t - \tau),$$

and a sealed end condition at each end of the interval $[0, L]$.

13.1.2.3 *Depolarization voltage is not Markovian*

We see from (13.1.17) that the solution for the depolarization voltage $V(x,t)$ has a deterministic mean component $av(x,t;x_0)$ and a zero mean random process $u(x,t;x_0)$ with

$$\langle V(x,t) \rangle = av(x,t;x_0)$$

since $W(t)$ is of zero mean so that $\langle u(x,t;x_0) \rangle = 0$. For the random process $u(x,t;x_0)$, we see from differentiating the expression for $u_n(t)$ in (13.1.16) (or taking the expectation of both sides of (13.1.14)) that

$$\frac{du_n}{dt} + \lambda_n^2 u_n = \phi_n(x_0)W(t). \tag{13.1.21}$$

Therefore, each $u_n(t)$ process is an Ornstein–Uhlenbeck process (O.U.P.). Thus each component process of the random part of the $V(x,t)$ in the eigenfunction (aka *normal mode*) decomposition (13.1.12) is an O.U.P. and therefore Markovian (given that knowledge of the process at any instant τ in time completely determines the process for all subsequent time $t > \tau$ by the IVP (13.1.21) for $u_n(t)$).

However, even if the individual constituent $u_n(t)$ processes that constitute $V(x,t)$ are Markovian, the depolarization voltage itself is not necessarily Markovian. The value $V(y,\tau)$ at a point y along the nerve cylinder and an instant τ does not determine $V(y,t)$ for $t > \tau$ at that point. We see this from the solution of the IBVP defined by

$$\frac{\partial V_I}{\partial t} = \frac{\partial^2 V_I}{\partial x^2} - V_I, \quad V(x,t_0) = U(x) \tag{13.1.22}$$

and the sealed end conditions (13.1.7). With $a = W(t) = 0$ in (13.1.6), the Fourier series solution for the new problem can similarly be found to be

$$V_I(x,t) = \sum_{n=0}^{\infty} \bar{V}_n e^{-\lambda_n^2(t-t_0)} \phi_n(x)$$

with

$$\bar{V}_n = \int_0^L U(z)\phi_n(z)dz.$$

Upon re-arranging the solution $V_I(x,t)$ as

$$V_I(x,t) = \sum_{n=0}^{\infty} \left\{ \int_0^L U(z)\phi_n(z)dz \right\} e^{-\lambda_n^2(t-t_0)}\phi_n(x)$$

$$= \int_0^L \left\{ \sum_{n=0}^{\infty} e^{-\lambda_n^2(t-t_0)}\phi_n(x)\phi_n(z) \right\} U(z)dz,$$

we obtain the following expression for the solution of the pure IVP in terms of the Green's function $G(x, z; t - t_0)$ as previously defined in (13.1.20):

$$V_I(x, t) = \int_0^L G(x, z; t - t_0)U(z)dz. \tag{13.1.23}$$

with

$$G(x, z; t - t_0) = \sum_{n=0}^{\infty} e^{-\lambda_n^2(t-t_0)}\phi_n(x)\phi_n(z). \tag{13.1.24}$$

In the form (13.1.23), it can be seen that prescription of a data point at (y, t_0) is not sufficient to determine $V(y, t)$ at the same spatial location for all $t > t_0$. We need to know the initial data for all point x over the span of the nerve cylinder $[0, L]$ to specify even just the voltage at time t for the same spatial point where the initial voltage data has been prescribed. Physically, this is understandable because the potential at all points along the nerve cylinder influences and contributes to the potential at any single spatial location of the neuron at a later time. Mathematically, this dependence is a consequence of the Green's function representation (13.1.23) for the solution of the pure initial value problem (13.1.22)–(13.1.7). Hence, space-time depolarization processes is not a Markov process if we simply extend the "short term memory" or "memoryless" property to apply to both spatial and temporal coordinates. In fact, the solution for our simple pure IVP suggest that it is not meaningful to talk about the transition probability density for $V_I(x, t)$ at a fixed x. The same Green's function representation however shows that specifying the voltage process for the entire cable at any instant of time does determine the depolarization for all points in the cable thereafter. As such, the depolarization voltage process has been classified as a new type of s.p. to be known as *Markov random fields*.

13.1.2.4 *Steady state depolarization mean and variance*

Mean depolarization   Returning to the original problem of a white noise current injection with drift $a > 0$ at a point x_0, the expected value of the depolarization voltage is given by (13.1.18). As $t \to \infty$, that expression tends to a unique steady state

$$\lim_{t \to \infty} E[V(x.t)] = a \sum_{n=0}^{\infty} \frac{1}{\lambda_n^2}\phi_n(x_0)\phi_n(x). \tag{13.1.25}$$

The series solution for the steady state behavior (as well as the series (13.1.18) for the time-dependent solution) converges rather slowly at (and near) the point of application x_0 of the external stimulus. It is therefore extremely useful from a computational viewpoint to have an alternative expression for the expected steady state depolarization that does not involve an infinite series. To find such an expression, we take expectation of both sides of (13.1.6) and (13.1.7) to get

$$\langle V \rangle_{,t} = \langle V \rangle_{,xx} - \langle V \rangle + a\delta(x - x_0),$$

$$\langle V(0,t) \rangle_{,x} = \langle V(L,t) \rangle_{,x} = 0,$$

where we have used the subscript notations to denote partial derivatives.

In steady state, the limiting voltage would be independent of time,

$$\lim_{t \to \infty} \langle V(x,t) \rangle = \bar{v}(x),$$

so that the PDE and sealed end conditions for the mean voltage simplify to

$$\bar{v}'' - \bar{v} = -a\delta(x - x_0), \quad \bar{v}'(0) = \bar{v}'(L) = 0.$$

The exact solution of this boundary value problem (BVP) is easily determined to be in terms of elementary functions:

$$\bar{v}(x) = \begin{cases} \dfrac{a\cosh(L - x_0)\cosh(x)}{\sinh(L)} & (0 \le x \le x_0) \\[2ex] \dfrac{a\cosh(x_0)\cosh(L - x)}{\sinh(L)} & (x_0 \le x \le L). \end{cases} \tag{13.1.26}$$

There is no infinite series to compute and hence no problem of slow convergence. It is not difficult to verify that the expression (13.1.25) is in fact the eigenfunction expansion of (13.1.26) and is not as well suited for evaluating the steady state expected voltage at the point of the injected current x_0.

b) *Variance:* To find an expression for the variance of $V(x,t)$, we use the normal mode decomposition (13.1.12) in the expression

$$Var[V(x,t)] = \langle V^2(x,t) \rangle - \langle V(x,t) \rangle^2$$

to get formally

$$Var[V(x,t)] = \sum_{n=0}^{\infty} \sum_{m=0}^{\infty} \{ \langle V_n(t)V_m(t) \rangle - \langle V_n(t) \rangle \langle V_m(t) \rangle \} \phi_n(x)\phi_m(x)$$

$$= D \sum_{n=0}^{\infty} \sum_{m=0}^{\infty} \frac{1 - e^{-(\lambda_n^2 + \lambda_m^2)t}}{\lambda_n^2 + \lambda_m^2} \phi_n(x)\phi_m(x)\phi_n(x_0)\phi_m(x_0)$$

It is not difficult to show by an integral test that the double infinite series diverges at $x = x_0$ while the convergence of the double infinite series for $x \ne x_0$ has been established in [74]. Extensive numerical results for mean and variance of the depolarization voltage process have been obtained there. In addition to demonstrating the divergence of $Var[V(x,t)]$ at x_0, they also show how the standard deviation and coefficient of variation of depolarization voltage at the soma (cell body) are much more sensitive to the position of the current injection when it is close to the soma.

13.1.3   Cable model neuron with distributed input current

Owing to the point source nature of the input current in the problem previously solved, the Green's function representation of the solution (13.1.19)–(13.1.20) is not representative of the response to a general forcing. For a more typical representation, suppose $F(x,t) = E(x)f(t)$ where $f(t)$ is a stochastic process and $E(x)$ is a known *envelop function* characterizing the distribution of the synapses along the equivalent nerve cylinder. In that case, the term $\{a + W(t)\}\,\delta(x - x_0)$ in (13.1.6) is replaced by $E(x)f(t) = F(x,t)$. After the Fourier series decomposition, the PDE for depolarization $V(x,t)$, (13.1.13), now takes the form

$$\sum_{n=0}^{\infty} \left(\frac{dV_n}{dt} + \lambda_n^2 V_n \right) \phi_n(x) = E(x)f(t).$$

Applications of the orthogonality among the eigenfunctions give

$$\frac{dV_n}{dt} + \lambda_n^2 V_n = E_n f(t), \quad V_n(0) = 0$$

where

$$E_n = \int_0^L E(x)\phi_n(x)dx.$$

Upon obtaining the solutions $\{V_n(t)\}$, we use them to write the solution of the original IBVP in the form of a Green's function representation:

$$V(x,t) = \int_0^t \int_0^L G(x,z,t-\tau)E(z)f(\tau)dzd\tau = \int_0^t \int_0^L G(x,z,t-\tau)F(z,\tau)dzd\tau,$$

with $G(x,z,t-\tau)$ given by (13.1.24). It is straightforward to verify the same integral representation applies to a general random forcing.

For simplicity, we assume that the forcing $F(x,t) = E(x)f(t)$ is of zero mean and focus on the determination of the second moment of the response process given by

$$\langle V^2(x,t) \rangle = \int_0^t \int_0^t G(x,z,t-\xi)G(x,y,t-\eta)\langle F(z,\xi)F(y,\eta)\rangle dzdyd\xi d\eta$$

$$= \int_0^t \int_0^t \int_0^L \int_0^L G(x,z,t-\xi)G(x,y,t-\eta)E(x)E(y)R(\xi,\eta)dzdyd\xi d\eta$$

where

$$R(\xi, \eta) = \langle f(\xi)f(\eta) \rangle$$

is the autocorrelation function of the $f(t)$.

Since the input process is of zero mean, the main statistics of interest is the variance of the depolarization voltage. For a zero mean white noise process, $f(t) = W(t)$, we have $R(\xi, \eta) = D\delta(\xi - \eta)$ so that

$$\langle V^2(x,t) \rangle = D \int_0^t \int_0^L \int_0^L G(x, z, t - \xi)G(x, y, t - \xi)E(x)E(y)dzdyd\xi.$$

For the special case of an Ornstein–Uhlenbeck process generated by

$$\frac{df}{dt} = -\alpha f + W(t), \quad f(-\infty) = 0,$$

with a zero mean white noise process $W(t)$, we have the following integral representation

$$f(t) = e^{-\alpha t} \int_{-\infty}^t e^{\alpha \tau} W(\tau)d\tau.$$

For $\xi > 0$ and $\eta > 0$, the correlation function of $f(t)$ is given by

$$\langle f(\xi)f(\eta) \rangle \equiv F(\xi, \eta) = e^{-\alpha(\xi+\eta)} \int_{-\infty}^{\xi} \int_{-\infty}^{\eta} e^{\alpha(s+u)} D\delta(s-u)dsdu$$

$$= De^{-\alpha(\xi+\eta)} \int_{-\infty}^{\min(\xi,\eta)} e^{\alpha z}dz = \frac{D}{\alpha}e^{-\alpha(\xi+\eta)} \left[e^{2\alpha \min(\xi,\eta)}\right] = \frac{D}{\alpha}e^{-\alpha|\xi-\eta|}.$$

This expression can then be used in the Green's function representation for $V(x,t)$ to determine its second order moment:

$$\langle V^2(x,t) \rangle = \frac{D}{\alpha} \int_0^t \int_0^t \int_0^L \int_0^L G(x,z,t-\xi)G(x,y,t-\eta)E(x)E(y)e^{-\alpha|\xi-\eta|}dzdyd\xi d\eta.$$

Had the initial state of $V(x,t)$ not been the resting state 0, we would have an additional term associated with $V(x,t_0) = U(x)$ so that

$$V(x,t) = \int_0^L G(x,z;t-t_0)U(z)dz + \int_{t_0}^t \int_0^L G(x,z,t-\tau)F(z,\tau)dzd\tau.$$

13.2 The Method of Spatial Correlations

13.2.1 The zero mean problem

The method of eigenfunction expansions are effective for reducing the IBVP for the linear stochastic PDE to solving a sequence of IVP for relatively simple ODE.

Combining the solution for these problems led to a Green's function representation for the solution of the original IBVP with which we can compute the statistics of the unknown stochastic solution such as mean, variance and correlation. For more complicated linear PDE, it is not always possible to obtain the needed eigen-pairs in terms of known functions (or even determining their existence). This would be the case when the system parameters (such as resistance and capacitance of the model neuron) vary with time and/or location. (They may also vary with the depolarization voltage level. In that case, we would have a nonlinear PDE for which the method of eigenfunctions is not applicable). For linear PDE with variable coefficients subject to random forcing, we need a different approach to the solution statistics, and most likely some numerical method of solution after reduction to a deterministic problem.

For the case of a general system of linear PDE in the form

$$\mathbf{u}_{,t} = L[\mathbf{u}] + \mathbf{f}(\mathbf{x}.t), \quad (\ )_{,z} = \frac{\partial(\ )}{\partial z}$$

where $\mathbf{u}$ and $\mathbf{f}$ are vector functions of the time variable t and space variable $\mathbf{x}$, the latter may be a scalar or vector depending on the number of spatial dimensions. When the unknown $\mathbf{u}$ is a vector function, $L[\cdot]$ would generally be a matrix whose elements are linear differential operators involving only spatial derivatives of the components of unknown vector response and coefficients possibly dependent on the time and spatial variables. Some examples of such PDE systems can be found in [68] where an effective general method of solution was formulated for such problems. We begin a description of this method for the simpler scalar PDE

$$u_{,t} = L_x[u] + f(x,t) \quad (0 < x < \ell, \ t > 0) \tag{13.2.1}$$

where the subscript x in L_x denotes the fact that the differential operator involves only spatial derivatives. For the equation of heat conduction, we have $L_x[u] = u_{,xx}$. For the cable model neuron, we have $L_x[u] = u_{,xx} - u$. The given PDE is general augmented by appropriate boundary and an initial conditions. For illustrative purposes, we take them to be the homogeneous (Dirichlet) end conditions

$$u(0,t) = u(\ell,t) = 0, \quad (t > 0)$$

and a vanishing initial state at $t = 0$

$$u(x,0) = 0 \quad (0 \le x \le \ell).$$

With the forcing term $f(x,t)$ being the only stochastic process with limited known statistics, we often have to settle for similarly limited statistical information about the output. For example, we may have a white noise excitation with drift applied at a point x_0,

$$f(x,t) = [a + W(t)] \, \delta(x - x_0)$$

where a is a known constant. In that case, we have

$$\langle f(x,t) \rangle = a\delta(x - x_0) \equiv \mu_f(x,t),$$

$$\langle f(x,t)f(y,s) \rangle = [a^2 + D\delta(t - s)] \, \delta(x - x_0)\delta(y - x_0) \equiv C_f(x,t;y,s).$$

In other cases, $f(x,t)$ may be known only as a second order stochastic process with known mean and correlation

$$\langle f(x,t) \rangle = \mu_f(x,t), \quad \langle f(x,t)f(y,s) \rangle = C_f(x,t;y,s).$$

Upon taking the expected value of both sides of the PDE, we get a deterministic PDE for the mean response $\mu(x,t) = \langle u(x,t) \rangle$

$$\mu_{,t} = L_x[\mu] + \mu_f(x,t),$$

$$\mu(0,t) = \mu(\ell,t) = \mu(x,0) = 0.$$

This is a familiar deterministic problem for which we know how to solve analytically or numerically. To the extent that the PDE is linear, so that superposition principles apply, and solving for the mean $\mu(x,t)$ is straightforward, we will henceforth assume that the forcing $f(x,t)$ is of zero mean so that the response $u(x,t)$ is also of zero mean. In this way, we may focus our attention to the second order statistics such as variance and correlation of $u(x,t)$.

13.2.2   The adjoint Green's function

In the life sciences, system characteristics do change with time over the life span of a biological organism. For many problems of interest, the system properties (such as the structure of the wing imaginal disc of a Drosophila) often vary with spatial location. For these cases, the coefficients of the linear spatial differential operator L would be function of spatial and/or temporal variable so that the method of eigenfunctions are generally not applicable. For a different kind of method of solution for these problem, we need to develop a Green's function representation for the solution of IBVP in the absence of eigen-pairs for its construction (as was done in the previous chapter).

When $f(x,t)$ is a known function, the solution of the linear IBVP has been found to have a Green's function representation at least for the cable model neuron in the last chapter. As such, we generally expect there be a similar integral representation for the more general case. Once we have the Green's function representation similar to that for the cable model for a neuron, the response statistics can be calculated as ordinary integrals of the input statistics. The key then is to determine an appropriate Green's function for the problem on hand (with or without eigenfunctions).

We illustrate the development for determining the needed Green's function and integral representation for the following IBVP:

$$u_{,t} = L_x[u] + f(x.t) \quad (0 < x < \ell, \ t > 0) \tag{13.2.2}$$

with

$$u(0, t) = u(\ell, t) = 0 \quad (t > 0), \quad u(x, 0) = u^o(x) \quad (0 \le x \le \ell) \tag{13.2.3}$$

where

$$L_x[u] = (p(x, t)u_{,x})_{,,x} - q(x, t)u \tag{13.2.4}$$

(with appropriate conditions on the continuity, differentiability and positiveness of the known functions p and q similar to those in the standard Sturm–Liouville theory). For most problems in application, p and q are usually independent of the time variable. However, real life problems with p and q depending on t and/or x do exist. It is only a matter whether or not the time variation involved is relevant to the duration of the event of interest, i.e., whether or not there is significant variation within the time frame of the phenomenon being investigated.

For the problem (13.2.2)–(13.2.3), we define an *adjoint Green's function* $G^*(x, t; y^*, s^*)$ to be the solution of the *terminal value problem*

$$-G^*_{,t}(x, t; y^*, s^*) = L_x[G^*] + \delta(x - y^*)\delta(t - s^*) \quad (0 < x, y^* < \ell, \ t > 0)$$

subject to the auxiliary conditions

$$G^*(0, t; y^*, s^*) = G^*(\ell, t; y^*, s^*) = 0, \quad (t < T),$$
$$G^*(x, t; y^*, s^*) = 0 \quad (0 \le x \le \ell, \ t > s^*).$$

Upon integrating the bilinear form

$$G^*\{u_{,t}(x, t) - L_x[u]\} - u(x, t)\{-G^*_{,t} - L_x[G^*]$$

over $0 < x < \ell$ and $0 < t < T$ for $T > s^*$, we get after integration by parts and applications of the homogeneous auxiliary conditions

$$u(y^*, s^*) = \int_0^\ell \int_0^{s^*} G^*(x, t; y^*, s^*)f(x, t)dtdx + \int_0^\ell G^*(x, 0; y^*, s^*)u^o(x)dx \tag{13.2.5}$$

where we have made use of the (causality) condition $G^*(x, t; y^*, s^*) = 0$ for $0 \le x \le \ell$ and $t > s^*$.

13.2.3   The Green's function representation

While (13.2.5) is an integral representation for the unknown $u(x,t)$, it involves solving a backward heat equation type terminal value problem which is unconventional and undesirable. We will work to transform it into an integral representation involving the actual Green's function defined to be the solution of the PDE

$$G_{,t}(x,t;y,s) = L_x[G] + \delta(x-y)\delta(t-s) \quad (0 < x, y < \ell,\ t > 0) \tag{13.2.6}$$

and the auxiliary conditions

$$G(0,t;y,s) = G(\ell,t;y,s) = 0, \quad (t < T), \tag{13.2.7}$$

$$G(x,t;y,s) = 0 \quad (0 \le x \le \ell,\ t \le s). \tag{13.2.8}$$

We begin by applying the adjoint Green's function representation to the problem for $G(x,t;y,s)$ to get the following reciprocity relation:

Proposition 49. $G(y^*, s^*; y, s) = G^*(y, s; y^*, s^*).$

Proof. Apply (13.2.5) to the problem that defines the Green's function to get

$$G(y^*, s^*; y, s) = \int_0^\ell G^*(x,t;y^*,s^*)\delta(x-y)\delta(t-s)dtdx$$

$$= G^*(y, s; y^*, s^*)H(s^* - s)$$

keeping in mind (13.2.7)–(13.2.8). □

Note that the reciprocity relation preserves causality for both G and G^*, i.e.,

$$G(x,t;y,s) = 0 \quad (t < s), \quad G^*(x,t;y,s) = 0 \quad (t > s).$$

Corollary 16. The IBVP for $u(x,t)$ has the following Green's function representation:

$$u(x,t) = \int_0^\ell \int_0^t G(x,t;y,s)f(y,s)dsdy + \int_0^\ell G(x,t;y,0)u^\circ(y)dy \tag{13.2.9}$$

Proof. Apply Proposition 49 to the representation (13.2.5) and obtain (13.2.9) after a change of notation. □

We will also need the following property for the Green's function:

Proposition 50. $G(x, s^+; y, s) = \delta(x - y).$

To prove this property, we let $Q(x, t; y, s)$ be the solution of the IBVP

$$Q_{,t} = L_x[Q] \quad (0 < x < \ell, \ t > s)$$

with

$$u(0, t) = u(\ell, t) = 0, \quad (t > s), \quad u(x, s^+) = \delta(x - y) \quad (0 \le x \le \ell).$$

We now set

$$G(x, t; y, s) = Q(x, t; y, s)H(t - s).$$

Lemma 13. $G(x, t; y, s) = Q(x, t; y, s)H(t-s)$ *is the Green's function for the IBVP* (13.2.6)–(13.2.7).

Proof. $G(x, t; y, s)$ as given in the hypotheses satisfies the causality condition $G(x, t; y, s) = 0$ for $t < s$. It satisfies the two boundary conditions at $x = 0$ and $x = \ell$ because Q does. It remains to show that G satisfies the PDE (13.2.6). But

$$G_{,t}(x, t; y, s) - L_x[G] = Q(x, t; y, s)\delta(t - s)$$

$$= Q(x, s^+; y, s)\delta(t - s) = \delta(x - y)\delta(t - s). \qquad \square$$

Proposition (50) now follows from the lemma above given

$$G(x, s^+; y, s) = [Q(x, t; y, s)H(t - s)]_{t=s^+} = Q(x, s^+; y, s) = \delta(x - y).$$

13.3 Mean and Correlation

13.3.1 *The storage problem*

When the forcing term $f(x, t)$ in (13.2.2) is a stochastic process $F(x, t)$ with known statistics, the Green's function representation allows us to compute the statistics of the resulting response s.p., now denoted by $U(x, t)$, in terms of the statistics of $F(x, t)$. For examples, we have for cases with $u^o(x) = 0$

$$\langle U(x, t) \rangle = \int_0^\ell \int_0^t G(x, t; \xi, \tau)\langle F(\xi, \tau) \rangle d\xi d\tau,$$

$$C(x, t; y, s) = \int_0^\ell \int_0^s \int_0^\ell \int_0^t G(x, t; \xi, \tau)G(y, s; \eta, \sigma)\langle F(\xi, \tau)F(\eta, \sigma) \rangle d\xi d\tau d\eta d\sigma,$$

etc. While evaluating the multiple integrals may be manageable for the corresponding ODE problem, computing the response statistics by way of the Green's function representation is often not practical for PDE problems. The operation counts and storage requirement (particularly the latter) are excessive even for the illustrative example of a single scalar equation in one spatial dimension such as

(13.2.2)–(13.2.3). The corresponding requirements for vector unknowns in higher spatial dimensions are often beyond available computing capacity today. Whether or not we have sufficient storage capacity or can do the calculation, it is always desirable to find a more efficient method of solution.

To reduce the storage requirement, we can also formulate conventional IBVP for the response statistics to be solved numerically by a suitable method. Our experience with the ODE case (in Chapter 11) suggests that we should start by computing the "variance" of the solution (having already found the first moment, either by calculating it or by knowing that the forcing is of zero mean).

13.3.2 Correlation function

As we proceed to formulate an efficient method for finding the response statistics (by the approach used for ODE), it soon becomes clear that for PDE problems, the first step in this process is *not* to calculate the response variance which, assuming zero mean response, would be $\langle U^2(x,t)\rangle$. Instead, we should calculate the spatial correlation function $S(x,y,t) = \langle U(x,t)U(y,t)\rangle$. To see why, we start with an IBVP for the correlation function $C(x,t;y,s) = \langle U(x,t)U(y,s)\rangle$ by observing that it satisfies the PDE

$$\frac{\partial C(x,t;y,s)}{\partial t} = L_x[C] + C_{Fu}(x,t;y,s) \tag{13.3.1}$$

where

$$C_{FU}(x,t;y,s) = \langle F(x,t)U(y,s)\rangle = C_{UF}(y,s;x,t)$$

can be expressed in terms of the forcing function with the help of the Green's function representation

$$C_{FU}(x,t;y,s) = \int_0^s \int_0^\ell G(y,s,\xi,\tau)\langle F(\xi,\tau)F(x,t)\rangle d\xi d\tau. \tag{13.3.2}$$

However, numerical evaluation of the double integral is unattractive and inefficient. As we have learned from the ODE case, we could avoid evaluating this integral if $F(x,t)$ is temporally uncorrelated.

13.4 Temporal White Noise Excitation

13.4.1 Temporally uncorrelated excitation

When $F(x.t)$ is temporally uncorrelated (so that it is white noise in time) with

$$\langle F(x.t)F(y,s)\rangle = r_{FF}(x,y)\delta(t-s). \tag{13.4.1}$$

we have the following key observation for the development of an efficient method of solution:

Proposition 51. *If the random forcing is temporally uncorrelated so that* (13.4.1) *holds, then*

$$(i) \quad C_{FU}(x,t;y,s) = 0 \quad (t > s) \tag{13.4.2}$$

and therewith

$$(ii) \quad \frac{\partial C(x,t;y,s)}{\partial t} = L_x[C] \quad (0 < x < \ell, \ 0 < s < t). \tag{13.4.3}$$

Proof. With (13.4.1), the expression (13.3.2) for the cross-correlation $C_{FU}(x,t;y,s)$ reduces to

$$C_{FU}(x,t;y,s) = \int_0^\ell G(y,s,\xi,t) r_{FF}(\xi,x) d\xi = 0 \quad (s < t)$$

which is (i). Application of (i) in (13.3.1) gives (ii). □

To have an IBVP for $C(x,t;y,s)$, we need to augment the PDE (13.4.3) with appropriate auxiliary conditions. The boundary conditions

$$C(0,t;y,s) = C(\ell,t;y,s) = 0$$

follow from the given boundary conditions for $U(x,t)$. Since the PDE only holds for $t > s$, we need an initial condition for $C(x,t;y,s)$ at $t = s$ in the form

$$C(x,s;y,s) = \langle U(x,s)U(y,s)\rangle = S(x,y,s)$$

where $S(x,y,s)$ is known as the *spatial correlation function* of the response process $U(x,t)$. Our task is then to find $S(x,y,t)$ that, for the PDE problem, plays the role of the variance $\langle U^2(x,t)\rangle$ in the new efficient method of Chapter 11 for the second order statistics for the ODE case.

13.4.2 *Spatial correlations*

To formulate a deterministic problem for the spatial correlation function, we differentiate $S(x,y,t)$ partially with respect to t to get

$$S_{,t}(x,y,t) = \langle U_{,t}(x,t)U(y,t)\rangle + \langle U(x,t)U_{,t}(y,t)\rangle$$
$$= L_x[S] + \langle F(x,t)U(y,t)\rangle + L_y[S] + \langle U(x,t)F(y,t)\rangle \tag{13.4.4}$$

where the first two terms on the right side resulted from using the PDE (13.2.2) to eliminate $U_{,t}(x,t)$ from the expression $\langle U_{,t}(x,t)U(y,t)\rangle$ while the last two terms come from doing the same with $U_{,t}(y,t)$ in $\langle U(x,t)U_{,t}(y,t)\rangle$. Upon writing

$$C_m(x,y,t) \equiv \langle F(x,t)U(y,t)\rangle + \langle U(x,t)F(y,t)\rangle = C_m(y,x,t),$$

we obtain the following result for $C_m(x,y,t)$:

Lemma 14. *Suppose that the forcing process is (zero mean and) delta correlated in time with autocorrelation function (13.4.1). Then $C_m(x,y,t) = r_{FF}(x,y)$ for the IBVP (13.2.2)–(13.2.3) with $u^o(x)$ is either independent of $F(x,t)$ or $= 0$.*

Proof. By the Green's function representation, the term $\langle U(x,t)F(y,t)\rangle$ becomes

$$\langle U(x,t)F(y,t)\rangle = \int_0^\ell \int_0^t G(x,t,\xi,\tau)\langle F(\xi,\tau)F(y.t)\rangle d\tau d\xi$$

$$= \int_0^\ell \int_0^t G(x,t,\xi,\tau)r_{FF}(\xi,y)\delta(t-\tau)d\tau d\xi$$

$$= \frac{1}{2}\int_0^\ell G(x,t,\xi,t)r_{FF}(\xi,y)d\xi$$

$$= \frac{1}{2}\int_0^\ell \delta(x-\xi)r_{FF}(\xi,y)d\xi = \frac{1}{2}r_{FF}(x,y).$$

The other term $\langle F(x,t)U(y,t)\rangle$ can be similarly calculated to arrive at the same result to complete the proof of the lemma. □

With the help of this lemma, we obtain the following principal result for the determination of the spatial correlation function:

Theorem 43. *The spatial correlation function of the response $U(x,t)$ of the IBVP (13.2.2)–(13.2.3) with $u^o(x)$ independent of $F(x,t)$ (or $= 0$) is determined by the IBVP*

$$S_{,t}(x,y,t) = L_x[S] + L_y[S] + r_{FF}(x,y) \quad (0 < x,y < \ell,\ t > 0) \quad (13.4.5)$$

$$S(0,y,t) = S(\ell,y,t) = S(x,0,t) = S(x,\ell,t) = 0 \quad (0 < x,y < \ell,\ t > 0) \quad (13.4.6)$$

$$S(x,y,0) = 0 \quad (0 < x,y < \ell). \quad (13.4.7)$$

Proof. The PDE is a consequence of Lemma 14 and equation (13.4.4). The auxiliary conditions are consequence of the definition of $S(x,y,t)$ and the auxiliary conditions (13.2.3). □

The IBVP for $S(x,y,t)$ stated in Theorem 43 is a deterministic problem that can be solved quite simply by known numerical methods, even if analytical techniques do not apply. At each instant $t > 0$, only the values of $S(x,y,t)$ at the grid points

(x_i, y_j) of the $[0, \ell] \times [0, \ell]$ region of the x, y-plane are calculated and stored in order to move forward (possibly for two consecutive time steps for an implicit method). In particular, the solution values at the diagonal grid points give the second moment $\langle U^2(x_j, t) \rangle = S(x_j, x_j, t)$ (equal to the corresponding variance since the solution is of zero mean).

Example 75. For the IBVP (13.2.2)–(13.2.3) $p(x, t) = 1$ and $q(x, t) = 0$ in (13.2.4) (so that $L_x[U] = U_{,xx}$), solve by eigenfunction expansions the IBVP with $u^o(x) = 0$ and a temporally uncorrelated forcing function $F(x, t)$.

By the method of separation of variables, it is straightforward to find the (normalized) eigen-pair for the spatially two-dimensional differential operator to be

$$\{\phi_{mn}(x, y), \lambda_{mn}^2\} = \left\{ \frac{2}{\ell} \sin\left(m\pi\frac{x}{\ell}\right) \sin\left(n\pi\frac{y}{\ell}\right), (m^2 + n^2)\left(\frac{\pi}{\ell}\right)^2 \right\},$$

$$(m, n = 1, 2, 3, \ldots)$$

with the orthogonality conditions

$$\int_0^\ell \int_0^\ell \phi_{mn}(x, y)\phi_{ij}(x, y)dxdy = 0, \quad (i \neq m \ \text{ or } \ j \neq n)$$

and

$$\int_0^\ell \int_0^\ell \phi_{mn}^2(x, y)dxdy = 1.$$

Let

$$\{S(x, y, t), r_{ff}(x, y)\} = \sum_{m=1}^\infty \sum_{n=1}^\infty \{S_{mn}(t), r_{mn}\} \phi_{mn}(x, y)$$

(with the series for $S(x, y, t)$ satisfying the four edge conditions in (13.4.6)) and substitute into PDE (13.2.2) to obtain

$$S'_{mn}(t) + \lambda_{mn}^2 S_{mn}(t) = r_{mn} \quad (m, n = 1, 2, 3, \ldots)$$

where

$$r_{mn} = \int_0^\ell \int_0^\ell r_{ff}(x, y)\phi_{mn}(x, y)dxdy.$$

The ODE for $S_{mn}(t)$ is supplemented by the initial condition

$$S_{mn}(0) = 0 \quad (m, n = 1, 2, 3, \ldots).$$

Altogether, the solution for $S(x, y, t)$ is then found to be

$$S(x, y, t) = \sum_{m=1}^{\infty} \sum_{n=1}^{\infty} \frac{1 - e^{-\lambda_{mn}^2 t}}{\lambda_{mn}^2} r_{mn} \phi_{mn}(x, y)$$

with the variance $\langle U^2(x, t) \rangle = S(x, x, t)$ or

$$\langle U^2(x, t) \rangle = \frac{4}{\ell^2} \sum_{m=1}^{\infty} \sum_{n=1}^{\infty} \frac{1 - e^{-\lambda_{mn}^2 t}}{\lambda_{mn}^2} r_{mn} \sin\left(m\pi \frac{x}{\ell}\right) \sin\left(n\pi \frac{x}{\ell}\right).$$

13.4.3 *IBVP for $S(x, y, t)$ by numerical methods*

The usefulness of the spatial correlation method is for problems when eigenfunction expansions are not applicable or not practical. For such cases, we can still obtain the variance and correlation function of the response process of the SDE by the new method much more efficiently than by other numerical approaches. We will use the simple scalar SDE problem above to illustrate the gain in storage requirements (and the amount of computing).

We discretize the IBVP for $S(x, y, t)$ for an equally spaced mesh on the square $[0, \ell] \times [0, \ell]$, taking $\ell = 1$ for simplicity, with the mesh points (x_k, y_j) given by

$$x_k = k\Delta x = \frac{k}{M}, \quad y_j = j\Delta y = \frac{j}{N} \quad (k = 0, 1, 2, \ldots, M; \ j = 0, 1, 2, \ldots, N).$$

Let $S_{ij}(t) = S(x_i, y_j, t)$ and store $\{S_{ij}(t)\}$ as an $M \times N$ matrix $[S_{ij}(t)]$. After central differencing the first and second derivatives of $S(x, y, t)$, we obtain from the IBVP for $S(x, y, t)$ (with $u^o(x) = 0$) the following ODE system for the matrix $S(t) = [S_{ij}(t)]$

$$S' = AS + SB + [r_{ff}(x_i, y_j)], \quad S(0) = O,$$

where A and B are an $M \times M$ and $N \times N$ tri-diagonal matrix, respectively, with:

$$A = \begin{bmatrix} -2 & 2 & 0 & \cdot & \cdot & 0 & 0 \\ 1 & -2 & 1 & 0 & \cdot & \cdot & 0 \\ 0 & 1 & \cdot & \cdot & \cdot & \cdot & \cdot \\ \cdot & \cdot & \cdot & \cdot & \cdot & 0 & \cdot \\ \cdot & \cdot & \cdot & \cdot & \cdot & 1 & 0 \\ 0 & \cdot & \cdot & \cdot & 1 & -2 & 1 \\ 0 & 0 & \cdot & \cdot & 0 & 2 & -2 \end{bmatrix}$$

and B having a similar structure. This is an IVP for the matrix ODE solved in Chapter 11 with minimal computing and storage requirement. Similar gains in computing and storage requirements are possible for general linear PDE system with random forcing via the spatial correlation method.

13.5 Linear Stochastic PDE Systems

13.5.1 *An example*

If instead of the parabolic IBVP (13.2.2)–(13.2.3), we are interested in the hyperbolic type IBVP

$$U_{,tt} = L_x[U] + F(x.t) \qquad\qquad (0 < x < \ell,\ t > 0), \qquad (13.5.1)$$

$$U(0,t) = U(\ell,t) = 0 \qquad\qquad (t > 0) \qquad\qquad (13.5.2)$$

$$U(x,0) = u^o(x), \quad U_{,t}(x,0) = v^o(x) \quad (0 \le x \le \ell) \qquad (13.5.3)$$

where L_x is again the second order differential operator as defined in (13.2.4). For this problem, we can develop an IBVP for the spatial correlation function $S(x,y,t)$ for the unknown s.p. $U(x,t)$ following the steps taken for the parabolic problem with suitable modification. However, it is more efficient to re-cast the new problem into a form similar to that of the parabolic case. This would allow us to apply a vector version of all the results obtained in the previous section with minimal modification.

To recast the new problem (13.5.1)–(13.5.3) into parabolic form (but not in substance), we set $U_1 = U(x,t)$ and $U_2(x,t) = U_{,t}(x,t)$. With $\mathbf{U}(x,t) = (U_1, U_2)^T$, we may rewrite the new IBVP as

$$\mathbf{U}_{,t}(x,t) = \mathbb{L}_x[\mathbf{U}] + \mathbf{F}(x,t) \quad (0 < x < \ell,\ t > 0) \qquad (13.5.4)$$

$$B_0[\mathbf{U}(0,t)] = B_\ell[\mathbf{U}(\ell,t)] = 0 \quad (t > 0) \qquad (13.5.5)$$

$$\mathbf{U}(x,0) = \mathbf{U}^o(x) \equiv (u^o(x), v^o(x))^T \quad (0 < x < \ell), \qquad (13.5.6)$$

where

$$\mathbf{F}(x,t) = \begin{pmatrix} F_1(x,t) \\ F_2(x,t) \end{pmatrix} = \begin{pmatrix} 0 \\ F(x,t) \end{pmatrix}, \quad \mathbb{L}_x[\mathbf{U}] = \begin{bmatrix} 0 & 1 \\ L_x & 0 \end{bmatrix} \begin{pmatrix} U_1 \\ U_2 \end{pmatrix}$$

$$B_0[\mathbf{U}(0,t)] = \begin{bmatrix} 1 & 0 \\ 0 & 0 \end{bmatrix} \begin{pmatrix} U_1(0,t) \\ U_2(0,t) \end{pmatrix}, \quad B_\ell[\mathbf{U}(\ell,t)] = \begin{bmatrix} 1 & 0 \\ 0 & 0 \end{bmatrix} \begin{pmatrix} U_1(\ell,t) \\ U_2(\ell,t) \end{pmatrix}.$$

Now that both $\mathbf{U}$ and $\mathbf{F}$ are vectors, there are several new correlation functions associated with the different components of these vectors. For example, the spatial correlation function of $\mathbf{U}$ is now

$$\langle \mathbf{U}(x,t)\mathbf{U}^T(y,t)\rangle = [\langle U_i(x,t)U_j(y,t)\rangle] = [S_{ij}(x,y,t)] \tag{13.5.7}$$

with $S_{21}(x,y,t) = S_{12}(y,x,t)$. Calculations similar to the scalar case leads to the following IBVP for $S(x,y,t)$

$$S_{,t} = \mathbb{L}_x[S] + (\mathbb{L}_y[S])^T + \begin{bmatrix} 0 & 0 \\ 0 & r_{ff}(x,y) \end{bmatrix} \quad (0 < x,y < \ell,\ t > 0)$$

with the auxiliary conditions (13.4.6) and (13.4.7).

13.5.2   *The general case*

For the general case where $\mathbf{U}(x,t)$ is a k component vector, a similar set of calculations lead to the following PDE

$$S_{,t} = \mathbb{L}_x[S] + \left(\mathbb{L}_y[S^T]\right)^T + R_{FF}(x,y) \quad (0 < x,y < \ell,\ t > 0) \tag{13.5.8}$$

for the $k \times k$ spatial correlation matrix function

$$S(x,y,t) = \langle \mathbf{U}(x,t)\mathbf{U}^T(y,t)\rangle = [S_{ij}(x,y,t)], \tag{13.5.9}$$

where $R_{FF}(x,y) = [r_{ij}(x,y)]$ is a $k \times k$ matrix function associated with the temporally uncorrelated vector random forcing

$$\langle \mathbf{F}(x,t)\mathbf{F}^T(y,s)\rangle = [r_{ij}(x,y)\delta_{ij}(t-s)]. \tag{13.5.10}$$

To show this, we note that

$$S_{,t} = \langle \mathbf{U}(x,t)\mathbf{U}^T(y,t)\rangle_{,t} = \langle \mathbf{U}_{,t}\,\mathbf{U}^T\rangle + \langle \mathbf{U}(\mathbf{U}^T)_{,t}\rangle$$

with the first term easily seen to be

$$\langle \mathbf{U}_{,t}\,\mathbf{U}^T\rangle = \mathbb{L}_x[S] + \langle \mathbf{F}(x,t)\mathbf{U}^T(y,t)\rangle.$$

For the second term, we have

$$\langle \mathbf{U}(x,t)\left\{\mathbf{U}^T(y,t)\right\}_{,t}\rangle = \langle \mathbf{U}(x,t)\left\{\mathbf{U}_{,t}(y,t)\right\}^T\rangle$$

$$= \langle \mathbf{U}(x,t)\left\{\mathbb{L}_y\left[\mathbf{U}(y,t)\right] + \mathbf{F}(y,t)\right\}^T\rangle$$

$$= \left(\mathbb{L}_y[S^T]\right)^T + \langle \mathbf{U}(x,t)\mathbf{F}^T(y,t)\rangle.$$

The result of the following exercises will be needed for completing the derivation of (13.5.8):

Exercise 61. Show $G(x,s_+;y,s) = I\delta(x-y)\ (s_+ > s)$ (with $G(x,s;y,s) = \frac{1}{2}I\delta(x-y)$).

(Hint: Let $W(x,t)$ be the solution of $W_{,t} = \mathbb{L}_x[W(x,t)]$, $(t > t')$ with $W(x,t') = I\delta(x - x')$ and satisfy the appropriate boundary conditions for the problem. Show $G(x,t;x',t') = H(t - t')W(x,t)$.)

To see the relevance and importance of this result, we re-write the wave equation $U_{,tt} = U_{,xx} + f(x,t)$ for a deterministic forcing $f(x,t)$ as a system of two equations by setting $V = U_{,t}$ and $\mathbf{U}(x,t) = (U, V)^T$. In this way, the single second order PDE in time is equivalent to the following system that is first order in time:

$$\mathbf{U}_{,t} = \begin{bmatrix} 0 & 1 \\ \dfrac{\partial^2}{\partial s^2} & 0 \end{bmatrix} \mathbf{U} + \begin{pmatrix} 0 \\ F(x,t) \end{pmatrix} = \mathbb{L}_x[\mathbf{U}] + \mathbf{F}(x,t).$$

For a system initially at rest, the Green's function representation for this system is

$$\mathbf{U}(x,t_+) = \int_0^{t_+} \int_0^\ell G(x,t_+;\xi,\tau)\mathbf{F}(\xi,\tau)d\xi d\tau.$$

To see this, we differentiate both sides with respect to t to get for t_+ just above t

$$\mathbf{U}_{,t}(x,t) = \int_0^t \int_0^\ell \{\mathbb{L}_x[G(x,t;\xi,\tau)] + I\delta(x-\xi)\delta(t-\tau)\}\mathbf{F}(\xi,\tau)d\xi d\tau$$

$$+ \int_0^\ell G(x,t;\xi,t)\mathbf{F}(\xi,t)d\xi$$

$$= \mathbb{L}_x\left[\int_0^t \int_0^\ell G(x,t;\xi,\tau)\mathbf{F}(\xi,\tau)d\xi d\tau\right] + I\mathbf{F}(x,t)$$

$$= \mathbb{L}_x[\mathbf{U}(x,t)] + \mathbf{F}(x,t).$$

keeping in mind $G(x,t;\xi,t) = \frac{1}{2}I\delta(x-\xi)$ with I being the identity matrix.

Exercise 62. For the IBVP (13.5.4)–(13.5.6). show $\langle \mathbf{F}(x,t)\mathbf{U}^T(y,t)\rangle + \langle \mathbf{U}(x,t)\mathbf{F}^T(y,t)\rangle = R_{FF}(x,y)$.

The PDE (13.5.8) is augmented by the auxiliary conditions (13.4.6) and (13.4.7) now for the $k \times k$ spatial correlation matrix function (13.5.9). Once $R_{FF}(x,y)$ is specified, the IBVP for $S(x,y,t)$ can be solved by available analytical or numerical methods.

13.5.3   *The correlation matrix*

Recall that the method of spatial correlation was motivated by the need for an initial condition for the determination of the correlation function of the stochastic solution for the IBVP with random forcing when the excitation is temporally uncorrelated. For a vector excitation, the correlation matrix for the vector unknown is given by $\langle \mathbf{U}(x,t)\mathbf{U}^T(y,s)\rangle = C(x,t;y,s)$ where $C_{ij}(x,t;y,s) = \langle U_i(x,t)U_j(y,s)\rangle$. For a vector unknown with two components, we have

$$C(x,t;y,s) = \begin{bmatrix} C_{11}(x,t;y,s) & C_{12}(x,t;y,s) \\ C_{21}(x,t;y,s) & C_{22}(x,t;y,s) \end{bmatrix}$$

with $C_{21}(x,t;y,s) = C_{12}(y,s;x,t)$. For an n-vector $\mathbf{U}(x,t)$ function, the PDE for the correlation (matrix) function is obtained by post-multiplying the given PDE for $\mathbf{U}(x,t)$ by $\mathbf{U}^T(y,s)$ and ensemble averaging the resulting equation to get

$$C_{,t}(x,t;y,s) = \langle \mathbf{U}_{,t}(x,t)\mathbf{U}^T(y,s)\rangle$$
$$= \langle \{\mathbb{L}_x[\mathbf{U}] + \mathbf{F}(x,t)\}\,\mathbf{U}^T(y,s)\rangle$$
$$= \mathbb{L}_x[C] + M(x,t;y,s)$$

where

$$M(x,t;y,s) = \left[\langle \mathbf{F}(x,t)\mathbf{U}^T(y,s)\rangle\right] = [\langle F_i(x,t)U_j(y,s)\rangle] = [M_{ij}(x,t;y,s)]$$

with $M_{ij}(x,t;y,s) = M_{ji}(y,s;x,t)$.

We may now proceed as in the scalar case by specializing the forcing to be temporally uncorrelated with

$$M^T(x,t;y,s) = \int_0^s \int_0^\ell G(y,s,\xi,\tau)\langle \mathbf{F}(\xi,\tau)\mathbf{F}^T(x,t)\rangle d\xi d\tau.$$

$$= \int_0^s \int_0^\ell G(y,s,\xi,\tau)[R_{ij}(\xi,x)\delta_{ij}(\tau - t)]d\xi d\tau = O \quad (t > s)$$

to get

$$C_{,t}(x,t;y,s) = \mathbb{L}_x[C] \quad (t > s). \tag{13.5.11}$$

The deterministic PDE for $C(x,t;y,s)$ is augmented by the auxiliary conditions

$$B_{0x}[C(0,t;y,s)] = B_{\ell x}[C(\ell,t;y,s)] = 0 \quad (t > s) \tag{13.5.12}$$

and

$$C(x,s;y,s) = \langle \mathbf{U}(x,s)\mathbf{U}^T(y,s)\rangle \equiv S(x,y,s) \quad (0 < x,y < \ell,\ t > s) \tag{13.5.13}$$

which are consequences of the given auxiliary conditions on $\mathbf{U}(x,s)$ and the definition of the correlation function. In (13.5.12), $B_{\alpha x}$ is the boundary operator B_α, $\alpha = 0$ or ℓ (see (13.5.5)), operating in the (x,t) space with (y,s) as parameters.

Note that the spatial correlation function $S(x, y, t)$ is now a matrix function

$$S(x, y, t) = \langle \mathbf{U}(x, t)\mathbf{U}^T(y, t) \rangle = \langle [U_i(x, t)U_j(y, t)] \rangle$$

that can be determined by the spatial correlation method formulated in this section.

Proposition 52. *Given the spatial correlation matrix function $S(x, y, \tau)$, the correlation matrix function $C(x, t; y, \tau)$ is determined by the IBVP (13.5.11)–(13.5.13).*

Proof. We have already established the proposition for $t > \tau$. The result for $t < \tau$ is obtained with the help of the symmetry condition $C(x, t; y, \tau) = C^T(y, \tau; x, t)$. ☐

13.5.4 *Correlated excitations*

If the random excitation $\mathbf{F}(x, t)$ is not delta-correlated in time, it often is (or can be approximated by) a filtered white noise. That is, $\mathbf{F}(x, t)$ is in the stochastic response of some SDE with temporally uncorrelated input. This is similar to the what has been discussed for ODE with random excitation except the response sought is determined by one or more PDE. For the scalar case such as (13.2.1), the random forcing $f(x, t)$ would be the output of the differential equation

$$f_{,t} = D_x[f(x, t)] + \Phi(x, t) \tag{13.5.14}$$

where $\Phi(x, t)$ is temporally uncorrelated so that

$$\langle \Phi(x, t)\Phi(y, s) \rangle = r_{\Phi\Phi}(x, y)\delta(t - s),$$

and D_x is a differential operator involving only spatial derivatives. In that case, the SDE (13.5.14) for $F(x, t)$ may be appended to the original PDE for $u(x, t)$ to get the vector PDE

$$\mathbf{U}_{,t}(x, t) = \mathbf{L}_x[\mathbf{U}] + \mathbf{F}(x, t)$$

(similar in form to (13.5.4)) with

$$\mathbf{U}(x, t) = (u(x, t), f(x, t))^T, \quad \mathbf{F}(x, t) = (0, \Phi(x, t))^T$$

and

$$\mathbf{L}_x[\mathbf{U}(x, t)] = \begin{bmatrix} \mathbb{L}_x & 1 \\ 0 & D_x \end{bmatrix} \mathbf{U}.$$

13.6 Cable Model Neuron with O-U Input Current

It was pointed out in the section on cable model neuron that the effects of the spatial span of neurons such as spinal cord motoneurons can not be completely

reproduced by point models such as the (Stein type) leaky integrate-and-fire model
excited synaptically by a homogeneous Poisson process $N(t)$ (see [52, 74]):

$$C\frac{dV}{dt} = -\frac{V}{R} + I, \quad \frac{dI}{dt} = -\alpha I + A\frac{dN}{dt}. \tag{13.6.1}$$

The model involves rapid and significant changes in the output postsynaptic membrane potential and irregular interspike intervals. While its output exhibits finite
rise times as observed experimentally at the soma (the main body of the neuron
from which long dendrite and dendritic trees emanate), the other factor strongly
influencing the rise time to be included in the model is the electrotonic distance of
the activated synapse(s) from the soma. Early efforts to gain insight to the effects
of spatial span between signal activation and signal received at the neuron's soma
include the study of a cable model neuron subject to a white noise current injection
at a point along the span of the neuron reported in the first section of this chapter.
A more appropriate model that capture both type of effects on rise time would be
to replace the leaky integrator equation by the nerve cylinder equation for depolarization voltage and to approximate the synaptic current in the point model (13.6.1)
by a filtered white-noise current injection [65]:

$$V_{,t} = V_{,xx} - V + I\delta(x - x_0), \quad I_{,t} = -\alpha I + a + \sigma W(t) \tag{13.6.2}$$

with $\langle W(t)W(s)\rangle = \sigma^2\delta(t - s)$. With $\mathbf{U}(x, t) = (V(x, t), I(t))^T$, the above system
may be written as

$$\mathbf{U}_{,t} = L_x[\mathbf{U}] + \begin{pmatrix} 0 \\ a + \sigma W(t) \end{pmatrix} \tag{13.6.3}$$

where

$$L_x[\mathbf{U}] = \begin{bmatrix} \frac{\partial^2}{\partial x^2} - 1 & \delta(x - x_0) \\ 0 & -\alpha \end{bmatrix} \mathbf{U}. \tag{13.6.4}$$

The cable equation (13.6.3) is again augmented by appropriate end condition and
an initial condition which we take to be

$$V_{,x}(0, t) = V_{,x}(\ell, t) = 0, \quad V(x, 0) = 0. \tag{13.6.5}$$

For simplicity, we work with the initial state

$$I(0) = 0. \tag{13.6.6}$$

13.6.1 Eigenfunction expansions

The problem provides an opportunity to show the finer details and the more subtle
aspects in the application of the new method. However, given the relatively simple
structure of the spatial differential operator, we first obtain the solution of the IBVP

by eigenfunction expansion in order to point out why we need more appropriate model for the problem than the simple white noise current investigated in the earlier section. For this purpose, we begin by taking advantage of the linearity of the IBVP and determine separately the mean and second moments of the depolaritzation voltage.

13.6.1.1 *The mean*

By taking the expectation of all the relations in the IBVP, we obtain the following deterministic IBVP for the mean response $\mu_V(x, t)$:

$$\mu_{V,t} = \mu_{V,xx} - \mu_V + \mu_I \delta(x - x_0),$$

$$\mu_{V,x}(0, t) = \mu_{V,x}(\ell, t) = 0 \quad (t > 0), \quad \mu_V(x, 0) = 0$$

and

$$\mu_{I,t} = -\alpha \mu_I + a, \quad \mu_I(0) = 0.$$

The solution of the deterministic IVP for the mean of $I(t)$ is

$$\mu_I(t) = \frac{a}{\alpha}\left(1 - e^{-\alpha t}\right).$$

Exercise 63. Obtain the following solution of the IBVP for the mean $\mu_V(x, t)$ by the method of eigenfunction expansions:

$$\mu_V(x, t) = \frac{a}{\alpha} \sum_{n=0}^{\infty} v_n(t) \phi_n(x) \phi_n(x_0)$$

where

$$v_n(t) = \begin{cases} \dfrac{1}{\lambda_n^2} - \dfrac{e^{-\alpha t}}{\lambda_n^2 - \alpha} + \dfrac{\alpha e^{-\lambda_n^2 t}}{\lambda_n^2(\lambda_n^2 - \alpha)} & (\lambda_n^2 \neq \alpha) \\[2ex] \dfrac{1}{\alpha}\left(1 - e^{-\alpha t}\right) - t e^{-\alpha t} & (\lambda_n^2 = \alpha). \end{cases}$$

13.6.1.2 *The variance*

For the variance $v(x, t) = E[\{V(x, t) - \mu_V(x, t)\}^2]$, we make use of the superposition principle and take $a = 0$ so that $\mu_V(x, t) = 0$ and $v(x, t) = E[\{V(x, t)\}^2]$. With

$$V(x, t) = \sum_{n=0}^{\infty} V_n(t) \phi_n(x)$$

where the relevant eigenfunctions $\{\phi_n(x)\}$ are as given in (13.1.9) and (13.1.10) with L replaced by ℓ. Orthogonality among the eigenfunctions reduces the PDE for

$V(x,t)$ to the system of ODE

$$\frac{dV_n}{dt} + \lambda_n^2 V_n = I(t)\phi_n(x_0), \quad V_n(0) = 0 \tag{13.6.7}$$

where

$$\frac{dI}{dt} = -\alpha I + \sigma W(t), \quad I(0) = 0 \tag{13.6.8}$$

or

$$I(t) = \sigma \int_0^t e^{-\alpha(t-s)} W(s)\,ds$$

with $\langle W(t)W(s)\rangle = \delta(t-s)$. For a fixed n, the IVP (13.6.7)–(13.6.8) for $\{V_n(t), I(t)\}$ is of the form addressed in the last section of the chapter on linear stochastic ODE with random forcing.

Exercise 64. With $a = 0$ (so that $I(t)$ and $V(x,t)$ are of zero mean), show that the variance $v(x,t)$ of $V(x,t)$ determined by the IBVP (13.6.2), (13.6.5) and (13.6.6) is given by

$$v(x,t) = \frac{\sigma^2}{2\alpha} \sum_{m=0}^{\infty} \sum_{n=0}^{\infty} v_{mn}(t)\phi_m(x)\phi_m(x_0)\phi_n(x)\phi_n(x_0)$$

where for $\alpha \neq 1$

$$v_{mn}(t) = \frac{\rho_{mn}}{\beta_{mn}} - \frac{e^{-2\alpha t}}{\gamma_m \gamma_n} + \frac{2\alpha e^{-\kappa_m t}}{\kappa_m \gamma_m \gamma_n} + \frac{2\alpha e^{-\kappa_n t}}{\kappa_n \gamma_n \gamma_m} - \frac{2\alpha e^{-\beta_{mn}t}}{\beta_{mn}\gamma_m \gamma_n}$$

with

$$\kappa_j = \lambda_j^2 + \alpha, \quad \gamma_j = \lambda_j^2 - \alpha, \quad \beta_{mn} = \lambda_m^2 + \lambda_n^2, \quad \rho_{mn} = \frac{1}{\kappa_m} + \frac{1}{\kappa_n}.$$

For the exceptional cases of $\alpha = 1$, we have instead

$$v_{00}(t) = \frac{1}{2}\left(1 - e^{-2t}\right) - (t + t^2)e^{-2t}$$

$$v_{m0}(t) = v_{0m}(t) = \frac{1}{\kappa_m}\left(\frac{1}{2} + \frac{1}{\kappa_m}\right) - \frac{e^{-2t}}{2\gamma_m}(1 + 2t) + \frac{e^{-\kappa_m t}}{\kappa_m \gamma_m}(1 + 2t)$$

with the other quantities remain unchanged.

13.6.2 *White noise vs. O-U noise forcing*

It is now clear from the results for the mean and variance of the response depolarization process that omitting the correlation in the input current process, approximating $I(t)$ as white noise) may lead to rather large differences in the subthreshold voltages for realistic choices of the parameters of both the neuron and the input current. Among the more striking differences is the variance of the depolarization voltage at the point of application of the input current. While the value of the variance at that point is unbounded for the temporally uncorrelated input current, the corresponding value for the OUP input current is finite.

The differences between the statistical properties of the membrane potential for the temporally uncorrelated (white noise) type input current process and those for the more realistic model with an OUP as the input current are considerable enough to make the earlier purely white noise approach sometimes questionable if accurate estimate are required. Based on the examples considered above, allowing for the exponential decay of the synaptic input current appears to be important if α is less than $O(10^{-1})$. More specific numerical differences between the outputs of the two types of input current can be found in [65].

13.6.3   The method of spatial correlation

As indicated previously, it would also be instructive to see the details of the spatial correlation method for this relative simple example for which we have an exact solution. For this problem, we have

$$S(x, y, t) = \left\langle \begin{pmatrix} V(x.t) \\ I(t) \end{pmatrix} (V(y, t), I(t)) \right\rangle = [S_{ij}(x, y, t)]$$

with

$$S_{11} = \langle V(x, t)V(y, t)\rangle \equiv S_V(x, y, t), \quad S_{22} = \langle I^2(t)\rangle$$
$$S_{12} = \langle V(x, t)I(t)\rangle \equiv S_I(x, t), \qquad S_{21} = S_I(y, t).$$

The PDE for the matrix spatial correlation function $S(x, y, t)$ for the cable model with an input current of zero mean is

$$S_{,t} = L_x[S] + \{L_y[S^T]\}^T + R(x, y) \tag{13.6.9}$$

with partial differential operator L_x given in(13.6.4) and $\mathbf{F}(x, t) = (0, \sigma W(t))^T$ so that

$$L_x[\cdot] = \begin{bmatrix} \frac{\partial^2}{\partial x^2} - 1 & \delta(x - x_0) \\ 0 & -\alpha \end{bmatrix}, \quad R(x, y) = \begin{bmatrix} 0 & 0 \\ 0 & \sigma^2 \end{bmatrix}.$$

The four scalar PDE corresponding to the four elements of matrix PDE (13.6.9) are

$$S_{V,t} = S_{V,xx} + S_{V,yy} - 2S_v + S_I(x, t)\delta(y - x_0) + S_I(y, t)\delta(x - x_0), \tag{13.6.10}$$

$$S_{I,t} = S_{I,xx} - (1 + \alpha)S_I + \langle I^2(t)\rangle\delta(x - x_0), \tag{13.6.11}$$

$$S_{I,t} = S_{I,yy} - (1 + \alpha)S_I + \langle I^2(t)\rangle\delta(y - x_0), \tag{13.6.12}$$

$$\langle I^2\rangle' = -2\alpha\langle I^2\rangle + \sigma^2, \tag{13.6.13}$$

for $0 < x, y < \ell$ and $t > 0$, with

$$S_V(x, y, 0) = S_I(z, 0) = 0 \quad (z = x, y), \quad \langle I^2 \rangle_{t=0} = 0,$$

$$S_{I,z}(z, t) = 0 \quad (z = 0, \ell), \quad S_{V,n}(x, y, t) = 0 \quad (x, y = 0 \text{ and } \ell)$$

where $S_{V,n}$ is the normal derivatives of S_V along the boundary of the square $(0, \ell) \times (0, \ell)$.

From the IVP for $\langle I^2 \rangle$, we have immediately

$$\langle I^2 \rangle = \frac{\sigma^2}{2\alpha} \left(1 - e^{-2\alpha t} \right).$$

The exact solution for the remaining IBVP for $S_I(z, t)$ and $S_V(x, y, t)$ can be obtained by eigenfunction expansions with $S_I(z, t)$ given by

$$S_I(z, t) = \frac{\sigma^2}{2\alpha} \sum_{n=0}^{\infty} K_n(t) \phi_n(z) \phi_n(x_0)$$

with

$$K_n(t) = \frac{1}{\lambda_n^2 + \alpha} - \frac{e^{-2\alpha t}}{\lambda_n^2 - \alpha} + \frac{2\alpha}{\lambda_n^4 - \alpha^2} e^{-(\lambda_n^2 + \alpha)t}.$$

The solution for $S_V(x, y, t)$ can be obtained similarly with the result for $y = x$ as found in Exercise 64.

Exercise 65. Obtain (13.6.10) by forming $S_{V,t} = \langle V_{,t}(x, t) V(y, t) \rangle + \langle V(x, t) V_{,t}(y, t) \rangle$ and applying the relations in (13.6.2).

Exercise 66. Obtain (13.6.11) by forming $S_{I,t} = \langle V_{,t}(x, t) I(t) \rangle + \langle V(x, t) I_{,t}(t) \rangle$ and applying the relations in (13.6.2).

Exercise 67. Obtain (13.6.13) by forming $\langle I^2(t) \rangle_{,t} = 2 \langle I(t) I_{,t}(t) \rangle$ and applying the relations in (13.6.2).

Chapter 14

Dpp Gradient in the Wing Imaginal Disc of Drosophila

Morphogens are molecular substances that bind to cell surface receptors and other molecules. The gradients of morphogen-receptor complex concentrations (or *signaling gradients* for brevity) are known to be responsible for the patterning of biological tissues during the developmental phase of the host biological organism. For a number of morphogen families, it is known experimentally that the signaling gradients are formed by morphogens transported from a localized production site and bound to surface receptors of cells near and away from the production site (see references cited in [28]). A theoretical basis for diffusion as a mechanism of morphogen transport was addressed in [28, 29] by analyzing appropriate PDE that model the known morphogen activities in the wing imaginal disc of Drosophila fruit flies. The results show that, when observations and data are correctly interpreted, diffusive transport offers a possible and likely mechanism for signaling gradient formation. We summarize in the sections below the simplest extracellular diffusive model of [29] and use it to investigate some stochastic effects on signaling gradients.

14.1 An Extracellular Morphogen Gradient Model

14.1.1 *Model formulation*

As in [28, 29], we simplify the development of the wing imaginal disc of a Drosophila fly as a one-dimensional phenomenon, ignoring variations in the ventral-dorsal and apical-basal directions since the inclusion of developments in these directions are straightforward (see [30] for example). To investigate the consequence of spatially distributed morphogen production, we work with an extracellular formulation of [29] where the results for such a model have been re-interpreted for a model where morphogen-receptor complexes internalize (through endocytosis) before degradation.

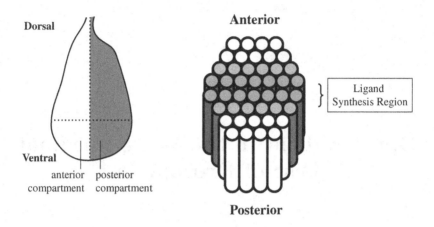

Fig. 14.1.   The Drosophila wing imaginal disc structure.

Let $[L(X,T)]$ be the concentration of the diffusing morphogen *Decapentaplegic* (Dpp) at time T and location X in the span from the midpoint of the morphogen production region $X = -X_{\min}$ to the edge of the posterior compartment of the wing disc at $X = X_{\max}$, with morphogens produced only in $-X_{\min} < X < 0$. Let $[R(X,T)]$ and $[LR(X,T)]$ be the concentration of unoccupied receptors and morphogen-occupied receptors, respectively. For the underlying biological processes of the development described in [28, 29], we add to Fick's second law for diffusive transport of Dpp ($\partial [L]/\partial T = D \partial^2 [L]/\partial X^2$, D being the diffusion coefficient) terms that incorporate the rate of morphogen-receptor binding, $-k_{on}[L][R]$, and dissociation, $k_{off}[LR]$, respectively. In living tissues, molecules that bind receptors do not simply stay bound or dissociate; they also (endocytose and) degrade through a degradation rate term with a rate constant k_{deg}. There is also a separate accounting of the time rate of change of the concentration of unoccupied receptors as they are being synthesized and degrade continuously in time with a degradation rate constant k_R. In this way, we obtain the following reaction-diffusion system for the evolution of three concentrations $[L], [LR]$ and $[R]$:

$$\frac{\partial[L]}{\partial T} = D\frac{\partial^2[L]}{\partial X^2} - k_{on}[L][R] + k_{off}[LR] + V(X,T) \qquad (14.1.1)$$

$$\frac{\partial[LR]}{\partial T} = k_{on}[L][R] - k_{off}[LR] - k_{\deg}[LR] \qquad (14.1.2)$$

$$\frac{\partial[R]}{\partial T} = V_R(X,T) - k_{on}[L][R] + k_{off}[LR] - k_R[R] \qquad (14.1.3)$$

where X lies in the interval $(-X_{\min}, X_{\max})$. The quantity $X_{\max}$ is the span of the wing imaginal disc in the distal direction from the edge of the ligand synthesis zone. The production rate of ligands takes place only in the narrow region $(-X_{\min}, 0)$ with the synthesis rate given in terms of the *Heaviside unit step function* $H(z)$:

$$V(X,T) = \bar{V}_L H(-X) = \begin{cases} \bar{V}_L & (-X_m \leq X < 0) \\ 0 & (0 < X \leq X_{\max}). \end{cases}$$

The three differential equations are augmented by the following boundary conditions

$$X = -X_{\min}: \quad \frac{\partial [L]}{\partial X} = 0, \quad X = X \max: \quad [L] = 0, \qquad (14.1.4)$$

all for $t > 0$. For the wing imaginal disc, the no flux condition at the compartment border being a consequence of assumed symmetry of the two compartments of the wing imaginal disc. The killed end condition at the edge, $X = X_{\max}$, reflects the assumption of an absorbing end (which we may occasionally take to be infinitely far away to avoid making such an assumption). Until morphogens being generated at $t = 0$, the biological system is in quiescence so that we have the initial conditions

$$t = 0: \quad [L] = [LR] = 0, \quad [R] = [R_0(X)] \qquad (14.1.5)$$

for some initial (unoccupied) receptor distribution.

14.1.2 *Uniform receptor synthesis rate*

The model formulated above was first introduced and analyzed in [29] for *Dpp* gradient formation in the posterior compartment of a *Drosophila* wing imaginal disc and extended in [26] to investigate feedback mechanisms for the robustness of such gradients. The model is generally applicable to other morphogen gradient systems with similar characteristics, at least for some insight to the qualitative behavior of such gradients. The idealization of activities in the wing imaginal disc leading to this basic model has already been described in [28, 29].

For most purposes, it suffices to consider receptor synthesis to be at a uniform rate $\bar{V}_R$ in both time and space (throughout the spatial domain) with

$$v_R(x,t) = \frac{\bar{V}_R/R_0}{D/X_{\max}^2} = \bar{v}_R. \qquad (14.1.6)$$

For receptors degrading at a rate proportional to its concentration with a degradation rate constant k_R prior to the onset of ligand synthesis (see (14.1.3)), the receptors should be in a steady state concentration determined by (14.1.3) to be

$$[R] = R_0 = \frac{\bar{V}_R}{k_R} \qquad (14.1.7)$$

for $t < 0$.

For efficient analysis and computation, we normalize the three principal concentrations $\{[L], [LR], [R]\}$ by the initial (steady state) receptor concentration R_0 to get the three dimensionless concentrations

$$\{a, b, r\} = \frac{1}{R_0} \{[L], [LR], [R]\}. \tag{14.1.8}$$

In terms of a, b and r, the three principal PDE (14.1.1)–(14.1.3) become

$$\frac{\partial a}{\partial t} = \frac{\partial^2 a}{\partial x^2} - h_0 ar + f_0 b + \bar{v}_L H(-x), \tag{14.1.9}$$

$$\frac{\partial b}{\partial t} = h_0 ar - (f_0 + g_0)b, \quad \frac{\partial r}{\partial t} = \bar{v}_R - h_0 ar + f_0 b - g_R r, \tag{14.1.10}$$

with

$$x = \frac{X}{X_{\max}}, \quad x_m = \frac{X_{\min}}{X_{\max}}, \quad t = \frac{D}{X_{\max}^2} T, \quad \bar{v}_L = \frac{\bar{V}_L / R_0}{D / X_{\max}^2} \tag{14.1.11}$$

and

$$\{h_0, f_0, g_L, g_0, g_R\} = \frac{X_{\max}^2}{D} \{k_{on} R_0, k_{off}, k_L, k_{\deg}, k_R\}. \tag{14.1.12}$$

The composite parameter $D / X_{\max}^2$ has the dimension of $(\text{time})^{-1}$ and is known as the *time constant* for the particular morphogen.

The three differential equations are augmented by the following (normalized) boundary conditions

$$x = -x_m: \quad \frac{\partial a}{\partial x} = 0, \quad x = 1: \quad a = 0, \tag{14.1.13}$$

all for $t > 0$. Until morphogens being generated at $t = 0$, the biological system is in quiescence so that we have the (normalized) initial conditions

$$t = 0: \quad a = b = 0, \quad r = 1, \tag{14.1.14}$$

keeping in mind

$$r(x, 0) = \frac{\bar{v}_R}{g_R} = \left(\frac{\bar{V}_R / R_0}{D / X_{\max}^2}\right) \left(\frac{k_R}{D / X_{\max}^2}\right)^{-1} = \left(\frac{\bar{V}_R / R_0}{k_R}\right) = 1. \tag{14.1.15}$$

The initial-boundary value problem (IBVP) defined by (14.1.9), (14.1.10), (14.1.13) and (14.1.14) has been analyzed mathematically and computationally in [26, 28, 29] and elsewhere.

14.1.3   Time independent steady state

With both synthesis rates V_L and V_R being time-invariant, a diffusive system
such as (14.1.9)–(14.1.10) typically tends to a time-independent steady state
$\{a_s(x), b_s(x), r_s(x)\}$ as $t \to \infty$. In the limit, the system (14.1.9)–(14.1.10) tends to

$$\frac{d^2 a_s}{dx^2} - h_0 a_s r_s + f_0 b_s + \bar{v}_L H(-x) = 0, \tag{14.1.16}$$

$$0 = h_0 a_s r_s - (f_0 + g_0)b_s, \quad 0 = \bar{v}_R - h_0 a_s r_s + f_0 b_s - g_R r_s. \tag{14.1.17}$$

The two equations in (14.1.17) may be solved for b_s and r_s in terms of a_s

$$b_s(x) = \frac{a_s(x)}{\alpha_0 + \zeta_0 a_s(x)}, \quad r_s(x) = \frac{\alpha_0}{\alpha_0 + \zeta_0 a_s(x)}, \tag{14.1.18}$$

where

$$\zeta_0 = \frac{g_0}{g_R}, \quad \alpha_0 = \frac{g_0 + f_0}{h_0}. \tag{14.1.19}$$

These expressions may then be used to eliminate b_s and r_s from (14.1.16) to get a
single ODE for $a_s(x)$:

$$a_s'' - \frac{g_0 a_s}{\alpha_0 + \zeta_0 a_s} + \bar{v}_L H(-x) = 0. \tag{14.1.20}$$

The ODE (14.1.20) is augmented by the two boundary conditions

$$a_s'(-x_m) = 0, \quad a_s(1) = 0. \tag{14.1.21}$$

An exact solution of the BVP (14.1.20)–(14.1.21) has been obtained in [29]. However, for a signaling gradient to give rise to tissue patterning, a simpler form of the
solution is more relevant. Both exact and approximate time-independent steady
state solutions are discussed in [29].

14.1.4   Low receptor occupancy solution

For a morphogen gradient to induce a distinctive biological tissue pattern, the gradient of the signaling morphogen concentration must change gradually from the
edge of the ligand source, $x = 0$, to the edge of the posterior compartment. It
cannot be too flat across much of the span between the two ends of that interval $0 < X < X_{\max}$. In Fig. 14.2, Graphs 1 and 2 do not induce complex tissue
patterns and generally not biologically useful for that purpose. Any distinct pattern requires signaling gradients like Graph 3. Such gradients are typically attained
when $\zeta_0 a_s \ll \alpha_0$ and $g_0/\alpha_0 = O(1)$ corresponding to a morphogen system with
relatively low free ligand concentration $[L]$ compared to the initial receptor concentration R_0. For such a system, we may neglect terms involving $\zeta_0 \bar{a}_s$ in (14.1.20) to

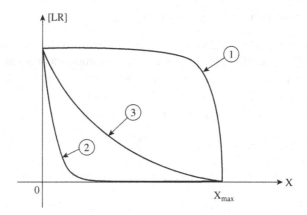

Fig. 14.2.   Signaling Morphogen gradients.

get an approximate set of solutions $\{A_s(x), B_s(x), R_s(x)\}$ to be determined by

$$A_s'' - \mu^2 A_s + \bar{v}_L H(-x) = 0, \quad \mu^2 = \frac{g_0}{\alpha_0} \tag{14.1.22}$$

$$A_s'(-x_m) = 0, \quad A_s(1) = 0. \tag{14.1.23}$$

The exact solution for $A_s(x)$ is

$$A_s(x) = \begin{cases} \dfrac{\bar{v}_L}{\mu^2}\{1 - \dfrac{\cosh(\mu)}{\cosh(\mu(1+x_m))}\cosh(\mu(x+x_m))\} & (-x_m \le x \le 0) \\[2mm] \dfrac{e\bar{v}_L}{\mu^2}\dfrac{\sinh(\mu x_m)}{\cosh(\mu(1+x_m))}\sinh(\mu(1-x)) & (0 \le x \le 1), \end{cases} \tag{14.1.24}$$

with

$$b_s(x) \simeq B_s(x) = \frac{A_s(x)}{\alpha_0}, \quad r_s(x) \simeq R_s(x) = 1. \tag{14.1.25}$$

For $\mu \gg 1$, the expression for $A_s(x)$ in the signaling range of $0 \le x < 1$ is effectively a boundary layer adjacent to $x = 0$, steep near $x = 0$ and dropping sharply to $\bar{v}_L/\mu^2$ (which is rather small) away from $x = 0$. Thus, even if a morphogen system is in a steady state of low occupancy, its signaling gradient may not be biologically useful for patterning if the condition $\mu^2 = O(1)$ is not met.

14.2   Transient States

14.2.1   Low receptor occupancy state

When the receptor synthesis rate is high compared to the ligand synthesis rate, there should be relatively few free morphorgrens (not bound to receptors) in the system.

Mathematically, this corresponds to $v_L(x,t) = \varepsilon \tilde{v}_L(x,t)$ for $\varepsilon \ll 1$ with $\tilde{v}_L = O(1)$. In that case, we have to a good approximation $\{a, b, r\} \approx \{\varepsilon a_1, \varepsilon b_1, 1 + \varepsilon \rho_1\}$. With this assumption, the nonlinear system of three PDE is linearized to take the form

$$\frac{\partial a_1}{\partial t} = \frac{\partial^2 a_1}{\partial x^2} - h_0 a_1 + f_0 b_1 + \tilde{v}_L(x,t), \tag{14.2.1}$$

$$\frac{\partial b_1}{\partial t} = h_0 a_1 - (f_0 + g_0) b_1, \qquad \frac{\partial \rho_1}{\partial t} = -h_0 a_1 + f_0 b_1 - g_R \rho_1 \tag{14.2.2}$$

with the boundary conditions

$$x = -x_m: \quad \frac{\partial a_1}{\partial x} = 0, \quad x = 1: \quad a_1 = 0, \tag{14.2.3}$$

all for $t > 0$. Until the onset of morphogen expression at $t = 0$, the biological system is in quiescence so that we have the (normalized) initial conditions

$$t = 0: \quad a_1 = b_1 = \rho_1 = 0. \tag{14.2.4}$$

with the receptor concentration in its uniformly distributed steady state of $r(x,t) = 1$ for all $t < 0$. (A formal derivation of this approximate system by parametric series expansions in a small amplitude parameter ε will be given in a later subsection.)

The simplified system for $\{a_1, b_1, \rho_1\}$ is known as a Low Receptor Occupancy (LRO) state for the morphogen system. The simplification de-couples the original three coupled PDE into two subsystems. One is a coupled system of two equations for the two main unknowns $a_1(x,t)$ and $b_1(x,t)$ which may be written in terms of a vector unknown $\mathbf{u}(x,t) = (a_1, b_1)^T$ as

$$\mathbf{u}_{,t} = L_x[\mathbf{u}] + \mathbf{f}(x,t)$$

where

$$L_x[\mathbf{u}] = \begin{bmatrix} \dfrac{\partial^2}{\partial x^2} - h_0 & f_0 \\ h_0 & -(g_0 + f_0) \end{bmatrix} \begin{pmatrix} a_1 \\ b_1 \end{pmatrix},$$

$$\mathbf{f}(x,t) = \begin{pmatrix} \tilde{v}_L(x,t) \\ 0 \end{pmatrix} \equiv \begin{pmatrix} \tilde{v}_L \\ 0 \end{pmatrix} H(-x)$$

with the boundary conditions

$$B_0[\mathbf{u}] = \left[\begin{pmatrix} \frac{\partial}{\partial x} & 0 \\ 0 & 0 \end{pmatrix} \begin{pmatrix} a_1(x,t) \\ b_1(x,t) \end{pmatrix} \right]_{x=0}, \quad B_\ell[\mathbf{u}] = \begin{pmatrix} 1 & 0 \\ 0 & 0 \end{pmatrix} \begin{pmatrix} a_1(1,t) \\ b_1(1,t) \end{pmatrix},$$

and the initial condition

$$\mathbf{u}(x,0) = \mathbf{0}.$$

The secondary subsystem consists of a single ODE

$$\frac{\partial \rho_1}{\partial t} + g_R \rho_1 = f_0 b_1 - h_0 a_1, \quad \rho_1(x,0) = 0.$$

Once the solution for the primary system has been found, the third unknown is given by

$$\rho_1(x,t) = e^{-g_R t} \int_0^t e^{g_R \tau} \left\{ f_0 b_1(x,\tau) - h_0 a_1(x,\tau) \right\} d\tau.$$

14.2.2 Eigenfunction expansions

From the spatial differential operator $L_x[\cdot]$, the eigen-pairs $\{\lambda_n^2, \phi_n(x)\}$ for the problem is determined by

$$\phi_n'' + \left(\lambda_n^2 - h_0\right)\phi_n = 0, \quad \phi_n'(-x_m) = \phi_n(1) = 0,$$

so that, with $\mu_n^2 = \lambda_n^2 - h_0$,

$$\phi_n(x) = \sin\left(\mu_n\left(x - 1\right)\right)), \quad \mu_n = \frac{\left(n + \frac{1}{2}\right)\pi}{1 + x_m}, \quad (n = 0, 1, 2, \ldots)$$

and

$$\lambda_n^2 = h_0 + \left(\frac{2n+1}{1+x_m}\right)^2 \left(\frac{\pi}{2}\right)^2.$$

The completeness of the eigenfunctions enables us to work with the eigenfunction expansions

$$\{a_1(x,t), b_1(x,t), \rho_1(x,t)\} = \sum_{k=0}^{\infty} \left\{ a_1^{(k)}(t), b_1^{(k)}(t), \rho_1^{(k)}(t) \right\} \phi_k(x) \qquad (14.2.5)$$

and to solve deterministic problems for $\{a_1(x,t), b_1(x,t), \rho_1(x,t)\}$ (see [47] for example).

14.2.3 Formal perturbation solution

A state of *low receptor occupancy* (LRO) occurs typically when the morphogen synthesis rate is small compared to the receptor synthesis rate. We have previously characterized this state by setting $v_L(x,t) = \varepsilon \tilde{v}_L(x,t;\varepsilon)$. The state of low receptor occupancy may be taken as the leading term approximation of a formal asymptotic solution by a regular perturbation expansion in the small parameter ε:

$$\{z(x,t;\varepsilon), r(x,t;\varepsilon)\} = \{0,1\} + \sum_{n=1}^{\infty} \{z_n(x,t), \rho_n(x,t)\} \varepsilon^n. \qquad (14.2.6)$$

for $z(x, t; \varepsilon) = a(x, t; \varepsilon)$, $b(x, t; \varepsilon)$ and $\tilde{v}_L(x, t; \varepsilon)$. The requirement that the governing equations be satisfied identically in ε results in a sequence of simpler problems defined by

$$\frac{\partial a_n}{\partial t} = \frac{\partial^2 a_n}{\partial x^2} - h_0 \rho_{n-1} a_n + f_0 b_n + \tilde{v}_n(x, t), \qquad (14.2.7)$$

$$\frac{\partial b_n}{\partial t} = h_0 \rho_{n-1} a_n - (f_0 + g_0) b_n, \quad \frac{\partial \rho_n}{\partial t} = -h_0 \rho_{n-1} a_n + f_0 b_n - g_R \rho_n \qquad (14.2.8)$$

for $n = 1, 2, 3, \ldots$ with $\rho_0 = 1$. The PDE are supplemented by the auxiliary conditions

$$x = -x_m: \quad \frac{\partial a_n}{\partial x} = 0, \quad x = 1: \quad a_n = 0, \quad (n = 1, 2, 3, \ldots) \qquad (14.2.9)$$

all for $t > 0$. Until the onset of morphogen synthesis at $t = 0$, the biological system is in quiescence so that we have the (normalized) initial conditions

$$t = 0: \quad a_n = b_n = \rho_n = 0, \quad (0 < x < 1, \quad n = 1, 2, 3, \ldots). \qquad (14.2.10)$$

Evidently, the process of perturbation expansions implemented above facilitates the determination of higher order terms beyond the LRO approximation. Applications of eigenfunction expansions for each order of the perturbation series (14.2.6) lead to a solution of the form

$$\{a(x, t), b(x, t), \rho(x, t)\} = \{0, 0, 1\} + \sum_{n=1}^{\infty} \sum_{k=0}^{\infty} \left\{ a_n^{(k)}(t), b_n^{(k)}(t), \rho_n^{(k)}(t) \right\} \phi_k(x) \varepsilon^n$$

with $\left\{ a_n^{(k)}(t), b_n^{(k)}(t), \rho_n^{(k)}(t) \right\}$ for each pair of (n, k) determined by an IVP for a first order ODE system as seen in the previous section.

14.3 Noisy Morphogen Synthesis Rate

14.3.1 *The LRO problem*

In this volume, we are mainly concerned with biological phenomena in a noisy environment. For morphogen gradient formation in Drosophila wing imaginal discs, there are a number of sources for randomness. We consider here the case of the morphogen synthesis rate with a random component of the form

$$\tilde{v}_L(x, t; \varepsilon) = \varepsilon \left\{ \bar{v}_1 + \sigma W(t) \right\} H(-x)$$

where $W(t)$ is a zero mean white noise with $\langle W(t) W(s) \rangle = \delta(t - s)$. The LRO solution (or the leading term perturbation solution) $\mathbf{u}(x, t) = (a_1, b_1)^T$ is determined by

$$\mathbf{u}_{,t} = L_x[\mathbf{u}] + \mathbf{F}(x, t) \qquad (14.3.1)$$

where

$$
L_x[\mathbf{u}] = \begin{bmatrix} \dfrac{\partial^2}{\partial x^2} - h_0 & f_0 \\ h_0 & -(g_0 + f_0) \end{bmatrix} \begin{pmatrix} a_1 \\ b_1 \end{pmatrix}, \quad \mathbf{F}(x,t) = \begin{pmatrix} \bar{v}_1 + \sigma W(t) \\ 0 \end{pmatrix} H(-x)
$$

(14.3.2)

with the boundary conditions

$$
B_0[\mathbf{u}] = \begin{bmatrix} \begin{pmatrix} \dfrac{\partial}{\partial x} & 0 \\ 0 & 0 \end{pmatrix} \begin{pmatrix} a_1(x,t) \\ b_1(x,t) \end{pmatrix} \end{bmatrix}_{x=0}, \quad B_\ell[\mathbf{u}] = \begin{bmatrix} \begin{pmatrix} 1 & 0 \\ 0 & 0 \end{pmatrix} \begin{pmatrix} a_1(x,t) \\ b_1(x,t) \end{pmatrix} \end{bmatrix}_{x=1}.
$$

(14.3.3)

and the initial condition

$$
\mathbf{u}(x,0) = \mathbf{0}.
$$

(14.3.4)

14.3.2 The method of eigenfunction expansions

The developments of the previous section suggest that we apply (14.2.5) to the present problem to get from the PDE (14.3.1)

$$
\frac{da_1^{(k)}}{dt} + \mu_k^2 a_1^{(k)} - f_0 b_1^{(k)} + \gamma^{(k)}\{\bar{v}_1 + \sigma W(t)\} = 0, \quad a_1^{(k)}(0) = 0, \quad (14.3.5)
$$

$$
\frac{db_1^{(k)}}{dt} - h_0 a_1^{(k)} + (g_0 + f_0)b_1^{(k)} = 0, \quad b_1^{(k)}(0) = 0, \quad (14.3.6)
$$

where

$$
\gamma^{(k)} = \frac{4}{\pi} \frac{\cos(\mu_k)}{2k+1}
$$

(14.3.7)

for $k = 0, 1, 2, \ldots$.

Once the solution for this primary system has been found, the third unknown

$$
\rho_1(x,t) = \sum_{k=0}^{\infty} \rho_1^{(k)} \phi_k(x)
$$

is determined by

$$
\frac{d\rho_1^{(k)}}{dt} + g_R \rho_1^{(k)} + h_0 a_1^{(k)} - f_0 b_1^{(k)} = 0, \quad \rho_1^{(k)}(0) = 0.
$$

for $k = 1, 2, \ldots$.

We now embark on the task of obtaining the solution for the primary subsystem, particularly $b_1(x,t)$, the bound morphogen gradient that determines the fate of the wing imaginal disc. Orthogonality enables us to de-couple the infinite system to solving

$$\frac{da_1^{(k)}}{dt} = -\mu_k^2 a_1^{(k)} + f_0 b_1^{(k)} - \gamma^{(k)} \{\bar{v}_1 + \sigma W(t)\}, \qquad a_1^{(k)}(0) = 0,$$

$$\frac{db_1^{(k)}}{dt} = h_0 a_1^{(k)} - (g_0 + f_0) b_1^{(k)}, \qquad b_1^{(k)}(0) = 0.$$

We can use the second ODE above to write

$$a_1^{(k)} = \frac{1}{h_0} \left\{ \frac{db_1^{(k)}}{dt} + (g_0 + f_0) b_1^{(k)} \right\}$$

and eliminate $a_1^{(k)}$ from the first to get

$$\frac{d^2 b_1^{(k)}}{dt^2} + (g_0 + f_0 + \lambda_k^2) \frac{db_1^{(k)}}{dt} + \{\lambda_k^2 (g_0 + f_0) - f_0 h_0\} b_1^{(k)} = h_0 \gamma^{(k)} \{\bar{v}_1 + \sigma W(t)\}$$

with

$$b_1^{(k)}(0) = \left[\frac{db_1^{(k)}}{dt} \right]_{t=0} = 0.$$

By the method of variation of parameters, the exact solution of the IVP for $b_1^{(k)}(t)$ is

$$b_1^{(k)}(t) = \frac{h_0 \gamma^{(k)}}{\Lambda_2^{(k)} - \Lambda_1^{(k)}} \int_0^t \left\{ e^{-\Lambda_1^{(k)}(t-s)} - e^{-\Lambda_2^{(k)}(t-s)} \right\} \{\hat{v}_1 + \sigma W(s)\} \, ds \qquad (14.3.8)$$

where

$$\begin{pmatrix} \Lambda_2^{(k)} \\ \Lambda_1^{(k)} \end{pmatrix} = \frac{1}{2} \left\{ \omega_k \pm \sqrt{\omega_k^2 - 4 \{\lambda_k^2 (g_0 + f_0) - f_0 h_0\}} \right\}. \qquad (14.3.9)$$

with $\omega_k = g_0 + f_0 + \lambda_k^2$. Note that

$$\lambda_k^2 (g_0 + f_0) - f_0 h_0 = \left\{ h_0 + \frac{(2k+1)^2}{(1+x_m)^2} \left(\frac{\pi}{2} \right)^2 \right\} (g_0 + f_0) - f_0 h_0 > 0$$

and

$$\omega_k^2 - 4 \{\lambda_k^2 (g_0 + f_0) - f_0 h_0\} = (g_0 + f_0 - \lambda_k^2)^2 + 4 f_0 h_0 > 0.$$

so that

$$\Lambda_2^{(k)} > \Lambda_1^{(k)} > 0.$$

14.3.2.1   *Mean*

Take the expectation of both sides of (14.3.8) and observe that $W(t)$ is of zero mean to obtain

$$E[b_1^{(k)}(t)] = \bar{v}_1 h_0 \gamma^{(k)} \int_0^t \Psi^{(k)}(t,s) ds, \tag{14.3.10}$$

$$= \frac{\bar{v}_1 h_0 \gamma^{(k)}}{\Lambda_1^{(k)} \Lambda_2^{(k)}} \left\{ 1 - \frac{\Lambda_2^{(k)} e^{-\Lambda_1^{(k)} t} - \Lambda_1 e^{-\Lambda_2^{(k)} t}}{\Lambda_2^{(k)} - \Lambda_1^{(k)}} \right\}$$

where

$$\Psi^{(k)}(t,s) = \frac{1}{\Lambda_2^{(k)} - \Lambda_1^{(k)}} \left\{ e^{-\Lambda_1^{(k)}(t-s)} - e^{-\Lambda_2^{(k)}(t-s)} \right\}. \tag{14.3.11}$$

The actual mean signaling morphogen gradient is given by

$$E[b_1(x,t)] = \sum_{k=0}^{\infty} E[b_1^{(k)}(t)] \phi_k(x). \tag{14.3.12}$$

It tends to the steady state solution of (14.1.24) as $t \to \infty$ given $e^{-\Lambda_j^{(k)} t} \to 0$ as $t \to \infty$, $j = 1, 2$. The expression (14.3.12) itself is the eigenfunction expansion form of the previously found in [26, 29, 47].

From the second equation of (14.3.1), we obtain

$$E[a_1(x,t)] = \frac{1}{h_0} \left\{ \frac{d}{dt} \{E[b_1]\} + (g_0 + f_0) E[b_1] \right\}$$

$$= \frac{1}{h_0} \sum_{k=0}^{\infty} \left\{ \frac{d}{dt} \left(E[b_1^{(k)}] \right) + (g_0 + f_0) E[b_1^{(k)}] \right\} \phi_k(x)$$

with $E[b_1^{(k)}]$ given by (14.3.10).

14.3.2.2   *Second moment*

In terms of the relevant eigenfunction expansions, the second moments of $b_1(x, t)$ is given by

$$E[\{b_1(x,t)\}^2] = \sum_{k=0}^{\infty} \sum_{i=0}^{\infty} E[b_1^{(k)}(t) b_1^{(i)}(t)] \phi_k(x) \phi_i(x)$$

with

$$E\left[b_1^{(k)}(t) b_1^{(i)}(t) \right] = h_0^2 \gamma^{(k)} \gamma^{(i)} \int_0^t \int_0^t \Psi^{(k)}(t,\xi) \Psi^{(i)}(t,\eta) \left\{ \hat{v}_1^2 + \sigma^2 \langle W(\xi) W(\eta) \rangle \right\} d\xi d\eta$$

$$= E[b_1^{(k)}(t)] E[b_1^{(i)}(t)] + \sigma^2 h_0^2 \gamma^{(k)} \gamma^{(i)} \int_0^t \Psi^{(k)}(t,\xi) \Psi^{(i)}(t,\xi) d\xi$$

given $E[W(t)] = 0$ and $E[W(\xi)W(\eta)] = \delta(\xi - \eta)$. It follows that

$$E\left[b_1^2(x,t)\right] = \{E[b_1(t)]\}^2 + \sigma^2 h_0^2 \sum_{k=0}^{\infty}\sum_{i=0}^{\infty} \gamma^{(k)}\gamma^{(i)} \int_0^t \Psi^{(k)}(t,\xi)\Psi^{(i)}(t,\xi)d\xi$$

where $\Psi^{(\ell)}(t,s)$ is as given in (14.3.11).

Exercise 68. Show

$$\lim_{t \to \infty} \int_0^t \Psi^{(k)}(t,\xi)\Psi^{(i)}(t,\xi)d\xi = \frac{\Lambda_2^{(k)} + \Lambda_1^{(k)} + \Lambda_2^{(i)} + \Lambda_1^{(i)}}{\Pi_{m=1}^{m=2}\Pi_{n=1}^{n=2}\left(\Lambda_m^{(k)} + \Lambda_n^{(i)}\right)} > 0.$$

It is then a straightforward calculation to evaluate the last integral and obtain for $\bar{v}_1 = 0$

$$\lim_{t\to\infty} E\left[b_1^2(x,t)\right] = \sigma^2 h_0^2 \sum_{k=0}^{\infty}\sum_{i=0}^{\infty} \frac{\gamma^{(k)}\gamma^{(i)}\left(\Lambda_2^{(k)} + \Lambda_1^{(k)} + \Lambda_2^{(i)} + \Lambda_1^{(i)}\right)}{\Pi_{m=1}^{m=2}\Pi_{n=1}^{n=2}\left(\Lambda_m^{(k)} + \Lambda_n^{(i)}\right)} > 0.$$

$$(14.3.13)$$

From the dependence of $\Lambda_2^{(k)} > \Lambda_1^{(k)} > 0$ on the eigenvalues $\{\lambda_k^2\}$ shown in (14.3.9) with

$$\lambda_k^2 = h_0 + \left(\frac{2k+1}{1+x_m}\right)^2 \left(\frac{\pi}{2}\right)^2 > h_0,$$

we have

$$\lambda_k^2 \gg h_0 > g_0 + f_0$$

for $k \geq 1$ given that $h_0 = O(10)$, $g_0 = O(1)$ and $f_0 \ll 1$. The double infinite series for (14.3.13) for the limiting behavior of the variance of $b_1(x,t)$ should converge rather quickly. However, with

$$\Lambda_1^{(k)} = \frac{1}{2}\left\{(g_0 + f_0 + \lambda_k^2) - \sqrt{(g_0 + f_0 + \lambda_k^2)^2 - 4\{\lambda_k^2(g_0 + f_0) - f_0 h_0\}}\right\}$$

$$\simeq \frac{\lambda_k^2(g_0 + f_0) - f_0 h_0}{g_0 + f_0 + \lambda_k^2} \simeq (g_0 + f_0)\{1 + O(\lambda_k^{-2})\}$$

(and $g_0 + f_0 = O(1/5)$ for a typical Dpp system in wing imaginal disc), the rate of approach of both mean and variance of the signaling gradient $b_1(x,t)$ toward their respective steady state is rather slow.

14.3.3 The spatial correlation method

For comparison, let us now determine the second moment of the same signaling gradient $b_1(x,t)$ by the method of spatial correlations. We work with the spatial correlation matrix $S(x, y, t)$ defined by

$$S(x, y, t) = E[\mathbf{u}(x, t)\mathbf{u}^T(y, t)] = \begin{bmatrix} S_{11} & S_{12} \\ S_{21} & S_{22} \end{bmatrix}.$$

Since the forcing term $\mathbf{F}(x, t)$ in the linear PDE for the vector process $\mathbf{u}(x, t)$ is temporally uncorrelated, we may apply the general results of the spatial correlation method in (13.5.8) to our problem to get

$$S_{,t} = L_x[S] + \{L_y[S^T]\}^T + R_{FF}(x, y) \quad (0 < x, y < \ell, \ t > 0) \tag{14.3.14}$$

where

$$L_z[S] = \begin{bmatrix} \dfrac{\partial^2}{\partial z^2} - h_0 & f_0 \\ h_0 & -(g_0 + f_0) \end{bmatrix} \begin{bmatrix} S_{11} & S_{12} \\ S_{21} & S_{22} \end{bmatrix}$$

and $R_{FF}(x, y) = [r_{ij}(x, y)]$ is a $k \times k$ matrix associated with the temporally uncorrelated vector random forcing $\mathbf{F}(x, t)$. For morphogen gradients problem with ligand synthesized in a noisy environment, we have from (14.3.2)

$$\mathbf{F}(x, t) = \begin{pmatrix} \bar{v}_1 + \sigma W(t) \\ 0 \end{pmatrix} H(-x).$$

To determine the second moment of $b_1(x, t)$, we take (for simplicity) $\bar{v}_1 = 0$ to get

$$\langle \mathbf{F}(x, t)\mathbf{F}^T(y, s) \rangle = \begin{bmatrix} H(-x)H(-y)\delta(t - s) & 0 \\ 0 & 0 \end{bmatrix} = \begin{bmatrix} r_{11}(x, y)\delta(t - s) & 0 \\ 0 & 0 \end{bmatrix}$$

with

$$R_{FF}(x, y) = [r_{ij}(x, y)] = \begin{bmatrix} H(-x)H(-y) & 0 \\ 0 & 0 \end{bmatrix}.$$

Since the forcing term $R_{FF}(x, y)$ in (14.3.14) is time-invariant, we expect $S(x, y, t)$ to approach a time-invariant steady state $\bar{S}(x, y) = [\bar{S}_{ij}(x, y)]$ determined by

$$L_x[\bar{S}] + \{L_y[\bar{S}]\}^T + R_{FF}(x, y) = O \quad (-x_m < x, y < 1) \tag{14.3.15}$$

with the boundary conditions

$$\left[\frac{\partial \bar{S}_{ij}}{\partial z} \right]_{z=-x_m} = 0, \quad \bar{S}_{ij}(1, y) = \bar{S}_{ij}(x, 1) = 0, \quad (i, j = 1, 2) \tag{14.3.16}$$

for $z = x$ and y. The four components of the matrix equation (14.3.15) are:

$$0 = \left(\nabla^2 - 2h_0\right)\bar{S}_{11} + f_0\left(\bar{S}_{12} + \bar{S}_{21}\right) + \sigma^2 H(-x)H(-y), \qquad (14.3.17)$$

$$0 = \left\{\frac{\partial^2}{\partial x^2} - (h_0 + f_0 + g_0)\right\}\bar{S}_{12} + h_0\bar{S}_{11} + f_0\bar{S}_{22}, \qquad (14.3.18)$$

$$0 = \left\{\frac{\partial^2}{\partial y^2} - (h_0 + f_0 + g_0)\right\}\bar{S}_{21} + h_0\bar{S}_{11} + f_0\bar{S}_{22}, \qquad (14.3.19)$$

$$0 = h_0\left(\bar{S}_{12} + \bar{S}_{21}\right) - 2\left(f_0 + g_0\right)\bar{S}_{22}. \qquad (14.3.20)$$

14.3.3.1 Solution by eigenfunction expansions

Let

$$\left[\bar{S}_{ij}(x,y)\right] = \sum_{k=0}^{\infty}\sum_{j=0}^{\infty}\left[A_{ij}^{(kn)}\right]\phi_k(x)\phi_n(y) \qquad (14.3.21)$$

so that the four boundary conditions in (14.3.16) are satisfied. With the expressions (14.3.21) for $\left[\bar{S}_{ij}(x,y)\right]$, the PDE (14.3.15) becomes

$$-\sigma^2\gamma_k\gamma_n = -\left(\mu_k^2 + \mu_n^2\right)A_{11}^{(kn)} + f_0\left(A_{21}^{(kn)} + A_{12}^{(kn)}\right),$$

$$0 = -\left(\mu_k^2 + f_0 + g_0\right)A_{12}^{(kn)} + h_0 A_{11}^{(kn)} + f_0 A_{22}^{(kn)},$$

$$0 = -(\mu_n^2 + g_0 + f_0)A_{21}^{(kn)} + h_0 A_{11}^{(kn)} + f_0 A_{22}^{(kn)},$$

$$0 = h_0\left(A_{12}^{(kn)} + A_{21}^{(kn)}\right) - 2(g_0 + f_0)A_{22}^{(kn)}.$$

For each (k, n) pair, the four equations above are linear algebraic equations for the four unknowns $\{A_{12}^{(kn)}, A_{21}^{(kn)}, A_{11}^{(kn)}, A_{22}^{(kn)}\}$. The steady state variance of the bound morphogen concentration $b_1(x, t)$ is given by

$$\lim_{t \to \infty}\{\text{Var}\left[b_1(x,t)\right]\} = \bar{S}_{ij}(x,x) = \sum_{k=0}^{\infty}\sum_{n=0}^{\infty}A_{ij}^{(kn)}\phi_k(x)\phi_n(x).$$

The convergence of this double infinite series is moderate since

$$A_{22}^{(kn)} = O\left(\left(k^2 + n^2\right)^{-1}\right).$$

14.3.3.2 Solution by the method of finite difference

Let us examine now the computational requirements by the method of finite difference for an accurate approximate solution for the same system (14.3.17)–(14.3.20). For illustrative purposes, we denote the unit square $[0, 1] \times [0, 1]$ by D_{11} and cover this solution domain for the problem by an $N \times N$ mesh with equally spaced mesh points at (x_ℓ, y_m) located at

$$x_\ell = (\ell - 1)\Delta, \quad y_m = (m - 1)\Delta, \quad \Delta = \frac{1}{N}, \quad (\ell, m = 1, 2, \ldots, N).$$

Let the value of the unknown matrix $Z(x, y)$ at a mesh point (x_ℓ, y_m) approximated by $Z^{(\ell m)}$ and approximate derivatives of a function $g(x, y)$ by its central difference approximations

$$\left[\frac{\partial g}{\partial x}\right]_{(x_\ell, y_m)} = \frac{g_{\ell+1.m} - g_{\ell-1,m}}{2\Delta}, \quad \left[\frac{\partial^2 g}{\partial x^2}\right]_{(x_\ell, y_m)} = \frac{g_{\ell+1.m} - 2g_{\ell,m} + g_{\ell-1,m}}{\Delta^2},$$

$$\left[\frac{\partial g}{\partial y}\right]_{(x_\ell, y_m)} = \frac{g_{\ell.m+1} - g_{\ell,m-1}}{2\Delta}, \quad \left[\frac{\partial^2 g}{\partial y^2}\right]_{(x_\ell, y_m)} = \frac{g_{\ell.m+1} - 2g_{\ell,m} + g_{\ell,m-1}}{\Delta^2}.$$

The four matrix equations (14.3.17)–(14.3.20) in their finite difference form are

$$-\sigma^2 \left[H^{(\ell m)}\right] = A\left[S_{11}^{(\ell m)}\right] + \left[S_{11}^{(\ell m)}\right] A^T - 2h_0 \left[S_{11}^{(\ell m)}\right] + f_0 S^{(\ell m)}, \tag{14.3.22}$$

$$0 = A\left[S_{12}^{(\ell m)}\right] - (h_0 + f_0 + g_0)\left[S_{12}^{(\ell m)}\right] + h_0 \left[S_{11}^{(\ell m)}\right] + f_0 \left[S_{22}^{(\ell m)}\right], \tag{14.3.23}$$

$$0 = \left[S_{21}^{(\ell m)}\right] A^T - (h_0 + f_0 + g_0)\left[S_{21}^{(\ell m)}\right] + h_0 \left[S_{11}^{(\ell m)}\right] + f_0 \left[S_{22}^{(\ell m)}\right], \tag{14.3.24}$$

$$0 = h_0 S^{(\ell m)} - 2(f_0 + g_0)\left[S_{22}^{(\ell m)}\right] \tag{14.3.25}$$

where

$$S^{(\ell m)} = \left[S_{12}^{(\ell m)}\right] + \left[S_{21}^{(\ell m)}\right],$$

and

$$A = \begin{bmatrix} -2 & 2 & 0 & & & \\ 1 & -2 & 1 & & & \\ 0 & 1 & -2 & 1 & & \\ . & . & . & . & . & \\ 0 & & 0 & 1 & -2 & 1 \\ 0 & 0 & . & & 1 & -2 \end{bmatrix}.$$

For a compact and efficient solution scheme for this algebraic system for the for matrix unknowns $\left[S_{11}^{(\ell m)}\right]$, $\left[S_{12}^{(\ell m)}\right]$, $\left[S_{21}^{(\ell m)}\right]$ and $\left[S_{22}^{(\ell m)}\right]$, we diagonalize the matrix A and A^T by setting

$$A = P\Lambda P^{-1}, \quad A^T = Q\Lambda Q^{-1} \tag{14.3.26}$$

where $\Lambda = [\Lambda_\ell \delta_{\ell m}]$ is the diagonal matrix of the eigenvalues of the matrix A and A^T and P and Q are the modal matrix of A and A^T, respectively. With (14.3.26), the system (14.3.22)–(14.3.25) becomes

$$-\sigma^2 \left[\tilde{H}^{(\ell m)}\right] = \Lambda \left[\tilde{S}_{11}^{(\ell m)}\right] + \left[\tilde{S}_{11}^{(\ell m)}\right] \Lambda - 2h_0 \left[\tilde{S}_{11}^{(\ell m)}\right] + f_0 \tilde{S}^{(\ell m)}, \tag{14.3.27}$$

$$0 = \Lambda \left[\tilde{S}_{12}^{(\ell m)}\right] - (h_0 + f_0 + g_0) \left[\tilde{S}_{12}^{(\ell m)}\right] + h_0 \left[\tilde{S}_{11}^{(\ell m)}\right] + f_0 \left[\tilde{S}_{22}^{(\ell m)}\right], \tag{14.3.28}$$

$$0 = \left[\tilde{S}_{21}^{(\ell m)}\right] \Lambda - (h_0 + f_0 + g_0) \left[\tilde{S}_{21}^{(\ell m)}\right] + h_0 \left[\tilde{S}_{11}^{(\ell m)}\right] + f_0 \left[\tilde{S}_{22}^{(\ell m)}\right], \tag{14.3.29}$$

$$0 = \tilde{S}^{(\ell m)} - 2 (f_0 + g_0) \left[\tilde{S}_{22}^{(\ell m)}\right] \tag{14.3.30}$$

with

$$\tilde{S}^{(\ell m)} = \tilde{S}_{12}^{(\ell m)} + \tilde{S}_{21}^{(\ell m)}, \quad \left[\tilde{Z}^{(\ell m)}\right] = P^{-1} \left[Z^{(\ell m)}\right] Q$$

where $Z^{(\ell m)}$ is any of $S_{ij}^{(\ell m)}$ or $H^{(\ell m)}$. The new system (14.3.27)–(14.3.30) are in fact a de-coupled system for each of the matrix components in the form

$$-\sigma^2 \tilde{H}^{(\ell m)} = (\Lambda_\ell + \Lambda_m - 2h_0)\tilde{S}_{11}^{(\ell m)} + f_0 \tilde{S}^{(\ell m)},$$

$$0 = \{\Lambda_\ell - (h_0 + f_0 + g_0)\} \tilde{S}_{12}^{(\ell m)} + h_0 \tilde{S}_{11}^{(\ell m)} + f_0 \tilde{S}_{22}^{(\ell m)},$$

$$0 = \tilde{S} \{\Lambda_m - (h_0 + f_0 + g_0)\} \tilde{S}_{21}^{(\ell m)} + h_0 \tilde{S}_{11}^{(\ell m)} + f_0 \tilde{S}_{22}^{(\ell m)} 0,$$

$$0 = h_0 \tilde{S}^{(\ell m)} - 2 (f_0 + g_0) \tilde{S}_{22}^{(\ell m)}.$$

For a fixed pair of (ℓ, m), the system above consists of four scalar equations for four scalar unknowns $\left\{\tilde{S}_{ij}^{(\ell m)}\right\}$ that can be solved very simply (and even symbolically) by generally available mathematical computing software such as MatLab, Mathematica and Maple.

Evidently, the spatial correlation method would be more efficient if the eigenfunction expansions converge so slowly to require summing over more than N terms in each series.

14.4 Effects of Noisy Environment on System Properties

In addition to inducing random fluctuation on the morphogen synthesis rate, a noisy environment also has an impact on other activities in the course of biological development. Returning to the LRO system (14.3.1) but now for the usual uniform synthesis rate in $(-x_m, 0)$ and a s.p. $h(t)$ instead of a constant binding rate h_0:

$$\mathbf{u}_{,t} = L_h[\mathbf{u}] + \mathbf{f}(x, t) \tag{14.4.1}$$

where

$$L_h[\mathbf{u}] = \begin{bmatrix} \frac{\partial^2}{\partial x^2} - h(t) & f_0 \\ h(t) & -(g_0 + f_0) \end{bmatrix} \begin{pmatrix} a_1 \\ b_1 \end{pmatrix}, \quad \mathbf{f}(x, t) = \begin{pmatrix} \bar{v}_1 \\ 0 \end{pmatrix} H(-x). \tag{14.4.2}$$

Suppose the noisy environment is impacting the binding rate in such a way that

$$h(t) = h_0 + \sigma W(t)$$

where h_0 is the known constant basic (or mean) normalized binding rate per unit concentration of the previous sections and $W(t)$ is a (zero mean) white noise process with unit variance. Let $\mathbf{V}(x, t) = (V_1, V_2)^T = \mathbf{u}(x, t) - (E[a_1(t)], E[b_1(t)])^T$ and re-write (14.4.1) as

$$\mathbf{V}_{,t} = L_x[\mathbf{V}] + \mathbf{F}(\mathbf{V}, x, t)W(t) \tag{14.4.3}$$

where

$$\mathbf{F}(\mathbf{V}, x, t) = \sigma \{V_1(x, t) + E[a_1]\} \begin{pmatrix} -1 \\ 1 \end{pmatrix}. \tag{14.4.4}$$

Note that the linear differential operator $L_x[\cdot]$ is as given in (14.3.2). Hence, the same eigenfunction expansions are available for the new problem. With

$$\mathbf{V}(x, t) = \sum_{k=0}^{\infty} \mathbf{V}^{(k)}(t) \phi_k(x)$$

orthogonality enables us to deduce from the PDE (14.4.3) the following set of SDE for $\{\mathbf{V}^{(k)}(t)\}$

$$\frac{dV_1^{(k)}}{dt} = -\mu_k^2 V_1^{(k)} + f_0 V_2^{(k)} - \sigma W(t) \left\{ V_1^{(k)}(t) + E[a_1^{(k)}(t)] \right\},$$

$$\frac{dV_2^{(k)}}{dt} = h_0 V_1^{(k)} - (g_0 + f_0) V_2^{(k)} + \sigma W(t) \left\{ V_1^{(k)}(t) + E[a_1^{(k)}(t)] \right\},$$

$$V_1^{(k)}(0) = 0, \quad V_2^{(k)}(0) = 0,$$

for $k = 0, 1, 2, \ldots$. For each k, the two SDE for $V_1^{(k)}(t)$ and $V_2^{(k)}(t)$ form an Ito system

$$\frac{d\mathbf{V}^{(k)}}{dt} = C_k \mathbf{V}^{(k)} + \mathbf{G}(\mathbf{V}^{(k)}.t)W(t)$$

where

$$C_k = \begin{bmatrix} -\mu_k^2 & f_0 \\ h_0 & -(f_0 + g_0) \end{bmatrix}, \quad \mathbf{G}(\mathbf{V}^{(k)}.t) = \sigma \left\{ V_1^{(k)}(t) + E[a_1^{(k)}(t)] \right\} \begin{pmatrix} -1 \\ 1 \end{pmatrix}.$$

The mathematical tools developed in Chapter 12 are available for extracting useful probabilistic information about $\mathbf{V}^{(k)}(t)$.

The same method of solution also applies to stochastic effects on other system parameters such as the normalized degradation rate constant g_0 for the signaling gradient, normalized dissociation rate constant f_0 and normalized degradation rate for receptors g_R. In all cases, the problems can be reduced to Ito type SDE systems as long as a complete set of eigen-pairs exists for the spatial dynamics of the problems.

Part 6
First Exit Time Statistics

Chapter 15

First Exit Time

15.1 Threshold for Action Potential in Nerve Axon

Many problems in the life sciences involve finding the time or location for a phenomenon to exceed some threshold. An example is an early approach to neuron firing by way of the Leaky-Integrator model. For this model, the transmembrane potential is determined by a simple IVP for $t > 0$ that can be taken in the form

$$\frac{dV}{dt} = -\alpha V + I(t), \quad V(0) = 0.$$

For the case of a uniform input current of finite duration,

$$I(t) = \begin{cases} 0 & (t < 0) \\ I_0 H(\tau - t) & (t > 0), \end{cases}$$

where $H(z)$ is the Heaviside unit step function and τ is some positive constant, the exact solution of the IVP is

$$V(t) = \begin{cases} \dfrac{I_0}{\alpha} (1 - e^{-\alpha t}) & (0 < t < \tau) \\ \dfrac{I_0}{\alpha} (1 - e^{-\alpha \tau}) e^{-\alpha(t-\tau)} & (t > \tau). \end{cases}$$

For a sufficiently large τ and I_0/α, $V(t)$ exceeds the threshold θ for the initiation of an action potential at some time $T < \tau$. We can determine the time of firing T by solving the relation

$$\theta = \frac{I_0}{\alpha} (1 - e^{-\alpha T})$$

for T to get

$$T = -\frac{1}{\alpha} \ln \left(1 - \frac{\alpha \theta}{I_0} \right).$$

Note that it is necessary to have $\alpha\theta < I_0$ for the existence of some $T < \tau$. As such, we have an example of a *threshold crossing* problem with the first crossing time T also known as the *first exit time*.

Exercise 69. Determine $V(t)$ and $T(\theta; v_0, t_0) < \tau$ for the same ODE but now with $V(t_0) = v_0$.

Here we are interested in cases where there is uncertainty in the problem. For such problems, we would want to know some basic statistics of the first exit time such as the mean and variance of T. A stochastic problem for neuronal threshold crossing arises naturally when the input current I_0 is a random variable. We also consider a different stochastic problem when the threshold parameter θ is a random variable with prescribed statistics such as the pdf $p_0(\theta)$ for θ. The situation arises when the environment for the neuron is noisy and the threshold for action potential may be affected by the noise surrounding the neuron. For the simple model above, we have from the transformation theory developed in Chapter 9 that the pdf for the first exit time $p_T(T)$ is given in term of $p_0(\theta)$ by

$$p_T(T) = p_0(\theta) \left| \frac{d\theta}{dT} \right| = p_0(\theta)(I_0 - \alpha\theta).$$

The mean exit time is easily calculated from

$$E[T] = \int_{-\infty}^{\infty} T p_T(T) dT = \int_{-\infty}^{\infty} T p_0(\theta) d\theta$$

$$= -\frac{1}{\alpha} \int_{-\infty}^{\infty} \ln\left(1 - \frac{\alpha\theta}{I_0}\right) p_0(\theta) d\theta.$$

Note that $E[T]$ is positive since $0 < 1 - \alpha\theta/I_0 < 1$. Higher moments of T can be calculated similarly.

15.2   The Moments of First Exit Time

We saw in the previous section that transformation theory developed in Chapter 9 applies when the solution for the stochastic problem can be expressed in terms of known functions and the prescribed stochastic input. Since we are interested mainly in the time when the response process $\mathbf{X}(t)$ reaches a certain threshold, the conditional density function $p(\mathbf{x}, t \mid \mathbf{x}_0, t_0)$ for $\mathbf{X}(t)$ is one useful piece of probabilistic information for this threshold crossing time. When the SDE is an Ito type equation,

$$d\mathbf{X}(t) = \mathbf{f}(\mathbf{X}(t), t)dt + G(\mathbf{X}(t), t)d\mathbf{B}(t), \quad \mathbf{X}(t_0) = \mathbf{A} \qquad (15.2.1)$$

this pdf has been shown in Chapter 12 to satisfy the Fokker–Planck or Kolmogorov forward equation,

$$\frac{\partial p}{\partial t_0} = -\sum_{j=1}^{n} f_j(\mathbf{x}, t) \frac{\partial p}{\partial x_j} + \frac{1}{2} \sum_{j=1}^{n} \sum_{i=1}^{n} (GDG^T)_{ij} \frac{\partial^2 p}{\partial x_i \partial x_j}, \qquad (15.2.2)$$

with $[p(\mathbf{y}, t \,|\, \mathbf{x}, t_0)]_{t=t_0, \mathbf{y}=\mathbf{x}} = p_0(\mathbf{x}, t_0)$ and p vanishing at infinity in all directions. In (15.2.1), $\mathbf{X}(t)$ and $\mathbf{f}(\mathbf{X}, t)$ are n-vectors, $G = [G_{ij}]$ is an $n \times m$ matrix, and $\mathbf{W}(t) = d\mathbf{B}(t)/dt$ is an m-vector white noise process with covariance matrix $D = [D_{ij}]$. It turns out that the corresponding Kolmogorov (backward) equation for $p(\mathbf{x}, t \,|\, \mathbf{x}_0, t_0)$ is more directly useful for the first exit time problem.

To keep things simple at the start, we again begin with the scalar case

$$dX = f(X, t)dt + dB(t), \quad X(0) = A$$

for which the conditional density function $p(y, t|x, t_0)$ is the solution of the scalar *Kolmogorov (backward) equation*

$$\frac{\partial p}{\partial t_0} = -f(x, t_0)\frac{\partial p}{\partial x} - \frac{1}{2}\sigma^2(x, t_0)\frac{\partial^2 p}{\partial x^2}$$

with

$$p(y, t; \pm\infty, t_0) = 0, \quad p(y, t_0|x, t_0) = \delta(x - y),$$

$$\left\langle \frac{dB(t)}{dt}\frac{dB(\tau)}{d\tau} \right\rangle = \langle W(t)W(\tau) \rangle = D\delta(t - \tau) \equiv \sigma^2\delta(t - \tau).$$

We are interested in the special case of a stationary process $X(t)$ with $f(x, t_0) = f(x)$ and $\sigma^2(x, t_0) = \sigma^2(x)$. Even more relevant to a method for the first exit time statistics is the corresponding equation for the Laplace transform of $p(y, t \,|\, x, t_0) = p(y, t - t_0|x)$ in $\tau = t - t_0$:

$$\hat{p}(y, s \,|\, x) = \int_0^\infty e^{-s\tau} p(y, \tau|x)d\tau.$$

After observing

$$\frac{\partial p}{\partial t_0} = -\frac{\partial p}{\partial t},$$

given $p(y, t \,|\, x, t_0) = p(y, t - t_0| x)$, we get the following ODE for the Laplace transform $\hat{p}(y, s \,|\, x)$ of $p(y, t - t_0| x)$

$$s\hat{p} = f(x)\frac{d\hat{p}}{dx} + \frac{1}{2}\sigma^2(x)\frac{d^2\hat{p}}{dx^2} \quad (x \neq y). \tag{15.2.3}$$

For a stationary process governed by a scalar Ito SDE

$$dX = f(X)dt + G(X)dB(t), \quad X(t_0) = A \tag{15.2.4}$$

where $X(t_0) = A < \theta$ with θ known as an *absorbing barrier* for the response process, the *first exit time* is a random variable T_θ for the process $X(t)$ to reach (and cross) the magnitude level θ from below for the first time. We let $F_\theta(t|x)$ be the probability distribution of T_θ,

$$F_\theta(t\,|\,x) = \text{Prob}\,\{T_\theta(x) \le t\},$$

with

$$f_\theta(t\,|\,x) = \frac{\partial F_\theta(t\,|\,x)}{\partial t}, \quad \hat{f}_\theta(x.s) = \int_{t_0}^{\infty} e^{-st} f_\theta(t\,|\,x) dt = \int_{-\infty}^{\infty} e^{-st} f_\theta(t\,|\,x) dt$$

since $f_\theta(t\,|\,x) = 0$ for $t < t_0$. We state the following lemma without proof as the proof does not offer useful information on, or insight to the SDE or its response processes. Readers interested in the details of the proof are referred to [7] and [53]

Lemma 15. $\hat{f}_\theta(x.s)$ *satisfies the same ODE (15.2.3) as* $\hat{p}(y, s\,|\,x)$ *with* $\sigma^2(x) = G^2 D$.

The lemma is useful in proving the following theorem on the moments of the first exit time:

Theorem 44. *If $f(X,t)$, $G(X,t)$ and $\sigma^2(x,t)$ are all independent of t, the moments $T_n(x)$ of the first exit time $T_\theta(x)$ associated with the temporally homogeneous response process $X(t)$ of the Ito SDE (15.2.4) satisfy the following system of differential equations*

$$\frac{1}{2}\sigma^2(x)\frac{d^2 T_n}{dx^2} + f(x)\frac{dT_n}{dx} = -nT_{n-1}, \quad (n = 1, 2, 3, \ldots) \tag{15.2.5}$$

with $T_0(x) = 1$ and

$$\left[\frac{dT_n}{dx}\right]_{t\to-\infty} = T_n(\theta) = 0, \quad (n = 1, 2, 3, \ldots). \tag{15.2.6}$$

Proof. Expanding formally $\hat{f}_\theta(x.s)$ as a Taylor series about $s = 0$,

$$\hat{f}_\theta(x.s) = \int_{-\infty}^{\infty} \sum_{n=9}^{\infty} \frac{(-s)^n}{n!} t^n f_\theta(t\,|\,x) dt = \sum_{n=0}^{\infty} \frac{(-s)^n}{n!} T_n(x)$$

we get from the ODE (15.2.3)

$$\sum_{n=1}^{\infty} \frac{(-s)^n}{n!} \left\{ nT_{n-1}(x) + f(x)\frac{dT_n}{dx} + \frac{1}{2}\sigma^2(x)\frac{d^2 T_n}{dx^2} \right\} = 0.$$

From this follows the system

$$\frac{1}{2}\sigma^2(x)\frac{d^2 T_n}{dx^2} + f(x)\frac{dT_n}{dx} = -nT_{n-1}(x), \quad (n = 1, 2, 3, \ldots, N)$$

when the moments of T_θ exist up to order N. The boundary conditions (15.2.6) also follow from substitution of the Taylor expansion of $\hat{f}_\theta(x.s)$ into the end condition

$$\left[\frac{dT_\theta}{dx}\right]_{x\to-\infty} = T_\theta(\theta) = 0.$$

$\square$

The BVP (15.2.5) and (15.2.6) will be applied to a number of problems in different areas of the life sciences in the rest of this chapter.

15.3 Input Variability and Neuronal Firing

In previous sections, we have formulated the Leaky Integrator as a simple approximate model for the evolution of subthreshold depolarization voltage $X(t)$ of a nerve cell receiving synaptic input currents. When these currents are random, the IVP for the output voltage in a diffusion approximation may be taken to be

$$dX = [-\alpha X + \mu]\,dt + \sigma dB, \quad X(0) = A \tag{15.3.1}$$

where α, μ and σ are positive constant, $B(t)$ is a standard Wiener process (with zero mean and variance t) so that $dB/dt = W(t)$ is formally a zero mean white noise with unit variance. In the development of this section, we re-scale time (as well as μ and σ) so that t is measured in units of the membrane time constant. After re-scaling (by setting $\tau = \alpha t$), the SDE (15.3.1) may be rewritten as

$$dX = [-X + \mu]\,dt + \sigma dB, \quad X(0) = A$$

where we have kept the same notation for t, μ and σ after having divided them by α.

We know already that the stochastic IVP for depolarization voltage $X(t)$ is of limited applicability for a number of reasons including the fact it is only a diffusion approximation of more realistic neuronal models. Another limitation is that the model is probably only appropriate for low threshold level, i.e., $\theta = O(\mu)$. Nevertheless, it was used in early efforts for investigating the first exit time problem. We describe in this section one such early effort to illustrate the applications of the BVP for the moments to determine the mean and variance of the first exit time. A limitation not previously noted is the scarcity of statistical information on the synaptic input excitation. It is generally unrealistic to expect a complete knowledge of the pdf for the synaptic input (other than an outright assumption). More realistic expectation would be having reasonable estimates of the mean and the variance of the input. In that case, the more appropriate and realistic approach for estimating the first exit time would be to calculate its mean and variance by the moment equations developed in the previous section (since that approach makes use of only these two pieces of information on the input). In addition, it will be shown that this alternative method is still advantageous even when the transformation theory is applicable.

For the stochastic IVP for $X(t)$ above, the first exit time moment equations are specialized to

$$\frac{1}{2}\sigma^2 \frac{d^2 T_n}{dx^2} + (\mu - x)\frac{dT_n}{dx} = -nT_{n-1}(x), \quad (n = 1, 2, 3, \ldots, N).$$

The boundary conditions (15.2.6) also follow from substitution of the Taylor expansion of $\hat{f}_\theta(x.s)$ into the end condition

$$\left[\frac{dT_n}{dx}\right]_{x \to -\infty} = T_n(\theta) = 0.$$

Before proceeding with further analysis, we re-scale the various relevant quantities by the setting

$$z = \frac{x}{\theta}, \quad \varepsilon = \frac{\sigma}{\theta}.$$

The ODE system for T_n then becomes

$$\frac{1}{2}\varepsilon^2 \frac{d^2 T_n}{dz^2} + (\gamma - z)\frac{dT_n}{dz} = -nT_{n-1}(z), \quad (-\infty < z < 1, \ n = 1, 2, 3, \ldots)$$

$$(15.3.2)$$

augmented by the auxiliary conditions

$$\left[\frac{dT_n}{dz}\right]_{z \to -\infty} = T_n(1) = 0. \tag{15.3.3}$$

The second order ODE for T_n is effectively a first order ODE for dT_n/dz and the exact solution is immediate.

For example, the BVP for the first moment $T_1(z)$ is

$$\frac{1}{2}\varepsilon^2 T_1'' + (\gamma - z)T_1' = -1, \quad T_1'(-\infty) = T_1(1) = 0. \tag{15.3.4}$$

The exact solution of this BVP is

$$T_1(z; \gamma, \varepsilon) = \frac{2}{\varepsilon^2} \int_z^1 e^{I(s)} \int_{-\infty}^s e^{-I(y)} dy\, ds, \quad I(z) = \frac{1}{\varepsilon^2}(\gamma - z)^2. \tag{15.3.5}$$

Similar exact solution can also be obtained for higher order moments. It should be evident that even after evaluating the quadratures involved, numerically or otherwise, the results are not particularly informative about the moments of the first exit time $\{T_n\}$, not to mention the first exit time itself. One of the purpose of this section is to show that more informative solution is possible.

15.3.1 *Low threshold and input variability*

Consider the simplest case of a low threshold value so that $\gamma = \mu/\theta > 1$ (and hence $\gamma > z$) and low input invariability (with σ of the synaptic input small compared to θ) so that $\varepsilon^2 \ll 1$. For this case, we work with parametric expansions of the moments $\{T_n(z)\}$ in powers of ε^2:

$$T_0 = 1, \quad T_n(z; \gamma, \varepsilon) \sim \sum_{k=0}^{\infty} T_n^{(k)}(z; \gamma) \varepsilon^{2k}. \quad (n = 1, 2, 3, \ldots). \qquad (15.3.6)$$

Upon substitution of these expansions into the ODE (15.3.2)–(15.3.3) and collecting coefficients of the same power of ε^2, we obtain for $k = 0$

$$(\gamma - z)\frac{dT_n^{(0)}}{dz} = -nT_{n-1}^{(0)}, \quad T_n^{(0)}(1) = \left[\frac{dT_n^{(0)}}{dz}\right]_{z \to -\infty} = 0 \qquad (15.3.7)$$

and

$$(\gamma - z)\frac{dT_n^{(k+1)}}{dz} = -\frac{1}{2}\frac{d^2T_n^{(k)}}{dz^2} - nT_{n-1}^{(k+1)}, \quad T_n^{(k)}(1) = \left[\frac{dT_n^{(k)}}{dz}\right]_{z \to -\infty} = 0. \qquad (15.3.8)$$

15.3.1.1 *The mean exit time*

The exact solution for the two-point BVP for $n = 1$ and $k = 0$ is

$$T_1^{(0)}(z) = \ln\left(\frac{\gamma - z}{\gamma - 1}\right).$$

Remarkably, the solution of the first order ODE in (15.3.7) actually satisfies both auxiliary conditions for the problem (without the need of a boundary layer solution at one of the boundaries). The exact solution for $n = k = 1$ determined by

$$(\gamma - z)\frac{dT_1^{(1)}}{dz} = -\frac{1}{2}\frac{d^2T_1^{(0)}}{dz^2} = \frac{1}{2}\frac{1}{(\gamma - z)^2}$$

$$T_1^{(1)}(1) = \left[\frac{dT_1^{(1)}}{dz}\right]_{z \to -\infty} = 0$$

is also easily obtained so that we have altogether

$$E[T_\theta(x)] = T_1(z) = \ln\left(\frac{\gamma - z}{\gamma - 1}\right) - \frac{\varepsilon^2}{4}\left[\frac{1}{(\gamma - 1)^2} - \frac{1}{(\gamma - z)^2}\right] + O\left(\varepsilon^4\right).$$

Exact solutions for $T_1^{(k)}(z)$ for $k > 1$ can also be obtained. But they are small of order ε^2 compared to what we have already found and will not be reported here. Evidently, the parametric series solution for $T_1(z)$ is simple and from it the change of first exit time with the initial state x is easily discerned (when compared to the more complicated exact solution (15.3.5)).

15.3.1.2 *The second moment and variance*

The exact solution of the two-point BVP for $n = 2$ and $k = 0$ is determined by

$$(\gamma - z)\frac{dT_2^{(0)}}{dz} = -2T_1^{(0)} = -2\ln\left(\frac{\gamma - z}{\gamma - 1}\right), \quad T_2^{(0)}(1) = \left[\frac{dT_2^{(0)}}{dz}\right]_{z \to -\infty} = 0,$$

giving

$$T_2^{(0)}(z) = \left[\ln\left(\frac{\gamma - z}{\gamma - 1}\right)\right]^2.$$

Again, the solution actually satisfies both auxiliary conditions for the problem (without the need of a layer solution at one of the boundaries). The exact solution for BVP for $n = 2$ and $k = 1$ can be similarly obtained so that we have altogether

$$T_2(z) = \left[\ln\left(\frac{\gamma - z}{\gamma - 1}\right)\right]^2 - \frac{\varepsilon^2}{2}\left[\ln\left(\frac{\gamma - z}{\gamma - 1}\right) - 1\right]\left[\frac{1}{(\gamma - 1)^2} - \frac{1}{(\gamma - z)^2}\right] + O\left(\varepsilon^4\right).$$

The variance of the first exit time,

$$Var\left[T_\theta(x)\right] = T_2(x) - \left[T_1(x)\right]^2$$

is easily shown to be

$$Var\left[T_\theta(x)\right] = \varepsilon^2\left[\frac{1}{(\gamma - 1)^2} - \frac{1}{(\gamma - z)^2}\right],$$

with

$$z = \frac{x}{\theta} \le 1, \quad \gamma = \frac{\mu}{\theta} > 1.$$

The advantages of the parametric series solutions are now seen to be much more substantial. It is simply impractical to try to extract useful information about the second moment and variance of the first exit time from the exact solution for (15.3.2)–(15.3.3).

15.3.1.3 *Coefficient of variation*

A statistical measure for quantifying the variability of a random variable is the *coefficient of variation*, denoted by CV or ρ. It is defined as the ratio of standard deviation of the random variable to its mean. For our problem, we have

$$CV = \rho = \frac{\sqrt{T_2(x) - [T_1(x)]^2}}{T_1(x)} \sim \frac{\varepsilon/\sqrt{2}}{\ln\left(\frac{\gamma - z}{\gamma - 1}\right)}\sqrt{\frac{1}{(\gamma - 1)^2} - \frac{1}{(\gamma - z)^2}} + O(\varepsilon^2)$$

with $z = x/\theta$ and $\gamma = \mu/\theta > 1$. Of particular interest is

$$[CV]_{x=0} = \rho(0) \simeq \frac{\varepsilon}{\sqrt{2}\ln\left(\gamma/(\gamma - 1)\right)}\frac{\sqrt{2\gamma - 1}}{\gamma(\gamma - 1)}.$$

15.3.2 *High threshold value and low input variability*

15.3.2.1 *Outer solutions*

For the more complicated case of high threshold level so that $\gamma = \mu/\theta < 1$, the analysis of mean and variance is more involved. The factor $|\gamma - z|$ is now small compared to ε when z is sufficiently close to γ whatever $\varepsilon > 0$ may be. The value $z = \gamma$ for which $\gamma - z = 0 < \varepsilon$ is known as a *turning point* of the ODE (15.3.2) for T_n. Clearly, the approximating IVP (15.3.8) is not applicable for z in a sufficiently small neighborhood of γ. Still we may take

$$T_1^{(0)}(z) = \begin{cases} c_\ell + \ln\left(\dfrac{\gamma - z}{1 - \gamma}\right) & (z < \gamma) \\ c_r + \ln(z - \gamma) & (z > \gamma) \end{cases} \tag{15.3.9}$$

away from the turning point. The two constants c_ℓ and c_r are available for bridging these pieces with an appropriate solution for z near γ (to render $T_1^{(0)}(z)$ at least continuous for the solution domain).

15.3.2.2 *Layer solutions*

To satisfy the threshold condition $T_1^{(0)}(1) = 0$, we introduce a new independent variable

$$y = \frac{1 - z}{\varepsilon^2}$$

and write the ODE for $T_1^{(0)}(z)$ as

$$\frac{1}{2}\frac{d^2 T_1}{dy^2} - (\gamma - 1 + \varepsilon^2 y)\frac{dT_1}{dy} = -\varepsilon^2.$$

Parametric series solutions

$$T_1(z; \gamma, \varepsilon) \sim \sum_{k=0}^{\infty} \bar{T}_1^{(k)}(y; \gamma)\varepsilon^{2k}. \quad (n = 1, 2, 3, \ldots),$$

in powers of ε^2 and independent variable y is now appropriate. The leading term solution is

$$\bar{T}_1^{(0)}(y; \gamma) = a_r \left\{ 1 - e^{-2(1-\gamma)y} \right\} \quad (\gamma < z \lesssim 1) \tag{15.3.10}$$

where we have specified one of the constants of integration so that $T_1^{(0)}(z = 1) = 0$. The second constant of integration a_r is available for matching the layer solution (15.3.10) to the solution for $\gamma \ll z < 1$ in (15.3.9). The method of match asymptotic expansions [21] requires

$$a_r = c_r.$$

On the other hand, the solution in (15.3.9) for $z < \gamma$ satisfies the limiting condition at $z = -\infty$. Hence there is no need for an layer solution at that end.

15.3.2.3 The turning point solution

The two yet unspecified constants c_r and c_ℓ are needed to connect the solution pieces in (15.3.9) smoothly with an appropriate *turning point solution* for a sufficiently small interval containing the turning point $z = \gamma$. In that small interval, there is no possible simplification for the IVP (15.3.4) for T_1. To simplify the exact solution found in (15.3.5) for the purpose of matching with (15.3.9), we let

$$\xi = \frac{z - \gamma}{\varepsilon}, \quad \tilde{T}(\xi) = T_1(z; \gamma, \varepsilon),$$

and re-write the ODE for $T_1(z; \gamma, \varepsilon)$ as

$$\frac{1}{2}\frac{d^2\tilde{T}_1}{d\xi^2} - \xi\frac{d\tilde{T}_1}{d\xi} = -1. \tag{15.3.11}$$

The solution for the first order ODE for $\tilde{T}_1'$ is

$$e^{-\xi^2}\frac{d\tilde{T}_1}{d\xi} = c_1 - 2\int_{-\infty}^{\xi} e^{-\eta^2}\,d\eta = c_1 + \frac{e^{-\xi^2}}{\xi} - 2\int_{-\infty}^{\xi} e^{-\eta^2}\,d\eta. \tag{15.3.12}$$

For $|\xi| \gg 1$, we have

$$e^{-\xi^2}\frac{d\tilde{T}_1}{d\xi} = c_1 + \frac{e^{-\xi^2}}{\xi} + \int_{-\infty}^{\xi} \frac{e^{-\eta^2}}{\eta^2}\,d\eta = c_1 + \frac{e^{-\xi^2}}{\xi}\left\{1 + O(\xi^{-1})\right\}$$

and therewith

$$\tilde{T}_1'(\xi) = c_0 + c_1\int_{-\infty}^{\xi} e^{\eta^2}\,d\eta + \ln(\xi)\left\{1 + O(\xi^{-1})\right\}$$

$$= c_0 + \left\{c_1\frac{e^{\xi^2}}{\xi} + \ln(\xi)\right\}\left\{1 + O(\xi^{-1})\right\} \quad |\xi| \gg 1. \tag{15.3.13}$$

15.3.2.4 The leading term mean exit time

Matching of the turning point solution (15.3.13) with the solution pieces in (15.3.9) by the method of Langer or matched asymptotic expansions [21] carried out in [75] results in

$$T_1^{(0)}(z) \simeq \begin{cases} C\left\{1 - e^{-2(1-\gamma)(1-z)/\varepsilon^2}\right\} & (0 \le 1 - z = O(\varepsilon)) \\[2mm] C + \ln\left(\dfrac{z-\gamma}{1-\gamma}\right) - \dfrac{\sqrt{\pi}\varepsilon}{z-\gamma}e^{(z-\gamma)^2/\varepsilon^2} & (\varepsilon \ll z - \gamma < 1 - \gamma) \\[2mm] C & |z - \gamma| = O(\varepsilon) \\[2mm] C + \ln\left(\dfrac{\gamma-z}{1-\gamma}\right) & (\gamma - z \gg \varepsilon) \end{cases}$$

where, except for negligibly small terms of order ε or small,

$$C \sim \frac{\sqrt{\pi}\varepsilon}{1-\gamma}e^{(1-\gamma)^2/\varepsilon^2} \gg 1. \qquad (15.3.14)$$

In listing the solution pieces, we have omitted those in intervals (such as the one around the turning point γ) not particularly relevant to our interest in the mean and variance of the first exit time since we are principally concerned with an initial state at the resting state of the neuron, namely for $z = x/\theta = 0$ with

$$T_1^{(0)}(0) \simeq C + \ln\left(\frac{\gamma}{1-\gamma}\right).$$

15.3.2.5   Variance and coefficient of variation

By similar calculations, we obtain also

$$T_2^{(0)}(z) \simeq \begin{cases} 2C^2\left\{1 - e^{-2(1-\gamma)(1-z)/\varepsilon^2}\right\} & (0 \le 1-z = O(\varepsilon)) \\ [T_1(z)]^2 + \left\{C^2 - 2C\ln\left(\frac{1-\gamma}{\varepsilon}\right)\right\} & (\varepsilon \ll z - \gamma < 1-\gamma) \\ 2C[C - \ln((1-\gamma)/\varepsilon)] & |z-\gamma| = O(\varepsilon) \\ [T_1(z)]^2 + C^2 - 4C\ln\left(\frac{1-\gamma}{\varepsilon}\right) & (\gamma - z \gg \varepsilon) \end{cases}$$

so that

$$Var\,[T_\theta(z)] \simeq \begin{cases} C^2\left\{1 - e^{-2(1-\gamma)(1-z)/\varepsilon^2}\right\} & (0 \le 1-z = O(\varepsilon)) \\ C^2 - 2C\ln\left(\frac{1-\gamma}{\varepsilon}\right) & (\varepsilon \ll z - \gamma < 1-\gamma) \\ C[C - 2\ln((1-\gamma)\varepsilon)] & |z-\gamma| = O(\varepsilon) \\ C^2 - 4C\ln\left(\frac{1-\gamma}{\varepsilon}\right) & (\gamma - z \gg \varepsilon) \end{cases} \quad .$$

Of particular interest is

$$Var\,[T_\theta(0)] \simeq C^2 + 2C\ln\left(\frac{\gamma}{1-\gamma}\right).$$

Correspondingly, the leading term coefficient of variation for $x = 0$ is given by

$$[CV]_{x=0} = \rho(0) \simeq 1 + \frac{2}{C}\ln\left(\frac{\gamma}{1-\gamma}\right).$$

Evidently, the first exit time of the response process $X(t)$ for $\gamma/\theta < 1$ is considerably more variable than the previous high threshold case of $\gamma/\theta > 1$ when the input variability is itself low with $\varepsilon = \sigma/\theta < 1$, at least for z away from a small neighborhood of the turning point.

Before moving on to the case of high input variability, it is noted here that for low input variability, we only reported results for a certain range of initial state. In particular, no results on cases where the initial depolarization state is very close to the exit threshold with $|z - \gamma| = O(\varepsilon)$. For those interested, the ranges not covered here can be found in [75].

15.3.3 *High input variability*

15.3.3.1 *Exact solution*

When the input variability is high with $\varepsilon = \sigma/\theta > 1$, the determination of the moments of the first exit time is generally also challenging computationally. The governing ODE (15.3.11) for

$$\tilde{T}_1(\xi) = T_1(z; \gamma, \varepsilon), \quad \xi = \frac{z - \gamma}{\varepsilon},$$

applies generally. With $\xi = 0$ being an ordinary point of that ODE, we have the following Taylor series solution for $\tilde{T}(\xi)$ about the origin:

$$\tilde{T}_1(\xi) = B_0 + B_1 g_o(\xi) + g_e(\xi)$$

where B_0 and B_1 are two constants of integration and

$$g_o(\xi) = \xi + \frac{1}{3}\xi^3 + \frac{1}{10}\xi^5 + \frac{1}{42}\xi^7 + \cdots \equiv \frac{1}{2\sqrt{\pi}} g(\xi),$$

$$g_e(\xi) = -\left(\xi^2 + \frac{1}{3}\xi^4 + \frac{4}{45}\xi^6 + \cdots\right).$$

It is straightforward to verify that

$$g_o(\xi) = \int^{\xi} e^{\eta^2} d\eta, \quad g_e(\xi) = -2 \int^{\xi} e^{\eta^2} \left(\int^{\eta} e^{-s^2} ds\right) d\eta$$

(see also the quadrature solution (15.3.12) of the same ODE). The task of determining the two constants B_0 and B_1 by the two auxiliary conditions

$$T_1(z = 1) = \left[\frac{dT_1}{dz}\right]_{z \to -\infty} = 0$$

is much more challenging and will not be done here. Instead, we provide presently a simpler asymptotic solution for the same range of σ.

15.3.3.2 Asymptotic solution

For an asymptotic solution, it is natural to attempt a perturbation solution

$$T_n(x) = \sum_{k=0}^{\infty} t_n^{(k)}(x)\sigma^{-2k-2}$$

with $t_n^{(0)}(x, y; \mu)$ determined by

$$\frac{d^2 t_n^{(0)}}{dx^2} = -n t_{n-1}^{(0)} \quad (n = 1, 2, 3, \ldots).$$

This approach leads to

$$\frac{d^2 t_1^{(0)}}{dx^2} = -1, \quad \frac{dt_1^{(0)}}{dx} = d_1 - \frac{x}{\sigma^2}.$$

For the case where there is only one threshold at θ_v for the initiation of a action potential, we need a sealed end at $-\infty$ that cannot be met by any choice of the constant of integration d_1.

For a more appropriate approach taken in [62] to ensure a sealed end at $-\infty$, we set

$$T_1(x) = e^{p(x)} t(x), \quad p(x) = \frac{1}{2\sigma^2}(\mu - x)^2.$$

The ODE equation for the first moment $T_1(x)$ is then reduced to

$$\frac{\sigma^2}{2} t'' + \left[\frac{1}{2} - \frac{1}{2\sigma^2}(\mu - x)^2 \right] t = -e^{-p(x)}.$$

A parametric series for $t(x; \sigma^2)$ in powers of σ^{-2},

$$t(x; \sigma^2) = \sum_{k=0}^{\infty} t_k(x)\sigma^{-2k-2}$$

now gives for the leading term solution $t_0'' = -2e^{-p(x)}/\sigma^2$ or

$$t_0'(x) = -\frac{2}{\sigma^2} \int_{-\infty}^{x} e^{-(\mu-\eta)^2/2\sigma^2} d\eta = \frac{2\sqrt{2}}{\sigma} \int_{(\mu-x)/\sqrt{2\sigma^2}}^{\infty} e^{-s^2} ds$$

$$= \frac{\sqrt{2\pi}}{\sigma} \left[1 - \mathrm{erf}\left(\frac{\mu - x}{\sqrt{2}\sigma} \right) \right],$$

having chosen the constant of integration to satisfy the sealed end condition at $-\infty$. One more integration leads to

$$t_0(x) = c_0 - \sqrt{\frac{2\pi}{\sigma^2}} \left[x - \int^{x} \mathrm{erf}\left(\frac{\mu - \eta}{\sqrt{2}\sigma} \right) d\eta \right].$$

The threshold condition $t_0(\theta) = 0$ specializes $t_0(x)$ to

$$t_0(x) = \sqrt{\frac{2\pi}{\sigma^2}} \left[(\theta - x) - \int_x^\theta \text{erf} \left(\frac{\mu - \eta}{\sqrt{2}\sigma} \right) d\eta \right].$$

The leading term approximation for the mean first exit time for a process starting at the resting state is given by

$$T_1(0) \sim \sqrt{\frac{2\pi}{\sigma^2}} e^{\mu^2/2\sigma^2} \left[\theta - \int_0^\theta \text{erf} \left(\frac{\mu - \eta}{\sqrt{2}\sigma} \right) d\eta \right].$$

For $|z| \ll 1$, $\text{erf}(z)$ may be approximated by

$$\text{erf}(z) \simeq \sqrt{\frac{2}{\pi}} \left(z - \frac{z^3}{3} + \cdots \right).$$

Hence $T_1(0)$ can be evaluated relatively simply when the input mean is small relative to the input variability, i.e., $\mu/\sigma \ll 1$. For this range of parameter values, the coefficient of variation has been calculated and given in [75].

15.4   Stochastic HIV-1 Models

15.4.1   *Multi-component HIV model*

Relatively simple models of HIV virus growth in human that are amenable to mathematical analysis involve three or four populations. They typically consist of a normal human CD4+T cell population $X(t)$, an HIV virion population $V(t)$ and an infected CD4+T cell population $I(t)$. The early three component model of Herz *et al.* and its modification in [59] (see [72] and references therein) consist of the following three ODE

$$X' = \lambda - \mu X - k_1 XV, \quad I' = k_1 XV - \alpha I, \quad V' = cI - \gamma V - k_2 XV, \quad (15.4.1)$$

with the modification of the original model of Herz *et al.* being the addition of the last term in the third equation. The various rate constants to be estimated from available clinical data are

λ — synthesis rate of uninfected CD4+T cells
μ — net death rate of uninfected CD4+T cells per cell
α — death rate of active infected CD4+T cells per cell
c — rate of HIV virion emission by active infected CD4+T cells per cell
γ — net death (or clearance) rate of HIV virions per virion
k_1 — infection rate per HIV virion per unit CD4+T cell
k_2 — virion loss rate per unit virion per unit uninfected CD4+T cell

A fourth equation is added in a number of ways. In [59], the additional equation characterizes the impotent infected CD4+T cells. In [56] and [20], the fourth equation characterizes latently infected CD4+T cells, and there are others. Many of these models have been modified to allow for randomness in the infection process. An example is the following three-component version of the Tuckwell–LeCorfec stochastic model:

$$dX = (\lambda - \mu X - k_1 XV)\, dt - \sqrt{k_1 XV}\, dB_1,$$

$$dI = (k_1 XV - \alpha I)\, dt + \sqrt{k_1 XV}\, dB_2,$$

$$dV = (cI - \gamma V - k_2 XV)\, dt - \sqrt{k_2 XV}\, dB_3,$$

where $\{B_i(t)\}$ are standard Wiener process (with zero mean and variance t at time t).

The three-component model can be simplified to the following two-component model for early stages of the infection process:

$$dI = (k_x V - \alpha I)\, dt + \sqrt{k_x V}\, dB_2, \tag{15.4.2}$$

$$dV = (cI - \gamma_k V)\, dt - \sqrt{k_v V}\, dB_3, \tag{15.4.3}$$

where

$$k_x = k_1 \lambda/\mu, \quad k_v = k_2 \lambda/\mu, \quad \gamma_k = \gamma + k_v.$$

At an early stage of the infectious process, there is an abundance of uninfected CD4+T cells so that $X(t)$ remains more or less unchanged from the steady state population λ/μ. Such simplification has been validated by clinical observations to be applicable for about the first 15 days after the initial infection.

The multi-component HIV models above are stationary process governed by the Ito type vector SDE

$$d\mathbf{X} = \mathbf{f}(\mathbf{X})dt + G(\mathbf{X})d\mathbf{B}(t), \quad \mathbf{X}(t_0) = \mathbf{A} \tag{15.4.4}$$

where $\mathbf{A}$ is not at any of the *absorbing barriers* of the response process and $[\langle \Delta B_i(t)\Delta B_j(t)\rangle] = [D_{ij}\Delta t] = D\Delta t$. The backward Kolmogorov equation for the transition pdf $p(\mathbf{x}, t|\, \mathbf{a}, t_0)$ for the vector process is [50]

$$\frac{\partial p}{\partial t_0} = -\sum_{j=1}^{n} f_j(\mathbf{a}, t)\frac{\partial p}{\partial a_j} - \frac{1}{2}\sum_{j=1}^{n}\sum_{i=1}^{n} (GDG^T)_{ij}\frac{\partial^2 p}{\partial a_i \partial a_j}, \tag{15.4.5}$$

The corresponding equations for the moments of first exit time was found to be the following multi-dimensional version of (15.2.5) [51, 53]:

$$\frac{1}{2}\sum_{j=1}^{n}\sum_{i=1}^{n} (GDG^T)_{ij}\frac{\partial^2 T_n}{\partial a_i \partial a_j} + \sum_{j=1}^{n} f_j(\mathbf{a}, t)\frac{\partial T_n}{\partial a_j} = -nT_n. \tag{15.4.6}$$

15.4.2 Early stage two-component HIV model

One problem of clinical interest is the time to a virus threshold level θ_V starting from some initial set of cell and virus populations. Since the infection process has a random component, the time T_θ for the HIV virion population $V(t)$ to reach (and cross) the threshold level θ_V for the first time is a random variable. The case $\theta_V = \theta_I = 0$ provides some indication of the patient's progress toward recovery. The possibility of viral extinction is supported by clinical evidence that disease is not always fatal among individuals exposed to HIV [35]. At the other end of population levels, we expect an upper limit $\theta_I = \theta_i$ on infected cells and/or $\theta_V = \theta_v$ on HIV virions beyond which the patient would not survive.

To determine the probabilistic and statistical information on T_θ, we need the vector version of the first passage time theory obtained for a scalar process developed in Section 2 of this chapter. For a derivation of the more general theory, readers are referred to [12, 55]. Here our discussion will be limited to the early stage two-component model with $\mathbf{X}(t) = (I(t), V(t))^T$. For $D = [D_{ij} = 1]$, the (backward) Kolmogorov equation (15.4.5) for the transition probability density function $p(\mathbf{x}, t; \mathbf{a}, t_0)$ for $t > t_0$ is

$$\frac{\partial p}{\partial t_0} = -(k_x v - \alpha i)\frac{\partial p}{\partial i} - (ci - \gamma_k v)\frac{\partial p}{\partial v} - \frac{v}{2}\left\{ k_x \frac{\partial^2 p}{\partial i^2} + k_v \frac{\partial^2 p}{\partial v^2} - 2\sqrt{k_x k_v}\frac{\partial^2 p}{\partial v \partial i} \right\}$$

where the forward and backward vector variables $\mathbf{x}$ and $\mathbf{a}$ are, respectively, (x_I, x_V) and $(a_I, a_V) \equiv (i, v)$.

The corresponding PDE for the moments of first exit time are the following specialization of the vector process version of (15.2.5) for Ito type SDE (15.2.4) (see [12, 55]):

$$\frac{v}{2}\left\{ k_x \frac{\partial^2 T_n}{\partial i^2} + k_v \frac{\partial^2 T_n}{\partial v^2} - 2\sqrt{k_x k_v}\frac{\partial^2 T_n}{\partial v \partial i} \right\}$$
$$+ (k_x v - \alpha i)\frac{\partial T_n}{\partial i} + (ci - \gamma_k v)\frac{\partial T_n}{\partial v} = -n T_{n-1}$$

$n = 1, 2, 3, \ldots$. Solutions for this system (or those for models with more components) have been obtained for different sets of rate constants by numerical simulation upon specification of the threshold level(s) for infected CD4+T cells and HIV virions (see [56] for example). Stochastic simulations (by Monte-Carlo, Euler Maruyama and other methods) of the original Ito SDE leading to approximate determination of the mean exit time can be found in [20] and elsewhere.

As auxiliary conditions for the PDE for T_n, the usual threshold conditions,

$$T_n(i, \theta_V) = T_n(\theta_I, v) = 0,$$

are not appropriate for the nth moment of time to the patient's demise. The actual thresholds virion and CD4+T cell populations, θ_v and θ_i, are well above the ranges for application of the early stage model. For moderate values of $\theta_V > v$ and $\theta_I > i$, the corresponding mean exit time T_1 could serve to delimit the continued application of our two-component model. On the other hand, moderate thresholds with $0 \lesssim \theta_V < v$ and $0 \lesssim \theta_I < i$ (together with sealed end conditions at some large positive values of i and v) may offer some indication pertaining to a patient's prospect of not developing AIDS after having been exposed to HIV, a prospect supported by available clinical data [35].

15.4.3 *Virion population and first exit time*

15.4.3.1 *A one-component model*

An even simpler stochastic HIV model is concerned with the virion population alone. While it may be formulated in several different ways including a one-dimensional diffusion approximation to a branching process, here we merely limit our attention to the early stage of infection when the infected cell population is very small compared to uninfected population. For a good period of time after contracting the HIV virus, we may neglect any change in $I(t)$ in the SDE (15.4.2) and use the resulting relation to eliminate I(t) from (15.4.3) to get a one-component model. Written in the notation of [60], the single SDE in this model is

$$dV = mpV\,dt + m\sqrt{pV}\,dB_3, \quad V(0) = v_0 \tag{15.4.7}$$

where $B_3(t)$ is a zero mean Wiener process with correlation $E[B_3(t)B_3(t+\tau)] = \tau$. The two parameters m and p are related to the parameters in (15.4.2) and (15.4.3) by

$$\sqrt{p} = \frac{ck_x - \alpha\gamma_k}{c\sqrt{k_x} - \alpha\sqrt{k_v}}, \quad m = \frac{\left(c\sqrt{k_x} - \alpha\sqrt{k_v}\right)^2}{ck_x - \alpha\gamma_k}.$$

Starting with an initial population v_0,

$$V(0) = v_0,$$

we are interested in the first exit time statistics for virus population $V(t)$ reaching some threshold. Given the growth dynamics (15.4.7), the origin is an exit threshold of interest. With

$$T_n(0) = 0 \tag{15.4.8}$$

as a threshold condition, $T_1(v_0)$ gives some indication when the patient may become virus free (and hence recovered). At the other end, the patient may not survive an excessively large virion population θ (that would deplete the CD4+T cell to an unacceptably low level). For the threshold condition

$$T_n(\theta) = 0, \tag{15.4.9}$$

$T_1(v_0)$ offers some indication on the mean time prior to the patient's demise for an initial virion population $v_0 \ll \theta$.

15.4.3.2 *The transition probability density*

The transition probability density function $q(v, t; v_0, 0)$ for virion population growth satisfies the Fokker–Planck or Kolmogorov forward equation

$$\frac{\partial q}{\partial t} = -mp\frac{\partial (vq)}{\partial v} + \frac{1}{2}m^2 p\frac{\partial^2 (vq)}{\partial v^2}, \quad (0 < v < \infty).$$

The exact solution is

$$q(v, t; v_0, 0) = \rho\sqrt{\frac{M(t)}{v}}e^{-\rho(v+M)}I_1(2\rho\sqrt{vM})$$

where

$$\rho = \frac{1}{m(e^{mpt} - 1)}, \quad M(t) = v_0 e^{mpt}$$

and

$$I_1(x) = \sum_{n=0}^{\infty} \frac{(x/2)^{2n+1}}{n(n+1)!}$$

is a modified Bessel function of the first kind.

15.4.3.3 *Mean and variance of $V(t)$*

From the solution for $q(v, t; v_0, 0)$, we can calculate the mean and variance of the virion population at time t to be

$$E[V(t)] = M(t) = v_0 e^{mpt}, \quad Var[V(t)] = 2me^{mpt}\left(e^{mpt} - 1\right).$$

It is of some interest that the coefficient of variation is

$$C.V. = \frac{\sqrt{Var[V(t)]}}{E[V(t)]} = \frac{1}{v_0}\sqrt{2m\left(1 - e^{-mpt}\right)}$$

with

$$\lim_{t\to\infty}[C.V.] = \sqrt{2m}/v_0 \quad (m > 0).$$

For $m < 0$, the coefficient of variation increases with t asymptotic to

$$C.V. \sim \frac{\sqrt{2|m|}}{v_0}e^{|m|pt/2} \quad (m < 0).$$

15.4.3.4 *Mean first exit time*

From the (backward) Kolmogorov equation, we obtain the system of differential equations for the first time moments:

$$\frac{1}{2}m^2 p v_0 \frac{d^2 T_n}{dv_0^2} + m p v_0 \frac{dT_n}{dv_0} = -T_{n-1} \quad (n = 1, 2, 3, \ldots) \tag{15.4.10}$$

with $T_0(v_0) = 1$. While we may examine the two possible threshold problems (with (15.4.8) and (15.4.9)) separately, we consider here both virion clearance as well as an upper limit for virion population size by taking the two required auxiliary conditions for (15.4.10) to be

$$T_n(0) = T_n(\theta) = 0.$$

The mean first exit time is then determined by

$$\frac{1}{2}m^2 p v_0 \frac{d^2 T_1}{dv_0^2} + m p v_0 \frac{dT_1}{dv_0} = -1, \quad T_1(0) = T_1(\theta) = 0.$$

The exact solution of this BVP is

$$T_1(v_0) = \frac{2}{m^2 p}\left\{\frac{1 - e^{-2mv_0}}{1 - e^{-2m\theta}}\int_0^\theta \int_0^x \frac{e^{2(x-z)/m}}{z}dzdx - \int_0^{v_0}\int_0^x \frac{e^{2(x-z)/m}}{z}dzdx\right\}.$$

The unimodal $T_1(v_0)$ provides the mean time to patient recovery or patient demise depending on the initial virion population.

15.5 First Exit Time with a Moving Threshold

15.5.1 *A threshold moving with time*

For many threshold crossing phenomena, the threshold or barrier changes with time. For example, after the initiation and propagation of an action potential in a neuron, there is usually a short time interval, known as a *refractory* (or an absolutely refractory) *period* during which no action potential can be generated. In effect, the threshold for an action potential is very high during the refractory period. But shortly after a refractory period when action potential is again possible, the threshold for initiation effectively decreases with time. The change is seen as the inhibitory effects of the previous action potential gradually decline to a sufficiently low level for another action potential to be more likely. As such, the threshold is time-varying, monotone decreasing during and after an refractory period in the case of initiation of a neuronal action potential.

One approach to solving a first exit time problem with a time-varying threshold characterized by a threshold function $Y(t)$ is to introduce the rate of change relation

$$\frac{dY}{dt} = \phi(Y, t), \quad Y(0) = y.$$

We are interested in cases where either $Y(t)$ is known (and thus so is $\phi(Y,t)$) or determined by a prescribed $\phi(Y,t)$. This additional scalar ODE is to be coupled with an Ito type SDE such as (15.4.4) that governs the dynamics of the phenomenon of interest. Together they allow us to apply the method of first exit time moments to the modified problem. To illustrate, we consider a first exit time problem for a single scalar Wiener process with drift $X(t)$ that satisfies the SDE

$$dX = \mu dt + \sigma dB, \quad X(0) = x$$

where threshold $Y(t)$ for first exit is increasing linearly with time. In that case, the threshold relation is

$$dY = kdt, \quad Y(0) = y(> x).$$

The following equations for the moments of the first exit time $T(x, y)$ for this problem are special cases of (15.4.6):

$$\frac{\sigma^2}{2}\frac{\partial^2 T_n}{\partial x^2} + \mu\frac{\partial T_n}{\partial x} + k\frac{\partial T_n}{\partial y} = -nT_{n-1}, \quad (n = 1, 2, 3, \ldots) \tag{15.5.1}$$

for all $x < y$ with the threshold condition

$$T_n(y, y) = 0. \tag{15.5.2}$$

Note that the region $x > y$ is not a part of the solution domain. In other words, the solution domain for $T_n(x, y)$ consists of all (x, y) pairs for which $x < y$ and $|y| < \infty$. Since $T_n(x, y) > 0$ for all $x < y$ pair, we must have a sealed end condition far away from the line $x = y$ (with $x < y$). Given the auxiliary conditions, it seems more natural to work with two alternative variables $\xi = x - y$ and $\eta = x + y$. A change of independent variables from (x, y) to (ξ, η) transforms the equations for $\{T_n\}$ from (15.5.1) to

$$L_{\xi\eta}[T_n] + 2\mu_\xi\frac{\partial T_n}{\partial \xi} + 2\mu_\eta\frac{\partial T_n}{\partial \eta} = -\frac{2}{\sigma^2}nT_{n-1}, \quad (-a < \xi < 0, |\eta| < \infty)$$

for $n = 1, 2, 3, \ldots$, with $T_0 = 1$ and

$$L_{\xi\eta}[T_n] = \frac{\partial^2 T_n}{\partial \xi^2} + 2\frac{\partial^2 T_n}{\partial \xi\partial \eta} + \frac{\partial^2 T_n}{\partial \eta^2}$$

$$\mu_\xi = \sigma^{-2}(\mu - k), \quad \mu_\eta = \sigma^{-2}(\mu + k).$$

The boundary conditions for these equations are

$$T_n(0, \eta) = \left[\frac{\partial T_n}{\partial \xi}\right]_{\xi \to -\infty} = \left[\frac{\partial T_n}{\partial \eta}\right]_{\eta \to \pm\infty} = 0. \tag{15.5.3}$$

The first of these is (15.5.2) in terms of the new independent variables. The other three are appropriate when there is only a single exit threshold at $x = y$.

For an exact solution of this BVP, we first solve the problem with the second condition replaced by

$$\left[\frac{\partial T_n}{\partial \xi}\right]_{\xi=-a} = 0$$

and then let $a \to \infty$ in the solution. Given the sealed end conditions at $\eta \to \pm\infty$, we consider a particular solution independent on η so that the equation for $T_1(\xi, \eta)$ becomes

$$\frac{d^2 T_1}{d\xi^2} + 2\mu_\xi \frac{dT_1}{d\xi} = -\frac{2}{\sigma^2}$$

or, after one integration

$$\frac{dT_1}{d\xi} + 2\mu_\xi T_1 = d_1 - \frac{2\xi}{\sigma^2}.$$

The exact solution is

$$T_1 = d_2 e^{-2\mu_\xi \xi} + \frac{1}{2\mu_\xi}\left(d_1 + \frac{2}{\mu_\xi \sigma^2} - \frac{2\xi}{\sigma^2}\right)$$

which trivially satisfies the two sealed end conditions for $\eta \to \pm\infty$.

The two constants of integration are determined by the two boundary conditions in the ξ variable in (15.5.3) to result in the solution

$$T_1(x, y) = \frac{e^{-2\mu_\xi a}}{2\mu_\xi^2 \sigma^2}\left(1 - e^{-2\mu_\xi \xi}\right) - \frac{\xi}{\mu_\xi \sigma^2}. \tag{15.5.4}$$

For a process that starts at $x = 0 (< y)$, the mean exit time is

$$T_1(0, y) = \frac{e^{-2\mu_\xi a}}{2\mu_\xi^2 \sigma^2}\left(1 - e^{2\mu_\xi y}\right) + \frac{y}{\mu_\xi \sigma^2}.$$

For the limiting case of $a \to \infty$, we have

$$\lim_{a \to \infty} [T_1(x, y)] = \frac{y - x}{\mu - k}. \tag{15.5.5}$$

Evidently, we need $\mu > k$ in order for the process to be able to pass through the threshold. If the barrier rises faster than the drift, the process that starts below the initial threshold is not likely to catch up with the faster moving barrier at a later time. Mathematically, $e^{-2\mu_\xi a}$ would tend to ∞ if $\mu < k$ and the solution (15.5.4) becomes unbounded as $a \to \infty$.

Similar calculations lead to

$$\lim_{a \to \infty} [T_2(x, y)] = \frac{\sigma^2 (y - x)}{(\mu - k)^2} + \left(\frac{y - x}{\mu - k}\right)^2. \tag{15.5.6}$$

Exercise 70. Explain the predicted outcomes of the solution for the same problem but with $k = 0$ (corresponding to the case of a time-invariant barrier).

15.5.2 Firing time of an OUP-model neuron

The more complex problem of action potential initiation with a time dependent threshold function $Y(t)$, monotone as well as oscillatory, has also been investigated (see [58] and references therein). In the simplest setting, the nerve cell potential $X(t)$ is taken to be an Ornstein–Uhlenbeck process

$$dX = (\mu - X)dt + \sigma dB, \quad X(0) = x$$

with time measured in unit of the cell membrane time constant. When $X(t)$ reaches a time-varying threshold value $Y(t)$, the neuron fires an action potential. This diffusion approximation to the underlying discontinuous process is known as the Stein model [54]. We are interested in the case of an initially resting cell with x set to 0. To illustrate, we consider the case of an exponentially decreasing threshold $Y(t)$ characterized by

$$\frac{dY}{dt} = -k_1 Y + k_2, \quad Y(0) = y$$

for a very small k_1. This corresponds to a slowly moving threshold

$$Y(t) = \left(y - \frac{k_2}{k_1} \right) e^{-k_1 t} + \frac{k_2}{k_1}.$$

The moments for the first exit time are determined by

$$\frac{\sigma^2}{2} \frac{\partial^2 T_n}{\partial x^2} + (\mu - x) \frac{\partial T_n}{\partial x} + (k_2 - k_1 y) \frac{\partial T_n}{\partial y} = -n T_{n-1}, \quad (x < y), \tag{15.5.7}$$

for $n = 1, 2, 3, \ldots$, with $T_0 = 1$. The threshold condition is again

$$T_n(y, y) = 0 \tag{15.5.8}$$

and the condition of no other threshold crossing is the sealed end condition at infinity in any direction with $x < y$. Given the variable coefficients $(\mu - x)$ and $(k_2 - k_1 y)$, the solution for T_n can no longer be a function of only one variable (x or y) or a single combination of x and y. It is of some interest to note that the second order linear PDE with variable coefficients is separable and has solutions in the product form $T_n^{(c)}(x, y) = \phi(x)\psi(y)$ with the x-dependent part $\phi(x)$ being solutions of the Hermite equation,

$$\frac{\sigma^2}{2} \frac{d^2\phi}{dx^2} + (\mu - x) \frac{d\phi}{dx} + \lambda \phi = 0,$$

for different choices of the constant λ. However, the task of determining solutions of this ODE that allow us to satisfy the sealed end condition at $-\infty$ is decidedly nontrivial and the final results on the first exit time moments are not informative. (It is also worth mentioning that the usual Hermite polynomial solutions are inappropriate for this purpose.) A more helpful approach would be to seek asymptotic solutions for small and large values of σ.

For $\sigma^2 \ll 1$, we may seek an outer (asymptotic expansion) solution for $T_n(x, y)$ in parametric series of σ^2:

$$T_n(x, y; \sigma^2) = \sum_{m=0}^{\infty} T_n^{(m)}(x, y)\sigma^2.$$

The leading term approximation $T_1^{(0)}(x, y)$ for the mean first exit time is then determined by the first order linear PDE

$$(\mu - x)\frac{\partial T_1^{(0)}}{\partial x} + (k_2 - k_1 y)\frac{\partial T_1^{(0)}}{\partial y} = -1 \tag{15.5.9}$$

supplemented by suitable auxiliary conditions consisting of the threshold condition (15.5.8) and appropriate sealed end condition(s) far away from the single (solution domain) boundary at $x = y$. This PDE for $T_1^{(0)}(x, y)$ may be solved by the method of characteristics with the characteristic ODE

$$\frac{dy}{dx} = \frac{k_2 - k_1 y}{\mu - x}, \quad \frac{dT_1^{(0)}}{dx} = -\frac{1}{\mu - x}.$$

The exact solution for this system that satisfies the threshold condition, now rephrased as

$$x(y_0) = y_0, \quad T_1^{(0)}(y_0, y_0) = 0,$$

is given parametrically by

$$T_1^{(0)}(x, y) = \ln\left(\frac{\mu - x}{\mu - y_0}\right), \quad \frac{k - y}{k - y_0} = \left(\frac{\mu - x}{\mu - y_0}\right)^{k_1}. \tag{15.5.10}$$

Evidently, we need to restrict $y_0 < \mu$ and $y_0 < k = k_2/k_1$ for our solution to be well-defined. In principle, the second relation determines y_0 as a function of x and y; the expression $y_0 = g(x, y)$ may then be used to eliminate y_0 from the first expression for $T_1^{(0)}$ to give the mean first exit time as a function of x and y.

Also, for (15.5.10) to be an approximate solution of the original problem, it needs to satisfy the sealed end condition far away from the only finite solution domain boundary at $x = y$. However, this is met by the solution (15.5.10) since we have from the first order PDE (15.5.9)

$$\lim_{y \to \infty} \left[\frac{\partial T_1^{(0)}}{\partial y} \right] = - \lim_{y \to \infty} \left[\frac{\mu - x}{k_2 - k_1 y} \frac{\partial T_1^{(0)}}{\partial x} + \frac{1}{k_2 - k_1 y} \right] = 0,$$

$$\lim_{x \to -\infty} \left[\frac{\partial T_1^{(0)}}{\partial x} \right] = - \lim_{x \to -\infty} \left[\frac{k_2 - k_1 y}{\mu - x} \frac{\partial T_1^{(0)}}{\partial x} + \frac{1}{\mu - x} \right] = 0,$$

given that $0 < \mu - x \to \infty$ and $-\infty < k_2 - k_1 y < 0$ are required of the range of x and y for possible exit through the moving barrier.

With the solution in parametric form, the mean first exit time for a process that started at $x = 0$ is given by

$$T_1^{(0)}(0, y) = \ln \left(\frac{\mu}{\mu - y_0} \right) \equiv \tilde{T}_1^{(0)}(y_0), \quad \frac{k - y}{k - y_0} = \left(\frac{\mu}{\mu - y_0} \right)^{k_1}.$$

With the second relation written as

$$y = k - (k - y_0) \left(\frac{\mu}{\mu - y_0} \right)^{k_1} \equiv \tilde{Y}(y_0), \quad k = \frac{k_2}{k_1} \gg y = Y(0),$$

the two relations $\tilde{T}_1^{(0)}(y_0)$ and $\tilde{Y}(y_0)$ may be plotted parametrically in a phase portrait to show the graph of $T_1^{(0)}(0, y)$.

As seen from the development above, the approximate solution by the leading term outer (asymptotic expansion) solution for small σ can be expressed in terms of elementary functions that yield quantitative information and qualitative insight to the mean first exit time pertaining to the initiation of an action potential for an OUP model neuron. The corresponding asymptotic solution for large σ is less informative and more complicated to derive. For this reason, numerical solutions of the original BVP for $T_1(0, y)$ were obtained in [58] for a range of values of y and several sets of values for (μ, σ, k_1, k_2). Readers interested in a matched asymptotic expansion solution are referred to the same solution process implemented for a more realistic neuronal model with high input variability ($\sigma \gg 1$) in the next chapter as it also applies to the present problem.

15.5.3 *Gene fixation in a changing population*

In a model of fluctuating gene frequency when selection varies randomly, the number of genes of a certain type is described by a Malthusian law in the form of a Stratonovich differential corresponding to

$$dX = \left(\mu + \frac{\sigma^2}{2} \right) X dt + \sigma X dB, \quad X(0) = x$$

in the Ito formalism. The total number of genes in the population is assumed to grow at a linear rate

$$\frac{dY}{dt} = kY, \quad Y(0) = y.$$

We have what is known as *gene fixation* when the gene frequency X/Y reaches unity. In other words, there would be only one gene type left in the population at that point.

Upon specialization of the general result (15.4.6), the moments of fixation time are solutions of

$$\frac{\sigma^2 x^2}{2} \frac{\partial^2 T_n}{\partial x^2} + \left(\mu + \frac{\sigma^2}{2}\right) x \frac{\partial T_n}{\partial x} + ky \frac{\partial T_n}{\partial x} = -nT_{n-1} \quad (x < y)$$

with $T_n(x, y)$ required to satisfy the threshold condition

$$T_n(y, y) = 0.$$

Unlike previous cases such as the initiation of an action potential in nerve cells, we are interested here in an initial gene population of $x > 0$ (with $y > x > 0$). In that case, we may set

$$\xi = \ln(x), \quad \eta = \ln(y)$$

and transform the PDE for T_n into one with ξ and η as independent variables:

$$\frac{\sigma^2}{2} \frac{\partial^2 T_n}{\partial \xi^2} + \mu \frac{\partial T_n}{\partial \xi} + k \frac{\partial T_n}{\partial \eta} = -nT_{n-1} \quad (\xi < \eta).$$

The new PDE is of constant coefficients. In fact, it is identical to (15.5.1) for a different problem with x and y as independent variables there. For a solution that satisfies the threshold condition at $x = y$ and a sealed end condition far away from the only barrier for crossing, we have from (15.5.5) and (15.5.6) without further calculations

$$T_1 = \frac{\eta - \xi}{\mu - k} = \frac{1}{\mu - k} \ln\left(\frac{y}{x}\right),$$

$$T_2 = \frac{\sigma^2 (\eta - \xi)}{(\mu - k)^2} + \left(\frac{\eta - \xi}{\mu - k}\right)^2 = \frac{\ln(y/x)}{(\mu - k)^2} \left\{\sigma^2 + \ln\left(\frac{y}{x}\right)\right\}.$$

Chapter 16

The Hodgkin–Huxley Model Neuron

16.1 The Hodgkin–Huxley Model

16.1.1 *The HH model*

The Hodgkin–Huxley (HH) model for nerve axon is arguably the most well known mathematical model in the biological sciences. It is effectively the linear cable model but now with specific voltage dependent current components associated with trans-membrane currents for different gated ion channels. The *full* or *complete (nonlinear) Hodgkin–Huxley model* adequate for investigating the generation of action potentials is essentially the cable model

$$c_m \frac{\partial V}{\partial t} = \frac{1}{r_i} \frac{\partial^2 V}{\partial x^2} + I, \quad I = I_K + I_{Na} + I_\ell + I_A \tag{16.1.1}$$

where the four components of the current I are the current I_K associated with the transport of potassium ions, I_{Na} associated with sodium ions, the *leakage current* I_ℓ associated with other ions (such as chloride) and I_A, the synaptic input currents received at various synaptic sites along the axon. The first three transmembrane current components are proportional to the difference between the depolarization voltage (above the resting potential) and the Nernst potential:

$$I_K = G_K(V_k - V), \quad I_{Na} = G_N(V_N - V), \quad I_\ell = G_\ell(V_\ell - V). \tag{16.1.2}$$

(The Nernst potential for any given ionic species is the membrane potential at which the ionic species is in equilibrium; i.e., there is no net movement of the ion across the membrane.) The parameters G_K and G_N are known to depend on the certain activation and inactivation variables of the relevant ions normalized to the range $[0, 1]$. More specifically, we have

$$G_K = g_K^* n^4, \quad G_N = g_N^* m^3 h.$$

While G_N depends on both activation and inactivation variables $m(t)$ and $h(t)$, the compliance G_K associated with sodium ion transport depends only on its activation variable $n(t)$. The three activation and inactivation variables evolve with time according to

$$\frac{dw}{dt} = \alpha_w(V)(1 - w) - \beta_w(V)w, \quad (w = n, m, h) \tag{16.1.3}$$

where the dependence of $\alpha_w(V)$ and $\beta_w(V)$ on $V(t)$ has been found empirically to be approximately given by

$$\alpha_n(V) = \frac{10 - V}{100 \left[e^{(10-V)/10} - 1 \right]}, \quad \beta_n(V) = \frac{1}{8} e^{-V/80},$$

$$\alpha_m(V) = \frac{25 - V}{10 \left[e^{(25-V)/10} - 1 \right]}, \quad \beta_m(V) = 4 e^{-V/18},$$

$$\alpha_h(V) = \frac{7}{100} e^{-V/20}, \quad \beta_h(V) = \frac{1}{e^{(30-V)/10} + 1}.$$

Upon using (16.1.2) to eliminate I_K, I_{Na} and I_ℓ from (16.1.1), we obtain

$$C_m \frac{\partial V}{\partial t} = \frac{a}{2\rho_i} \frac{\partial^2 V}{\partial x^2} + g_K n^4 (V_K - V) + g_N m^3 h (V_N - V) + g_\ell (V_\ell - V) + \frac{I_A}{2\pi a} \tag{16.1.4}$$

where $\{V_K, V_N\}$ are the Nernst potential for potassium and sodium, respectively, a is the radius of the axon cylinder, $\rho_i = \pi a^2 r_i$ and

$$\{C_m, g_K, g_N, g_\ell\} = \frac{1}{2\pi a} \{c_m, g_K^*, g_N^*, G_\ell\}.$$

The PDE (16.1.4) and the three ODE (16.1.3) supplemented by initial and boundary conditions constitute the full Hodgkin–Huxley model. The publication of the theory and analysis of this model earned the authors the 1963 Nobel Prize in Physiology or Medicine. A great deal of research has been done on this rich and complex model on a single nerve cell and its generalizations since then. A review of the importance of the contributions of Hodgkin and Huxley can be found in [6].

16.1.2   *The space-clamped HH model*

The full Hodgkin–Huxley model is mathematically challenging. One of the directions for further research in this area is to simplify the model by eliminating some of the features not needed in certain environment. For example, in laboratory preparations the voltage can be made practically uniform over a patch of nerve membrane. The phenomenon is known as a *space clamp*. In that case, any spatial derivative of V would be zero over the patch so that the diffusive term in the one PDE in the Hodgkin–Huxley model can be omitted when the synaptic input current I_A depends only on t and not on the spatial variable x. The full Hodgkin–uxley (HH) model then simplifies to the so-called space-clamped HH model consisting the four ODE

$$C_m \frac{dV}{ct} = g_K n^4 (V_K - V) + g_N m^3 h(V_N - V) + g_\ell(V_\ell - V) + \frac{I_A}{2\pi a}$$

(16.1.5)

$$\frac{dw}{dt} = \alpha_w(V)(1 - w) - \beta_w(V)w, \quad (w = n, m, h).$$

(16.1.6)

The system is augmented by four initial conditions

$$V(0) = V_0, \quad n(0) = n_0, \quad m(0) = m_0, \quad h(0) = h_0$$

with the activation and inactivation variables being their steady state values

$$w(0) = \frac{\alpha_w(V_0)}{\alpha_w(V_0) + \beta_w(V_0)} \quad (w = n, m \text{ and } h).$$

The space-clamped model is obviously more tractable in that it is easy to extract useful information from the ODE system and gain insight on the neuronal behavior. There are many directions for new research on this space-clamped model as can be seen from what have been reported in [6, 53, 54]. One of them is to learn more about neuronal behavior in a noisy environment. As a natural extension of the previously investigated white noise current injection at a point of the nerve axon, we consider the case

$$I_A(t) = 2\pi a \{\mu + \sigma W(t)\}$$

where $W(t)$ is a zero mean white noise with unit variance. In that case, the space-clamped HH system is a vector Ito SDE for the four component vector process $\mathbf{X}(t) = (V, n, m, h)^T$

$$d\mathbf{X} = \{\mathbf{f}(\mathbf{X}) + \mathbf{g}(\mathbf{X})W(t)\}\, dt$$

with

$$\mathbf{f}(\mathbf{X}) = \begin{pmatrix} C_m^{-1} \{g_K n^4 (V_K - V) + g_N m^3 h(V_N - V) + g_\ell(V_\ell - V) + \mu\} \\ \alpha_n(V)(1 - n) - \beta_n(V)n \\ \alpha_m(V)(1 - m) - \beta_m(V)m \\ \alpha_h(V)(1 - h) - \beta_h(V)h \end{pmatrix},$$

and

$$\mathbf{g}(\mathbf{X}) = \begin{pmatrix} \sigma \\ 0 \\ 0 \\ 0 \end{pmatrix}.$$

The mathematical results developed in Chapters 9 and 12 now apply to the Ito SDE for the four component process $\mathbf{X}(t)$.

16.2   The Fitzhugh–Nagumo Model

16.2.1   *FHN simplifications of the HH model*

While the space-clamped model is an order of magnitude simpler to analyze than the original HH model, it is still difficult to extract useful information from the analytical and numerical results for such a model, especially for stochastic problems. Hence, it is desirable to simplify the HH model in different ways in order to be able to obtain more insight to various neuronal activities.

The Fitzhugh–Nagumo (FHN) model formulated in [10] and [37] is one outcome of such effort. It involves only a voltage variable $v(t)$ and a recovery variable $w(t)$ so that the dynamics is only second order in time. In his original paper, Fitzhugh noted that the behavior of the computed voltage of the space-clamped HH model is qualitative similar to the Van der Pol oscillator. This observation motivated him to modify its dynamics and add a second variable $w(t)$ that simulates the slower kinetics of the potassium activation variable $n(t)$ and sodium inactivation variable $h(t)$. We have then the following set of two ODE (for a space-clamped FHN model):

$$\frac{dv}{dt} = f(v) - w + I, \quad \frac{dw}{dt} = \varepsilon(v - \gamma w), \tag{16.2.1}$$

where $f(v)$ is a cubic function of v typically of the form

$$f(v) = v(1 - v)(v - a). \tag{16.2.2}$$

It is clear from the dynamics modeled by (16.1.6) that $w(t)$ in the FHN model is not $n(t)$ or $h(t)$, but some variation or combination of these slowly varying variables. In the original formulation of (16.2.1), it was noted that "the path of an action potential, plotted in the (n, h) plane, can be fitted to within 0.1 by the line $h + n = 0.85$" [10] (with 0.85 somehow changed to 0.8 by others who made use of the FHN model); hence we only need to include one of the two variables in the model. However, since "the curves of n and $-h$ versus t during an action potential have similar shapes, n and $-h$ can be replaced by their average $w = 0.5(n - h)$ to give a simplified model." To get the final form of the ODE for dw/dt, it was further noted that the null-cline of ODE for dn/dt is effectively a linear relation between V and n; this led to the choice of the second equation in (16.2.1).

On the other hand, it is not so obvious or straightforward to argue how the dynamics of the depolarization voltage could be simplified to the first equation in (16.2.1). With n as one variable and the corresponding specified rate of change in (16.1.6), the rate of change of $V(t)$ can be reduced to

$$C_m \frac{dV}{dt} = g_K n^4 (V_K - V) + g_N m^3 (0.85 - n)(V_N - V) + g_\ell (V_\ell - V) + I_0, \tag{16.2.3}$$

where $I_0 = I_A/2\pi a$. Even with the slowly varying $m(t)$ taken to be a constant, this relation on the rate of change of $V(t)$ is rather different from its counterpart in the space-calmped FHN model, the first ODE in (16.2.1). In [10], the modified transmembrane voltage equation adopted was justified by the observation that the null-clines for $dv/dt = 0$ as $v(I)$ curves for different values of w are N-shaped curves that intersect the I axis at three locations and these (I, v) "curves resemble qualitatively the corresponding theoretical (I, V) curves of the (full) HH model." With the Van der Pol oscillator also exhibiting similar qualitative features, the Van der Pol equation was suitably adjusted to give the first equation in (16.2.1). Correspondingly, the second equation for the slow or "recovery" variable is taken to be the second ODE in (16.2.1). Together, they constitute the simplified model now known as the (space-clamped) FHN model with $f(v)$ being a cubic having three real roots. The considerable leap from (16.2.3) to the first equation in (16.2.1) notwithstanding, it is customary to identify $v(t)$ with the depolarization voltage $V(t)$.

To the extent that the system (16.2.1) evolves from the space-clamped HH model, it is reasonable to think of it as a space-clamped specialization of a diffusive model with

$$\frac{\partial v}{\partial t} = \frac{\partial^2 v}{\partial^2 x} + f(v) - w + I, \quad \frac{dw}{dt} = \varepsilon(v - \gamma w). \tag{16.2.4}$$

Supplemented by suitable boundary and initial conditions, the system (16.2.4) has been called the general FitzHugh–Nagumo model and a body of research has been accumulated for both the general FHN model and its space-clamped specialization (see [54] and references cited therein).

16.2.2 *Stochastic space-clamped FHN*

While there is now a substantial body of research results for the HH model and its various modifications, our interest in this volume is in stochastic phenomena. To illustrate progress in this direction, we consider the space-clamped FHN model with random synaptic input current injection:

$$dV = \{f(V) - W + \mu\} dt + \sigma dB(t), \quad V(0) = v_0,$$

$$dW = \varepsilon(V - \gamma W) dy, \quad W(0) = w_0,$$

where $B(t)$ is a standard Wiener process (of zero mean and variance t). We are interested in the first exit time $T(v_0, w_0)$ when $v(T) = \theta_v$. Given the complexity of $f(V)$ and two dimensional nature of the problem in the state variables V and W, the most effective method of solution is likely to be numerical simulations. Here, we obtain several approximate solutions by analytical methods to illustrate possible approaches to analytical solutions. While they may not always be adequate for an accurate description of the solution characteristics, they are valuable for validating approximate solutions by stochastic numerical simulations.

Upon specializing the general result in (15.4.6), the system of equations for the first exit time moments for this problem is

$$\frac{\sigma^2}{2} \frac{\partial^2 T_1}{\partial x^2} + \{f(x) - y + \mu\} \frac{\partial T_1}{\partial x} + \{\varepsilon(x - \gamma y)\} \frac{\partial T_1}{\partial y} = -1,$$

$$\frac{\sigma^2}{2} \frac{\partial^2 T_2}{\partial x^2} + \{f(x) - y + \mu\} \frac{\partial T_2}{\partial x} + \{\varepsilon(x - \gamma y)\} \frac{\partial T_2}{\partial y} = -2T_1,$$

etc., with a change of notations from (v_0, w_0) to (x, y). The exit time of interest typically pertains to the voltage threshold θ_a for the initiation of an action potential so that

$$T_n(\theta_a, y) = 0 \quad (y_\ell < y < y_h). \tag{16.2.5}$$

If we are only interested in the initiation of an action potential, then we would have no other exit threshold so that

$$\left[\frac{\partial T_n(x, y)}{\partial x} \right]_{x \to -\infty} = 0, \quad (y_\ell < y < y_h). \tag{16.2.6}$$

(If we are interested in the possibility that $V(t)$ may reach the lower limit θ_h of hyper-depolarization when the transmembrane potential reverses direction and increases toward the resting potential, we would have a second threshold condition $T_n(\theta_h, y) = 0$ as well, with $T_n(x, y) > 0$ for $\theta_h < x < \theta_a$. This case has been investigated in [62, 64] and elsewhere.)

To the extent that the recovery variable $W(t)$ has no direct relation to the initiation of action potential or reversal after reaching hyper-depolarization, it seems reasonable to prescribe a sealed-end condition at the two end points of y, say $y = y_\ell$ and $y = y_h$:

$$\left[\frac{dT_n(x, y)}{dy} \right]_{y = y_\ell} = \left[\frac{dT_n(x, y)}{dy} \right]_{y = y_h} = 0, \quad (\theta_h < x < \theta_a). \tag{16.2.7}$$

However, the PDE for $T_n(x, y)$ are only first order in the y variable; hence only one auxiliary condition is appropriate. While an appropriate single auxiliary condition may be formulated on the basis that $w(0)$ should be a steady state value consistent with $v(0)$, it will be seen that the sealed-end conditions are satisfied by the asymptotic solution for low input variability to be discussed in the next subsection.

With $f(x)$ typically a cubic function of x, accurate solutions for these (deterministic) BVP have been obtained mainly by numerical simulations (see [62] for example). With the sealed end conditions (16.2.6) and (16.2.7), it is possible to obtain an exact analytical solution for the problem as shown in the next subsection.

16.2.3   First passage at an early stage

Since $w(t)$ is a slowly varying function compared to $v(t)$, it is essentially unchanged for a period after the start of the process. Therefore, we may take to a good approximation $y(t) = y_0$ and consider the solution $T_n(x, y) = T_n(x, y_0)$ with y fixed at y_0. In that case, the equations for the moments of first exit time simplify to

$$\frac{\sigma^2}{2} \frac{\partial^2 T_1}{\partial x^2} + \{f(x) - y_0 + \mu\} \frac{\partial T_1}{\partial x} = -1,$$

$$\frac{\sigma^2}{2} \frac{\partial^2 T_2}{\partial x^2} + \{f(x) - y_0 + \mu\} \frac{\partial T_2}{\partial x} = -2T_1,$$

etc. The ODE for $T_1(x)$ is essentially a first order ODE in dT_1/dx so that an exact solution is immediate. Note that the two auxiliary conditions (16.2.7) are satisfied by $T_n(x, y_0)$ since these moments are not dependent on y in this early stage approximate model.

Exercise 71. Obtain the exact solution for $T_1(x, y_0)$ for $f(x) - y_0 + \mu. = -(x - x_1)(x - x_2)(x - x_3)$.

The correct exact solution obtained in the exercise should be simpler than normally expected. To confirm that solution, consider the case where $\sigma^2 \ll 1$ so that we can neglect the second derivative term in the ODE for $T_n(x, y_0)$ leaving us with

$$\{f(x) - y_0 + \mu\} \frac{\partial T_1}{\partial x} = -1, \quad \{f(x) - y_0 + \mu\} \frac{\partial T_2}{\partial x} = -2T_1. \tag{16.2.8}$$

The exact solution for T_1 that satisfies the threshold condition (16.2.5) is

$$T_1(x, y_0) = c_1 \ln\left(\frac{x - x_1}{\theta_v - x_1}\right) + c_2 \ln\left(\frac{x - x_2}{\theta_v - x_2}\right) + c_3 \ln\left(\frac{x - x_3}{\theta_v - x_3}\right) \tag{16.2.9}$$

where $\{c_k\}$ are the coefficients in the partial fraction

$$\frac{-1}{f(x) - y_0 + \mu} = \frac{c_1}{x - x_1} + \frac{c_2}{x - x_2} + \frac{c_3}{x - x_3}.$$

The expression (16.2.9) is the solution for T_1 provided the initial voltage v_0 is to the right of (i.e., $>$) the largest root among the three zeros $\{x_k\}$ of the coefficient $F(x) = f(x) - y_0 + \mu$ in the ODE (16.2.8). Otherwise, $F(x)$ would vanish at least once in the interval (v_0, θ_v) and it would be necessary to address the turning point problem for the correct solution. This is similar to what we have encountered previously in the early stage model for HIV infection. More discussion about asymptotic solutions for both small and large input variability can be found in [64].

16.3   A Model for the Fast Variables

16.3.1   *A space-clamped model for the two fast variables*

To the extent that the time scale for significant changes is much faster for v and m compared to those for n and h, it is rather curious why the FHN model (and its variations) should pick one variable from each of the two groups to form their 2-component model. In the development above (and extensive numerical simulations in the references cited), the slow variable is effectively unchanged over a relatively long period immediately after the start of the process. During that period, we may ignore the changes in the slow variables in the FHN type models so that we have effectively a one-component model. While the simplification is a welcome relief for the purpose of mathematical analysis, the flip side is a loss of opportunity to capture more of the dynamics of the biological phenomenon by a different kind of two-component model. It is with this perspective that the new two-component model of [62] is examined below.

With $n(t)$ and $h(t)$ slowly varying, they may be considered unchanged at their initial value to a good approximation for a period of time near the start of the evolving process

$$n(t) = n_0, \quad h(t) = h_0.$$

During that same period, the rates of change for the $V(t)$ and $m(t)$ of the full HH model (16.1.4) and (16.1.3) become two equations for these two unknowns $\bar{V}$ and $\bar{m}$:

$$\frac{\partial \bar{V}}{\partial t} = \frac{a}{2C_m \rho_i} \frac{\partial^2 \bar{V}}{\partial x^2} + c_K n_0^4 (V_K - \bar{V}) + c_N \bar{m}^3 h_0 (V_N - \bar{V}) + c_\ell (V_\ell - \bar{V}) + \frac{I_0}{C_m}$$

$$(16.3.1)$$

$$\frac{d\bar{m}}{dt} = \alpha_m(\bar{V}) - \gamma_m(\bar{V})\bar{m}, \quad c_w = g_w/C_m \quad (w = K, N, \ell) \tag{16.3.2}$$

where

$$\alpha_m(\bar{V}) = \frac{25 - \bar{V}}{10\left[e^{(25-\bar{V})/10} - 1\right]}, \quad \gamma_m(\bar{V}) = \frac{25 - \bar{V}}{10\left[e^{(25-\bar{V})/10} - 1\right]} + 4e^{-\bar{V}/18}.$$

$$(16.3.3)$$

We refer to this new two-component model as the *Vm-model* for brevity.

Upon omitting the spatial derivative term in $\bar{V}(t)$ from this new *Vm*-model; we get a new space-clamped two-component model as an alternative to the FHN model. Naturally, we would like to compare $\bar{V}(t)$ from the space-clamped *Vm*-model and $V(t)$ from the space-clamped HH model. Such comparison was made in [62] for both deterministic input current and for a white noise input current computed. Numerical solutions for both show good agreement between the results from the two models for more than 1.5 electrotonic time units as long as the input variability is not large compared to the input mean in the stochastic cases. Here we investigate the first exit time problem as it relates to the firing of an action potential.

For the stochastic problem with a white noise input current, we may write the corresponding SDE as

$$d\bar{V} = P(\bar{V}, \bar{m})dt + \sigma dB(t), \quad d\bar{m} = Q(\bar{V}, \bar{m})dt,$$

$$\bar{V}(0) = v_0, \quad \bar{m}(0) = m_0$$

where

$$P(\bar{V}, \bar{m}) = c_K n_0^4(V_K - \bar{V}) + c_N \bar{m}^3 h_0(V_N - \bar{V}) + c_\ell(V_\ell - \bar{V}) + \mu$$

$$Q(\bar{V}, \bar{m}) = \alpha_m(\bar{V}) - \gamma_m(\bar{V})\bar{m}, \quad \frac{I_0}{C_m} = \mu + \sigma W(t),$$

and $W(t)dt = dB(t)$ with $B(t)$ being a standard Wiener process (of zero mean and variance t). Upon specializing the general result for Ito type SDE (15.4.6), the system of equations for the moments $T_n(x, y)$ of the first exit time (for the initiation of an action potential) is

$$\frac{\sigma^2}{2}\frac{\partial^2 T_n}{\partial x^2} + P(x, y; \mu)\frac{\partial T_n}{\partial x} + Q(x, y; \mu)\frac{\partial T_n}{\partial y} = -nT_{n-1} \quad (n = 1, 2, 3, \ldots)$$

with $T_0 = 1$.

The auxiliary conditions augmenting the system of PDE consist of the threshold conditions

$$T_n(\theta_v, y) = 0.$$

To be concrete, we are interested only in the threshold for initiating an action potential (and not the reversal after hyper-depolarization). In that case, we have as a second condition in the x variable the sealed end condition (16.2.6). Given the PDE for $T_n(x, y)$ are only first order in the y variable, only one auxiliary condition can be prescribed at some specific value of y. The appropriate auxiliary condition will be formulated as we investigate the solutions for low and high input variability problems keeping in mind $0 \le m(t) \le 1$ as noted in the formulation of the full HH model.

16.3.2 *Low input variability and method of characteristics*

For $\sigma \ll \mu$, outer (asymptotic expansion) solutions by expanding $\{T_n\}$ as parametric series in σ^2/μ,

$$T_n(x, y; \sigma^2, \mu) = \sum_{k=0}^{\infty} T_n^{(k)}(x, y; \mu)\left(\frac{\sigma^2}{\mu}\right)^k$$

have as their leading term $T_n^{(0)}(x, y; \mu)$ of the expansions determined by the first order PDE

$$P(x, y; \mu)\frac{\partial T_1^{(0)}}{\partial x} + Q(x, y; \mu)\frac{\partial T_1^{(0)}}{\partial y} = -1, \tag{16.3.4}$$

$$P(x, y; \mu)\frac{\partial T_2^{(0)}}{\partial x} + Q(x, y; \mu)\frac{\partial T_2^{(0)}}{\partial y} = -2T_1^{(0)} \tag{16.3.5}$$

etc. These PDE are augmented by the initial conditions

$$\left[T_n^{(0)}\right]_{x=\theta_v} = 0, \quad [y]_{x=v_0} = m_0 = \frac{\alpha_m(v_0)}{\gamma_m(v_0)}. \tag{16.3.6}$$

For the nerve cell initially at resting state taken to be $v_0 = 0$ with $m_0 = \alpha_m(0)/\gamma_m(0)$ generally taking to be 0.06 in the literature. These first order PDE can be recast into solving IVP for ODE by way of the method of characteristics.

In particular, the characteristics ODE for $T_1^{(0)}(x, y)$ are

$$\frac{dy}{dx} = \frac{Q(x, y; \mu)}{P(x, y; \mu)}, \quad \frac{dT_1^{(0)}}{dx} = -\frac{1}{P(x, y; \mu)} \tag{16.3.7}$$

augmented by the two auxiliary conditions (16.3.6) with $n = 1$. For the specific case of a process starting at $x(0) = v_0 = 0$ with the corresponding initial value $y(0) = m_0 = 0.06$, the first ODE can be solved (numerically if necessary) to give

$$y = Y(x, d_1; \mu)$$

where the constant of integration d_1 is determined by the initial condition

$$Y(v_0, d_1; \mu) = m_0 = 0.06.$$

The right hand side of the second ODE in (16.3.7) is now completely known and the ODE itself can be integrated to give

$$T_1^{(0)}(x) = d_2 - \int_0^x \frac{dz}{P(z, Y(z, d_1; \mu))}.$$

The constant of integration d_2 is determined by the threshold condition in (16.3.6) so that

$$T_1^{(0)}(x) = \int_x^{\theta_v} \frac{dz}{P(z, Y(z, d_1; \mu))}.$$

The mean time to threshold θ_v for the process starting at $v_0 = 0$ is then

$$T_1^{(0)}(0) = \int_0^{\theta_v} \frac{dz}{P(z, Y(z, d_1; \mu))}.$$

It is worth noting an unusual feature in the application of the method of characteristics to solve the first order PDE (16.3.4) and the auxiliary conditions (16.3.6). Unlike the usual method of solution, we did not obtain the solution surface for the PDE that passes through a given non-characteristic initial curve. All we obtained is a particular characteristic curve of the solution surface starting from a particular initial point $(v_0, m_0, 0)$. This is all we could obtain since we have no information about the solution T_1 at other (x, y) locations. On the other hand the single characteristics found is all we need for the solution for our problem.

In [62], numerical results for $T_1^{(0)}(0)$ were obtained for the range $5 < \mu < 20$ that are good approximations for results from stochastic simulations of both the present alternative 2-component model and the space-clamped HH model for $\sigma \ll 1$. Similar calculations can also be made to determine $T_1^{(0)}(x)$ and $T_2^{(0)}(x)$ for other x values.

16.3.3 High input variability

At the other end of the spectrum, solutions for the same problem for $\sigma \gg 1$ can also be obtained by asymptotic methods. The obvious perturbation solution

$$T_n(x,y) = \sum_{k=0}^{\infty} t_n^{(k)}(x,y)\sigma^{-2k-2}$$

with $t_n^{(0)}(x, y; \mu)$ determined by

$$\frac{d^2 t_n^{(0)}}{dx^2} = -n t_{n-1}^{(0)} \quad (n = 1, 2, 3, \ldots)$$

is inappropriate for the same reason shown previously for the simpler leaky integrator model in the last chapter. A more appropriate approach taken in [62] to ensure a sealed end at $-\infty$ sets

$$T_n(x,y) = e^{p(x,y)/\sigma^2} t_n(x,y),$$

where

$$p(x,y) = \frac{1}{2a_1(y)}[a_0(y) - a_1(y)x])^2 = \frac{a_1(y)}{2}[x - \zeta(y)]^2,$$

$$a_0(y) = c_N h_0 V_N y^3 + c_K n_0^4 V_K + c_\ell V_\ell + \mu, \quad a_1(y) = c_N h_0 y^3 + c_K n_0^4 + c_\ell.$$

The equation for the first moment $T_1(x, y)$ is then reduced to

$$\frac{\sigma^2}{2}\frac{\partial^2 t_1}{\partial x^2} - \frac{1}{2}\left[\frac{P^2}{\sigma^2} + \frac{\partial P}{\partial x} - 2\frac{Q}{\sigma^2}\frac{\partial p}{\partial y}\right]t_1 + Q\frac{\partial t_1}{\partial y} = -e^{-p(x,y)/\sigma^2}.$$

A parametric series for t_1 in powers of σ^{-2},

$$t_1(x,y) = \sum_{k=0}^{\infty} t_1^{(k)}(x,y)\sigma^{-2k-2},$$

now gives as the leading term solution

$$\frac{1}{2}\frac{\partial^2 t_1^{(0)}}{\partial x^2} = -e^{-p(x,y)/\sigma^2} = -e^{-a_1(x-\varsigma)^2/2\sigma^2}, \quad \varsigma = \frac{a_0(y)}{a_1(y)}.$$

Upon integration, we get

$$\frac{\partial t_1^{(0)}}{\partial x} = -2\int_{-\infty}^{x} e^{-a_1(z-\varsigma)^2/2\sigma^2}\,dz = -2\sigma\sqrt{\frac{2}{a_1}}\int_{-\infty}^{\xi} e^{-\eta^2}\,d\eta, \qquad (16.3.8)$$

with

$$\xi = \sqrt{\frac{a_1}{2\sigma^2}}\,(x-\varsigma).$$

In arriving at (16.3.8), the constant of integration has been chosen to satisfy the sealed end condition at $-\infty$. The integral in the expression for $\partial t_1^{(0)}/\partial x$ above is related to the error function erf(ξ) with

$$\int_{-\infty}^{\xi} e^{-\eta^2}\,d\eta = \frac{\sqrt{\pi}}{2} + \int_{0}^{\xi} e^{-\eta^2}\,d\eta = \frac{\sqrt{\pi}}{2}\{1 + \text{sgn}(\xi)\text{erf}\,(|\xi|)\}.$$

Upon integrating once more, we obtain

$$t_1^{(0)} = 2\sigma\sqrt{\frac{2}{a_1}}\int_{x}^{\theta_v}\int_{-\infty}^{\xi} e^{-\eta^2}\,d\eta dx = \frac{4\sigma^2}{a_1}\int_{\xi_x}^{\xi_\theta}\int_{-\infty}^{\xi} e^{-\eta^2}\,d\eta d\xi$$

$$\xi_x = \sqrt{\frac{a_1}{2\sigma^2}}\,(x-\varsigma), \quad \xi_\theta = \sqrt{\frac{a_1}{2\sigma^2}}\,(\theta_v-\varsigma)$$

with the second constant of integration chosen to satisfy the threshold condition at $x = \theta_v$. The leading term approximation for the mean first exit time for a process starting at $v_0 = 0$ (and the corresponding value for $y(v_0 = 0) = m_0$) is

$$t_1^{(0)}(0, m_0) = \left[\frac{4\sigma^2}{a_1}\int_{\xi_0}^{\xi_\theta}\int_{-\infty}^{\xi} e^{-\eta^2}\,dz d\xi\right]_{y=m_0}.$$

In terms of the erf($\cdot$), we have to a first approximation for $\sigma^2 \gg 1$

$$T_1(0, m_0) \sim e^{p(0,m_0)/\sigma^2} t_1^{(0)}(0, m_0)$$

$$= \phi(m_0)\left[\frac{4\sigma^2}{a_1}\frac{\sqrt{\pi}}{2}\int_{\xi_0}^{\xi_\theta}\{1 + \text{sgn}(\xi)\text{erf}\,(|\xi|)\}\,d\xi\right]_{y=m_0}$$

$$= \frac{4\sigma^2\phi(m_0)}{a_1(m_0)}\frac{\sqrt{\pi}}{2}\left\{[\xi_\theta(m_0) - \xi_0(m_0)] + \int_{\xi_0(m_0)}^{\xi_\theta(m_0)} \text{sgn}(\xi)\text{erf}\,(\xi)\,d\xi\right\},$$

where

$$\phi(m_0) = e^{a_0^2(m_0)/2\sigma^2 a_1(m_0)}.$$

With $\xi_0(m_0) = \xi_x = -\zeta\sqrt{a_1/2\sigma^2} < 0$, the effect of sgn($\xi$) needs to be enforced. This leading term solution has been shown in [62] to be in good or reasonable agreement with results for the expected first exit time from stochastic simulations of the SDE with $\sigma = 10$ for a range of input mean $\mu < \sigma = 10$.

Since much of the concrete numerical results for solutions of SDE in the literature have been obtained by numerical simulations, this volume will end with a brief description of the basic elements of numerical simulations of SDE (or, more briefly, stochastic simulation) in the next (last) chapter.

Chapter 17

Numerical Simulations

17.1 Stochastic Simulations

17.1.1 *Introduction*

It would have been most exhilarating had we ended this volume after the discussion in the previous chapter on a topic that had won its originators the 1963 Nobel Prize in Physiology or Medicine. But we would be remiss not to acknowledge the limitation of the innovative and elegant mathematical techniques expounded in the preceding chapters. It is a fact that many realistic mathematical models involving stochastic differential (and difference) equations cannot be analyzed by these and other available mathematical methods (without further simplifications). For other models amenable to these techniques, the actual applications of the relevant methods may be impractical. For these SDE models, numerical simulations are necessary for extracting useful insight and acceptable approximate quantitative information. It suffices to focus our discussion of numerical simulations on stochastic ODE. A stochastic PDE may be converted into a system of stochastic ODE by eigenfunction expansions, the method of lines in finite difference or some type of Ritz-Galerkin method.

Numerical simulations of SDE problems requires more than routine applications of existing algorithms for deterministic problems and their mathematical underpinning. They involve an added level of complexity that does not exist in the corresponding simulations for deterministic models. This chapter provides the readers some basic elements of stochastic simulation that would enable those interested to be able to further explore the literature on this topic.

In this volume, we have limited analyses of SDE models to those of the Ito type. For that reason, we will focus our discussion first to problems involving the following type of SDE:

$$d\mathbf{X} = \mathbf{f}(\mathbf{X}, t)dt + G(\mathbf{X}, t)d\mathbf{B}$$

where G is a matrix of appropriate dimensions and $\mathbf{B}(t)$ is a vector Wiener process. (We continue to use $B(t)$ for Wiener process to avoid confusion given that $W(t)$ is often used to denote white noise as it is done in this volume.) Since an increment of the Wiener process is involved, we need to be able to generate such increments in the simulation process. For that purpose, we need to know how to generate random numbers. Hence, we begin our discussion with some comments about generation of (pseudo-)random numbers on a computer and the existence of (computer software for) *Pseudorandom Number Generators* (PRNG). Since much of the historical and general information on PRNG can be found on the internet and in the literature on random number generation (including documentation by MatLab, Mathematica and Maple), references are cited below only when appropriate.

17.1.2 *Random (and Pseudorandom) number generation*

The earliest methods for generating random numbers included dice rolling, coin flipping, roulette wheels, shuffling of playing cards and scattering of yarrow stalks according to the Chinese Book of I Ching. Some are still used today but mainly in games and gambling as they tend to be too slow for most applications. Because of their mechanical nature, these techniques for generating large numbers of sufficiently random numbers require a lot of work and/or time and therefore are of limited utility.

A physical random number generator can also be based on an essentially random (most often atomic or subatomic) physical phenomenon. Sources of randomness, disorder and chaos (often group under the term entropy) include radioactive decay, thermal noise, shot noise, radio noise and other more technical phenomena in science and engineering as well as certain human behavior. Various imaginative ways of collecting this entropic information have been and continue to be devised. One drawback of many such "physical" methods is the speed at which entropy can be harvested from natural sources often dependent on the underlying physical phenomena being measured. Thus, sources of naturally occurring "true" entropy are said to be *blocking*; they are rate-limiting to varying degrees. Such blocking behavior often contributes to the impracticality of this type of entropy source.

With the advent of high speed computing, implementing a computational algorithm on a modern computer produces very quickly (in a fraction of a second or so) a large group of (long sequence) numbers with good randomness properties. However, such algorithms typically starts with some initial value known as a *seed*. For the same seed, the sequence eventually repeats. To illustrate, an early computer-based PRNG, known as the middle-square method, was suggested by John von Neumann in 1946. It is based on the following simple algorithm:

- *Take any 10 digit number, square it, remove the middle 10 digits of the resulting number as the "random number", then use that number as the seed for the next iteration. For example, squaring the 10 digit number "1111111111" yields "0123456790098 7654321" with the "0" added so that the same number of digits are deleted from the middle 10 to give "5679009876". Repeating the process yields a new (pseudorandom number) "1531717055", etc.*

This method suffers the same "repetition" defect as all such PRNG; all sequences eventually repeat themselves (after a "period"), some rather quickly. The method has since been supplanted by more elaborate generators that produce sufficiently high quality output through a sufficiently long period (before repetition) adequate for most but the most sophisticated applications (such as cryptography). As they do not generate truly random numbers, these generators are called *Pseudorandom Number Generators* (PRNG); the numbers they generate are known as *pseudorandom numbers*. Some of these PRNG with historical significance are the *Mersenne Twister* (the first to avoid major drawbacks), the Marsaglia family of *xorshift generators* (significantly improved speed) and the *WELL family* of generators (improving on Mersenne Twister issues on large state space and slow recovery from state spaces with a large number of zeros). At least one such acceptable PRNG is usually available on standard math software such as MatLab (*rand*), Mathematica (*Random*) and Maple (*rand*). As such, it is not necessary to go into more details about PRNG for the purpose of generating (pseudo-)random numbers when needed for applications to a stochastic simulation of a SDE model.

17.1.3 *Generating a wiener process*

As we are interested mainly in the numerical simulations of Ito type SDE models and the Wiener process is an important part of such SDE, we need to know how the Wiener process and Wiener increments are simulated numerically. For this purpose, we recall from Chapter 10 that a scalar Wiener process (aka Brownian motion) over the time interval $[0, T]$ is, for a fixed t, a random variable $B(t)$ that depends continuously on t in $[0, T]$ with $B(0) = 0$ and is normally distributed with mean μ and variance t, i.e., $B(t) \sim N(\mu, t)$. It follows from these properties that $B(t)$ satisfies the following three conditions:

- $B(t) - B(s)$, $0 \leq s < t \leq T$, is also normally distributed with mean 0 and variance $t - s$, i.e., $B(t) - B(s) \sim \sqrt{t - s} N(0, 1)$.
- $B(t) - B(s)$ and $B(v) - B(u)$, $0 \leq s < t \leq u < v \leq T$, are statistically independent.
- $E[B(t) - B(s)] = 0$ and $E[\{B(t) - B(s)\}^2] = t - s$ (exercise).

For numerical simulations, we work with a discretized version of the Brownian motion over a discrete set of $M + 1$ equally spaced points in the interval $[0, T]$ by setting

$$\Delta t = \frac{T}{M}, \quad t_0 = 0, \quad t_k = t_{k-1} + \Delta t \quad (k = 1, 2, \ldots, M).$$

Exercise 72. From the properties listed above, show $B(t_j) - B(t_{j-1}) \sim \sqrt{t_j - t_{j-1}} N(0, 1)$.

Given the hypothesis, we have

$$B(t_j) = B(t_{j-1}) + \sqrt{\Delta t} N(0,1)$$

where $\sqrt{\Delta t} N(0,1)$ is a realization of the random variable $\Delta B(t_j) = B(t_j) - B(t_{j-1})$. We need now to be able to generate a random number for ΔB, or to generate a pseudorandom number for a random variable that is $\sim N(0,1)$. Note that we are interested in a random number from a random variable with a specific probability distribution, not one from a uniform or any type other than a standard normal distribution. It cannot be one that is generated by the PRNG described in the previous section but can be generated using a (uniform distribution) PRNG and a function that relates the output to the specified distribution as illustrated below.

Suppose the realization of the random variable Y with the standard normal distribution $N(0,1)$ is y. Let

$$z = \frac{1}{\sqrt{2\pi}} \int_{-\infty}^{y} e^{-\xi^2/2} d\xi, \tag{17.1.1}$$

with $0 < z < 1$ for $-\infty < y < \infty$. If z is a pseudorandom number in the interval $(0,1)$ generated by a (quality) PRNG, say *rand* in MatLab or *Random* in Mathematica, then the corresponding value y in the relation above is evidently a pseudorandom number of the random variable $Y \sim N(0,1)$. In other words, the inverse of cumulative normal distribution with an ideal uniform PRNG input z in the range $(0,1)$ would produce a sequence of values with a standard normal distribution. It remains to formulate an effective scheme to determine y given z.

For $z > 1/2$, one possible approach would be to write (17.1.1) as

$$z = \frac{1}{2} + \frac{1}{\sqrt{2\pi}} \int_{0}^{y} e^{-\xi^2/2} d\xi$$

and differentiate z with respect to y to get

$$\frac{dy}{dz} = \sqrt{2\pi} e^{y^2/2}, \quad y\left(\frac{1}{2}\right) = 0.$$

Let $\Psi(z)$ be the solution of the IVP above for $z > 1/2$ (even a numerical solution would do). Then, given any pseudorandom number $z_r > 1/2$, the value $y_z = \Psi(z_r)$ is the pseudorandom number for y sought. For $z_r < 1/2$, we work with

$$z = \frac{1}{2} - \frac{1}{\sqrt{2\pi}} \int_{y}^{0} e^{-\xi^2/2} d\xi \quad (y < 0)$$

and proceed similarly.

The development above on possible approaches to generating pseudorandom numbers from a non-uniform probability distribution is only meant to provide a relatively simple description of the issues involved in such a task. More sophisticated and effective approaches have been formulated and implemented in practice. Adequate software for generating random numbers of normal distribution is available in

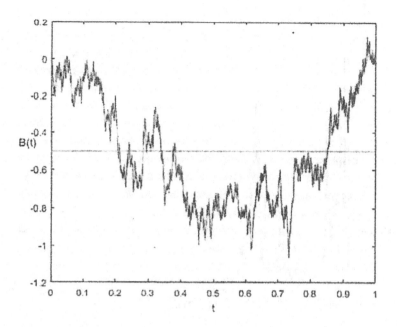

Fig. 17.1. Numerical simulations for the Wiener process.

MatLab, Mathematica, Maple and others. Fig. 17.1 gives is a sample output among homework papers submitted by a recent class of students who did the computation in MatLab using *randn* to generate normally distributed pseudorandom numbers for the calculation of a Wiener process for an equally space set of 5,000 discrete points over the time interval $[0, 1]$.

17.1.4 *The Euler-Maruyama algorithm*

We are now ready to describe a simple method for numerical simulations of Ito type SDE. The general approach is illustrated by the scalar problem

$$dX = f(X(t))dt + G(X(t))dB(t), \quad X(0) = A,$$

for some known initial state A. Treatments of more general cases (such as non-autonomous vector DE and random initial condition) can be found in [22, 67]. To the extent that the SDE is a shorthand notation for the stochastic integral

$$X(t) = A + \int_0^t f(X(s))ds + \int_0^t G(X(s))dB(s)$$

we may form the approximate relation

$$X(t_k) - X(t_{k-1}) \simeq f(X(t_{k-1}))\Delta t + G(X(t_{k-1})))\Delta B(t_k) \qquad (17.1.2)$$

for sufficiently small $\Delta t = t_k - t_{k-1}$. The approximate relation reduces to the simple Euler method (see [72] and references therein) in the absence of the noise term

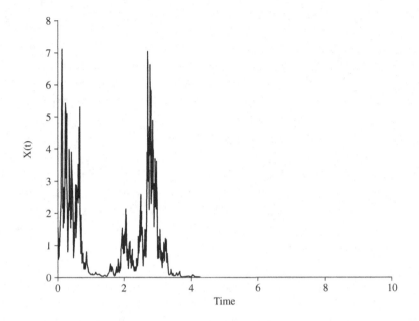

Fig. 17.2.   Numerical simulations for the stochastic process $X(t)$ defined by (17.1.3).

$dB(t)$. The recurrence relations (17.1.2) was proposed in [36] and often designated as the Euler–Maruyama method (see [18] and references therein). It may be used to provide an approximate characterization of the evolution of $X(t)$ over a set of M equally spaced points $\{t_k\} = \{k\Delta t\}$ in the time interval $[\Delta t, T]$ with $\Delta t = T/M$. To implement the algorithm starting with $X_0 = X(0) = A$, we need only to generate a sequence of normally distributed pseudorandom numbers for $\Delta B(t_k) = \sqrt{\Delta t}N(0.1)$ which we now know how to do. To implement the algorithm using MatLab, the needed sequence can be generated by *randn*. Fig. 17.2 is a sample output among homework papers submitted by a recent class of students. The computation was done with MatLab using *randn* to generate normally distributed pseudorandom numbers for the calculation of one realization of the solution of

$$dX = 2X(t)dt + 3X(t)dB(t), \quad X(0) = 1, \tag{17.1.3}$$

over an equally spaced set of $10,000$ discrete points for the time interval $[0, 10]$.

As in the deterministic case, a numerical method of solution does not ends with the development of an algorithm. The algorithm itself must be investigated for accuracy (rate of convergence), efficiency (speed in execution) and stability. Discussion of such issues as well as more sophisticated methods for SDE models (such as the Milstein method) is beyond the scope of this volume but can be readily found in the SDE literature starting with [18, 22].

17.2 Simpler Approaches for Less Information

In theory, we know a stochastic process by having the entire collection of joint pdf, each for the random variables associated with a different combination of (indexing) times. In practice, we often can only obtain the first and second order statistics (pdf or moments) of the process. For specific applications, it is often the case that we need (or would settle for) even less, possibly just the mean and variance of the process at a particular instant of time.

17.2.1 *First and second moments at one instant of time*

Take for example an evolving biomass of a population governed by the stochastic IVP

$$X' = f(X, t; C), \quad X(0) = A. \tag{17.2.1}$$

When C is a known constant c and A is random variable with a prescribed pdf $p_A(a)$. It is generally not possible to obtain an explicit solution of the IVP for the application of the transformation theory. Stochastic simulations that generate a large number of sample paths would allow us to calculate some desired statistics of $X(t)$ approximately (including estimates of its pdf). The situation with C being a random variable (whether or not A is also a random variable) is similar.

In specific applications, we may be interested mainly in the mean value and the variance of the output process at an instant of time. For example, we may be concerned for planning purposes only with the expected population size at some terminal time T, $E[X_T]$ with X_T being the random variable $X(T)$. In that case, a full scale stochastic simulation as described in the previous section (that generates thousands or millions of sample paths) would be over-killed. A much less computing-intensive approach suffices for that purpose. We describe an example of a more efficient method using the IVP (17.2.1) with C being a constant c and A a random variable with a prescribed pdf $p_A(a)$.

For simplicity, let $p_A(a)$ be with compact support so that it vanishes for $a < a_\ell$ and $a > a_u$. The suggested efficient method consists of the following steps:

(1) Solve the IVP to get the output $\{X^{(i)}(t)\}$ for a set of suitably chosen initial values $(a_1, a_2, \ldots, a_N)$ with most of the points $\{a_i\}$ inside the interval (a_ℓ, a_N).
(2) Interpolate the (deterministic) solutions $\{X^{(i)}(T)\}$ to get the interpolating function $X_I(T; a)$ giving a close estimate for X_T at different values of the initial value a that are not one of the interpolation points $\{a_k\}$.
(3) Use the usual definitions to calculate the first and second order statistics of X_T using the interpolation function $X_I(T; a)$ for $X(T; a)$ as a function of a for the full range of a. For example the mean value and second moment of X_T are

given by

$$E[X_T] \cong \int_{a_1}^{a_2} X_I(T; a) p_A(a) da,$$

$$E[X_T^2] \cong \int_{a_1}^{a_2} X_I^2(T; a) p_A(a) da.$$

17.2.2 *PDF at one instant of time*

We can also calculate quite efficiently the pdf of $X(t)$ if we only need it for one instant (or a few instants) of time. Suppose we want only the pdf of $X_T = X(T)$. An approach similar to the previous determination of $E[X_T]$ and $E[X_T^2]$ now requires some additional steps in order to take advantage of the transformation theory developed in Chapter 9. One such step involves solving a new IVP

$$V' = f_{,X}(X, t; c) V, \quad V(0) = 1 \tag{17.2.2}$$

where $f_{,X} = \partial f / \partial X$ and $V(t) = \partial X / \partial A$. The ODE and initial condition in (17.2.2) are obtained by differentiating the ODE and initial condition in (17.2.1) with respect to A. The suggested efficient approach to the new problem consists of the following steps:

(1) Solve the two IVP (17.2.1) and (17.2.2) simultaneous (using any numerical software on MatLab or Mathematica) to get the output $\{X^{(i)}(t)\}$ and $\{V^{(i)}(t)\}$ for a set of suitably chosen initial values $(a_1, a_2, \ldots, a_N)$ with most points inside the interval (a_ℓ, a_N).
(2) Interpolate the (deterministic) solutions $\{X^{(i)}(T)\}$ and $\{V^{(i)}(T)\}$ to get the interpolating function $X_I(T; a)$ and $V_I(T, a)$ giving a good estimate for X_T and V_T at different values of the initial data a that are not one of the interpolation points $\{a_k\}$.
(3) Invert the interpolation function $X_I(T; a)$ by working with a as the pre-image of $X(T)$ through $X_I(T; a)$ and denote the result by $a_I(X_T)$.
(4) Use Theorem 24 for the pdf of X_T to get

$$p_{X_T}(x_T) = p_A(a_I(x_T))/|V_I(T; a_I(x_T))|,$$

(allowing for possible multiple solutions of $X_I(T; a) = x_T$ as appropriate).

17.3 Genetic Instability and Carcinogenesis

In the remainder of this chapter, we discuss an unusual application of the expectation of a function $g(X)$ of a random variable X in terms of the pdf of X (see Theorem 25 in Chapter 10).

17.3.1 *Genetic mutation is a two-edge sword*

We learned in an earlier chapter that mutation is one of the important mechanisms in biological development including genetic evolution. It enables organisms to change from their existing stable (but not asymptotically stable) state in an otherwise Hardy–Weinberg environment. Not all mutations are beneficial for an organism. Here we undertake the study of a class of possibly deleterious mutation on the genetic development of organisms that involves the intertwining of the issues of evolution, instability, control and optimization: mutations that are carcinogenic, i.e., having the potential of initiating various types of cancers.

Homeostasis is a process for maintaining relative constancy in a biological system. In particular, homeostatic control of cells maintains a dynamic equilibrium (or acceptable variations) of cell populations in (organs of) multicellular biological organisms. Sometimes cells can break out of homeostatic control and enter a phase of abnormal expansion. Cancer is a manifestation of such uninhibited growth, which results when cell lineages lose the ability to maintain a sufficient rate of *apoptosis* (programmed cell death) to counterbalance the rate of *mitosis* (cell division). Mechanisms of homeostatic control can be disrupted by *genetic mutations*, to which every cell is prone to some extent. A highly elevated susceptibility to mutations is known in oncology as *genetic* (or *genomic*) *instability* and is closely related to carcinogenesis (see [24] and references therein).

Genetic instability is known to involve the following two (among other) competing effects on carcinogenesis: (1) an increased frequency of cell deaths resulting from deleterious mutations, and (2) an increased frequency of mutations producing cancerous cells. The former effect impedes cancer growth; the latter accelerates it. This gives rise to the question, *What "amounts" of instability are most favorable to the onset of cancer?* The first mathematical models for this class of problems were based on two conceptual premises, (a) the Darwinian view of a cell colony as a collection of phenotypes struggling for survival and predominance, and (b) the degree of genetic instability in the colony can be (hypothetically) set to any level and remain constant in time. A colony of cells was regarded as a species undergoing a birth-and-death process, interpreted as a Darwinian microevolution, and subject to genetic instability. The degree of genetic instability was defined quantitatively as the probability of genetic mutation of a specified type per cell, per mitosis. The "microevolutionary success" of the species was defined as the ability to produce a cancerous population of a given size M, and was measured as the inverse of the time required to reach the target population size. The question how fast (if ever) the target cancerous population size is reached was thus recast as a problem in optimization. The first results showed that "too much" instability impedes the growth of the colony by increasing the cell death rate (a result of too many deleterious mutations in cells), while "too little" instability impedes the growth by lowering the rate of acquisition of cancerous mutants. In the same paper, an optimal time-invariant level of genetic instability that maximizes the rate of cancerous progression was found. This level, defined as the probability of chromosomal loss per cell division, agreed well with available in vitro experimental measurements and turned out to be a robust mathematical result: it depended only logarithmically on the relevant parameter values, and its order of magnitude was consistent with available data (see [24] and references therein).

While it is expected (and of interest to show) that a "more optimal" result (i.e., shorter time to target cancerous population) can be attained by allowing the mutation rate to vary with time (see [24, 45, 73]), we consider here a more realistic issue in a different direction. To the extent that the mutation rate causing genetic instability takes place in a complex environment and influenced by many cellular activities, an important aspect in trying to understand carcinogenesis is to have some handle on the effects of these influences that are not known with certainty. As a first attempt to include the effects of such noisy environment, we let the optimal mutation rate obtained in the absence of the influence of noisy cellular activities be modified by a time-invariant random part. We further simplify the computational aspects of the problem by taking the deterministic part of the mutation to be induced by a one-step activation process as the activation of an oncogene analyzed in [73].

17.3.2 The mathematical model

A dynamical system model developed in [24] for the aforementioned one-step activation process consists of the two normalized ODE for the normal cell and mutated cancerous cell population, $x_1(t)$ and $x_2(t)$, normalized by the initial normal cell population size N and the target mutated cell M, respectively:

$$x_1' = -(\mu + u)x_1 + [1 - d(u)](1 - x_1)x_1 \equiv g_1(x_1, x_2, u), \qquad (17.3.1)$$

$$x_2' = \frac{1}{\sigma}(\mu + u)x_1 + [1 - d(u)](a - x_1)x_2 \equiv g_2(x_1, x_2, u), \qquad (17.3.2)$$

where $(\ )' = d(\ )/dt$, $\sigma = M/N \gg 1$, $a \geq 2$, and $0 < \mu \ll 1$ is the positive *basic mutation rate constant* (for maintaining normal homeostasis). Typically, we have $\sigma \geq 10$ and $\mu \leq 10^{-1} \ll 1$. For simplicity, we will focus our discussion herein to a specific death rate of the form

$$d(u) = 1 - (1 - u)^\alpha \qquad (17.3.3)$$

for some real parameter α. An in-depth discussion for the more complicated (but realistic) general $d(u)$ distinguished only by its convexity as a function of the *cancerous mutation rate constant* can be found in [24],[73].

The two ODE (17.3.1) and (17.3.2) are subject to the following three auxiliary conditions

$$x_1(0) = 1, \quad x_2(0) = 0, \quad x_2(T) = 1 \qquad (17.3.4)$$

where T is the (unknown) terminal time when the cancerous mutant cell population reaches the target size. Evidently, the time to target T depends on the choice of u which is normalized so that it is within the range

$$0 \leq u \leq 1. \tag{17.3.5}$$

(For the general case considered in [24, 73] (but not herein), the mutation rate constant varies with time with $u(t)$ constrained to be among the class of piecewise continuous functions on the interval $[0, T]$, denoted by Ω, and by the inequality (17.3.5).) Needless to say, the two normalized cell populations are also subject to the non-negativity constraints

$$x_1 \geq 0, \quad x_2 \geq 0. \tag{17.3.6}$$

17.3.3 *Shortest time by a constant mutation rate*

As indicated earlier, we are interested in learning about the effects of noisy cellular activities on the optimal mutation rate found in the absence of noise. Undertaking such a study also offers an opportunity to illustrate applications of the efficient method described in the previous section, for determining the effect of noise on the optimal constant u of [23]. However, in contrast to the probabilistic approach in [23], we determine here the optimal mutation rate prior to the introduction of environmental noise by way of a deterministic optimization problem. The addition of the influence of environmental noise on this optimal rate is modeled by adding a time-invariant noise term to the optimal rate.

For the problem without noise, we note that (17.3.1) and (17.3.2) are essentially uncoupled as (17.3.1) can be solved for $x_1(t)$. With the solution for $x_1(t)$, the second equation becomes just a first order linear ODE for $x_2(t)$ which can always be solved by an integrating factor. Equation (17.3.1) may look complicated, but it can also be solved exactly in two ways. It is separable and can be integrated after separating the variables and performing a partial fraction decomposition. However, it is simpler to write it as

$$x_1' = \ell(u)x_1 - [1 - d(u)]x_1^2 \tag{17.3.7}$$

with

$$1 - d(u) = (1 - u)^\alpha, \tag{17.3.8}$$

$$\ell(u) = -(\mu + u) + 1 - d(u) = -(\mu + u) + (1 - u)^\alpha. \tag{17.3.9}$$

While we could have kept the death rate $d(u)$ unspecified, it is taken to be that given in (17.3.3) to be concrete and for computing numerical results in specific cases below. In the form (17.3.7), the ODE for y is a Bernoulli equation and can be transformed into a linear equation

$$\frac{dy}{dt} + \ell(u)y = (1 - u)^\alpha, \tag{17.3.10}$$

for $y(t) = 1/x_1(t)$ with

$$y(0) = \frac{1}{x_1(0)} = 1. \tag{17.3.11}$$

The exact solution for the IVP (17.3.10)–(17.3.11) is

$$x_1(t) = \frac{\ell e^{\ell t}}{(1 - u)^\alpha (e^{\ell t} - 1) + \ell}. \tag{17.3.12}$$

With (17.3.12), the second ODE becomes

$$x_2' - (1 - u)^\alpha(a - x_1)x_2 = \frac{1}{\sigma}(\mu + u)x_1. \tag{17.3.13}$$

For the initial condition $x_2(0) = 0$, the exact solution for the IVP for $x_2(t)$ is found to be

$$x_2(t) = \frac{(\mu + u))\ell}{\sigma \{a(1 - u)^\alpha - \ell\}} \frac{e^{\{a(1-u)^\alpha - \ell\}t} - 1}{(1 - u)^\alpha - (\mu + u))e^{-\ell t}}. \tag{17.3.14}$$

The target cancer population is reached at time T with $x_2(T) = 1$:

$$1 = \frac{(\mu + u))\ell}{\sigma \{a(1 - u)^\alpha - \ell\}} \frac{e^{\{a(1-u)^\alpha - \ell\}T} - 1}{(1 - u)^\alpha - (\mu + u))e^{-\ell T}} \equiv C(T, u; a, \sigma, \alpha, \mu) \tag{17.3.15}$$

with $\ell(u; \alpha, \mu)$ given by (17.3.9). For a prescribed value of the mutation rate u (as well as all the other five parameters), the relation (17.3.15) determines the terminal time $T(u)$. The magnitude of T generally changes with u. We seek the optimal value of u_{op} that minimizes $T(u)$.

In principle, we may determine u_{op} by locating the stationary points of $T(u)$ (and look for the point that corresponds to the global maximum). This is accomplished by differentiating (17.3.15) with respect to u and solving the resulting relation (between T and u);

$$\left[\frac{dC}{du}\right]_{\frac{dT}{du}=0} = 0$$

simultaneously with (17.3.15) to get u_{op} and T_{min}. A second order test would be needed if there should be more than one stationary points. In practice, it may be simpler to do bisections on u to narrow down the minimum T. Pairs of (u_{op}, T_{min})

for typical values of two important parameters α and a are shown in the Table below:

Typical Pairs of (u_{op}, T_{min})
$(\mu = 0.1, \ u_m = 1, \ \sigma = 10)$

$a \setminus \alpha$	0.5	0.75	1.0	1.5
2	----	----	----	----
3	----	$(0.168, 2.370)$	----	----
5	----	$(0.110, 1.435)$	----	----

Exercise 73. Do the computation needed to fill in one or more blanks.

17.3.4 *Noisy environment and expected time to cancer*

Suppose the noisy cellular environment induces an additive noise component to the time-invariant control u so that

$$u = u_{op} + r \tag{17.3.16}$$

where r is a random variable with a prescribed pdf $p(r)$ and u_{op} is the optimal time-independent (normalized) optimal mutation rate constant when $r = 0$. We are interested in the effects of noise on the optimal time to cancer. For example, we would like to know the expected time to target T and the corresponding variance.

With the exact solution (17.3.12) and (17.3.14), now with the time-invariant mutation rate u given by (17.3.16), we still can determine the terminal time T as a function of r (with $u = u_{op} + r$) by way of (17.3.15). We can then calculate the expected value of T from

$$E[T] = \int_{R_\ell}^{R_u} T(r)p(r)dr \tag{17.3.17}$$

where R_ℓ and R_u are finite when $p(r)$ is with compact support; otherwise we have $R_\ell = -\infty$ and $R_u = \infty$. While we actually need $T(r)$ for all r in $[R_\ell, R_u]$, only $T(r)$ is needed for a sufficiently large number of r values adequate for interpolation. The following four $E[T]$ values have been so calculated for a typical pair of (u_{op}, T_{min}) to illustrate the relatively efficient solution process when only a small number of T values are needed for (17.3.17).

Expected Time to Target Cancerous Population
$(\mu = 0.1, \ \sigma = 10, \ \alpha = 0.75, R_u = -R_\ell = 0.05)$

$a \setminus p(r)$	$10\{H(r+0.05) - H(r-0.05)\}$	$50e^{-100|r|}/(1-e^{-5})$	$\delta(r)$
3	2.376	2.371	2.370
5	1.439	1.436	1.435

Appendix

It has been said that different people learn in different ways. However, it is the author's experience that most students learn better by an active learning process. The experience has led him to require graded weekly assignments as an important component of his courses and have the students' performance contribute significantly to their final course grade. The official policy on late homework calls for a 10% reduction of the actual score for every business day late but no more than a 50% reduction for submissions before the final exam week. A typical set of these weekly assignments is included in this appendix.

A.1 Assignment I

READING: Ch. 1 and Ch. 2
EXERCISES: Do all 10 problems.

1. On a small island, 250 people have been exposed to an infectious disease. Consider an experiment which consists of determining the number N of people on the island who have the disease.

   a) What is the sample space of the experiment?
   b) Will this space be an equiprobable space?
   c) If E is the event that $N \leq 50$, what is E^c, the complement of E?
   d) If A is the event $N \geq 40$, what are A^c, $A \cup E$ and $A \cap E$?

2. Show $P(E_1 \cup E_2) \leq P(E_1) + P(E_2)$ for any two events E_k, not necessarily elementary.

3. There are three fish A, B and C in a fish tank. When a food pellet is periodically dropped into the tank and the fishes compete to catch and swallow it, it is observed over a long period of time that either A or B gets the pellet 50% of the time and either A or C gets it 75% of the time.

   a) What is the probability that A gets the pellet when it is dropped?

b) Which fish is the best fed, i.e., most successful in catching dropped food pellets?

4. A study of 10, 000 people in the same age group for lung disease found that 4, 000 in the group have been steady smokers and among them 1, 800 have serious lung disorder. It also found the 1, 500 of the non-smoker also have serious lung disorders. Are smoking and lung disorder independent events?

5. If $\mathbf{x}(0) = \mathbf{p}$ is a probability vector and M is a probability matrix, prove that $\mathbf{x}(n) = M^n \mathbf{p}$ is also a probability vector.

6. Prove that the product of two probability matrices is a probability matrix and that any power of a probability matrix is a probability matrix.

7. If $M > O$, show $\mathbf{y} = M\mathbf{x} > \mathbf{0}$ for any probability vector $\mathbf{x}$. (A matrix $M > O$ means that all elements of M are positive. A vector $\mathbf{p} > \mathbf{0}$ means all components of the vector are positive.)

8. If M is the transition matrix of a regular MC and $\mathbf{x}(0) = \mathbf{p}$ is a probability vector, show that $\mathbf{x}(n) = M^n \mathbf{p}$ is a positive probability vector for a sufficiently large n.

9. Let $\mathbf{x}(n)$ be the solution of the initial value problem (IVP) for the first order linear system of difference equations $\mathbf{x}(n + 1) = M\mathbf{x}(n)$, with $\mathbf{x}(0) = \mathbf{p} = (p_1, 1 - p_1)^T$.

$$\tilde{M} = \begin{bmatrix} 0.5 & 0.25 \\ 0.5 & 0.75 \end{bmatrix}, \quad \hat{M} = \begin{bmatrix} 1 & 0.25 \\ 0 & 0.75 \end{bmatrix}$$

What is the difference in the behavior of the solution for large n between the transition matrices $M = \tilde{M}$ and $M = \hat{M}$?

10. Determine the eigen-pairs of the transition matrix for the social mobility problem,

$$M = \begin{bmatrix} 0.6 & 0.1 & 0.1 \\ 0.3 & 0.8 & 0.2 \\ 0.1 & 0.1 & 0.7 \end{bmatrix},$$

and use the results to solve the IVP $x(n + 1) = Mx(n)$ with $\mathbf{x}(0) = \mathbf{p} = (p_1, p_2, p_3)^T$, $p_1 + p_2 + p_3 = 1$.

A.2   Assignment II

READING: Ch. 3.

EXERCISES: Hand in Problems 1 - 6 and any two of the last three problems.

1. a) A math professor assigns 20 different grades to his/her 20 students. Because of a bug in the computer software, the grades were randomly recorded on the student transcripts.

    i) What is the probability that every student receives the correct grade?
    ii) What is the probability that exactly 19 students receive their correct grades?

b) A chimp is placed at a toy typewriter that has only the letters A, B, C, D and E. Suppose the chimp types four keys at random.

    i) What is the probability of the typed word being "BEAD"?
    ii) What is the probability that all four letters are the same?

2. a) Show that the Markov chain (MC) defined by the transition matrix A below is not a regular MC:

$$A = \begin{bmatrix} 1 & 1/2 & 0 & 0 & 0 \\ 0 & 0 & 1/2 & 0 & 0 \\ 0 & 1/2 & 0 & 1/2 & 0 \\ 0 & 0 & 1/2 & 0 & 0 \\ 0 & 0 & 0 & 1/2 & 1 \end{bmatrix}, \quad B = \begin{bmatrix} 0 & 0 & 1 \\ 1 & 0 & 0 \\ 0 & 1 & 0 \end{bmatrix}, \quad C = \begin{bmatrix} 1/2 & 1 & 0 & 0 \\ 1/2 & 0 & 1 & 0 \\ 0 & 0 & 0 & 1/2 \\ 0 & 0 & 0 & 1/2 \end{bmatrix}$$

b) Show that the Markov chains defined by the transition matrices B and C are neither regular nor absorbing.

3. There are different definitions of a recurrent state of a Markov chain though most are equivalent. Here, we define it in terms of the hitting time probability

$$H(x, y) = \sum_{n=1}^{\infty} H_n(x, x)$$

(summing over all positive integers) with $H_n(x, x)$ being the probability of the chain returning to state x for the first time after n steps (stages), i.e.,

$$H_n(x, x) = \mathrm{Prob}\{x(n) = x,\ x(k) \neq x, k = 1, 2, \ldots, n - 1 | x(0) = x\}$$

A state of the MC is a *recurrent state* if $H(x, y) = 1$ and is a transient state if it is not recurrent. Apply this definition to determine the recurrent and transient states of the MC defined by the transition matrix C in Problem 2. (You may want to make use of the weighted and directed graph of the MC constructed in class for this problem.)

4. In an examination consisting of 100 (true-false) questions, statistics show that if a question is answered correctly, the probability of a correct answer on the next question is $\frac{3}{4}$. On the other hand, if a question is answered incorrectly, the probability of getting the correct answer on the next question is only $\frac{1}{4}$. Formulate a Markov chain model and give an estimate for the average score on this examination.

5.

$$\hat{M} = \begin{bmatrix} 1 & m_{12} \\ 0 & m_{22} \end{bmatrix}$$

a) For the MC defined by the transition matrix above, we know that any initial distribution will evolve into state 1 with probability 1. Prove this by obtaining the absorption matrix for the problem (without re-arranging the transition matrix into the canonical form adopted in class).

b) Starting from the transition state of the MC, obtain the expected number of steps prior to absorption.

6. Suppose in selectively breeding show dogs, the breeder crossbreeds a dog of unknown genotype with a hybrid. One offspring chosen at random is mated with a hybrid and the same process is repeated in succeeding generations. The possible states (genotypes) of a randomly selected offspring are *dominant* (D_m, D_f), *hybrid* (D_m, R_f) or (R_m, D_f) and *recessive* (R_m, R_f)).

a) Construct the transition matrix M.

b) Obtain a limiting distribution for the Markov chain $\mathbf{x}(n+1) = M\mathbf{x}(n)$.

c) Without determining the solution of the IVP for the linear difference system (which you will do in the next problem), explain why is the equilibrium distribution is expected to be independent of the initial distribution.

7. For a prescribed initial distribution $\mathbf{x}(0) = \mathbf{p}$, solve the IVP problem of the Markov chain of Problem 6.

8. Suppose a randomly selected offspring from crossbreeding a dog of unknown genotype is mated with a dog of recessive genotype instead. A randomly selected offspring is again crossbred with a recessive and the same process continues in each generation. After n generations, what is the probability that an offspring chosen at random is recessive?

9. For the Equal Opportunity Base Substitution (or the Jukes–Cantor) model with parameter α characterizing is transition matrix, $M = J(\alpha)$, find γ to show that the product of two *Jukes–Cantor* matrices is also *Jukes–Cantor*, i.e., $J(\alpha)J(\beta) = J(\gamma)$.

10. An experiment must be repeated until two successful experiments have been completed. Suppose that the probability of a successful experiment on one trial is $1/3$.

a) Formulate the transition matrix for an absorbing MC of three possible states.

b) If the first experiment fails, what is the probability that the first success occurs on the fourth trial?

c) What is the expected number of times that the experiment will be repeated?

A.3   Assignment III

READING: Ch. 4.
EXERCISES: Hand in any 7 problems.

1. a) Formulate the transition matrix A for a one-dimensional random walk problem where a player has two marbles and loses one on each coin flip with probability p and gains one with probability $q = 1 - p$ with the game ending when the player has zero or four marbles.
   b) Obtain the absorption matrix A for $p = q = 1/2$ (by first re-ordering the states of the problem, if necessary, to get a new transition matrix B in a form suitable for our purpose).
   c) Determine the expected number of plays before the game ends for $p = q = 1/2$.

2. For a (time-invariant) MC with an $m \times m$ transition matrix M, denote by $M^{(n)} = \left[m_{ij}^{(n)} \right]$ the (power) matrix M^n. Prove the MC version of the Chapman–Kolmogorov equation:

$$M^{(n)} = M^{(n-k)} M^{(k)}, \quad m_{ij}^{(n)} = \sum_{\ell=1}^{m} m_{i\ell}^{(n-k)} m_{\ell j}^{(k)}.$$

3. When we combine the effects of both Selective Breeding II (recessive genotype not participating in reproduction) and mutation, the primary nonlinear difference equation for q_n is modified to

$$q_{n+1} = \frac{q_n + \alpha}{q_n + 1}$$

   a) Set $q_n = \sqrt{\alpha} + (a_n)^{-1}$ and transform the nonlinear difference equation into a first order linear difference equation for a_n.
   b) Solve the linear equation for a_n and obtain q_n.

4. The Kolmogorov ODE system for pure birth processes (with an initial population of size N) derived in class,

$$\frac{dP_k}{dt} = -k\lambda P_k + (k-1)\lambda P_{k-1}, \quad P_k(0) = 0,$$

   $k = N+1, N+2, \ldots$, can be solved sequentially and exactly in terms of elementary functions. With $P_N(t) = e^{-N\lambda t}$,

   a) Obtain P_{N+1}, $P_{N+2}(t)$ and P_{N+3}.
   b) Infer (or prove by induction) that $P_{N+j}(t)$ is as given by Exercise 15.

5. With similar underlying assumptions such as stationarity and independent increments, formulate and solve the Kolmogorov ODE system (and the initial conditions) for a pure death process.

6. Use the relevant IVP to show $\sum P_k(t) = 1$ for the Poisson process, the pure birth process and the pure death process.

7. Derive the corresponding *probability generating function* (pgf) for the birth and death process above.

8. As a continuous time MC of two states, consider a machine that works for an exponential amount of time having mean $1/\lambda$ before breaking down and that

takes an exponential amount of time with $1/\mu$ to repair. If the machine is working at $t = 0$, what is the probability that it will still be working at $t = 10$? (Hint: $\lambda_0 = 1$ and $\mu_1 = 1$ all other λ_k and μ_j are zero.)

9. Let $P_{ij}(t)$ be probability $P_i(t)$ of size i at time t with initial population size j and $P(t)$ the matrix $[P_{ij}(t)]$.

   a) Verify that

   $$P_k(t) = \sum_{j=1}^{\infty} P_{kj}(t) p_j^{(N)}(t), \quad \mathbf{p}^{(N)}(t) = \left(p_1^{(N)}(t), p_2^{(N)}(t), \cdots \right)^T, \quad p_j^{(N)}(0) = \delta_{jN}.$$

   b) Prove the (continuous time MC version of the) Chapman–Kolmogorov equation

   $$P_{ij}(t+s) = \sum_{k=1}^{\infty} P_{ik}(t) P_{kj}(t).$$

A.4 Assignment IV

READING: Ch. 5.
EXERCISES: Hand in any six problems.

1. a) Formulate the Kolmogorov ODE system and the initial conditions for the probability distribution $P_k(t)$ that allows for both birth and death with $P_k(t) = 0$ for $k < 0$.

   b) Show

   $$\sum_{k=0}^{\infty} P_k(t) = 1$$

2. Solve the PDE for the pgf for the birth and death process for $\lambda \neq \mu$.

3. Solve the same PDE for $\lambda = \mu$.

4. a) In a single queue service arrangement, let $P_k(t)$ be the probability of k customers waiting in the queue to be checked out. For a sufficiently small δt so that at most one customer arrives and enters the queue with probability $\lambda \delta t$ and at most one customer is served with probability $\mu \delta t$ and leaves the checkout counter. Formulate a system of Kolmogorov ODE for this $M/M/1$ queuing problem. (Hint: As usual, all terms that tend to zero faster than δt should be omitted to get a simple birth and death type theory.)

   b) Obtain the governing PDE for the generating function for probability distributions for this problem (but do not try to solve the PDE).

5. a) As t tends to infinity, determine the corresponding limiting steady state probability distribution $P_k(\infty)$ of Problem 4 in terms of λ, μ and k.

   b) Deduce from your steady state distribution $P_k(\infty)$ the relative magnitude of λ and μ in order for the queue to clear (or at least not to lengthen) eventually.

c) Obtain from $P_k(\infty)$ the expected value of the number of customers in the queue.

6. a) Formulate the Kolmogorov ODE for the same single queue problem above but now with finite queue capacity N (with only room for N customers waiting in the queue).

b) Determine the corresponding steady state distribution for $P_k(\infty)$.

7. Suppose the simple birth and death process is modified by immigration. With at most one addition in the small δt elapsed time with constant probability ν for one addition, formulate the Kolmogorov system of ODE for this process.

8. a) The one dimensional random walk problem is modified by allowing the unequal probability in taking a step in the two directions with $p_r \neq p_l$. Deduce in the limit of $\delta x \to 0$ the biased diffusion equation in the form of Eq. (6.15) in Ch. 4.

b) Show that the biased diffusion equation for $u(x,t)$ may be transformed in an unbiased diffusion equation for the function $w(x,t)$ related to $u(x,t)$ as given in the expression after Eq. (6.15).

9. a) Problem 6 is an extreme case of the following more realistic problem. Instead of not admitting more customers into the one queue when there are N customers in the queue, we have plenty of room but customers may decide not to join the queue when it is long. Suppose instead of joining with probability $\lambda\delta t$, they join with a probability $\lambda\alpha_n\delta t$ where α_n is a decreasing function of n. Formulate the Kolmogorov ODE system of a birth and death type model for this queueing problem.

b) Instead of a single queue, suppose there are k Poisson servers serving all arriving customers with the same efficient (same rate μ). The customer do not balk so that the arrival rate is at constant λ. Formulate the Kolmogorov ODE system for the probability $P_n(t)$ of having n customers in the queue at time t for this queueing problem. (Hint: The actual departure rate μ varies depending on whether or not n is bigger than k.)

A.5 Assignment V

READING: Ch. 6; Ch. 8: Sec. 1–3.
EXERCISES: Hand in any 7 problems.

1. The pgf for a special case of the branching process model for two-child policy (with $q(x) = q_0 + q_2x^2$, $q_0 + q_2 = 1$) was shown to be determined by the first order linear PDE

$$G_{,t} - q_2\mu(x-1)(x-x_1)G_{,x} = 0, \quad G_{,x} = \frac{\partial G}{\partial x}, \quad x_1 = \frac{q_0}{q_2}.$$

For the special case of $q_2 = q_0$, we have $x_1 = 1$.

a) Obtain the explicit solution for the pgf $G(x,t)$ given in course notes.

b) Determine the expected value of the population size as a function of t and its steady state value.

c) For any positive initial population $(m > 0)$, show that $0 < P_0(t) < 1$ for $t > 0$.

2. A random variable is uniformly distributed in the interval (a, b) if its density function is given by

$$p(x) = \begin{cases} \dfrac{1}{b-a} & (a < x < b) \\ 0 & (x < 0). \end{cases}$$

Determine $P(X \le z)$, $E[X]$ and $Var[X]$ for this density function.

3. A random variable is exponentially distributed with real-valued parameter $\lambda > 0$ if its density function is given by

$$p(x) = \begin{cases} \lambda e^{-\lambda x} & (x \ge 0) \\ 0 & (x < 0) \end{cases}.$$

a) Calculate $P(X \le z)$, $E[X]$ and $Var[X]$ for this density function.

b) Show $P(X \le \infty) = 1$.

4. a) Prove Proposition 20 of Chapter 8.

b) Prove Corollary 10 of Chapter 8.

5. The Poisson distribution can be discussed in the context of a pdf with the help of the Dirac delta function. The same is true for a random variable with a binomial distribution with real-valued parameter p, $0 < p < 1$, by taking its density function as

$$p(x) = \sum_{k-0}^{\infty} \binom{n}{k} p^k (1-p)^{n-k} \delta(x-k), \qquad \binom{n}{k} = \frac{n!}{k!(n-k)!}$$

where $\delta(z)$ is the Dirac delta function. Determine $P(X \le z)$, $E[X]$ and $Var[X]$ for $p(x)$.

6. a) Prove Proposition 34 of Chapter 8 (mean and variance of the average of a set of random variables).

b) Prove Proposition 35 of Chapter 8 (on derivatives of characteristic functions).

7. Two trains arrive at the same station some time during the interval $[0, T]$. The arrival times X and Y of the two trains are random and independent of each other. Let Z be the time interval between arrivals. Determine the probability density function of Z. (Hint: X and Y are uniformly distributed over the interval $[0, T]$ (and assume so if you cannot show this); then work first with $W = X - Y$ and then $Z = |W|$).

8. Show that the change of variable

$$u(x, t) = e^{(2bx - b^2 t)/2\alpha} w(x, t)$$

transforms the biased diffusion equation (5.17) for $u(x, t)$ into the conventional (unbiased) diffusion equation (5.7) for $w(x, t)$.

A.6   Assignment VI

READING: Ch. 8: Sec. $1-3, 5$; Ch. 9.
EXERCISES: Hand in any six problem.

1. $Y' = aY^2, \quad Y(0) = X, \quad p_X(x) = \lambda e^{-\lambda x} H(x)$

    a) Obtain the pdf $p_Y(y, t)$ for the response process $Y(t)$ of the IVP above in terms of a general pdf $p_X(x)$ for the r.v. X.

    b) Obtain the pdf $p_Y(y, t)$ if the pdf for X is (exponential) as given in the last assignment.

2. $Y = aX + b$

    a) Determine the (pre-image) range for $Y \le y$ for $a < 0$.

    b) Obtain the pdf and the Probability distribution $P_Y(y)$ in terms of the pdf of X.

3. $Y = |X|$ (known as full-wave rectifier in engineering)

    a) Determine the pdf $p_Y(y)$ and the distribution $P_Y(y)$ for the r.v. Y in terms of those for the r.v. X.

    b) Suppose X is normally distributed (say Gaussian with zero mean and unit variance). Sketch the graph of $p_Y(y)$.

4. $Y = \sin(X + \phi)$

    a) Determine the pdf $p_Y(y)$ and the distribution $P_Y(y)$ in terms of those for the r.v. X.

    b) Suppose X is uniformly distributed in the interval. Determine the $p_Y(y)$ and $P_Y(y)$.

    c) Sketch the graph of $p_Y(y)$ and $P_Y(y)$.

5. $Y = \begin{cases} -b & (X \le -b) \\ X & (|X| \le b) \\ b & (X \ge b) \end{cases}$ (known as a Limiter in engineering)

    a) Obtain the probability distribution $P_Y(y)$ in terms of the distribution of the r.v. X.

    b) Sketch $P_Y(y)$ for $P_X(x) = e^{-\lambda(x-c)} H(x - c)$ for some positive constant c.

    c) Is $P_Y(y)$ discontinuous?

    d) Obtain the pdf for Y.

6. $Y = \begin{cases} X + c & (X \le -c) \\ 0 & (|X| \le c) \\ X - c & (X \ge c) \end{cases}$

    a) Obtain the probability distribution $P_Y(y)$ for the r.v. Y in terms of of the r.v. X.

b) Show that $P_Y(y)$ is (multi-valued or) discontinuous at $y = 0$. Under what condition on $P_X(x)$ would $P_Y(y)$ be continuous at $y = 0$?

c) Sketch $P_Y(y)$ for $P_X(x) = [1 + \tanh(x)]/2$ and determine the probability of Y at zero (not the probability of $Y \leq 0$).

7. $$Y = \max(X, Y)$$

a) Sketch the region D_z in the x, y–plane for $Z \leq z$.

b) Determine the distribution $P_Z(z)$ in term of the joint distribution $P_{XY}(x, y)$.

c) Obtain the pdf $p_Z(z)$ from the distribution $P_Z(z)$.

d) Suppose X and Y are independent and exponential distribution with parameter α and β, respectively, with $\alpha \neq \beta$. Obtain $p_Z(z)$ and compare with the (previously found) result for $Y = \min(X, Y)$.

8. $$Y' = AY, \quad Y(0) = X$$
Suppose now both A and X are random variables. The determination of the pdf of $Y(t)$ with t as a parameter would involve the transformation $Y(t) = Xe^{At}$. We may make use of the known results for the simpler transformation $W = U + V$ by using some intermediate variables:

a) Determine the pdf $p_Z(z)$ of the random variable $Z = \ln(Y)$ in terms of the joint pdf $p_{AB}(a, b)$ for the two random variables A and $B = \ln(X)$.

b) Describe how you would get the pdf $p_{AB}(a, b)$ from $p_{AX}(a, x)$.

c) Describe how you would get the pdf $p_Y(y; t)$ for $Y(t)$ from the pdf $p_Z(z)$.

A.7 Assignment VII

READING: Ch. 10; Ch. 11: Sec. 1–5.
EXERCISES: Hand in any 7 problems.

1. For a fixed t, determine by way of the Liouville equation the pdf $p_Y(y)$ of the random variable Y, of the solution of the IVP $Y' = aY^2$, $Y(0) = Y_0$ where a is a known constant and Y_0 is a random variable with pdf $p_0(y_0)$.

2. For a fixed t, determine by way of the Liouville equation the pdf $p_Y(y; t)$ of the random variable $Y_t = Y(t)$ of the solution of the IVP $Y' = AY^2$, $Y(0) = y_0$ where y_0 is a known constant and A is a random variable with pdf $p_0(a)$.

3. a) Prove Proposition 31 of Chapter 8. (sum of m.s. convergent sequences)
   b) Prove Proposition 32 of Chapter 8. (uniqueness of m.s. convergence)

4. Prove Proposition 33 of Chapter 8. (interchangeability of $l.i.m.$ and $E[\cdot]$)

5. a) Prove $E[XY] = r\sigma_x\sigma_y$ for two jointly Gaussian (normal) random variables X and Y with a joint probability density function

$$p(x, y) = \frac{1}{2\pi\sigma_x\sigma_y\sqrt{1 - r^2}} \exp\left\{-\frac{q(x, y)}{\sqrt{1 - r^2}}\right\},$$

where

$$q(x,y) = \left(\frac{x-\mu_x}{\sigma_x}\right)^2 + \left(\frac{y-\mu_y}{\sigma_y}\right)^2 - 2r\left(\frac{x-\mu_x}{\sigma_x}\right)\left(\frac{y-\mu_y}{\sigma_y}\right).$$

b) For $\mu_x = \mu_y = 0$, obtain the joint characteristic function for the two jointly Gaussian r.v.

6. Show that the joint probability density function for the two jointly Gaussian random variable X and Y in Problem 5 above can be written as $p(x,y) = P_0 e^{-\mathbf{z}^T S \mathbf{z}}$ for suitable vector $\mathbf{z}$ with

$$S^{-1} = C = \begin{bmatrix} \sigma_x^2 & r\sigma_x\sigma_y \\ r\sigma_x\sigma_y & \sigma_y^2 \end{bmatrix}.$$

7. Let the stochastic process $X(t)$ be $\cos(t+Y)$ where Y is a random variable with a uniform density function on $[0, 2\pi]$. Show that $X(t)$ is a stationary process.

8. Suppose a fair coin is tossed every T seconds starting from $t = 0$. Let the second order stochastic process $X(t)$ be defined by

$$X(t) = \begin{cases} 1 & \text{if a head turns up at the } n^{\text{th}} \text{ toss} \\ -1 & \text{if a tail turns up at the } n^{\text{th}} \text{ toss} \end{cases}$$

for $(n-1) < t < nT$. Draw a sample path and show

$$(i)\ \ E[X(t)] = 0, \quad (ii)\ \ E[X^2(t)] = 1, \quad (iii)\ \ C_{xx}(t,s) = \begin{cases} 1 & |t-s| < T \\ 0 & \text{otherwise} \end{cases}$$

A.8 Assignment VIII

READING:　　Chapter 11; Ch.12: Sec. 1*, 2, 3.
EXERCISES:　Hand in any 6 problems.

1. Show that the s.p. $X(t)$ of Problem 7 in Assignment VII is m.s. differentiable.
2. Suppose A and B are two r.v. and $X(t) = A + Bt$.

   a) Obtain the $\mu(t)$ and $C_{XX}(t_1, t_2)$ (the autocorrelation, not the central moment) in terms of the moments of A and B.

   b) Suppose A and B are statistically independent, obtain the pdf $p_X(x, t)$ of the s.p. $X(t)$.

3. Let $P_X(x; t)$ be the probability distribution of a s.p. $X(t)$. Let the s.p. $Y(t)$ be defined by

$$Y(t) = \begin{cases} 1 & (X(t) > 0) \\ -1 & (X(t) \le 0) \end{cases} \quad \text{(so that the sample space of } Y(t) \text{ is } \{-1, 1\}).$$

   a) Show that $\text{Prob}\{Y(t) = 1\} = 1 - P_X(0; t)$ and $\text{Prob}\{Y(t) = -1\} = P_X(0; t)$.

   b) Obtain $E[Y(t)]$ and $E[Y(t+\tau)Y(t)]$.

4. The exact solution of the IVP $X' = -AX$, $X(0) = X_0$ is known to be $X(t) = X_0 e^{-At}$. Suppose A is a Gaussian random variable of zero mean and unit variance (and X_0 is a known constant).

   a) Determine the mean $\mu(t)$ and correlation function $C_{XX}(t_1, t_2)$ for the s. p. $X(t)$.

   b) Is the s. p. $X(t)$ weakly stationary?

5. a) Is the s.p. $X(t)$ of Problem 4 m.s. differentiable? Justify your answer.

   b) Obtain the pdf of $p_X(x, t)$ of the s.p. $X(t)$.

6. For a wide-sense stationary process (scalar or vector), its mean is independent of t and its autocorrelation $C(t, s)$ depends only of $\tau = t - s$ (and not on t and s individually). In that case, we can define the power spectrum (also known as the spectral density) $S(\omega)$ as the Fourier transform of the autocorrelation $C(\tau)$:

$$S(\omega) = \int_{-\infty}^{\infty} C(\tau) e^{i\omega\tau} d\tau.$$

Obtain the spectral density of the following autocorrelations of two different scalar wide-sense stationary stochastic processes:

   a) $C(\tau) = e^{-\alpha|\tau|}$ $(\alpha > 0)$, b) $C(\tau) = \left(1 - \dfrac{|\tau|}{T}\right) H(T - |\tau|)$ $(T > 0)$

   where $H(x)$ is the unit step function. [The s.p. associated with $C(\tau)$ of part a) is known as the random telegraph signal while the that of part b) is for random binary transmission. They are important processes in signal processing.]

7. For a weakly stationary s.p. $X(t)$, prove the following results:

   a) $C(-\tau) = C(\tau)$

   b) $C(0) \geq 0$

   c) $C(0) \leq C(\tau) \leq C(0)$,

   d) $S(-\omega) = S(\omega)$.

8. From the original discussion that led to the Poisson process, show that

   a) Probability $\{k$ fish caught in the time interval $t_2 < t < t_1\} = \lambda(t_2 - t_1)$.

   b) $E[X(t_1) - X(t_2)] = \lambda(t_1 - t_2)$.

   c) $E\left[\{X(t_1) - X(t_2)\}^2\right] = \lambda^2(t_1 - t_2)^2 + \lambda(t_1 - t_2)$.

A.9 Assignment IX

READING: Ch. 12; Ch. 13: Sec. 1, 2*, 3. (Also review Ch. 6: Sec. 3 − 4.)

EXERCISES: Hand in all problems.

1. $X_1' = -2X_1 + X_2$, $X_2' = X_1 - 2X_2 + f(t)$, $X_1(0) = X_2(0) = 0$.

The IVP, with the forcing function $f(t)$ is a white noise process of zero mean and unit variance, can be re-written as a SDE for the vector process $\mathbf{X}(t) = (X_1, X_2)^T$ with a constant coefficient matrix A.

a) Obtain the pair of unit eigenvectors $\{\mathbf{p}^{(1)}, \mathbf{p}^{(2)}\}$ for the coefficient matrix A and form the modal matrix so that $P^{-1} = P^T = P$. (Hint: The last equality is established by taking $\mathbf{p}^{(1)} = (1,1)^T / \sqrt{2}$.)

b) Derive the IVP for the covariance matrix function $V(t) = \langle \mathbf{X}(t)\mathbf{X}^T(t) \rangle = [V_{ij}(t)]$.

c) With $Re(\lambda_k) < 0, k = 1$ and 2, obtain the expected explicit steady state solution for $V(t)$.

2. a) Obtain the complete (time dependent) solution of the IVP in Problem 1 for $V(t)$.

b) Obtain the autocorrelation $C(t, s)$ for the vector s.p. of Problem 1 for $t > s$.

3. $$X'' - 2X/(1+t)^2 = f(t), \quad X(0) = X'(0) = 0.$$

For the s.p. generated by the stochastic ODE where $f(t)$ is a white noise process with zero mean and unit variance, set $\mathbf{X}(t) = (X, X')^T$ to transform it into a first order time-varying system for $\mathbf{X}(t)$.

a) Formulate the IVP the variance matrix $V(t)$ for $\mathbf{X}(t)$.

b) By setting $U(t) = E(t)V(t)E^T(t)$, obtain the IVP for the integrating factor $E(t)$ for the vector ODE that would transform the IVP for $V(t)$ into one for $U(t)$ of the form

$$U' = E(t)DE^T(t), \quad U(0) = O.$$

c) Determine the explicit solution for $E(t)$ with the initial condition $E(0) = I$.

4. Consider the vector stochastic IVP, $\mathbf{X}' = A(t)\mathbf{X} + \mathbf{F}(t)$, $\mathbf{X}(0) = \mathbf{0}$, with $\mathbf{F}(t) = E(t)\mathbf{W}(t)$ where $\mathbf{W}(t)$ is vector white noise process with $\langle \mathbf{W}(t)\mathbf{W}^T(s) \rangle = D\delta(t-s)$ for some constant matrix $D = [D_{ij}]$ and a known time dependent matrix envelop function $E(t)$.

a) Obtain the IVP for the variance $V(t)$ of the response process (mainly to see how $E(t)$ comes into the IVP for $V(t)$). Do NOT solve the IVP.

b) Obtain the IVP for the autocorrelation function $C_{XX}(t_1, t_2)$ for the response process $\mathbf{X}(t)$.

5. For a scalar process $Z(t)$ as the output of an IVP for the (Langevin) first order ODE $Z' = -\alpha Z + Y(t)$, $z(0) = 0$, where $Y(t)$ is not a white noise process but known to be of mean λ and with autocorrelation function $C_{YY}(t, s) = \lambda^2 + \lambda\delta(\tau)$.

a) Determine $\langle Z(t) \rangle$ for such a forcing s.p.

b) Without the restriction of $t > s$, formulate an IVP for the autocorrelation function $C_{ZZ}(t, s)$. (Note that it involves the cross-correlation function $C_{YZ}(\zeta, s) = \langle Y(\zeta)Z(s) \rangle$ which we do not have since $Y(t)$ is not white noise.)

c) Determine $C_{YZ}(\zeta, s)$ using the exact solution for $Z(s)$ in terms of $Y(\zeta)$ by way of the impulse response $h(s, \zeta) = $ where $H(x)$ is the unit step function.

6. As an alternative way for finding $C_{YZ}(\zeta, s)$, formulate an IVP for $C_{YZ}(t, s)$ and solve it exactly.

7. Obtain the solution for $V(t)$ keeping in mind that $Z(t)$ is not of zero mean. (Note: It is simpler not to solve an IVP for $V(t)$.)

8. a) Determine the eigen-pairs for the homogeneous BVP $\phi'' + \lambda\phi = 0$, $\phi(0) = \phi'(1) = 0$.

   b) With the explicit solutions in terms of trigonometric functions, show that the eigenfunctions for different eigenvalues are mutually orthogonal, i.e.,

$$\int_0^1 \phi_n(x)\phi_m(x)dx = 0 \quad (m \neq n).$$

   c) Prove the same orthogonality relation above without using the actual solution (but with the help of the ODE for $\{\phi_n(x)\}$).

   d) Let the solution $u(x, t)$ of the problem

$$u_{,t} = u_{,xx} + f(x, t), \quad u(0, t) = u_{,x}(1, t) = 0, \quad (t > 0), \quad u(t, 0) = 0 \ (0 \leq x \leq 1),$$

   be expressed in terms of the eigen-pairs $\{\lambda_n, \phi_n(x)\}$,

$$u(x, t) = \sum_{n=0}^{\infty} u_n(t)\phi_n(x).$$

Obtain the IVP for $\{u_n(t)\}$ (but do not try to solve it since we have not specified $f(x, t)$).

A.10 Assignment X

READING: Ch. 13; Ch. 14 and Ch. 17.
EXERCISES: Hand in any 7 problems or implement the stochastic simulations for a Weiner process and the IVP of Problem 1.

1. The simple scalar stochastic ODE $X' = -\alpha X + W(t)$, $X(0) = 0$ where $W(t)$ is a zero mean (Gaussian) white noise is known as Langevin's equation. We consider instead the ODE with the auxiliary condition $X(-\infty) = 0$.

   a) Obtain its autocorrelation function $C_{XX}(t, s) = \langle X(t)X(s)\rangle$ using in the expectation $\langle X(t)X(s)\rangle$ the exact solution

$$X(t) = \int_{-\infty}^{t} \Psi(t, s)W(s)ds, \quad \Psi(t, s) = e^{-\alpha(t-s)} \quad (t > s),$$

   b) Make use of $C_{WW}(t, s) = \langle W(t)W(s)\rangle = D\delta(t - s)$ to evaluate the double integral for $t > s$.

   c) Use the symmetry condition to determine $C_{XX}(t, s)$ for $t < s$.

2. Obtain the same autocorrelation function for the Langevin equation less directly by the process below:

   a) Derive the following IVP problem for the autocorrelation function of the response process of Langevin's equation with $X(-\infty) = 0$:

   $$\frac{\partial C_{XX}(t, s)}{\partial t} = -\alpha C_{XX}(t, s) + C_{WX}(t, s), \quad C_{XX}(-\infty, s) = 0.$$

   b) Derive the following IVP for $C_{WX}(t, s) = \langle W(t)X(s)\rangle$:

   $$\frac{\partial C_{WX}(t, s)}{\partial s} = -\alpha C_{WX}(t, s) + C_{WW}(t, s), \quad C_{WX}(t, -\infty) = 0.$$

   c) With $C_{WW}(t, s) = \langle W(t)W(s)\rangle = D\delta(t - s)$, solve the IVP for $C_{WX}(t, s)$.
   d) Use the result of c) to obtain an explicit solution for $C_{XX}(t, s)$.

3. $\qquad X'' + X = f(t), \quad X(0) = X'(0) = 0, \quad \langle f(t)f(s)\rangle = D\delta(t - s)$

   a) Derive or verify the following exact solution for the IVP above:

   $$X(t) = \int_0^t \Psi(t, s)f(s)ds, \quad \Psi(t, s) = \sin(t - s) \qquad \text{(A.10.1)}$$

   b) By setting $Y(t) = X'(t)$ and $\mathbf{Z}(t) = (X(t), Y(t))^T$, transform the problem for the second order ODE to a first order system for $\mathbf{Z}(t)$ by stating the matrix A and the vector forcing $\mathbf{F}(t)$.
   c) Formulate the IVP for the variance matrix $V(t) = \langle \mathbf{Z}(t)\mathbf{Z}^T(t)\rangle$. Express the forcing term $\Phi(t)$ of the matrix IVP for $V(t)$ in terms of the covariance functions $E[f(t)X(t)]$ and $E[f(t)X'(t)]$ and state the initial conditions.
   d) Use the formal exact solution (A.10.1) given above to show

   $$\Phi(t) = \begin{bmatrix} 0 & 0 \\ 0 & D \end{bmatrix}.$$

4. The eigen-pairs of the matrix A in Problem 3 are complex quantities. It is simpler to solve the IVP for $V(t) = [v_{ij}(t)]$ directly without matrix diagonalization:

   a) Obtain from the matrix ODE for $V(t)$ four scalar ODE:

   $$v'_{11} = v_{12} + v_{21}, \quad v'_{12} = v_{22} - v_{11} = v'_{21}, \quad v' = \cdots.$$

   b) Solve these ODE and their initial conditions for the three distinct components of $V(t)$.

5. Do Exercise 46 of Chapter 13 (on derivatives of moments).

6. Do Exercise 48 of Chapter 13 (on stochastic stability in the mean).
7. Do Exercise 50 of Chapter 13 (on stochastic stability in the mean square).
8. Do Exercise 53 of Chapter 14 (on $G(x, t_+; y, t)$).
9. Do Exercise 54 of Chapter 14 (on mean depolarization voltage).
10. Do Exercise 55 of Chapter 14 (on eigenfunction expansion solution of depolarization voltage).

Bibliography

1. L.F. Abbott, Lapique's introduction of the integrate-and-fire model neuron, *Brain Research Bulletin.*, Vol. 50(5/6), (1999), 303–304.
2. L.J.S. Allen, *An Introduction to Stochastic Processes with Applications to Biology*, 2nd ed., CRC Press, 2011.
3. R.H. Bartels and G.W. Stewart, Solution of the matrix equation $AX + XB = C$, *Comm. ACM*, Vol. 15, (1972), 820–826.
4. W.E. Boyce and R.C. DiPrima, *Elementary Differential Equations and Boundary Value Problems*, 7th Ed., John Wiley and Sons, New York, 2013.
5. A. Bryson and Y.C.Ho, *Applied Optimal Control*, Ginn and Company, Waltham MA, 1969.
6. W.A. Catterall, I.M. Raman, H.P.C. Robinson, T.J. Sejnowski and O. Paulsen, The Hodgkin-Huxley heritage: From channels to circuits, *J Neurosci.*, Vol. 32(41), (2012), 14064–14073.
7. D.A. Darling and A.J.F. Siegert, The first passage problem for a continuous Markov process, *Ann. Math. Statist.*, Vol. 24, (1953), 624–632.
8. G.A. Enciso, M. Tan and F.Y.M. Wan, Maximum expected terminal EB population at critical Chlamydia population size, submitted for publication (2019).
9. J. Feng, Is the integrate-and-fire model good enough? — A review, *Neural Comput.*, Vol. 13, (2001), 309–331.
10. R. Fitz-Hugh, Impulses and physiological states in theoretical models of nerve membrane, *Biophys. J.*, Vol. 1, (1961), 445–466.
11. W. Fuller, *An Introduction to Probability Theory and Its Applications*, Vol. I (2nd ed.), John Wiley and Sons, Inc., 1957.
12. Gihman and A.V. Skorohod, *Stochastic Differential Equations*, Springer-Verlag, Berlin, 1972.
13. C.S. Gillespie, Moment-closure approximations for mass-action models, *IET Systems Biology*, Vol. 3(1), (2009), 52–58.
14. N.S. Goel and N. Richter-Dyn, *Stochastic Models in Biology*, Elsevier, 2017.
15. G. Golub, S. Nash and C. Van Loan, A Hessenberg-Schur method for the problem $AX + XB = C$, *Trans. Automatic Control*, Vol. AC-24, (6), (1979), 909–913.
16. R. Guttman, L. Feldman and H. Lecar, Squid axon membrane response to white noise stimulation, *Biophys. J.*, Vol. 14, (1974), 941–955.
17. D. Hansel and G. Mato, Existence and stability of persistent states in large neuronal networks, *Phys. Rev. Lett.*, Vol. 86, (2001), 4175–4178.
18. D. Higham, An algorithmic introduction to numerical simulation of stochastic differential equations, *SIAM Review*, Vol. 43(3), (2001), 525–548.

19. A.L. Hodgkin and A.F. Huxley, A quantitative description of membrane current and its application to conduction and excitation in nerve, *J. Physiology*, Vol. 117(4), (1952), 500–544.

20. A. Kamina, R.W. Makuch and H. Zhao, A stochastic modeling of early HIV-1 population dynamics, *Math. Biosci.*, Vol. 170, (2001), 187–198.

21. J. Kevorkian and J.D. Cole, *Multiple Scale and Singular Perturbation Methods*, Springer-Verlag, New York, 1996.

22. P.E. Kloeden and E. Platen, *Numerical Solution of Stochastic Differential Equations*, Springer-Verlag, Berlin, 1999.

23. N.L. Komarova and D. Wodarz, The optimal rate of chromosome loss for the inactivation of tumor suppressor genes in cancer, *Proc. Natl. Acad. Sci.*, Vol. 101(18), (2004), 7017–7021.

24. N.L. Komarova, A. Sadovsky and F.Y.M. Wan, Selective pressures for and against genetic instability in cancer: A time-dependent problem, *J. Royal Society, Interface*, Vol. 5, (2008), 105–121.

25. F. Kozin, On the probability densities of the output of some random systems, *J. App. Mech.*, Vol. 28, (1961), 161–165.

26. T. Kushner, A. Simonyan and F.Y.M. Wan, A new approach to feedback for robust signaling gradients, *Studies in Appl. Math.*, Vol. 133, (2014), 18–51.

27. V. Kuznetsov, I. Makalkin, I.M. Taylor and A. Perelson, Nonlinear dynamics of immunogenic tumors: Parameter estimation and global bifurcation analysis, *Bull. Math. Biol.*, Vol. 56, (1994), 295–321.

28. A.D. Lander, Q. Nie and F.Y.M. Wan, Do morphogen gradients arise by diffusion? *Developmental Cell*, Vol. 2, (2002), 785–796.

29. A.D. Lander, Q. Nie and F.Y.M. Wan, Spatially distributed morphogen production and morphogen gradient formation, *Math. Bio. Sci.*, Vol. 2(2), (2005), 239–262.

30. A.D. Lander, Q. Nie, B. Vargas and F.Y.M. Wan, Size-normalized robustness of Dpp gradient in Drosophila wing imaginal disc, *JoMMS*, Vol. 6(1–4), (2011), 321–350.

31. L. Lapicque, Recherches quantitatives sur l'excitation electrique des nerfs traitee comme une polarization, *J. Physiol, Pathol. Gen.*, Vol. 9, (1907), 620–635.

32. P.E. Latham, B.J. Richmond, P.G. Nelson and S. Nirenberg, Intrinsic dynamics in neuronal networks I, Theory, *J. Neurophysiol.*, Vol. 83, (2000), 808–827.

33. J.K. Lee, G.A. Enciso, D. Boassa, C.N. Chander, T.H. Lou, S.S. Pairawan, M.C. Guo, F.YM, Wan, M.H. Ellisman, C. Sütterlin and M. Tan, Progressive decrease in Chlamydia cell size precedes differentiation, *Nat. Comm.*, Vol. 9, (2018), Article # 45 (doi:10.1038/s41467-017-02432-0).

34. R. Lefever and S. Garay, *Biomathematics and Cell Kinetics*. Elsevier, North-Hollan biomedical Press, 1978.

35. J.A. Levy, *HIV the Pathogenesis of AIDS*, 2nd ed., ASM, Washington, DC, 1998.

36. G. Maruyama, Markov proceesses and stochastic equations, *Nat. Sci. Rep.*, Ochanomizu University, Vol. 4 (1953), pp. 40–43. Zbl 0053.40802.

37. J. Nagumo, S. Arimoto and S. Yoshizawa, An active pulse transmission line simulating nerve axon, *Proc. IRE*, Vol. 50(10), (1962), 2061–2070.

38. Q. Nie, F.Y.M. Wan, Y.-T. Zhang and X.-F. Liu, Compact integration factor methods in high spatial dimension, *J. Comp. Phys.*, Vol. 227, (2008), 5238–5525.

39. R.M. Nisbet and W.S.C. Gurney, *Modelling Fluctuating Populations*, John Wiley & Sons, Chichester and New York, 1982.

40. L.S. Pontryagin *et al.*, *The Mathematical Theory of Optimal Control Processes*, Interscience Publishers, New York, 1962.

41. E.B. Postnikov and O.V. Titkova, A correspondence between the models of Hodgkin-Huxley and FitzHugh-Nagumo revisited, *Eur. Phys. J. Plus*, Vol. 131, (2016), Article 411.

42. W. Rall, Time constants and electrotonic length of membrane cylinders and neurons, *Biophys. J.*, Vol. 9, (1969), 1483–1508.

43. J. Rinzel, Excitation dynamics: Insights from simplified membrane models, *Fed Proc.*, Vol. 44(15), (1985), 2944–2946.

44. W. Rudin, *Principles of Mathematical Analysis*, 3rd ed. McGraw-Hill, New York, 1976.

45. C. Sanchez-Tapia and F.Y.M. Wan, Fastest time to cancer by loss of tumor suppressor genes, *Bull. Math Bio.*, Vol. 76, (2014), 2737–2784.

46. L. Segel, *Modeling Dynamic Phenomena in Molecular and Cellular Biology*, Cambridge University Press, 1984.

47. A. Simonyan and F.Y.M. Wan, Transient feedback and robust signaling gradients, *Int. J. Num. Anal. & Modeling*, Series B, Vol. 13(2), (2016), 175–200.

48. J.M. Smith, *The Theory of Evolution*, London, Penguin Books, 1958.

49. J.M. Smith, *Mathematical Ideas in Biology*, Cambridge University Press, 1968.

50. T.T. Soong, *Random Differential Equations in Science and Engineering*, Academic Press, New York and London, 1973.

51. H.C. Tuckwell, On the first-exit time problem for temporally homogeneous Markov processes, *J. Appl. Prob.*, Vol. 13, (1976), 39–48.

52. H.C. Tuckwell, Synaptic transmission in a model for stochastic neural activity, *J. Theor. Biol.*, Vol. 77, (1979), 65–81.

53. H.C. Tuckwell, *Introduction to Theoretical Neurobiology, Vol. 1 — Linear cable theory and dendritic structure*, Cambridge University Press, 1988.

54. H.C. Tuckwell, *Introduction to Theoretical Neurobiology, Vol. 2 — Nonlinear and stochastic theories*, Cambridge University Press, 1988.

55. H.C. Tuckwell, *Stochastic Processes in the Neurosciences*, SIAM, Philadelphia, 1989.

56. H.C. Tuckwell and E. Le Corlec, A stochastic model for early HIV-1 dynamics, *J. Theor. Biol.*, Vol. 195, (1998), 451–463.

57. H.C. Tuckwell and F.Y.M. Wan, The response of a nerve cylinder to spatially distributed white noise inputs, *J. Theo. Biol.*, Vol. 87, (1980), 275–295.

58. H.C. Tuckwell and F.Y.M. Wan, First passage time of Markov processes to moving barriers, *J. Appl. Prob.*, Vol. 21, (1984), 695–709.

59. H.C. Tuckwell and F.Y.M. Wan, Nature of equilibria and effects of drug treatments in some viral population dynamical models, *IMA J. Math. Appl. Med. & Biol.*, 17, (2000), 311–327.

60. H.C. Tuckwell and F.Y.M. Wan, First passage time to detection in stochastic population dynamical models for HIV-1, *Appl. Math. Letters*, Vol. 13, (2000), 79–83.

61. H.C. Tuckwell and F.Y.M. Wan, On the behavior of solutions in viral dynamical models, *BioSystems*, Vol. 73(3), (2004), 157–161.

62. H.C. Tuckwell and F.Y.M. Wan, Time to first spike in Hodgkin-Huxley stochastic systems, *Physica A — Stat. Mech. Applic.*, Vol. 351(2-4), (2005), 427–438.

63. H.C. Tuckwell, F.Y.M. Wan and R. Rodriguez, Analytical determination of firing times in stochastic nonlinear neural models, *Letters, Neurocomuting*, Vol. 48, (2002), 1003–1007.

64. H.C. Tuckwell, R. Rodriguez and F.Y.M. Wan, Determination of firing times for the stochastic Fitzhugh-Nagumo neuronal model, *Neural Comp.*, Vol. 15, (2003), 143–159.

65. H.C. Tuckwell, F.Y.M. Wan and J.-P. Rospars, A spatial stochastic neuronal model with Ornstein-Uhlenbeck input current, *Biol. Cybern.*, Vol. 86, (2002), 137–145.

66. H.C. Tuckwell, F.Y.M. Wan and Y.S. Wong, The interspike interval of a cable model neuron with white noise input, *Biol. Cybern.*, Vol. 49, (1984), 155–167.

67. N.G. van Kampen, *Stochastic Processes in Physics and Chemistry*, 3rd ed., Elsevier, 2008.

68. F.Y.M. Wan, A direct method for linear dynamical problems in continuum mechanics with random loads, *Studies in Appl. Math.*, Vol. 52, (1973), 259–276.

69. F.Y.M. Wan, *Mathematical Models and Their Analysis*, Harper and Row, 1989.

70. F.Y.M. Wan, *Introduction to the Calculus of Variations and Its Applications*, Chapman and Hall, 1995.

71. F.Y.M. Wan and G.A. Enciso, Optimal proliferation and differentiation of chlamydia trachomatis, *Studies in Applied. Math.*, 139(1), (2017), 129–178 (DOI: 10.1111/sapm.12175).

72. F.Y.M. Wan, *Dynamical System Models in the Life Sciences*, World Scientific, New Jersey, London, Singapore, 2017.

73. F.Y.M. Wan, A. Sadovsky and N.L. Komarova, Genetic instability in cancer: An optimal control problem, *Studies in Appl. Math.*, Vol. 125(1), (2010), 1–10.

74. F.Y.M. Wan and H.C. Tuckwell, The response of a spatially distributed neuron to a white noise current injection, *Biol. Cybern.*, Vol. 33, (1979), 39–55.

75. F.Y.M. Wan and H.C. Tuckwell, Neuronal firing and input variability, *J. Theo. Neurobiol.*, Vol. 1, (1982), 197–218.

76. S.C. Wecker, The role of early experience in habitat selection by the prairie deer mouse, Peromyscus maniculatus bairdii, *Ecol. Monogr.*, Vol. 33, (1963), 307–325.

77. S.C. Wecker, Habitat selection, *Sci. Amer.*, Vol. 211, (1964), 109–116.

78. H.F. Weinberger, *A First Course in Partial Differential Equations*, paperback ed., Courier Dover Publ., 1995.

79. D.J. Wilkinson, *Stochastic Modelling in Systems Biology*, Chapman-Hall/CRC Press, 2006.

Index

Printed in the United States
By Bookmasters